Seismic Geomorphology: Applications to Hydrocarbon Exploration and Production

Geological Society books refereeing procedures

The Society makes every effort to ensure that the scientific and production quality of its books matches that of its journals. Since 1997, all book proposals have been refereed by specialist reviewers as well as by the Society's Books Editorial Committee. If the referees identify weaknesses in the proposal, these must be addressed before the proposal is accepted.

Once the book is accepted, the Society Book Editors ensure that the volume editors follow strict guidelines on refereeing and quality control. We insist that individual papers can only be accepted after satisfactory review by two independent referees. The questions on the review forms are similar to those for *Journal of the Geological Society*. The referees' forms and comments must be available to the Society's Book Editors on request.

Although many of the books result from meetings, the editors are expected to commission papers that were not presented at the meeting to ensure that the book provides a balanced coverage of the subject. Being accepted for presentation at the meeting does not guarantee inclusion in the book.

More information about submitting a proposal and producing a book for the Society can be found on its website: www.geolsoc.org.uk.

It is recommended that reference to all or part of this book should be made in one of the following ways:

Davies, R. J., Posamentier H. W., Wood, L. J. & Cartwright, J. A. (eds) 2007. *Seismic Geomorphology: Applications to Hydrocarbon Exploration and Production.* Geological Society, London, Special Publications, **277**.

Sullivan, E. C., Marfurt, K. J. & Bhumentritt, C. 2007. Seismic geomorphology of palaeozoic collapse features in Fort Worth Basin (USA). *In*: Davies, R. J., Posamentier H. W., Wood L. J. & Cartwright, J. A. (eds) *Seismic Geomorphology: Applications to Hydrocarbon Exploration and Production.* Geological Society, London, Special Publications, **277**, 187–203.

GEOLOGICAL SOCIETY SPECIAL PUBLICATION NO. 277

Seismic Geomorphology: Applications to Hydrocarbon Exploration and Production

EDITED BY

R. J. DAVIES
Durham University, UK

H. W. POSAMENTIER
Anadarko, Canada

L. J. WOOD
Bureau of Economic Geology, University of Texas, USA

and

J. A. CARTWRIGHT
Cardiff University, UK

2007
Published by
The Geological Society
London

THE GEOLOGICAL SOCIETY

The Geological Society of London (GSL) was founded in 1807. It is the oldest national geological society in the world and the largest in Europe. It was incorporated under Royal Charter in 1825 and is Registered Charity 210161.

The Society is the UK national learned and professional society for geology with a worldwide Fellowship (FGS) of over 9000. The Society has the power to confer Chartered status on suitably qualified Fellows, and about 2000 of the Fellowship carry the title (CGeol). Chartered Geologists may also obtain the equivalent European title, European Geologist (EurGeol). One fifth of the Society's fellowship resides outside the UK. To find out more about the Society, log on to www.geolsoc.org.uk.

The Geological Society Publishing House (Bath, UK) produces the Society's international journals and books, and acts as European distributor for selected publications of the American Association of Petroleum Geologists (AAPG), the Indonesian Petroleum Association (IPA), the Geological Society of America (GSA), the Society for Sedimentary Geology (SEPM) and the Geologists' Association (GA). Joint marketing agreements ensure that GSL Fellows may purchase these societies' publications at a discount. The Society's online bookshop (accessible from www.geolsoc.org.uk) offers secure book purchasing with your credit or debit card.

To find out about joining the Society and benefiting from substantial discounts on publications of GSL and other societies worldwide, consult www.geolsoc.org.uk, or contact the Fellowship Department at: The Geological Society, Burlington House, Piccadilly, London W1J 0BG: Tel. +44 (0)20 7434 9944; Fax +44 (0)20 7439 8975; E-mail: enquiries@geolsoc.org.uk.

For information about the Society's meetings, consult *Events* on www.geolsoc.org.uk. To find out more about the Society's Corporate Affiliates Scheme, write to enquiries@geolsoc.org.uk

Published by The Geological Society from:
The Geological Society Publishing House, Unit 7, Brassmill Enterprise Centre, Brassmill Lane, Bath BA1 3JN, UK

(*Orders*: Tel. +44 (0)1225 445046, Fax +44 (0)1225 442836)
Online bookshop: www.geolsoc.org.uk/bookshop

British Library Cataloguing in Publication Data
A catalogue record for this book is available from the British Library.

ISBN-13: 978-1-86239-223-6

Typeset by Techset Composition Ltd, Salisbury, UK
Printed by MPG Books Ltd, Bodmin, UK

Distributors

North America
For trade and institutional orders:
The Geological Society, c/o AIDC, 82 Winter Sport Lane, Williston, VT 05495, USA
Orders: Tel +1 800-972-9892
Fax +1 802-864-7626
Email gsl.orders@aidcvt.com

For individual and corporate orders:
AAPG Bookstore, PO Box 979, Tulsa,
OK 74101-0979, USA
Orders: Tel +1 918-584-2555
Fax +1 918-560-2652
Email bookstore@aapg.org
Website http://bookstore.aapg.org

India
Affiliated East-West Press Private Ltd, Marketing Division, G-1/16 Ansari Road, Darya Ganj,
New Delhi 110 002, India
Orders: Tel. +91 11 2327-9113/2326-4180
Fax +91 11 2326-0538
E-mail affiliat@vsnl.com

Contents

Preface

This Geological Society Special Publication is the result of a highly successful joint SEPM and Geological Society conference held in February 2005 in Houston, Texas. The papers attempt to capture the development of a new discipline, 'Seismic Geomorphology'. The discipline of geomorphology has a long and illustrious history but in recent years an entirely new way of studying landscapes and seascapes has been developed. It involves the use of seismic data – specifically 3D seismic data. Just as CAT scans allow medical staff to view our anatomy in 3D, seismic data now allow Earth Scientists to do what the early geomorphologists such as William Morris Davis and Albrecht Penck could only dream of – and view 10's and 100's of square kilometres of the Earth's subsurface in 3D. Such evolving image technologies enable geoscientists to see more geological detail than before and how seascapes and landscapes have evolved through time. Seismic Geomorphology, when integrated with seismic and sequence stratigraphy, is also a powerful tool for the prediction of lithologies, stratigraphic architecture and processes in space and time. It therefore has a commercial significance for drilling lithologies suitable for hosting and sealing hydrocarbons. The book is divided into three main sections, (a) review of the basic assumptions in seismic data analysis of depositional systems and revisiting sequence stratigraphic models, (b) data interrogation strategies and (c) sedimentary environment case studies.

R. J. Davies
H. W. Posamentier
L. J. Wood
J. A. Cartwright

Acknowledgments

This Geological Society Special Publication is the result of the first joint Geological Society and SEPM conference, which has now become an annual event. We would to thank all the contributors – who made the meeting such a success. Vickey Sare (ChevronTexaco) is thanked for helping convene the meeting and we are extremely grateful to Päivi Heiniö for helping to coordinate the technical programme. Anadarko, BP, Paradigm Geophysical Ltd., The Bureau of Economic Geology, and the SEPM kindly sponsored the conference. This book would not have been possible without the expertise of the reviewers who kindly gave up their time to review papers. Published in collaboration with SEPM (Society for Sedimentary Geology).

1926 SEPM

Society for Sedimentary Geology

Seismic geomorphology – an overview

H. W. POSAMENTIER,[1] R. J. DAVIES,[2] J. A. CARTWRIGHT[3] & L. WOOD[4]

[1]*Anadarko Petroleum Corporation, 1201 Lake Robbins Drive, The Woodlands, TX 77380, USA (e-mail: Henry_Posamentier@anadarko.com)*

[2]*CeREES (Centre for Research into Earth Energy Systems), Department of Earth Sciences, University of Durham, Science Laboratories, Durham DH1 3LE, UK (e-mail: Richard.Davies@durham.ac.uk)*

[3]*3DLab, School of Earth, Ocean and Planetary Sciences, Main Building, Park Place, Cardiff University, Cardiff CF10 3YE, UK (e-mail: joe@ocean.cf.ac.uk)*

[4]*Bureau of Economic Geology, University of Texas at Austin, TX 78713-8924, USA (e-mail: lesli.wood@beg.utexas.edu)*

Abstract: Seismic geomorphology, the extraction of geomorphic insights using predominantly three-dimensional seismic data, is a rapidly evolving discipline that facilitates the study of the subsurface using plan view images. A variety of analytical techniques is employed to image and visualize depositional elements and other geologically significant features. This volume presents key technical papers presented at a recent research conference – the Seismic Geomorphology Conference (10–11 February 2005), co-convened by the Society for Sedimentary Geology and The Geological Society (London). These papers cover a broad range of topics, from detailed depositional element analysis to big picture regional issues, from lithology prediction to diagenetic modification of the stratigraphic section. This discipline is only in its early stages of development and will henceforth expand rapidly in response to the growing availability to researchers of high-quality three-dimensional seismic data.

The derivation of stratigraphic insights from seismic data has its origins in the early 1970s with the advent of improved quality two-dimensional (2D) seismic data. The assumption that seismic amplitude reflections approximate geological time lines was fundamental to the development of seismic stratigraphy (Vail *et al.* 1977). The discipline of seismic stratigraphy traces its roots to the landmark publication of AAPG Memoir 26, which summarized the work of Peter Vail and his colleagues at Exxon Production Research Company (Vail *et al.* 1977). Discrete seismic reflection packages, or depositional sequences, were defined by discontinuities shown in seismic data by the downlap, onlap, truncation or toplap of seismic reflections (Mitchum *et al.* 1977). Such seismic discontinuities were interpreted to represent stratigraphic discontinuities and unconformities. Inferences with regard to lithologies were based upon internal reflection character such as reflection amplitude and continuity. These 2D-based interpretations were then mapped and the spatial distribution of depositional systems with associated lithological predictions subsequently interpreted.

Seismic reflection technology underwent significant advances in the 1980s, making these data less expensive to acquire and hence more widely available. Three-dimensional (3D) seismic reflection data comprised acquisition of closely spaced 2D seismic lines with high precision navigation, which, when computationally manipulated, yielded true 3D coverage in *X–Y–Z* space. At first, such data were interpreted as a succession of parallel 2D seismic sections. Techniques included printing the lines on translucent vellum and interpreting each section while partially seeing the immediately adjacent section through the vellum. This approach made interpretation easier from the perspective of mapping horizons but did not truly take advantage of the 'third-dimension' inherent to the 3D volume. In essence this approach resulted in little more than a tightly spaced 2D seismic analysis. By the mid to late 1980s, computer-based display and visualization of 3D data began to take hold, making true 3D interpretations possible. Methods evolved for generating horizontal and flatted slices, arbitrary traverses, wavelet attribute extractions and mapping, and rapid analysis of large complex data volumes.

From: Davies, R. J., Posamentier, H. W., Wood, L. J. & Cartwright, J. A. (eds) *Seismic Geomorphology: Applications to Hydrocarbon Exploration and Production*. Geological Society, London, Special Publications, **277**, 1–14.
0305-8719/07/$15.00

Horizontal slices, flattened time slices, and proportional slices, derived from 3D volumes, provided plan view images of amplitude and other attribute distributions that strongly resembled depositional environments. Increasing computing power enabled large volumes to be manipulated and interpreted rapidly, and visualization software increased the ability for 3D visualization of surfaces. These volumes provided map views over geological time from which depositional elements and systems could be interpreted. Such plan view images yielded data that showed buried landforms; hence the discipline of seismic geomorphology came into being (Posamentier 2000). *Seismic geomorphology* may be defined as the application of analytical techniques pertaining to the study of landforms and to the analysis of ancient, buried geomorphical surfaces as imaged by 3D seismic data. Seismic geomorphology, when used in conjunction with seismic stratigraphy, represents the state of the art approach to extracting stratigraphic insights from 3D seismic data. The following section outlines workflows for optimizing this process.

Workflow

Most workflows designed to derive stratigraphic insights from 3D seismic data involve an initial reconnaissance step. Commonly the interpreter will quickly scan through a 3D seismic volume by in-line, crossline, and time. The objective is to identify anomalous seismic features referred to as FLTs (i.e. 'funny looking things'). Typical FLTs include local amplitude brights or dims, lineaments, or in general any features that might 'look geological'. Another reconnaissance approach involves opacity rendering, whereby the 3D volume is rendered transparent except for specific amplitude values associated with a particular target, such as a channel fill deposit. Usually it is the amplitude extremes that are rendered opaque, thus allowing the targeted opaque features to stand out.

Once an FLT is identified, it becomes the focus of further detailed analysis. This analysis can take the form of: (1) horizon picking and subsequent illumination; (2) amplitude extraction along specific horizons; (3) horizon slicing or stratal slicing, whereby the volume is flattened on a key horizon and then amplitude extractions are made from time slices parallel to the key horizon; (4) proportional horizon slicing, where an interval is bound between two mapped horizons and then is unproportionally sliced between those two horizons; (5) interval attribute analysis whereby an interval that brackets the FLT is defined and then characterized seismically; (6) voxbody picking; (7) extraction of horizon-based attributes such as dip magnitude, dip azimuth, curvature and roughness; (8) extraction of volume-based attributes such as phase, coherence, and impedance; and (9) volume co-rendering, whereby two or more volume attributes are displayed simultaneously within the same volume. Examples of each will be shown in the next section.

The key to each of these analyses is to look for and recognize geologically or geomorphologically meaningful patterns in plan view as well as in section view. Such patterns can take the form of fluvial or deep water channels, slumps and slides, shelf sediment ridges, and carbonate patch reefs, to name just a few. For the geomorphological approach to seismic interpretation to succeed, it is essential for the interpreter to have a broad experience base with respect to seismic plan-view and section-view expression of a variety of depositional elements. For example, if one is not familiar with the stratigraphy and geomorphology of fluvial point bars, then features like scroll bars in plan view, and lateral accretion sets in section view may pass by the interpreter unnoticed.

A critical step in the evaluation of any seismic feature is to iterate between section and plan views. A geological feature must have an expression that is scientifically reasonable in multiple dimensions. *Analyses of section view integrated with plan view images represent the integration of seismic stratigraphy with seismic geomorphology.*

An equally critical step in seismic geomorphologic analyses is the integration of borehole data if available. These data provide critical lithologic and sedimentologic ground truth for the seismic interpretation. Modern analogues or unequivocal ancient analogues can be used to enhance the credibility of an interpretation.

Examples of seismic analytical techniques

Horizon picking and illumination. 2D or 3D displays of seismic reflections can often significantly enhance the appearance of a depositional element's external morphology. For example, a channel can appear as a seismically expressed trough; a carbonate patch reef will be recognizable by its circular, positive relief. Once a horizon or horizons bounding such an element are interpreted, various attributes can then be draped on such a surface to further enhance the appearance of the depositional element. Figure 1 illustrates an unconformity surface mapped in time illustrating the current subsurface structure on that horizon. Figure 2 illustrates an amplitude extraction draped on a horizon. In some instances merely illuminating such a horizon from different lighting angles can provide significant stratigraphic insight (Fig. 3).

Horizon parallel or stratal slicing. Several techniques exist for slicing through a 3D seismic data volume. They include time slices, dipping planar

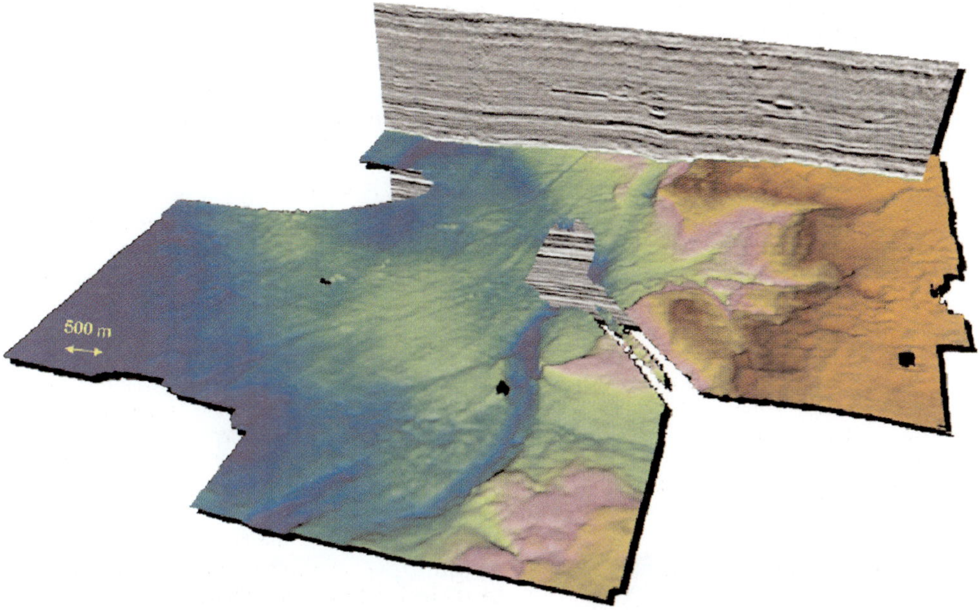

Fig. 1. Time structure draped on interpreted horizon. Base Cretaceous unconformity, western Canada basin, Alberta. Note the presence of fluvial channels on this surface.

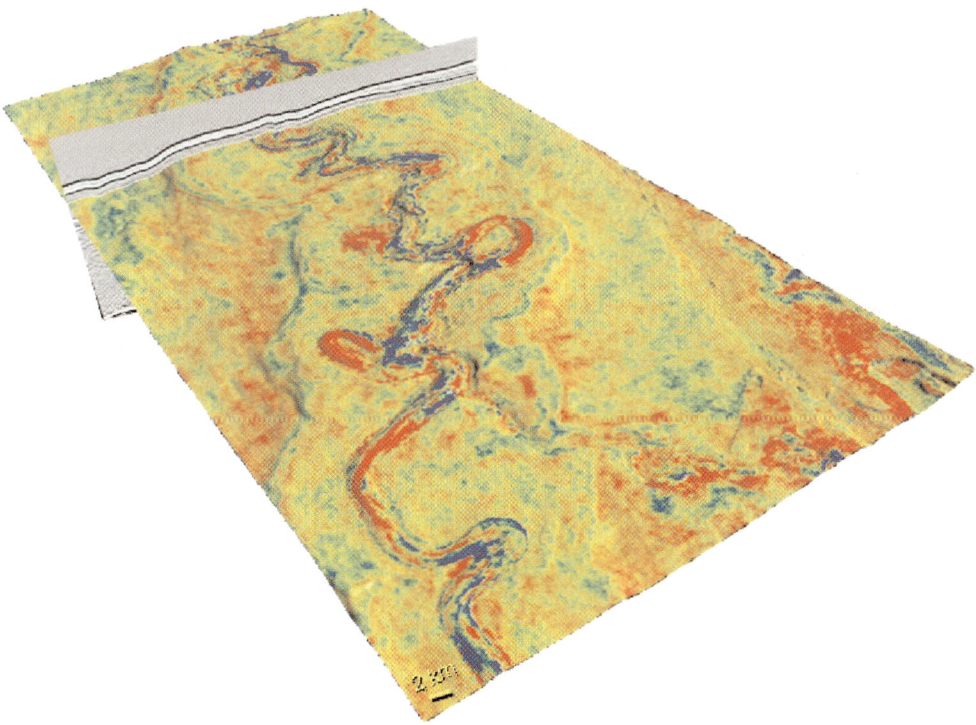

Fig. 2. Reflection amplitude draped on interpreted horizon suggesting presence of sand within deep-water turbidite channel, Gulf of Mexico.

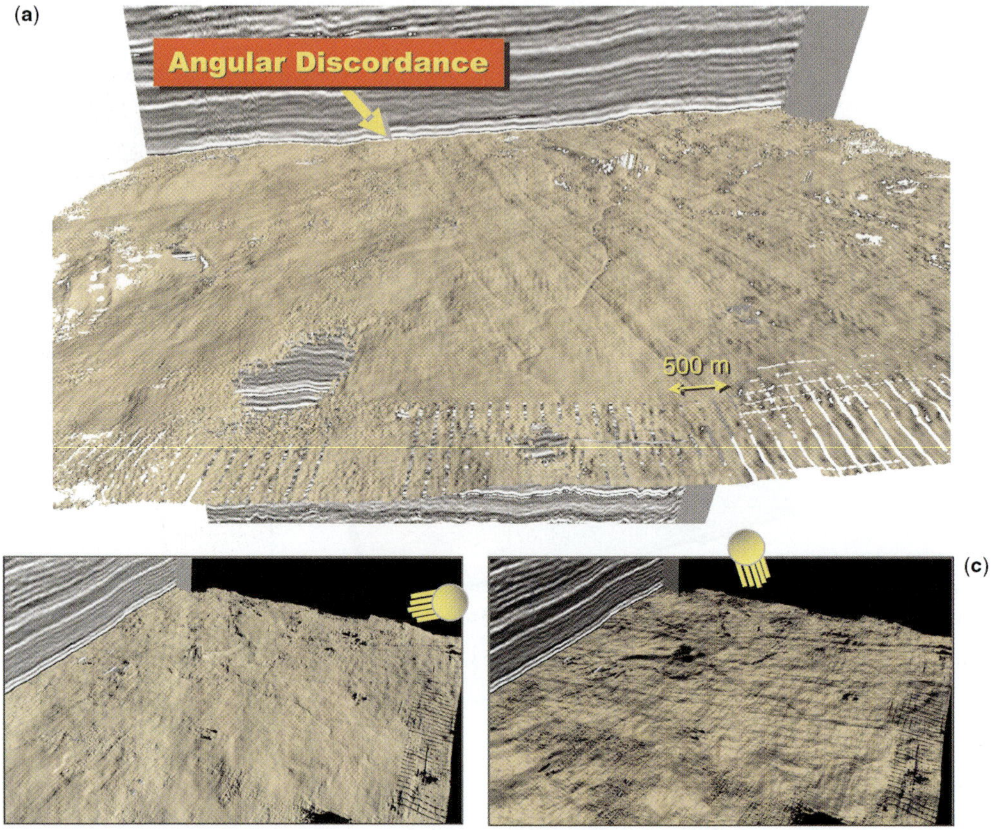

Fig. 3. (**a**) Illuminated unconformity surface in perspective view with channel and ridge-form erosional remnants (i.e. cuestas). (**b**) The same unconformity surface with lighting from the right – parallel to the trend of the ridge-form erosional remnants. (**c**) The unconformity surface with lighting from the top. Note that the ridge-form erosional remnants are far better visible when the lighting is orthogonal to the ridges.

slices, horizon-parallel slices (i.e. stratal slices), and proportional slices. Commonly, when looking for stratigraphic features, the best results are obtained by slicing as near as possible to the target interval and parallel to a well-mapped structural horizon. This technique will reduce the affects of structuring that might obscure the imaging of palaeo-geomorphic features. Where seismic reflections are parallel and nearly horizontal, then seismic time slices would suffice (Figs 4 and 5a). Where reflections are uniformly dipping, then dipping planar slices are appropriate (Fig. 5b). Where reflections are characterized by variable dip, then horizon parallel or horizon slices (i.e. stratal slices) yield the best results (Fig. 5c). In those instances where reflections are uniformly divergent, then proportional slicing would be ideal. This involves slicing between two non-parallel reflections, whereby the interval between the two reflections is proportionately divided into an equal number of slices (Fig. 6).

Horizon or slice amplitude extraction Amplitude extractions along horizons or along seismic slices can reveal the presence of depositional elements by virtue of the different impedance characteristics of the depositional element relative to the surrounding strata. For example, the deep-water crevasse splay shown in Figure 7 is apparent from the amplitude extraction along a horizon slice.

Interval attribute analysis. Seismically characterizing an interval that contains an FLT can sometimes yield superior results. Numerous interval attributes can be generated, such as maximum positive polarity amplitude, maximum absolute polarity amplitude, the ratio of positive maximum to negative maximum amplitude, total cumulative amplitude. Examples of interval attributes are illustrated in Figure 8. One interval attribute worthy of special mention is trace-shape or seismic-facies analysis. This interval attribute involves examining a sub-sample of traces

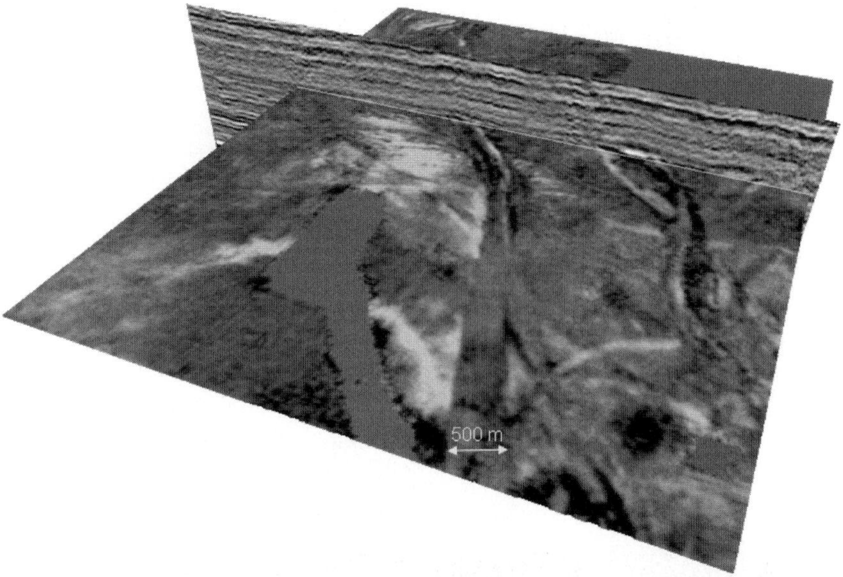

Fig. 4. Time slice amplitude extraction showing small fluvial channels, western Canada sedimentary basin, Alberta, Canada. Note that because the time slice is close to parallel with the seismic reflections, reasonable imagery of channels is achieved.

for a given area across a specific interval and then characterizing these traces according to shape. The interpreter arbitrarily determines how many classes are needed to characterize the variability of the interval in question, and then all traces are assigned to one of these classes. The result is a seismic facies map (Fig. 9a). An extension of this process is the generation of correlation maps, whereby each trace is correlated with a particular class that can yield further detail (Fig. 9b).

Voxbody picking. 3D seismic data are composed of numerous voxels, each voxel corresponding to a seismic sample along a seismic trace. Voxbody picking, also referred to as subvolume detection, involves selecting a voxel and then highlighting those connecting voxels that satisfy user-defined attribute values (most often amplitude). The connected highlighted voxels can help identify depositional elements such as channel fills or other geobodies of relatively similar attribute value (Fig. 10).

Horizon-based attribute mapping. Attributes calculated along horizons can bring to light subtleties of depositional elements not apparent in other displays. Figure 11 illustrates multiple horizon attributes calculated for a deep-water turbidite channel.

Volume-based attribute mapping. In instances where horizon slicing does not work well because no reference horizons useful for datuming can be interpreted with confidence close to the target interval, volume-based attribute analyses can be useful. One commonly used such attribute is coherence or discontinuity, whereby the similarity (or dissimilarity) between adjacent traces can be mapped. This technology, originally designed for mapping subsurface faults, is an excellent edge-detection tool and is especially useful for defining geobodies with sharply defined margins such as channels (Fig. 12).

Volume co-rendering. Where two attributes provide useful information regarding the lithofacies distribution within a geobody, co-rendering may provide the ideal display for extracting stratigraphic, geomorphic and depositional systems insights (Fig. 13). In some instances interpretations are facilitated by showing coherence in plan view and amplitude in section view (Fig. 14).

Summary

All of the techniques discussed above contribute to improving our understanding of a particular geomorphic feature or surface of interest. Critical to the success of this approach lies in the 'ground truth' calibration of lithofacies and depositional setting using borehole data. As with any other aspect of seismic interpretation, it is imperative

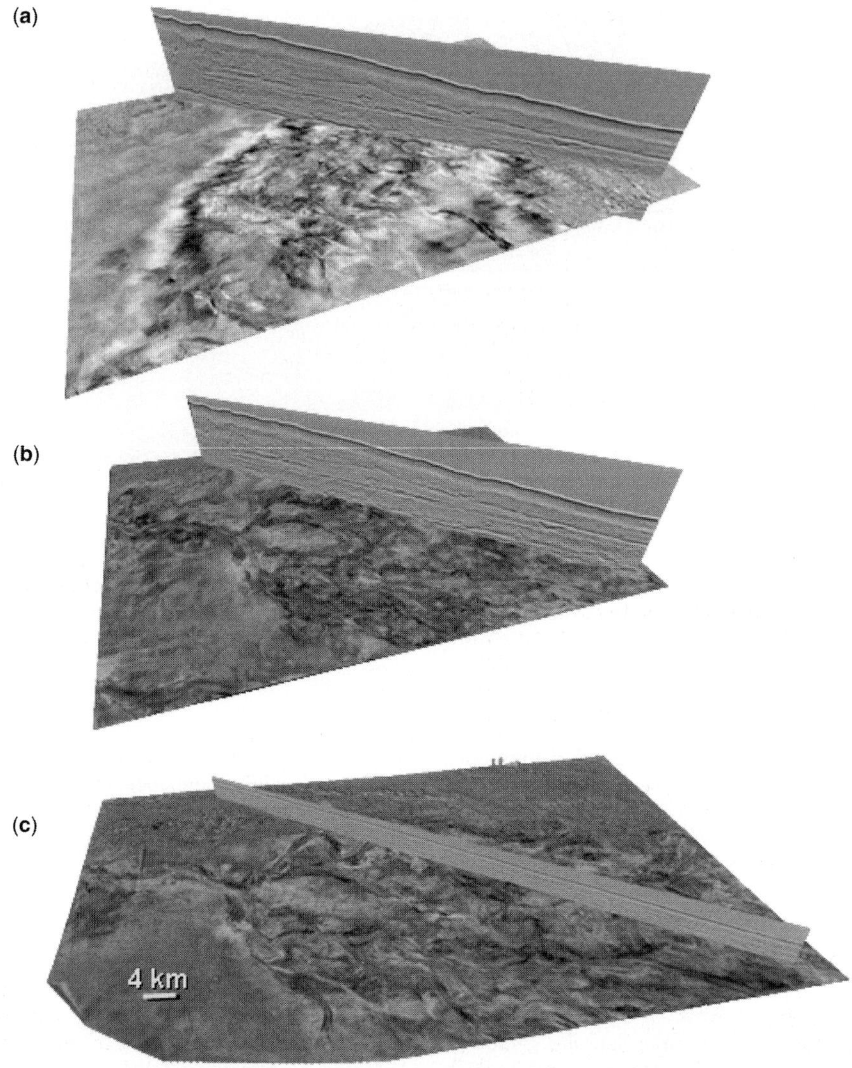

Fig. 5. (**a**) Seismic amplitude extraction along time slice through southeast-dipping reflections. Part of a frontal splay (i.e. lobe) turbidite system in the deep-water Gulf of Mexico is imaged. (**b**) Seismic amplitude extraction along dipping planar slice oriented approximately parallel to seismic reflections. More of the depositional system is imaged. (**c**) Seismic amplitude extraction along horizon-parallel slice, also referred to as horizon slice or stratal slice. This slice achieves the best possible image of the depositional system.

to be aware of pitfalls that model bias brings to the interpretation process. Moreover, the interpreter must be able to distinguish between seismic expression of actual depositional elements and geophysical data artifacts. As progressively more computer-driven analytical tools are employed, it is important that the interpreter be aware of the limitations of vertical and spatial resolution on their geologic interpretation, and how the various attributes derived and analysed in the workflow were calculated. Planform pattern recognition is a powerful addition to the interpreter's day-to-day toolkit that brings with it the need to be aware of the broader context of the features under scrutiny, and also to recognize that the planform imagery to a greater or lesser extent can be influenced by choices made by the interpreter in correlating any given horizon.

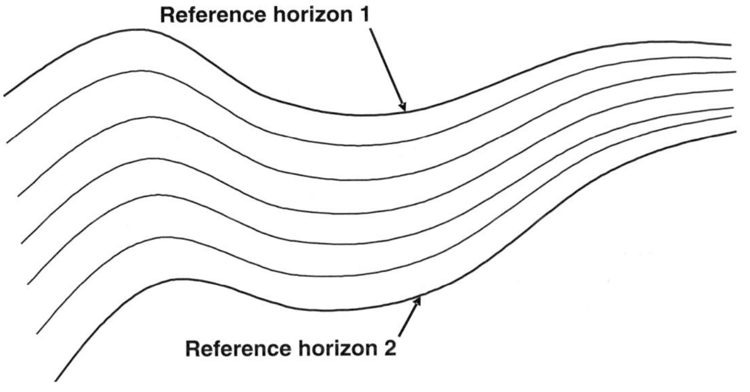

Fig. 6. Proportional slicing through divergent seismic reflections.

A look forward

Seismic geomorphology, based on interpretation of plan-view seismic images, is rapidly developing on several fronts. These are (a) understanding the development of seascapes and landscapes in clastic and carbonate settings, (b) advances in workflows directed towards lithological prediction through the integration of seismic stratigraphy and seismic geomorphology, (c) revising and improving sequence stratigraphic models, and (d) development of new and increasingly more sophisticated analytical techniques. These are all useful directions. How is seismic geomorphology poised to impact geoscience, compared for example with experimental analogue modelling and fieldwork? Where are the fundamental discoveries going to come from in the future – what new breakthroughs will be made?

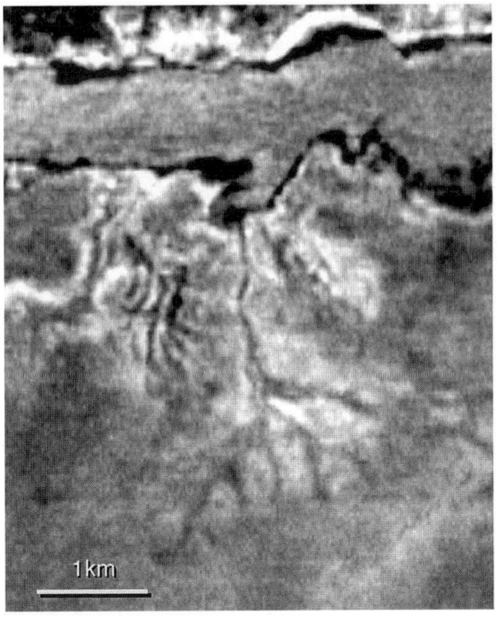

Fig. 7. Seismic amplitude extraction along a horizon slice through a deep-water crevasse splay (note the distributive channel pattern), Gulf of Mexico. Reflection amplitude suggests that the channels are sand-prone.

Palaeoceanography

Palaeoceanography is a discipline essential to the understanding of past climate change. Palaeocirculation patterns and water mass structure can be reconstructed through interpreting seismic data across contourite drift complexes. Seismic geomorphology studies of contourite drifts have been published recently (Knutz & Cartwright 2003; Hohbein & Cartwright 2006). As seismic data coverage increases in deep-water environments along continental margins, the role that these types of studies could play may become more important.

Palaeoclimatology

Newly developing techniques in morphometric analysis of seismic geomorphological features offer promise in better understanding the nature of palaeo-drainage and channel discharge over geological time. The potential for reconstructing a region's rainfall and discharge history through analysis of changes in channel orientation, size, width:depth ratios and sinuosity, for example, may eventually contribute significantly to a more detailed climate history for regions than is presently achievable. In addition, critical linkages between areas such as the South China Sea and the Indian Ocean may be better understood through more detailed seismic geomorphological-derived history of the region. Such ocean-to-ocean linkages are critical for understanding the palaeo-El Nino and La Nina

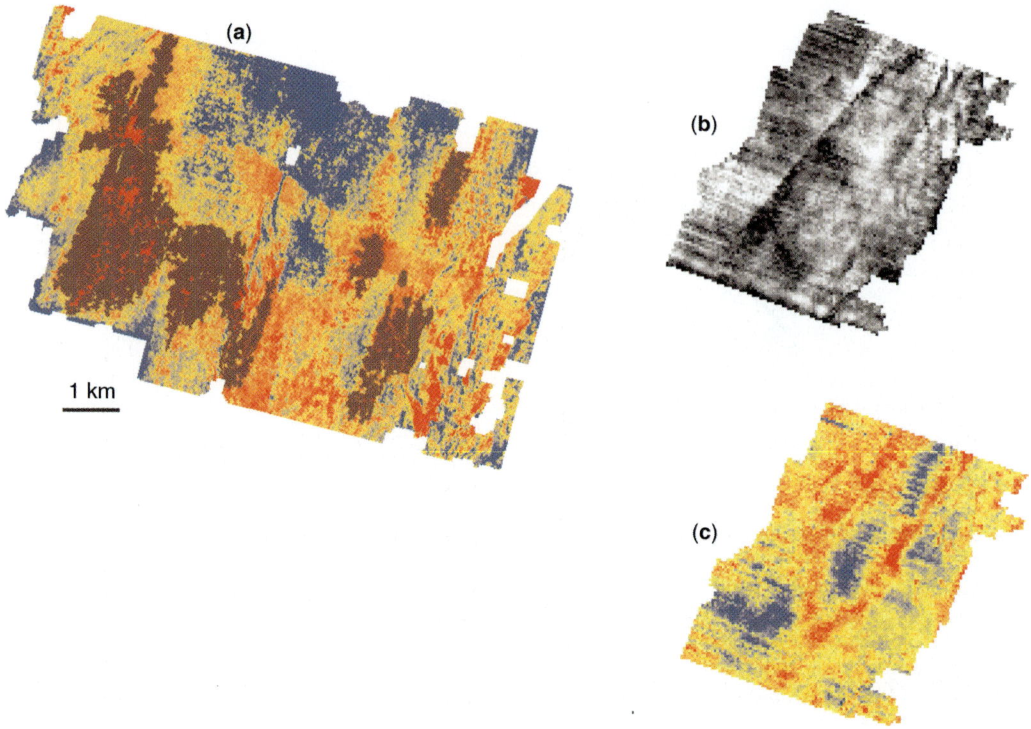

Fig. 8. Several interval attributes characterizing Miocene shelf sand ridges, offshore Northwest Java (modified from Posamentier 2002). (**a**) Interval attribute (cumulative amplitude) illustrating several sand shelf sand ridges. Note the presence of a 1 km-wide distributary channel that is imaged along with the sand ridges though it lies just beneath. (**b**) Detail of an isolated sand ridge delineated by a well-defined margin on one side (the edge of the ridge on the side that constitutes the leading edge) and a poorly defined margin on the other (i.e. the trailing edge). This attribute is the maximum negative polarity for an interval that brackets the sand ridge. (**c**) Detail of the same sand ridge as shown in (b). This attribute, a third derivative map for the same bracketing interval, highlights the presence of smaller sediment waves (large dunes?) superimposed on the larger sediment ridge.

events. A history of disruption of these linkages through sea level change is important for developing accurate palaeoclimate models.

Deep-water channel complexes

Most of the advances in understanding the stratigraphic architecture and lithofacies distributions of deep-water systems are driven by oil and gas exploration in offshore areas such as the deep Atlantic Ocean of West Africa, and the Gulf of Mexico. These systems are highly complex and variable. As seismic resolution continues to improve, some of this complexity will be more clearly imaged and consequently better understood. Such enhanced understanding is critical to the commercial success of exploration in these often very-high-cost environmental settings.

Biogeography

The growing coverage of 3D seismic data in our continental margins around the world affords us the unprecedented opportunity to understand the geomorphology of exposed shelves and shallow seaways of the world as never before. The history of exposure of regions used by palaeocultures as migration pathways, for example between mainland Asia and Indonesian archipelago, or between northern Asia and North America, is currently very poorly constrained. Seismic geomorphology (especially when applied to the near sea floor) offers an opportunity to redefine the history of lowstand shelves around the world and shed great insight into the migration routes of early cultures.

Anthropogenic hazards

The past 20 years of data collection, both seismic and borehole data, as a by-product of conventional energy exploration have brought to light a number of catastrophic processes active on margins around the world. Submarine slope failures have

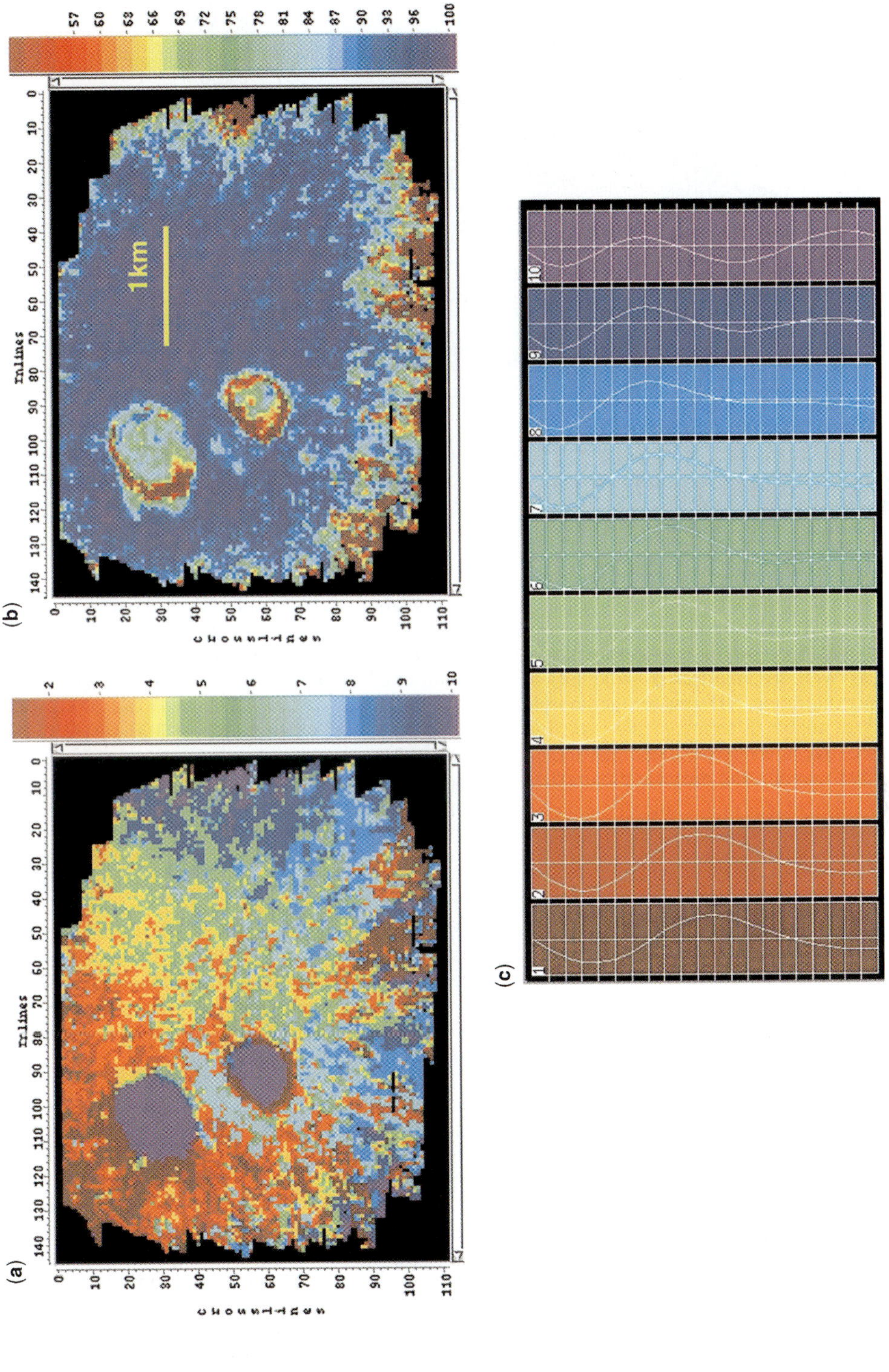

Fig. 9. Seismic facies mapping of two Devonian pinnacle reefs, western Canadian sedimentary basin. **(a)** Seismic facies map illustrating that all seismic traces within the reefs are assigned to a single facies class. **(b)** Seismic facies correlation map wherein each trace is compared with class 3 (c). Note the more detailed seismic and therefore geological expression of the reefs. **(c)** The neural-network-derived 10 seismic facies classes upon which the analysis was based (Stratimagic software by Paradigm Inc.).

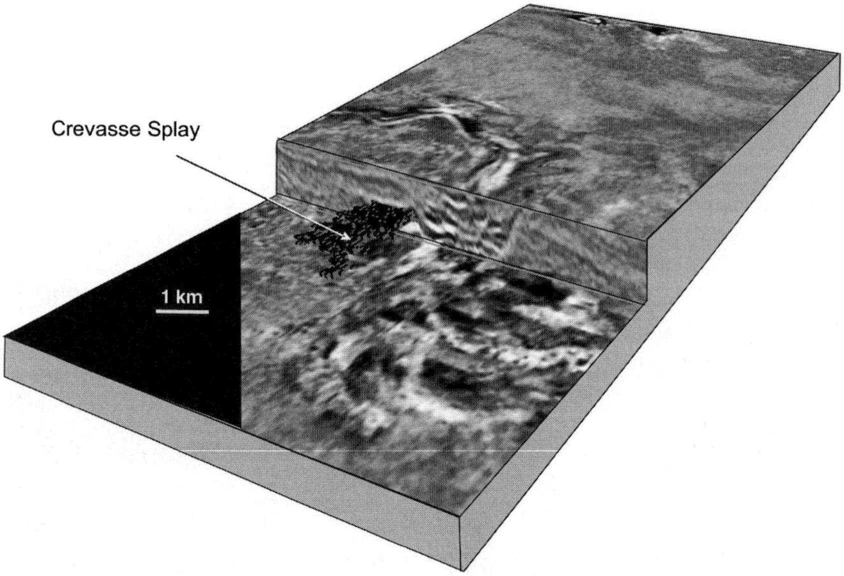

Fig. 10. Voxbody expression of deep-water crevasse splay, Pliocene of the eastern Gulf of Mexico.

been documented to generate large tsunamis in coastal regions. In addition, such mass slope failures can endanger ocean tele-communication equipment as well as engineered structures for hydrocarbon exploration. 3D seismic geomorphological analyses enable us to image and investigate these processes like never before, hopefully increasing our ability to predict and mitigate the impact of these events.

Tectonic geomorphology

This sub-discipline of geomorphology has evolved in the past decade as a result of increasing interest in Neotectonics, and the interplay of geodynamics and landscape evolution in the mountain belts (Burbank & Anderson 2001). Topics such as the structural control of river systems can easily be extended into the submarine and subsurface realm, for example, in the structural control of depositional systems on continental slopes. The seismic geomorphology of deformed surfaces is much more difficult to interpret in comparison with surfaces within simple layer-cake, unstructured successions, but with restoration techniques now available in three dimensions, the analysis of deformed surfaces is now tractable.

Igneous geomorphology

As exploration ventures into deep water, it is moving into regions of more stretched crustal type and even in some cases into basins floored by oceanic crust. Perhaps as much as 70% of divergent continental margin basins are of the volcanic type, exemplified by the northeast Atlantic margins of Greenland and Norway (Planke *et al.* 2000). Not surprisingly, therefore, more 3D surveys are acquired in areas of volcanic and intrusive activity, and this opens up potential for a wholly new approach to igneous geology, using a seismic geomorphological methodology. This approach has been defined as seismic volcanostratigraphy in the pioneering work of Planke *et al.* (2000). Studies of intrusive features such as mafic sills (Hansen *et al.* 2004) and extrusive features such as submarine volcanoes (Davies *et al.* 2002) and subaqueous lava flows (Trude 2004) are showing a host of features that have not been observed using traditional outcrop methodologies, and there is much potential for further advances in this field.

Contributions in this book

Back to basics

It is a healthy sign that, through the discipline of seismic geomorphology, some of the papers in this book go back to question some of the basic tenets of sequence stratigraphy, such as the correct placement of the sequence boundary relative to sea level. This revisited and sometimes contentious question is tackled by focusing on the late Quaternary

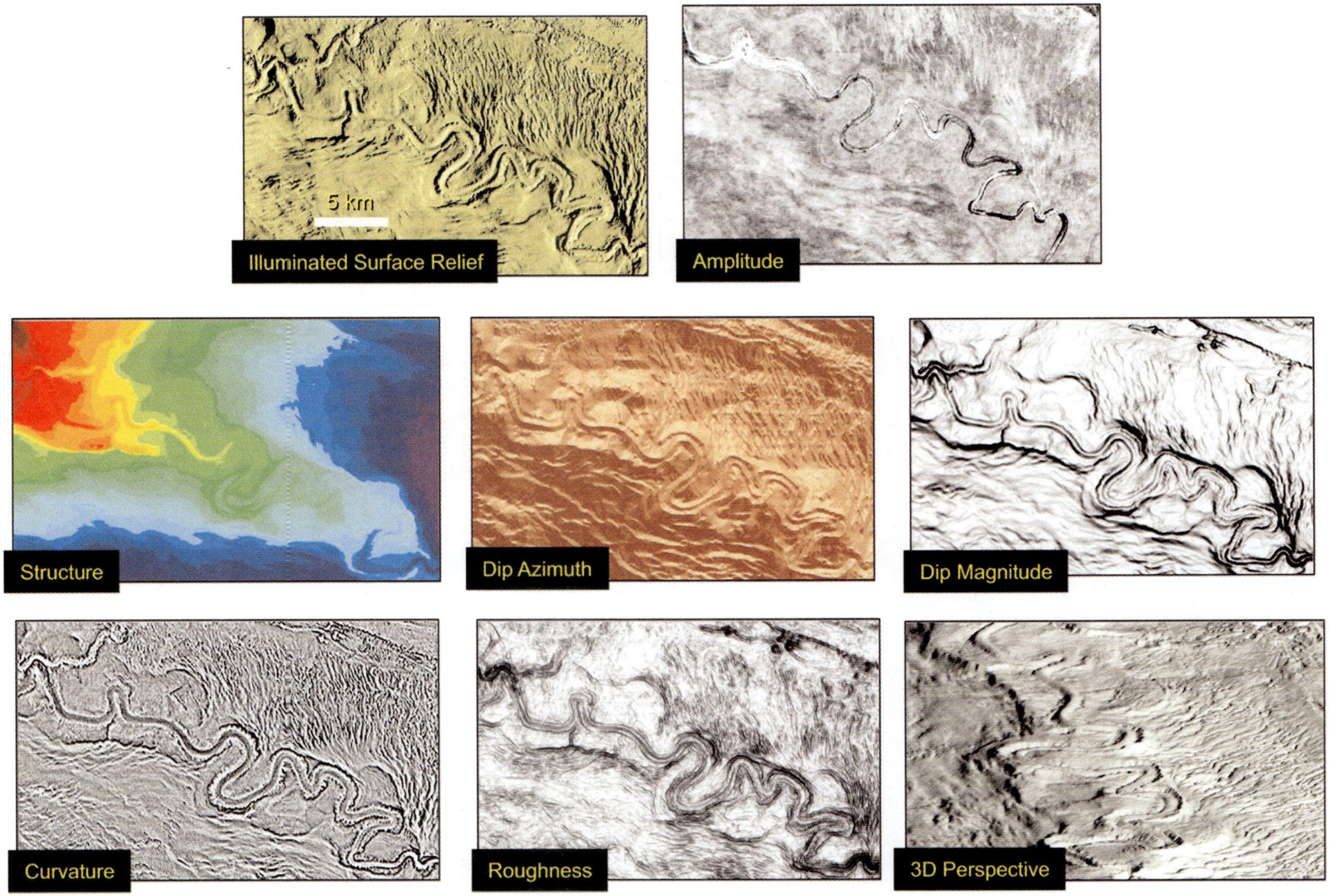

Fig. 11. Various horizon-based attributes extracted from the upper bounding surface of a deep-water turbidite channel, Gulf of Mexico. Each attribute affords the interpreter different and potentially valuable insights regarding depositional elements.

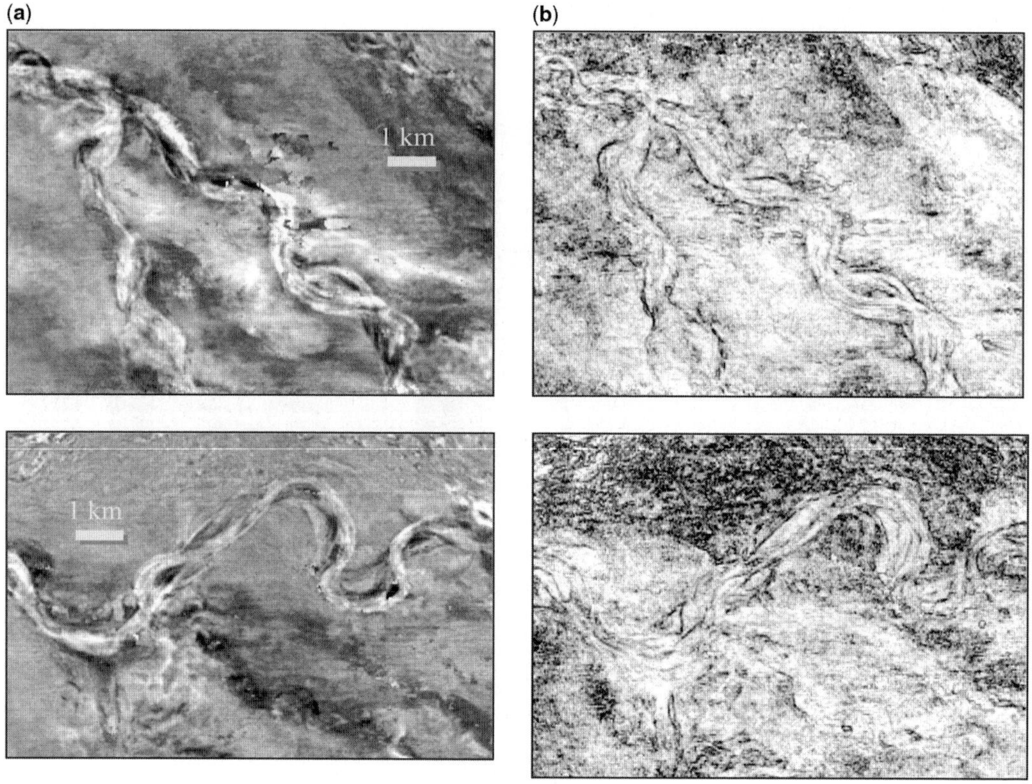

Fig. 12. Amplitude (**a**) and coherence (**b**) time slices illustrating a deep-water turbidite channel, Gulf of Mexico. The amplitude reflection attribute is an indicator of acoustic properties of the channel fill, whereas the coherence attribute highlights the edges of the channel.

geomorphology of a sequence boundary in the Gulf of Mexico, when sea-level history is well constrained through radiocarbon analyses and other chronological data (**Simms *et al.***). **Crumeyrolle *et al.*** also focus on the lowstand part of the sea level curve, using 2D and 3D seismic data to further refine sequence stratigraphic models. The datasets they employ allow them to be holistic in terms of depositional setting and cover fluvial to deepwater sediments in the Mahakam Shelf. Their approach is subsequently broadened through the use of outcrop analogues. **Hadler-Jacobsen *et al.*** take a geological long-term view, examining the stacking pattern of deepwater sediments over a 50 Ma window, and relating this to basin margin dynamics (adjustment, deformation and sediment deposition).

Data interrogation strategies

A book on seismic geomorphology would not be complete without consideration of seismic resolution, imaging quality and strategies for interrogating the data (**Zeng**). **Hart & Sagan** discuss clastic and carbonate settings from Precambrian to Late cenozoic in age, demonstrating the utility of techniques in curvature analysis. **Sullivan *et al.*** further discuss the use of seismic attributes to study karstification in the Pennsylvanian Marble Falls limestones.

Depositional elements – case studies

This book includes papers on a breadth of sedimentary environments. **Handford & Baria** illustrate how 3D seismic data are being used to re-think well-known exploration and development targets, such as the Smackover Formation, in the US Gulf Coast. They dispel the previous interpretations of the Smackover deposited as a homoclinal carbonate ramp, revealing clinoforms that they interpret to be oolitic carbonate shoreface deposits. **Rabelo *et al.*** look at fluvial depositional environments, parameterizing meander belts using 3D seismic data and seismic volumes. In the northern and the southern

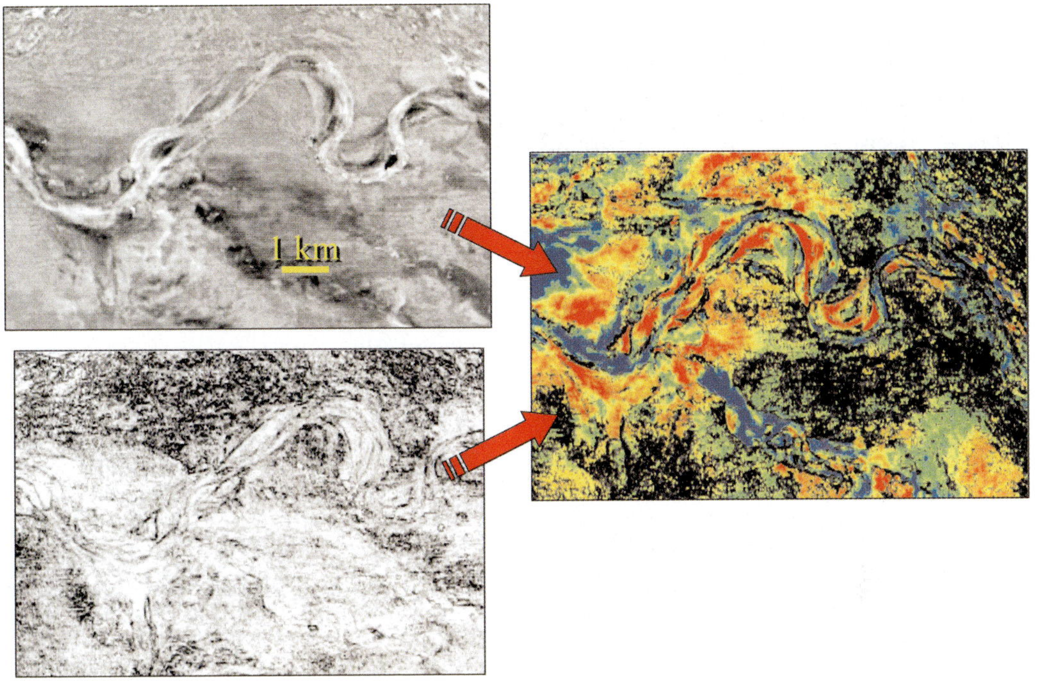

Fig. 13. Co-rendered amplitude and coherence image of a deep-water turbidite channel, Gulf of Mexico.

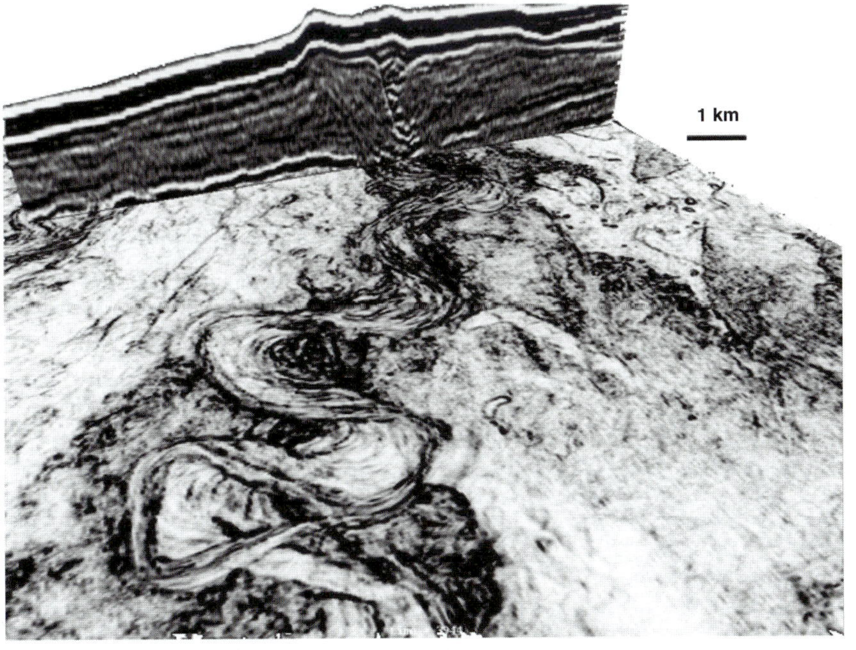

Fig. 14. Co-displayed coherence (in plan view, i.e. time slice) and reflection amplitude (in cross section view).

hemisphere the role of glacial processes in shaping landforms is particularly important. Such environments, in particular ice-streams, are investigated by **Andreassen *et al.*** using 3D seismic data from the Barents Sea.

The spectacular seismic data coming from offshore West Africa as well as other deep-water locations have revolutionized our understanding of deep-water sedimentary environments over the past 10–15 years. **Schwab *et al.*** use these data to full advantage in their analysis of a mixed turbidite-contourite system on the Mauritanian continental slope.

The scientific community is just beginning to understand the magnitude and impact of post-depositional compaction, diagenesis and sediment remobilization along continental margins. Increasingly, seismic investigations are uncovering extreme and repetitive sediment mobilization events. **Jackson** demonstrates the utility of 3D seismic data in understanding post-depositional remobilization of mud.

Conclusions

Pattern recognition, involving the interpreter being able to recognize geologically significant features in plan view on 3D the seismic data, is critical to the seismic geomorphological approach. In conjunction, it is also essential to cross reference plan view with section view images, thus integrating the geomorphology with the stratigraphy. Seismic geomorphology is a rapidly evolving discipline, benefitting from the rapidly accelerating widespread availability of 3D seismic data. Seismic geomorphological analyses address a broad range of disciplines ranging from igneous to sedimentary geology, focusing on questions ranging from lithology distribution to diagenesis to large-scale tectonic analyses. Although by no means an exhaustive treatise on this subject, this volume does present a representative cross section of applications and principles relevant to this rapidly evolving discipline.

The senior author wishes to thank Anadarko Petroleum Corporation for permission to publish this paper. We also thank VeritasDGC and Western Geco for permission to publish the seismic data shown here.

References

BURBANK, D. W. & R. ANDERSON. 2001. *Tectonic Geomorphology*. Blackwell Science, Oxford.

DAVIES, R., BELL, B., CARTWRIGHT, J. A. & SHOULDERS, S. 2002. Three-dimensional seismic imaging of dike-fed submarine volcanoes. *Geology*, **30**, 223–226.

HANSEN, D. M., CARTWRIGHT, J. A. & THOMAS, D. 2004. The development of intersecting igneous sills from 3D seismic. *In*: DAVIES, R. J., CARTWRIGHT, J., STEWART, S. A., UNDERHILL, J. R., LAPPIN, M. (eds) *3D Seismic Technology: Application to the Exploration of Sedimentary Basins*. Geological Society, London, Memoirs, Special Publications, **29**, 301–316.

HOHBEIN, M. & CARTWRIGHT, J. A. 2006. 3D seismic analysis of the West Shetland Drift: implications for late Neogene palaeoceanography of the NE Atlantic. *Marine Geology*, **230**, 1–20.

KNUTZ, P. & CARTWRIGHT, J. A. 2003. Seismic stratigraphy of the West Shetlands Drift: implications for North Atlantic circulation. *Palaeoceanography*, **18**, DOI: 10.1029/2002PA000786.

MITCHUM, R. M., JR, VAIL, P. R. & SANGREE, J. B. 1977. Seismic stratigraphy and global changes of sea level, part 6: stratigraphic interpretation of seismic reflection patterns in depositional sequences. *In*: PAYTON, C. E. (ed.) *Seismic Stratigraphy – Applications to Hydrocarbon Exploration*. American Association of Petroleum Geologists Memoirs, **26**, 117–133.

PLANKE, S., SYMONDS, P., ALVESTAD, E. & SKOGSEID, J. 2000. Seismic volcanostratigraphy of large-volume basaltic extrusive complexes on rifted margins. *Journal of Geophysical Research*, **105**, 19,335–19,351.

POSAMENTIER, H. W. 2000. Seismic stratigraphy into the next millennium; a focus on 3D seismic data. *American Association of Petroleum Geologists Annual Conference*, New Orleans, LA, 16–19 April, 2000, A118.

POSAMENTIER, H. W. 2002. Ancient shelf ridges – a potentially significant component of the transgressive systems tract: case study from offshore northwest Java. *American Association of Petroleum Geologists*, **86**, 75–106.

TRUDE, J. T. 2004. *Emplacement of igneous sills*. Ph.D. Thesis, Cardiff University.

VAIL, P. R., MITCHUM, R. M. JR & THOMPSON, S. III. 1977. Seismic stratigraphy and global changes of sea level, part 3: relative changes of sea level from coastal onlap. *In*: PAYTON, C. E. (ed.) *Seismic Stratigraphy – Applications to Hydrocarbon Exploration*. American Association of Petroleum Geologists Memoirs, **26**, 63–81.

Seismic imaging for seismic geomorphology beyond the seabed: potentials and challenges

H. ZENG

Bureau of Economic Geology, John A. and Katherine G. Jackson School of Geosciences, The University of Texas at Austin, University Station, Box X, Austin, TX 78713-8924, USA (e-mail: hongliu.zeng@beg.utexas.edu)

Abstract: A successful study of seismic geomorphology depends not only on knowledge of sedimentological, geomorphological principles and the local geological setting, but also on quality of the seismic geomorphological imaging. A thorough understanding of how seismic waves respond to geomorphology of depositional sequences and facies is essential prior to developing strategies and selecting tools for field seismic data interpretation. This is especially important in data that are of variable quality or lack marked amplitude anomalies. Studies presented in this paper show that it is more desirable to use stratal slices to display seismic information on geological time surfaces. Multi-slice or movie display of stratal slices is effective for the study of depositional process and is a good quality-control tool for avoiding seismic artifacts. Seismic wavelets adjusted to 90° phase help tie seismic traces to lithofacies with higher stratigraphic resolution. Seismic frequency bands of stratal slices should match the lithofacies thickness of interest for optimal facies imaging. Seismic facies analysis can be improved by automated geomorphological classification.

Seismic data from within a depositional basin fill present challenges for seismic geomorphology greater than those at seabed and shallower depth. With increased depth, seismic frequency decreases and seismic velocity increases, leading to reduced resolution. Seismic signal-to-noise ratio typically deteriorates with depth as well, and interpretation is further hampered at depth if the unit of interest is thin and below seismic resolution. In addition, reflections from multiple thin beds commonly interfere with each other and form merged events, making object- or volume-based imaging more problematic. Auto tracing can also be a poor choice for thinly interbedded sandstone/shale formations, which are less continuous and contain few strong seismic anomalies. In these situations, horizontal slices (time slices, horizon slices, stratal slices, etc.) are commonly applied to produce more desirable results. Studies in seismic geomorphology using slices have a great potential for providing new and vital information for basin analysis, exploration and production.

In this article, the author discusses lessons learned from the slice-based seismic geomorphologic studies in Miocene–Pliocene strata in Vermilion Block 50 and Tiger Shoal area, offshore Louisiana. Table 1 summarizes seismic data characteristics and basic geological interpretation from previous studies (Hentz & Zeng 2003; Zeng & Hentz 2004). The interpreted seismic data are in a 0.4 to 3.5 s section, with dominant frequency ranging from 40 (shallow) to 20 Hz (deep). Well- and seismic-interpreted depositional systems include highstand coastal fluvial plain (Pliocene), highstand coastal delta (upper Miocene), lowstand shelf-edge delta (middle Miocene) and lowstand incised valley (Pliocene-Miocene). Many depositional sequences, especially coastal-plain fluvial sequences, are characterized by discontinuous seismic events without significant amplitude anomalies that cannot be easily mapped by auto-tracing or geobody detection. Most of the sandstones are low in acoustic impedance (AI, relative to shale) and thin (3–40 m), and most are below seismic resolution. Previous studies detail high-frequency stratigraphy and depositional history of the strata using well logs and seismic data (e.g. Zeng *et al.* 2001; Hentz & Zeng 2003; Zeng & Hentz 2004). This paper is neither a detailed case study nor a comprehensive overview of the seismic interpretation techniques. Instead it is a discussion of some of the seismic imaging techniques that should be considered for a meaningful study of seismic geomorphology, with the strengths and weaknesses associated with each technique highlighted. The physical basis of seismic reflection data and the integration of geological concepts (models) with seismic applications are emphasized.

How to slice

For sediments within a basin fill, where targets are thin and hard to distinguish from the amplitude background, amplitude-anomaly- and volume-based analysis of seismic geomorphology is difficult to

From: DAVIES, R. J., POSAMENTIER, H. W., WOOD, L. J. & CARTWRIGHT, J. A. (eds) *Seismic Geomorphology: Applications to Hydrocarbon Exploration and Production*. Geological Society, London, Special Publications, **277**, 15–28.
0305-8719/07/$15.00

Table 1. *Characteristic seismic data and geological interpretation in Miocene–Pliocene strata in Vermilion Block 50 and Tiger Shoal area, offshore Louisiana (summarized from Hentz & Zeng 2003; Zeng & Hentz 2004)*

	Pliocene	Miocene
Typical seismic section	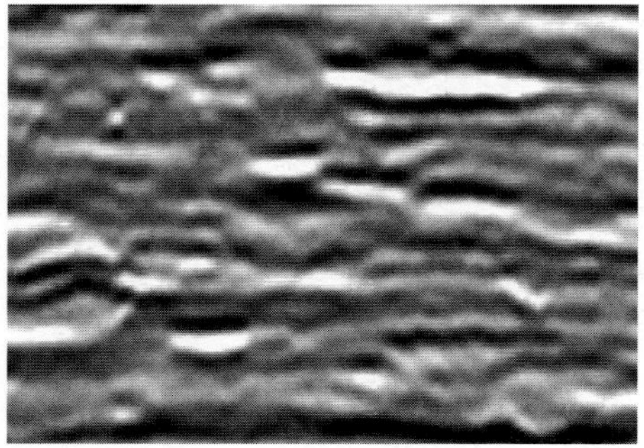	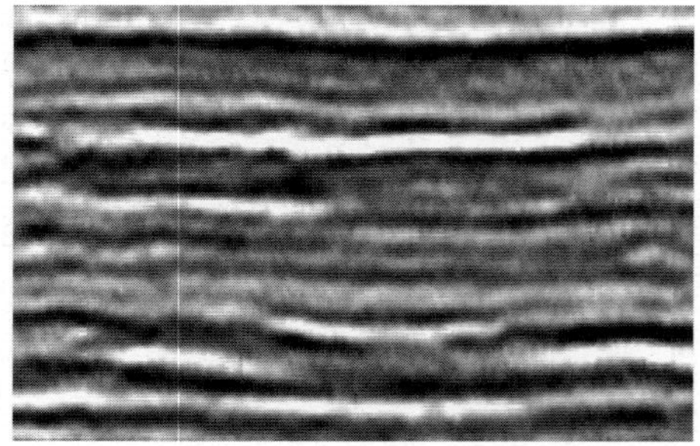
Depth range	400–1700 ms (time) 300–1800 m (depth)	1700–3500 ms (time) 1800–4400 m (depth)
Seismic frequency	30–40 Hz (dominant) 5–70 Hz (range)	20–30 Hz (dominant) 5–40 (range)
Sandstone thickness	3–30 m	3–40 m
Seismic facies	Discontinuous, variable amplitude	Variable continuity, variable amplitude
Depositional environment	Coastal fluvial plain (highstand), Incised valley (lowstand)	Coastal delta (highstand), Incised valley (lowstand), Shelf-edge delta (lowstand)

apply. Selecting the right slicing method for extracting geomorphic patterns properly from the 3D seismic data, without having to deal with event-picking problems related to inadequate seismic resolution, is therefore crucial. Three seismic slicing methods are available to the interpreter (Fig. 1):

1. Horizon slicing (Brown 1991) is the most widely used tool in seismic interpretation packages. Horizon slices are made by slicing parallel and away from a single reference seismic event (Fig. 1a). Because the method assumes a conformable depositional sequence without significant thickness changes, this approach is suitable for layer-cake-type successions represented by parallel seismic events or for successions near the reference event. The main problems with this method include the possibility of cutting obliquely through depositional units of different ages where significant lateral thickness changes occur and the nonuniqueness of slicing when the slices are made from different reference events.
2. Stratal slicing (Zeng 1994; Zeng *et al.* 1998*a*, *b*), or proportional slicing (Posementier *et al.* 1996), is increasingly being adopted by users and incorporated into software packages. These slices are made by linear sampling of seismic traces between two seismic events (Fig. 1b). The method improves geological time–surface extraction by adjusting to lateral thickness changes. Sediments among multiple reference events can be processed to generate a single stratal-slice volume for studying seismic geomorphology through time in a multi-slice mode or movie mode. The method is applicable to any data except those that are structurally complex (e.g. densely faulted or salt-diapir areas), in which good-quality seismic reference events are difficult to pick.
3. Seismic phase unwrapping (Stark 2004) is an attempt to build age (relative geological time) volume by unwrapping instantaneous phase traces (Fig. 1c). Volume sculpting using age volume generates seismic displays that follow defined, relative geological time. This approach assumes that all primary seismic reflections follow geological time surfaces that are intrinsically significant impedance boundaries. Good results are expected for high signal-to-noise ratio, high continuity and high-resolution data (e.g. Fig. 1c, left of section). However, the method is less suitable for formations that have lateral lithofacies changes characterized by seismic polarity reversals (e.g. from a low-AI channel sand to a high-AI floodplain shale). Also, age volume is less reliable for noisy data and seismically thin beds with amplitude and phase tuning (Fig. 1c), right of section], which can be falsely represented by phase artefacts.

Stratal slicing is considered a good choice in this study because of its ability to handle common and significant lateral thickness variations in Miocene–Pliocene strata. In this study, stratal slices were generated by using Recon StratalSlice™ software.

Stratal slicing, however, is still based on simplified geological models that can be violated in field data. The assumption that sedimentary rate and preserved thickness are linearly changing laterally is only a first-order approximation of subsurface formations and may not be correct in some situations. As a test, a stratal slice (Fig. 2b) of a fluvial system is not considered optimal because a slight adjustment (<10 ms or <10 m in magnitude, Fig. 2a) in surface picking in part of the survey area significantly improves channel imaging (channel a in the upper surface, Fig. 2c, and channel c in the lower surface, Fig. 2d). A future challenge is to develop a nonlinear slicing method that can account for lateral irregular sediment rate and thickness changes, thereby refining seismic geomorphological imaging whilst providing optimal ties to well data.

Seismic phase character

Seismic geomorphology can be an indicator of depositional elements and depositional processes. For example, a channel pattern seen in a coastal-plain sequence implies a fluvial process. However, it is also desirable for seismic data to be calibrated to lithology so that sediments of a depositional process can be evaluated and added to the analysis. A channel filled with sand or shale could imply an active channel or an abandoned channel, with completely different significance for hydrocarbon prospecting. With better correlation between amplitude and lithology, both sedimentary rocks and depositional processes can be studied using seismic data, and seismic sedimentology becomes possible.

Industrial standard seismic data is zero-phased. A zero-phase seismic trace is symmetrical to a single reflection surface, with the maximum amplitude approximating impedance contrast (Fig. 3a). Zero-phase data are ideal for lithofacies identification on an unconformity or in a thick bed. However, for seismically thin depositional units, amplitude traces from zero-phase data become asymmetrical and are more difficult to tie to lithology-indicative wireline logs (Fig. 3c). Worse, stratigraphic resolution of zero-phase data is low because of more significant side-lobe interference. As a result, the sign and strength of amplitudes observed on stratal slices do

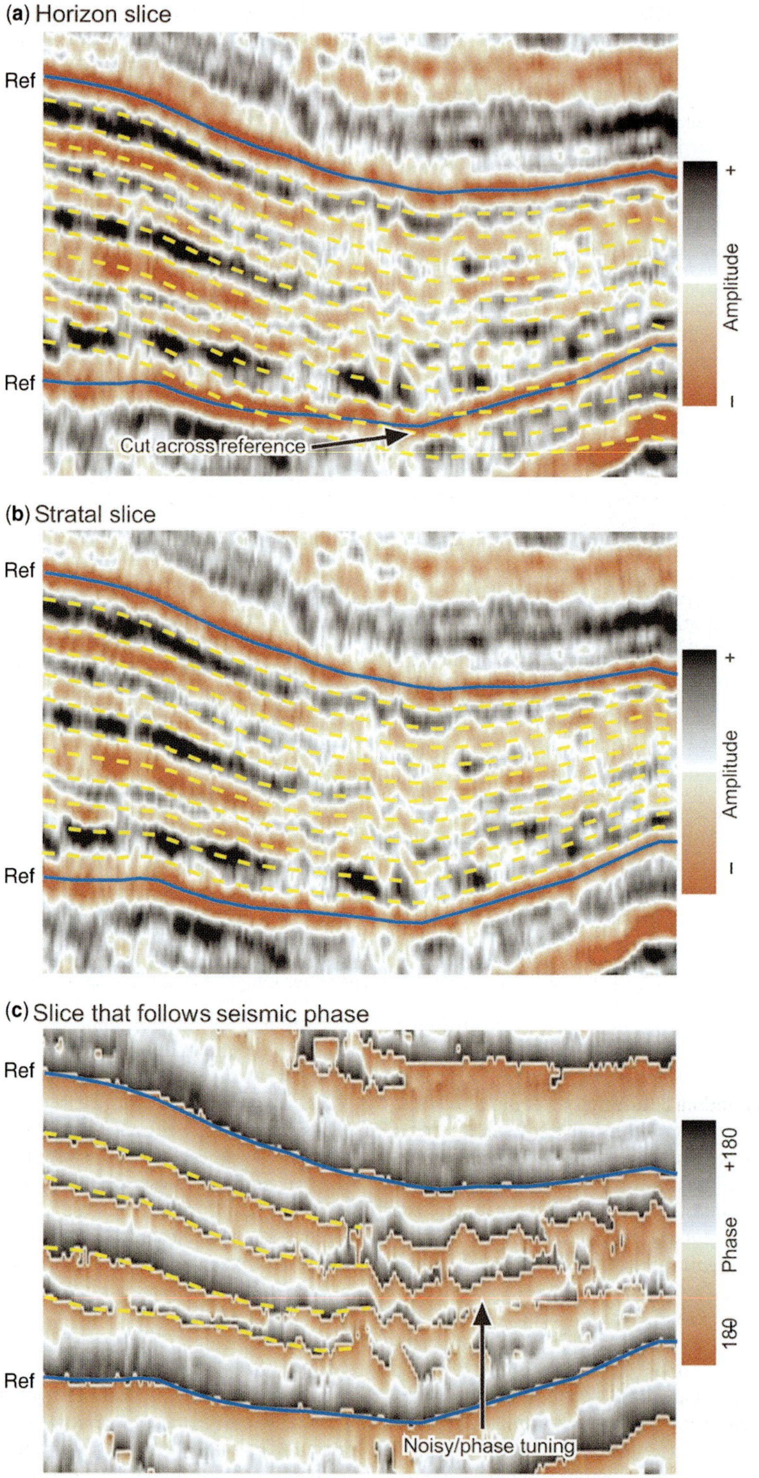

Fig. 1. Strategies of surface imaging for seismic geomorphology. Solid lines denote reference seismic events. Dashed lines are horizontal slices. Note that the horizon slice can cut cross-reference time events (**a**) and seismic phase unwrapping is unsuitable for noisy data or intervals with phase tuning of thin beds (**c**).

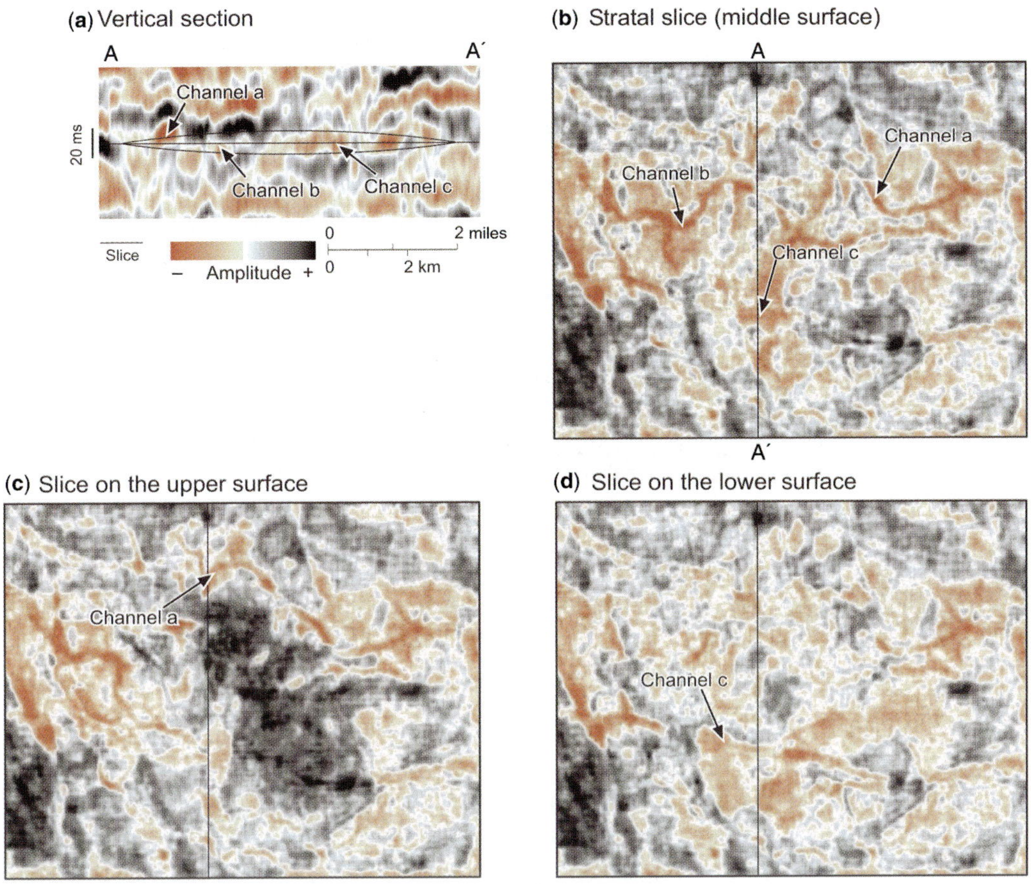

Fig. 2. Use of a nonlinear slicing tool to improve quality of stratal slices. (**a**) Vertical section A–A′, showing positions of the stratal slice picked at the middle surface on the basis of a linear sampling scheme and two slices, upper and lower, with corrections less than 10 ms (<10 m). (**b**) Stratal slice reveals complete image of channel b and incomplete images of channels a and c. (**c**) Upper slice has better image of channel a. (**d**) Lower slice has better image of channel c.

not necessarily indicate lithology (Zeng & Backus 2005*a*, *b*). The simplest and most effective way to improve interpretive value of amplitudes is to apply a 90° phase shift to the zero-phase data so that seismic traces are converted from reflectivity series to relative impedance series (Sicking 1982; Zeng & Backus 2005*a*, *b*). In 90° data maximum amplitude correlates to the centre of a thin bed (trough, Fig. 3d). A valuable byproduct of this process is that stratigraphic resolution of seismic data is improved. A seismic inversion fulfils a similar goal, but with nonuniqueness added to the data, however, because using post-stack seismic data can solve for either AI contrast or thickness, but not both, in a thin bed.

As a test, an attempt was made to tie spontaneous potential (SP) logs in wells to seismic data and to create a clear image of a sand-bearing depositional unit in an upper Miocene sequence (unit G in Fig. 4). As shown in a well log section (Fig. 4a, c), the F_2 Sand (17 m or 12 ms) is interpreted as an incised-valley-fill sandstone in shale, whereas the G Sand is a 13 m (9 ms), continuous sand body composed of lowstand incised-valley-fill sandstones and highstand sandy deposits of a coastal delta. A thin [9 m (6 ms)], transgressive shale separates the two sandy units (wells 218-54 and 218-14, Fig. 4a, c). Judging by well log evidence, the two sandy units are conformable in the area.

In standard zero-phase data (Fig. 4a), each sandy unit ties to a trough–peak couplet. The upper F_2 sandstone produces a strong, localized trough event in which the incised valley lies. The upper G sandstone is represented by a strong trough event that extends throughout the 3D survey, except where

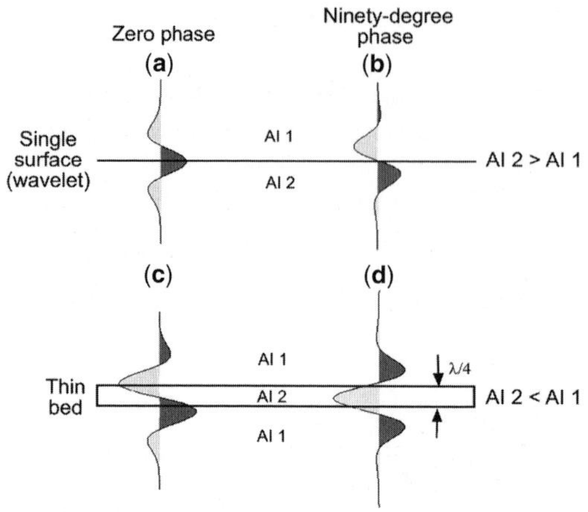

Fig. 3. Phase control of seismic waveform to a geological object. Corresponding to a (0° phase) Ricker wavelet, seismic reflection is symmetrical to a single surface (**a**), but is antisymmetrical to a seismically thin bed (λ/4 in this case, (**c**). In contrast, a 90° phase Ricker wavelet results in a reflection that is antisymmetrical to a single surface (**b**) but symmetrical to a thin bed (**d**). The 90° phase data are optimal for thin-bed lithological interpretation because the maximum amplitude (trough) points to the centre of the bed, seismic polarity approximates lithology, and stratigraphical resolution is higher (narrower side lobes).

the F_2 sandstone overlies it. In the area where F_2 developed, the seismic event associated with the G sandstone is of weaker amplitude. A stratal slice following the upper G trough event (Fig. 4b) shows a channel-like feature within the unit with a different amplitude level and polarity. This is interpreted as a seismic artefact because it does not match observation from well logs. In contrast, in 90° phase data (Fig. 4c) the two sand bodies can be tied to two individual seismic trough events. The F_2 sandstone is still expressed as a lenticular trough event, whereas the G sandstone is now a continuous event in the well-controlled area, even where the F_2 sandstone exists (Fig. 4d), showing no sign of the 'erosional' feature seen in the 0° phase slice (Fig. 4b). To the north, the feature still exists, but with a smaller extent (Fig. 4d), suggesting a local erosional event weaker than implied by the 0° phase slice (Fig. 4b). An interpretive pitfall is thus avoided.

Seismic frequency control

Seismic stratigraphy of thin depositional sequences is closely related to seismic frequency (Zeng & Kerans 2003). Similarly, seismic geomorphological patterns observed on stratal slices are controlled by seismic frequency content. In the frequency domain, seismic response is simply the product of wavelet spectrum and reflectivity spectrum.

$$S(\omega) = W(\omega) \times G(\omega) \quad (1)$$

where $\omega = 2\pi f$, with f denoting frequency in Hz, $W(\omega)$ is the wavelet spectrum, $G(\omega)$ is the reflection-coefficient (RC) spectrum and $S(\omega)$ is the seismic spectrum. These spectra are equally important because, in the seismic response, the frequency component that is strong in both spectra is relatively enhanced and the frequency component that is weak in either spectrum is relatively suppressed. For a fixed RC spectrum representing a specific geological profile, however, the wavelet spectrum is the sole cause of any changes in seismic reflections. In the time domain, seismic frequency control on seismic reflection is closely related to tuning phenomena. The lithofacies close to tuning thickness of a given wavelet is strengthened in amplitude, with other lithofacies suppressed in amplitude display. In short, seismic geomorphology is a function of wavelet frequency or tuning thickness.

Figure 5 illustrates a series of stratal slices imaged from the same interpreted geological time surface in the same 3D data volume. Each slice was extracted with a distinctive frequency band, the dominant frequency ranging from 15 to 45 Hz. Low-frequency slices (e.g. 15 Hz) reflect a field-scale lithofacies trend (fluvial sandstone belt *c.* 20 m thick at 2400 ms^{-1}). Midrange-frequency slices (e.g. 30 Hz) respond more to reservoir-scale lithofacies (channel sandstones *c.* 10 m thick). High-frequency slices (40 and 50 Hz) are related more to small channel sandstones and sub-reservoir-scale facies (levee, crevasse channel, etc. 6–8 m). In this case, the midrange, 30 Hz slice approximates the image made from original seismic data. Without frequency decomposition, information revealed in both low- and high-frequency bands is lost.

Tools for extracting and displaying more frequency components in seismic data are widely available. Frequency scan, which is based on panel filtering, is probably the easiest to use (Fig. 5). Spectral decomposition is preferred by some authors (e.g. Partyka *et al.* 1999) because it is capable of providing more accurate images from narrower frequency bands. In the author's experience, whether panel filtering or spectral decomposition is better depends on data and local geology. More tests should be documented to examine pros and cons.

The lesson learned from this example is that seismic data delivered to interpreters are not necessarily tuned to fit specific interpretation needs. Interpreters should be trained to be selective in utilizing seismic frequency bands on the basis of local geology and application purposes.

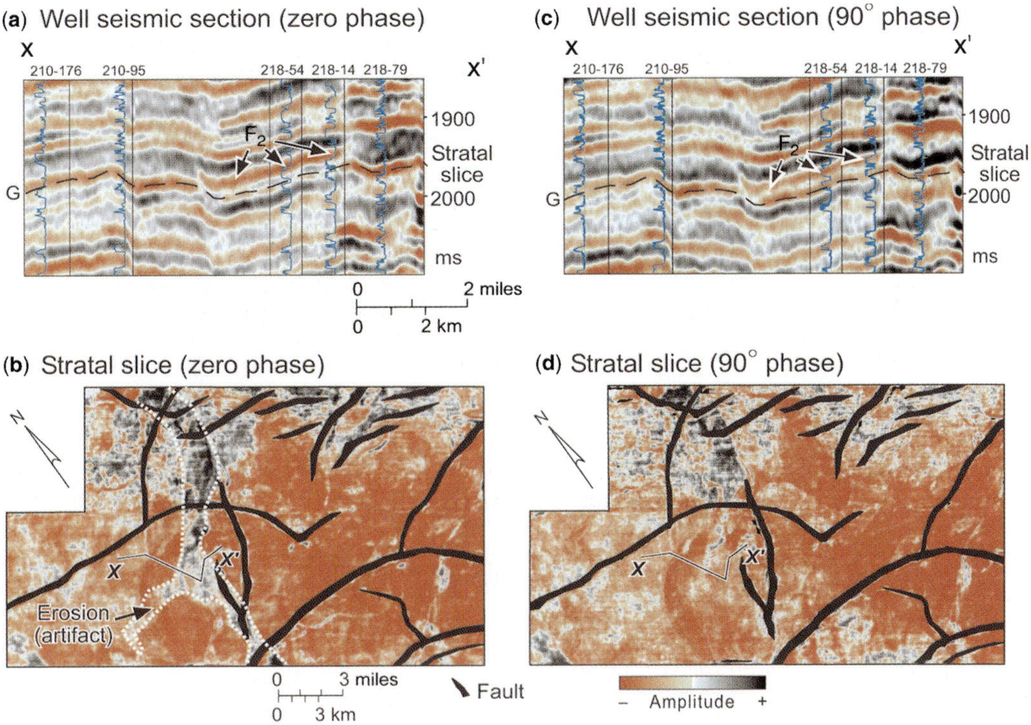

Fig. 4. Seismic interpretation example showing advantages of 90° phase seismic wavelets (modified from Zeng & Backus 2005*b*). (**a**) C_{clay} log section (X–X′) projected to zero-phase seismic data. Amplitude traces are not tied to lithology. A false, unconformable contact between the two sequences is visually apparent. (**b**) Stratal slice roughly following the seismic trough at upper G sandstone depicts a channel-like erosional feature wherever F_2 sandstone exists, creating an interpretive pitfall. (**c**) C_{clay} log section (X–X′) is tied to 90°-phase seismic data (red is sandstone; black is shale). The 'unconformable' contact observed in the zero-phase data disappears. (**d**) Stratal slice following seismic trough at the centre of G unit uncovers no seismic artefact.

Automation of geomorphology-based facies classification

Currently, facies classification using seismic geomorphology is done mostly by manually picking facies patterns on stratal slices. This process can be sped up and quality improved through geomorphology-based, automated, seismic facies classification.

Historically there are two different approaches to automatic pattern recognition of seismic data: waveform classification and texture mapping. Waveform classification is analysis of the shape of the amplitude trace at individual sample points and the relating of different trace shapes to different depositional facies. Currently, most widely used commercial software applications utilize this approach (e.g. StratiMagic™). Texture classification started with seismic facies analysis in seismic stratigraphy. Attempts have been made in recent years (e.g. West *et al.* 2001) to quantify seismic facies by grouping texture attributes derived from the gray-tone co-occurrence matrix. Seismic geomorphology has so far received little attention for automated facies analysis.

Zeng (2004) showed that the neural network can be used in supervised morphology classification. Each input node of the neural network takes on the value of one morphologic attribute, and each output node represents one facies class. Providing the neural network with morphologic attributes that can effectively describe different morphology groups (facies) in a statistical sense is key. Examples of morphologic attributes include (but are not limited to) size, orientation, geometric tendency (e.g. line or circle) and curvature. Supervised learning involves converting sample depositional facies (visually picked by geologists) into facies categories labelled arbitrarily (1, 2, 3, etc.). If results are satisfactory, the rule learned from the training can then be applied to the whole seismic data set for automated depositional facies mapping.

Figure 6 shows the morphology-based classification of some fluvial systems. On the basis of

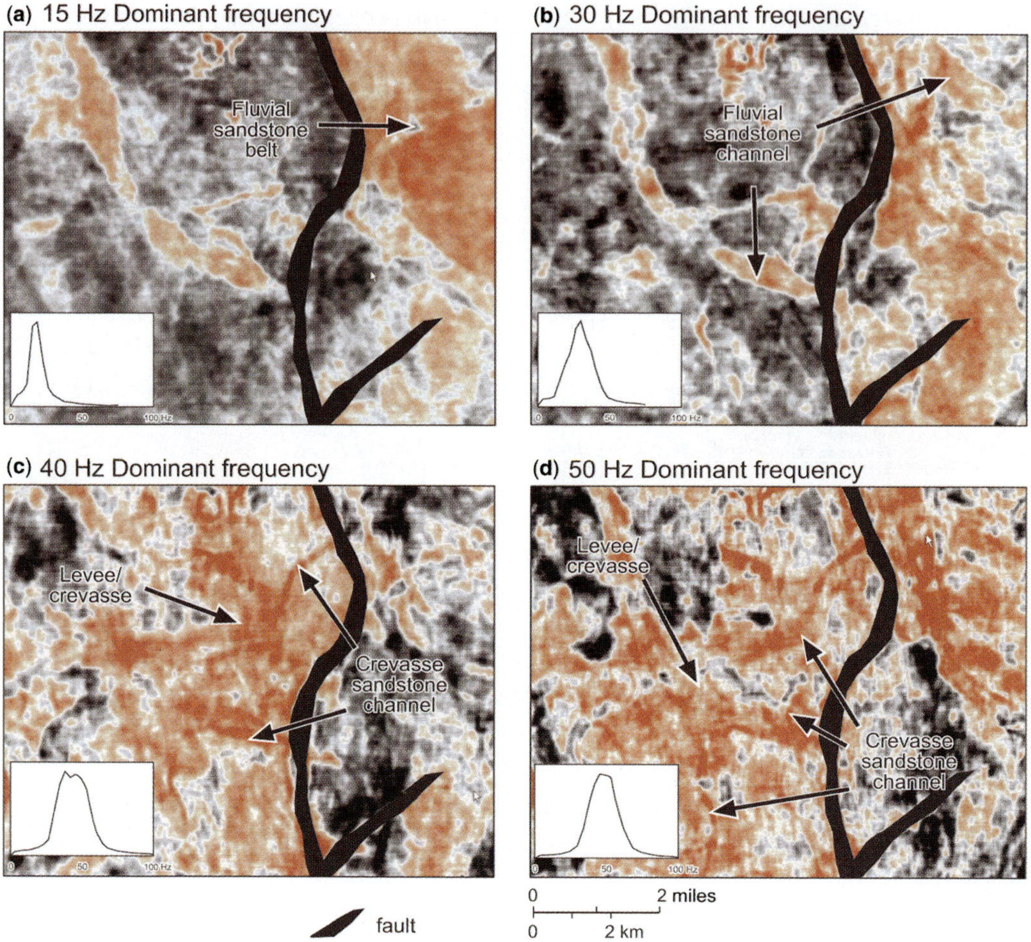

Fig. 5. Stratal slice in four-frequency band panel. A panel filtering that applied four bandpass filters with dominant frequencies ranging from 15 to 45 Hz (**a–d**) to the same 3D volume revealed geomorphological patterns that reflect different tuning thicknesses.

geomorphological patterns recognized on amplitude stratal slices (Fig. 6a), a neural network clearly identified systems having straight, sinuous, and anastomosing patterns as different fluvial facies (Fig. 6b). In the process, waveform (amplitude) is assumed to be insensitive to morphologic variations across facies spectra, and no texture attributes (e.g. grey-tone co-occurrence matrix) are used in analysis.

Showing the potential and feasibility of the automated method, the study is preliminary. More research should be encouraged on how to integrate all available seismic-pattern information – waveform, texture and geomorphology – for object-based, 3D facies classification. Another goal is to build an expert system for predicting depositional facies and lithofacies from seismic data by learning from geomorphology of modern depositional systems and seismic geomorphology of worldwide subsurface data.

A field-data example

A case history from a Pliocene coastal-plain sequence provides an example that requires special attention for slicing, phase character, and seismic frequency considerations for optimal depositional imaging. Being typical of what is being called 'less ideal' data, vertical seismic sections for the sequence (e.g. Fig. 7a) are dominated by laterally discontinuous seismic events, except for occasional traceable events (e.g. Ref. 1 and Ref. 2, Fig. 7a). Although few faults exist in the area, lateral depth and thickness

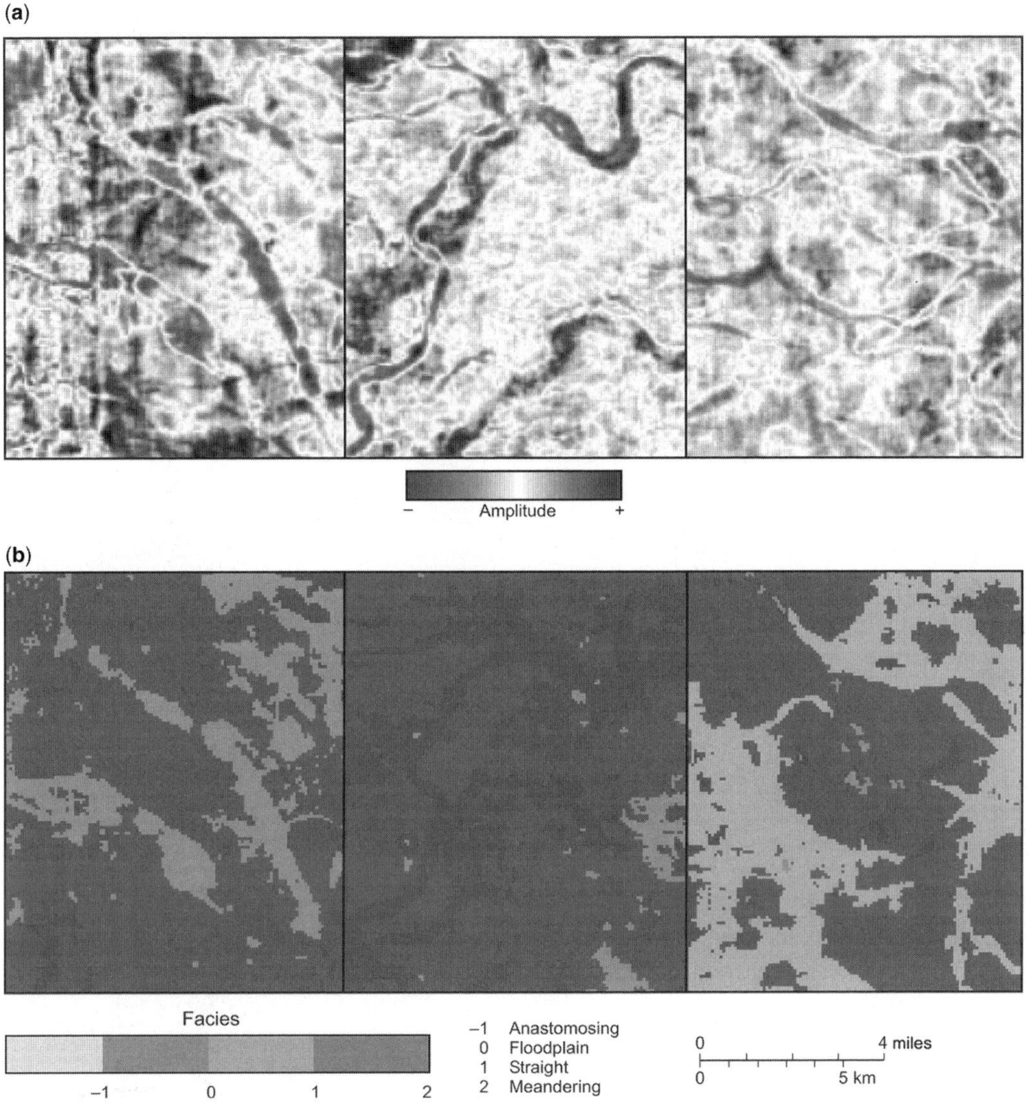

Fig. 6. Example of automated, geomorphology-based seismic facies classification (modified from Zeng 2004). (**a**) Amplitude stratal slices showing straight (left), meandering (middle), and anastomosing (right) fluvial systems. (**b**) Facies maps of straight (left), meandering (middle), and anastomosing (right) fluvial systems retained from a neural-network training of selected geomorphological attributes.

changes in sediments are common. Wells in adjacent areas penetrated fluvial sandstones ranging from 3 to 15 m in the sequence. In the 35 Hz dominant-frequency data, many of these sandstones are below seismic resolution ($\lambda/4 = 9$ m at 2400 m s^{-1}), and most of them are below tuning. Most commonly they are only marginally detectable in vertical section without resolving the tops and bases of the sand bodies, making a seismic facies analysis challenging (Fig. 7a). Non-parallel, lenticular events imply complex erosional–depositional relationships among different depositional units. Different channel systems and different lithofacies (e.g. channel axis v. overbank) have similar expressions on vertical seismic sections. They are characterized by the same amplitude range and the same seismic facies

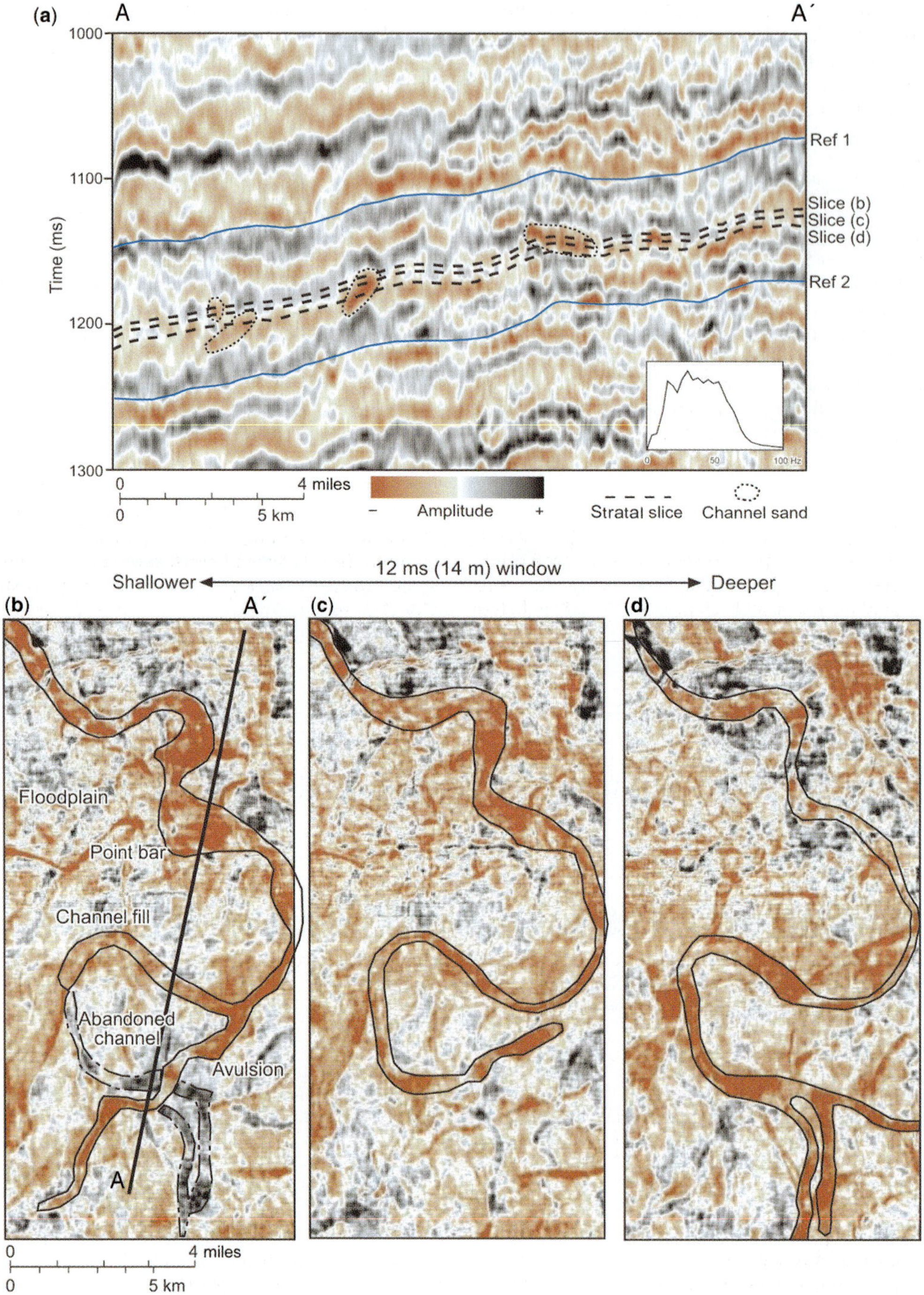

Fig. 7. Stratal slicing of a Pliocene meandering fluvial system, Offshore Louisiana. (**a**) A 90° phase seismic section of 35 Hz dominant frequency showing structurally simple but stratigraphically complex system. Highlighted lenticular channel bodies are indistinctive in amplitude and difficult to interpret without well control. (**b**) Stratal slice at the deeper end of a 12 ms (14 m) window. (**c**) Stratal slice at the middle of the window. (**d**) Stratal slice at the shallow end of the window. The stratal slices were generated by linear sampling of seismic amplitude traces between the two traceable maximum flooding surfaces in Ref. 1 and Ref. 2 in (a).

(i.e. discontinuous facies). In addition, seismic reflections from different lithofacies in different fluvial systems may merge to form new seismic events that are prone to being time transgressive, making a 3D correlation of those units – not to mention a study of fluvial processes – a formidable task.

Key to the study of this structurally simple but stratigraphically complex formation is good seismic geomorphological imaging. A set of amplitude stratal slices in the formation (Fig. 7b–d) reveals the presence of a meandering fluvial system that is superimposed on older fluvial plain deposits. Key depositional elements can be identified from images such as point bars, sandy channel fills, floodplains, channel avulsions and abandoned channels (mud plugs). Levee facies cannot be identified, probably having mixed with channel fill in the seismic images. Several tests were done to ensure that they are favourable seismic images for seismic sedimentologic study:

1. To correct lateral thickness variations in the fluvial depositional sequence, two reference events (Ref. 1 and Ref. 2, Fig. 7a) were applied for stratal slicing. Stratal slices better follow geological time surfaces (or narrow geological-time intervals) and typically offer more representative and complete depositional facies images. A horizon slice (Fig. 8a) that is equivalent to the stratal slice in Figure 7c was made by parallel-slicing off one of the reference events (Ref. 2, Fig. 7a). Even though the horizon slice is only 50 ms (60 m) or so away from the reference event, a segment of fluvial channel is completely missing, and channel images in other areas are more fragmented, showing poorer depositional surface approximation.
2. By careful adjustment of seismic phase to 90°, seismic amplitudes were tied to lithology, with negative amplitudes (red) indicating sandstones and positive amplitude (black) relating more to shale (Fig. 7). With a clear link to lithology, amplitudes in a channel pattern can be used to distinguish a sand-filled channel from mud plug in an abandoned channel (Fig. 7b). Lithology-labelled stratal slices typically show (cleaner) facies images (compare Fig. 7c with the zero-phase equivalent in Fig. 8b).
3. The original data (Fig. 7a) optimally tune to channel-fill sandstones (around 9 m), leading to a reliable interpretation of channel geometry (Fig. 7b–d). On a stratal slice made from a volume of higher-frequency (45 Hz dominant frequency, Fig. 8c), amplitudes tune to thinner marginal facies (levee, crevasse splay, etc.), making identification of the main channel more difficult. High-frequency data are not always better for facies imaging.
4. Other seismic attributes, when used appropriately, can help amplitude-based imaging and interpretation. For example, a continuity stratal slice (Fig. 8d) better images the boundaries of fluvial channels, especially small channels that seem blurred in amplitude displays (Fig. 7b–d). The slice also better reveals cut-and-fill relationships between different fluvial systems that provide helpful information in grouping depositional events and interpreting channel avulsion. The continuity attribute, however, has little lithological significance. Sand-filled channels look the same as mud-plugged channels (compare Figs 7b and 8d). Stratigraphic resolution is low owing to the winnowed nature of the attribute.

A succession of stratal slices allow the depositional process associated with fluvial systems to be studied in a multi-slice, historical mode. For example, in a 12 ms (14 m) interval, deeper (older) slices image a less-meandering channel system (Fig. 7d); in shallower (younger) slices, the channel becomes more sinuous (Fig. 7b and 7c); also in a shallower slice, channel avulsion is observed (Fig. 7b). With superimposition of channel patterns interpreted from slices (Fig. 9), a meandering-stream process emerges: through geological time, a meander channel became more sinuous, with migration of a channel downstream and occurrence of channel avulsion and abandonment. In this case, the time-series expression of the depositional process comes from detailed imaging of a narrow traveltime window equivalent to a single seismic event (Fig. 7a). The same precision and resolution of the depositional process study cannot be achieved by a volume-based amplitude detection of a geobody that is typically unable to track a geological event recorded in sediments thinner than a single seismic event.

Although it is somewhat subjective to make a judgement about whether a stratal slice follows geological time, a multi-slice (Fig. 7b–d) or movie display of stratal slices is also effective in reducing the risk of tracing a seismic artefact as a depositional element. Typically if a seismic geomorphological pattern is consistent or gradually changing through a stratal-slice series (e.g. Fig. 7b–d), it is more likely of geological-time significance; if the pattern simply cuts across seismic events, like time slices typically do, or parts of a depositional element appear on separate stratal slices (e.g. a segment of channel a that is apparent in Fig. 2c cannot be seen on Fig. 2c), an imperfect slicing may have occurred.

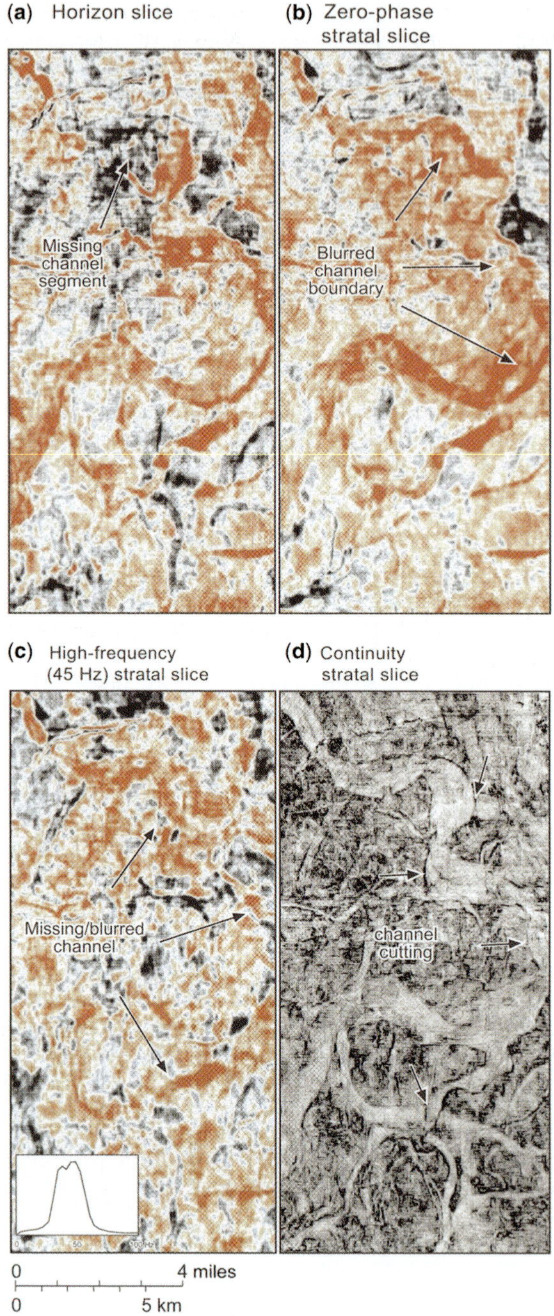

Fig. 8. Slices equivalent to the stratal slice in Figure 7(c), showing influence of the slicing method and imaging parameters on depositional facies imaging. (**a**) Horizon slice misses a segment of the main channel and makes the other part of the channel more fragmented. (**b**) On the zero-phase stratal slice amplitude is less correlated to lithology, and the image of the main channel becomes blurred. (**c**) A stratal slice generated from a volume of 45 Hz dominant frequency tunes amplitudes to thinner lithofacies, leading to the loss of main-channel identity. (**d**) Continuity stratal slice reveals cleaner channel images with recognizable cut-and-fill relationship between different fluvial systems, but fails to indicate lithology. Stratigraphic resolution is low, too, because of the time-window (12 ms in this case) required to generate the attribute.

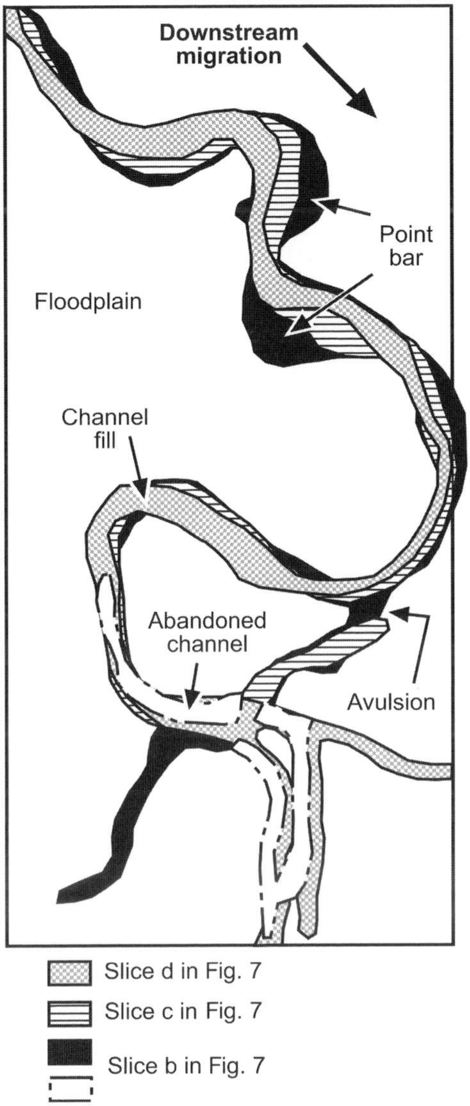

Fig. 9. Depositional process revealed by study of stratal slices in multi-slice or movie mode. Through geological time, the meandering fluvial system imaged in Figure 7 became more sinuous, migrated downstream and developed a channel avulsion.

Conclusions

1. A seismic geomorphological study using low-frequency seismic data from within a basin fill can be improved by analysing stratal slices that follow geological time surfaces (or narrow geological time windows).
2. Useful seismic calibrations for better geomorphological imaging include nonlinear correction to stratal slices, 90° phasing, and frequency scan.
3. Automated, geomorphology-based classification is a feasible and potentially useful tool.
4. As illustrated in a fluvial sequence, a multi-slice display of stratal slices through a depositional system is effective for study of depositional processes and for avoiding seismic artifacts caused by errors in stratal slicing.

The author acknowledges significant research contributions to this study by L. J. Wood, T. F. Hentz, M. V. DeAngelo, L. F. Brown Jr, A. C. Badescu, C. O. Kiliç and Y. Hu., C. Jackson and M. Gee reviewed and improved the presentation and are greatly appreciated. Figures were prepared by J. Coggin under the direction of J. L. Lardon. Lana Dieterich edited the text. Data are courtesy of ChevronTexaco. Landmark Graphics Corporation provided software for display of well log correlations and interpretation of seismic data via the Landmark University Grant Program. Partial support for this publication was received from the John A. and Katherine G. Jackson School of Geosciences and the Geology of Foundation of The University of Texas at Austin. Published by permission of the Director, Bureau of Economic Geology, John A. and Katherine G. Jackson School of Geosciences, The University of Texas at Austin.

References

BROWN, A. R. 1991. *Interpretation of Three-dimensional Seismic Data*, 3rd edn. American Association of Petroleum Geologists, Tulsa, OK, Memoirs, **42**.

HENTZ, T. F. & ZENG, H. 2003. High-frequency Miocene sequence stratigraphy, offshore Louisiana: cycle framework and influence on production distribution in a mature shelf province. *American Association of Petroleum Geologists Bulletin*, **87**, 197–230.

PARTYKA, G., GRIDLEY, J. & LOPEZ, J. 1999. Interpretational application of spectral decomposition in reservoir characterization. *The Leading Edge*, **18**, 353–360.

POSAMENTIER, H. W., DORN, G. A., COLE, M. J., BEIERLE, C. W. & ROSS, S. P. 1996. Imaging elements of depositional systems with 3-D seismic data: a case study. *In*: *Gulf Coast Section SEPM Foundation 17th Annual Research Conference*, Houston, TX, 213–228.

SICKING, C. J. 1982. Windowing and estimation variance in deconvolution. *Geophysics*, **47**, 1022–1034.

STARK, T. J. 2004. Relative geologic time (age) volumes – relating every seismic sample to a geologically reasonable horizon. *The Leading Edge*, **23**, 928–932.

WEST, B., MAY, S., EASTWOOD, J. & ROSSEN, C. 2001. Interactive seismic facies classification of stack and AVO data using textural attributes and neural networks. *In*: *Society of Exploration Geophysicists International Exposition and Seventy-Second Annual Meeting*, San Antonio, 9–14 September, CD-ROM.

ZENG HONGLIU. 1994. *Facies-guided 3-D seismic modeling and reservoir characterization*. Ph.D. dissertation, The University of Texas at Austin.

ZENG HONGLIU. 2004. Seismic geomorphology-based facies classification. *The Leading Edge*, **23**, 644–646.

ZENG HONGLIU & BACKUS, M. M. 2005*a*. Interpretive advantages of 90°-phase wavelets, part I: modeling. *Geophysics*, **70**, C7–C15.

ZENG HONGLIU & BACKUS, M. M. 2005*b*. Interpretive advantages of 90°-phase wavelets, part II: seismic applications. *Geophysics*, **70**, C17–C24.

ZENG HONGLIU & HENTZ, T. F. 2004. High-frequency sequence stratigraphy from seismic sedimentology: applied to Miocene, Vermilion Block 50, Tiger Shoal area, offshore Louisiana. *American Association of Petroleum Geologists Bulletin*, **88**, 153–174.

ZENG HONGLIU & KERANS, C. 2003. Seismic frequency control on carbonate seismic stratigraphy: a case study of the Kingdom Abo sequence, West Texas. *American Association of Petroleum Geologists Bulletin*, **87**, 273–293.

ZENG HONGLIU, BACKUS, M. M., BARROW, K. T. & TYLER, N. 1998*a*. Stratal slicing, part I: realistic 3-D seismic model. *Geophysics*, **63**, 502–513.

ZENG HONGLIU, HENRY, S. C. & RIOLA, J. P. 1998*b*. Stratal slicing, part II: real seismic data. *Geophysics*, **63**, 514–522.

ZENG HONGLIU, HENTZ, T. F. & WOOD, L. J. 2001. Stratal slicing of Miocene–Pliocene sediments in Vermilion Block 50-Tiger Shoal Area, offshore Louisiana. *The Leading Edge*, **20**, 408–418.

Geomorphology and age of the Oxygen isotope stage 2 (last lowstand) sequence boundary on the northwestern Gulf of Mexico continental shelf

A. R. SIMMS*, J. B. ANDERSON, K. T. MILLIKEN, Z. P. TAHA & J. S. WELLNER

Department of Earth Science, Rice University, 6100 S. Main, MS-126; Houston, TX 77005, USA

**Present address: Boone Pickens School of Geology, Oklahoma State University, 105 NRC, Stillwater, OK 74078, USA (e-mail: alex.simms@okstate.edu)*

Abstract: The sequence boundary associated with the last glacial–eustatic lowstand was mapped across the northwestern Gulf of Mexico continental shelf. The geomorphology of incised fluvial valleys varies widely across the shelf. These differences are due to differences in shelf physiography and the interval of the eustatic cycle the valleys were occupied. Incision begins during the falling limb of sea level and results in terraced valleys. Rivers that abandoned their valleys during the fall in sea level to cut new valleys during the lowstand generally have u-shaped profiles. Incised valleys connected to turbidite systems only occurred in two valleys (the Colorado and Rio Grande), but this may be because sea level did not fall below the shelf break during the last eustatic cycle. Some valleys deepen in an offshore direction, others become shallower. The timing of fluvial incision was constrained using radiocarbon dates so that incision can be tied directly to the sea-level curve for the last glacial–eustatic cycle. The results show that the fluvial incision occurred throughout the falling limb of sea level and lowstand; however, maximum incision occurred during the lowest position of sea level. The resulting surface has significant relief, extends across the shelf, and has time significance. The associated conformable surface, on the other hand, is much harder to recognize and occurs at different stratigraphic levels relative to different shelf-margin deltas.

Despite its predictive and organizational power, sequence stratigraphy has been scrutinized, debated, and revised in many papers since its inception. Some of these debates are new to stratigraphy while others are as old as the science itself. See Nystuen (1998) for an excellent review on this subject. One of the ongoing discussions within sequence stratigraphy revolves around which surface to use when separating sequences (Van Wagoner *et al.* 1988; Posamentier & Vail 1988; Galloway 1989). The 'Exxonian' view is that sequence boundaries separate depositional sequences, the sequence boundary being defined as an erosional unconformity and its correlative conformity (Mitchum *et al.* 1977; Vail *et al.* 1977*a–c*; Posamentier & Vail 1988; Van Wagoner *et al.* 1988). Galloway (1989) contends that the maximum flooding surface is a better bounding surface for defining genetic sequences. He argues that maximum flooding surfaces are easier to identify, span the transition in which marine and non-marine units inter-finger, and are marked by condensed sections that provide ample palaeontological material for dating.

Anderson *et al.* (2004) and Fillon *et al.* (2004) synthesized results from a sequence-stratigraphic analysis of late Quaternary strata of the northern Gulf of Mexico. Using high-resolution seismic data and platform borings and cores, they examined the nature of the three important bounding surfaces (sequence boundary, maximum flooding surface and transgressive surface). They found all three surfaces to be discernable and useful for sequence-stratigraphic analysis. The final conclusion from these prior works was that all three surfaces, the sequence boundary, maximum flooding surface and transgressive surface, have their value in stratigraphic analysis and each surface varies in its prominence on the shelf from one area to another. Sequence boundaries are more conspicuous where fluvial valleys exist but may be difficult to discern on seismic profiles in interfluve areas. Maximum flooding surfaces are most prominent where large deltas prograde across the shelf producing downlap surfaces. Transgressive surfaces are the most difficult to recognize in seismic data, but they may be quite distinct in cores, outcrop and, in certain settings, even on high-resolution seismic data.

Aside from the debates about methodology, of the many criticisms of sequence stratigraphy the most critical to its effectiveness as a stratigraphic

From: DAVIES. R. J., POSAMENTIER, H. W., WOOD, L. J. & CARTWRIGHT, J. A. (eds) *Seismic Geomorphology: Applications to Hydrocarbon Exploration and Production*. Geological Society, London, Special Publications, **277**, 29–46.
0305-8719/07/$15.00

tool is the time significance of the sequence boundary (Galloway 1989; Christie-Blick 1991; Jordan & Flemings 1991; Miall 1997; Catuneana *et al.* 1998; Pekar *et al.* 2003). It has even been argued that, if the sequence boundary is diachronous, then sequence stratigraphy is no more advanced than lithostratigraphy (Galloway 1989).

A key to determining the chronostratigraphic significance of the ('Exxonian') sequence boundary is its correct placement with respect to sea level. The dilemma is what age to assign to the sequence boundary: the time of initiation, greatest rate of change or ending of downcutting? Posamentier *et al.* (1988) and Jervey (1988) have argued for placing the sequence boundary at the inflection point of the sea-level curve, a time when sea level is falling at its maximum rate. Posamentier *et al.* (1992), Steckler *et al.* (1993), Miall (1997), Galloway (1989), and Posamentier & Allen (1999) have argued that fluvial incision and the formation of the sequence boundary occurs throughout the interval of sea-level fall. This interpretation is supported by Törnqvist *et al.* (2003), who used optical luminescence dating to constrain the timing of Quaternary fluvial deposits in the Rhine-Meuse incised valley. Hunt & Tucker (1992) and Abdulah *et al.* (2004) suggested that, while rivers begin to down cut when sea level starts to fall, the sequence boundary forms during the maximum lowstand. Part of the dilemma may be that not all rivers respond to sea-level fall in the same way at the same time. In addition, the placement of the sequence boundary may be dependent on factors such as amplitude and period of sea-level cycles, sediment flux, subsidence and the effectiveness of marine and nonmarine sediment transportation processes (Jordan & Flemings 1991).

Much of the controversy about bounding surfaces stems from the fact that these surfaces were first defined using seismic data with poor chonostratigraphic and independent eustatic constraints. This prevented detailed analysis of the actual nature of the bounding surfaces, such as the relief on these surfaces, their lateral extent and stratigraphic relationship with bounding strata. Another problem stems from the inferences that were made about the relationship between seismic units and their bounding surfaces and eustasy (i.e. Haq *et al.* 1987; Christie-Blick *et al.* 1988).

One way to address these problems is to focus on late Quaternary stratigraphic sequences because this is the interval of geological time when sea-level history is best constrained. Also, there are a number of high-resolution dating techniques, such as radiometric dating, oxygen-isotope stratigraphy and well-constrained biostratigraphy, which provide the chronostratigraphic resolution needed to relate the formation of a bounding surface to an independently established record of sea level (Anderson *et al.* 2004). This chronology allows a test of the time significance of the sequence boundary. Lastly, late Quaternary strata lie close enough to the sea floor to be imaged using high-resolution seismic methods, which provide outcrop-scale resolution. Thus, concerns about such issues as incised-valley extent and morphology can be addressed.

In this paper we present the results of a recent detailed study in which the geomorphology of the sequence boundary that formed during the last glacial eustatic lowstand was mapped in detail. This map shows the morphology of a sequence boundary and its incised valleys on the shelf and can be used as a template for mapping sequence boundaries and incised valleys in the subsurface. In addition to mapping, we used radiocarbon analyses and other chronological data to constrain the age of the sequence boundary and relate its formation to the most recent sea-level curves for the Gulf of Mexico. With this information we address three specific issues: (1) the geomorphology of the sequence boundary including its valleys that formed during the last fall in sea level, which culminated around 18 ka on the northwestern Gulf of Mexico; (2) the best placement of the sequence boundary; and (3) the time significance of the sequence boundary.

Study area

At the time scale of the last glacial–eustatic cycle (120 ka), subsidence and sediment supply are the dominant controls on shelf physiography (Anderson *et al.* 2004). Tectonic activity within the northwestern Gulf of Mexico occurs in the form of salt tectonics, growth faulting, and shale diapirism (Ewing 1991). These processes are more active on the middle and outer shelf and have a secondary influence on stratigraphic architecture (Anderson *et al.* 2004).

On the northwestern Gulf of Mexico margin, subsidence increases in a nearly linear fashion across the shelf (Anderson *et al.* 2004). Subsidence profiles for the late Quaternary can be estimated using the total thickness of strata deposited during the last glacial–eustatic cycle. Average rates of subsidence across the study area vary from a few centimetres/ky onshore to about a metre/ky at the shelf margin and increase in a more or less linear fashion across the shelf (Paine 1993; Anderson *et al.* 2004).

In general, long-term sediment supply to the margin has varied considerably across the shelf and is mainly related to the drainage-basin size of individual rivers. Climate has exerted a secondary effect on stratigraphic architecture by regulating the quantity and timing of sediment input from individual rivers (Anderson *et al.* 2004). Rivers with larger drainage basins, including the Brazos, Colorado, Rio Grande and western Louisiana rivers, formed extensive deltas on the shelf throughout the fall and rise in sea level to form broad, low-gradient shelves offshore south Texas, east Texas and western Louisiana (Abdulah *et al.* 2004; Banfield & Anderson 2004; Wellner *et al.* 2004; Fig. 1). In contrast, rivers with smaller drainage basins, such as the central Texas rivers, were unable to fill accommodation created by subsidence, resulting in a ramp-like profile (Eckles *et al.* 2004; Fig. 1). These differences in shelf physiography have strongly influenced fluvial incision during the falling stage of sea level and, therefore, resulted in differences in the nature of the sequence boundary across the shelf.

Eustatic record for the late Quaternary

Sea level has varied by as much as 120 m over the last 120 ka (Lambeck & Chappell 2001; Lambeck *et al.* 2002). The oxygen isotope curve (Shackleton 1987) provides a proxy for the trends of sea level over this period (Fig. 2), although the magnitude and, to some extent, direction of these trends vary from basin to basin (Lambeck *et al.* 2002; Potter & Lambeck 2004). Within the Gulf of Mexico, sea level was at a high, approximately 5 m above its present level, 120 ka ago during oxygen isotope stage (OIS) 5e (Paine 1993; Fig. 2). During the early fall in sea level (approximately 120–80 ka years ago) the shoreline episodically shifted across the inner shelf. Between 70 and 60 ka years ago, sea level fell to

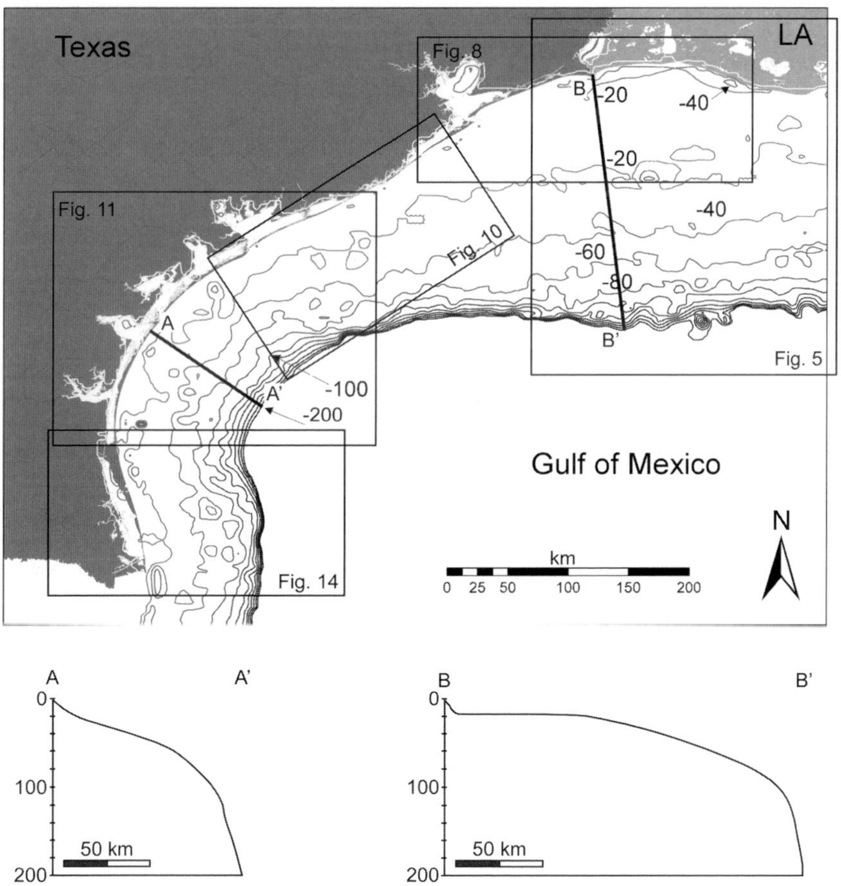

Fig. 1. Bathymetry and profiles of the Texas and western Louisiana continental shelf. Note that both the east Texas–Louisiana and south Texas shelves are broad and have low gradients and distinct shelf breaks whereas the central Texas shelf is more of a ramp setting. These differences reflect long-term sediment supply to these shelves. Bathymetry given in metres.

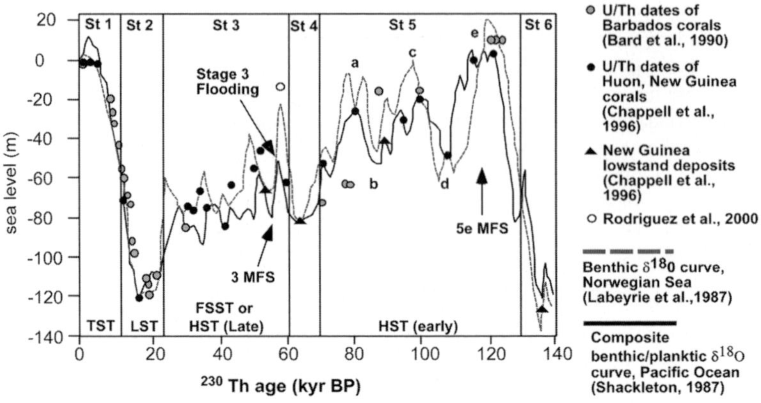

Fig. 2. The oxygen isotope curve is used as a proxy for sea level for the last 120 000 years. Also shown are the actual sea-level data that help constrain the curve.

approximately – 80 m, exposing most of the shelf (OIS4). This was followed by a rapid rise (OIS3) that culminated with sea level at approximately −15 m within the Gulf of Mexico (Rodriguez *et al.* 2000). During the Last Glacial Maximum (OIS2), between approximately 22 and 17 ka ago, the palaeoshoreline was situated near the shelf break (between −90 and −120 m water depth) in the northwestern Gulf (Curray 1960; Abdulah *et al.* 2004; Roberts *et al.* 2004). After approximately 17 ka, sea level began to rise, flooding the shelf. The rate of rise was relatively rapid until approximately 6 ka (Curray 1960).

Methods

Over the last 15 years, over 20,000 km of high-resolution seismic data were acquired across the northwestern Gulf of Mexico margin (Fig. 3). These data are augmented by oil company platform borings and cores and by high-resolution seismic data donated by Texaco Oil Company, the USGS, and by Fugro– McClellan Engineers. The Rice University data were collected aboard the *R/V Matagorda*, *R/V Lone Star*, and most recently *R/V Trinity*. The majority of the high-resolution seismic profiles were digitally collected using a 15 in^3 water gun. Chirper and boomer profiles were taken within bays. Several hundred pneumatic-hammer cores and vibracores were used to constrain the depth and age of the sequence boundary and transgressive surface on the inner shelf and within bays (http://gulf.rice.edu/coastal). In addition, several rotary cores (up to 25 m long) were acquired within the bays to complement chirper and boomer profiles.

The sequence boundary was identified within seismic profiles by truncation of reflections below and onlap above the surface (Mitchum *et al.* 1977). Within cores the sequence boundary was identified by the presence of an erosional surface marked by rip-up clasts and a landward shift in facies placing more proximal facies above more distal facies (Van Wagoner *et al.* 1990). Several different surfaces exhibited these properties in one or two locations, but only one was mappable throughout the region over the time scale of the last 120 ka.

A chronostratigraphic framework for the northwestern Gulf of Mexico was constructed using oxygen isotopes, radiometric dating, and biostratigraphy (Anderson *et al.* 2004; Fillon *et al.* 2004; Kohl *et al.* 2004). This chronostratigraphic framework constrains the timing of deposition and surface formation. The age of major flooding surfaces (MFS 5e and MFS 3) and the OIS4 erosional surface (higher-order sequence boundary), where present, was constrained using oxygen isotopic and biostratigraphic data. The time of formation of the OIS2 sequence boundary was constrained using radiocarbon ages from above and below this surface. Twenty-five radiocarbon dates provide age control on the sequence boundary, transgressive surface, and the strata they bound.

Geomorphology results

An important criterion for delineating sequence boundaries is based on the identification of incised valleys (Vail *et al.* 1977*a–c*; Van Wagoner *et al.* 1990). In our experience, proper identification of incised valleys, and hence sequence boundaries, can be done only with regional data sets that show more than one valley incised at the same stratigraphic level. Thus, examination of lowstand fluvial geomorphology and the response of different rivers to sea-level fall require a regional data set and robust chronostratigraphic framework. Figure 4 is a shaded elevation map of the OIS2 sequence boundary. This map was compiled from earlier studies (Abdulah *et al.* 2004; Banfield & Anderson 2004; Eckles *et al.* 2004; Wellner *et al.* 2004), and from more recent

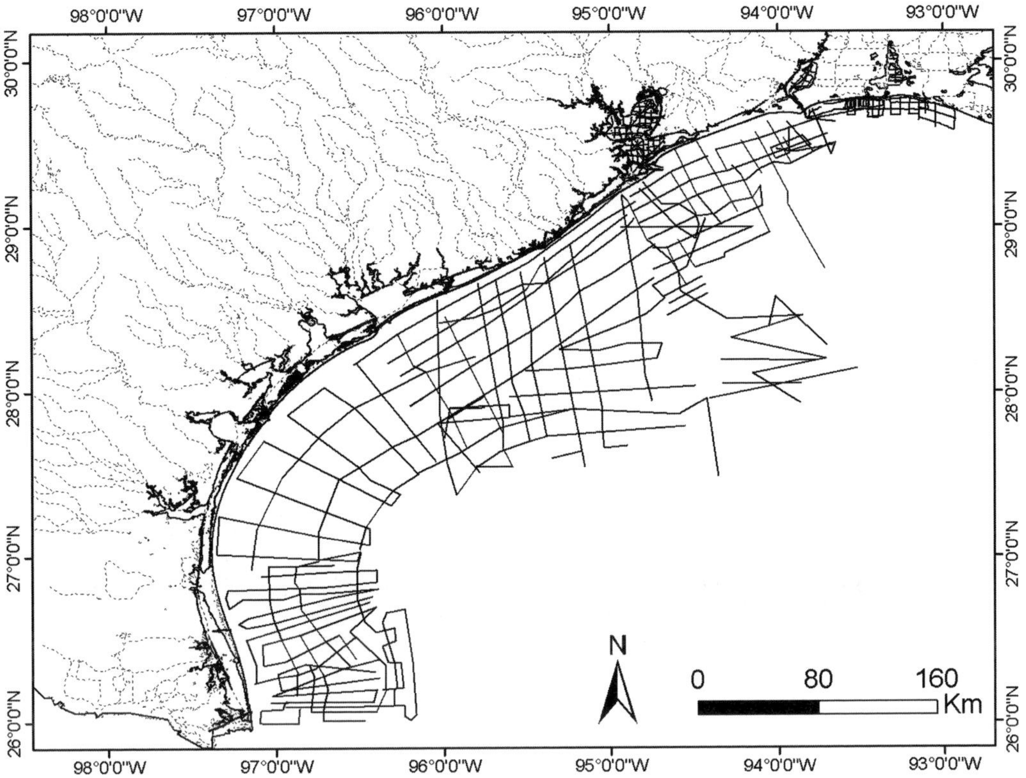

Fig. 3. Map showing *R/V Lone Star* cruise tracks on the continental shelf where high-resolution seismic data and cores used in the investigation were collected.

data (Tramp & Anderson 2004; Simms 2005). It is the focus of the following discussion.

Western Louisiana shelf

As sea level fell from 120 to 20 ka, western Louisiana was characterized by a complex fluvial drainage system (Berryhill *et al.* 1986). The Berryhill *et al.* (1986) map has been widely used to illustrate Quaternary incised valleys and shows many channels, some of which begin and end abruptly on the shelf (Fig. 5). The terminations are not the product of inadequate data coverage. The data set used by Berryhill and his colleagues was quite dense and our work has shown that some of the channels do indeed terminate both updip and downdip.

The chronostratigraphic control on the age of western Louisiana channels was minimal. Later work has shown that Berryhill *et al.*'s map includes valleys formed throughout the OIS5 falling limb of sea level into the OIS2 lowstand in sea level (Wellner *et al.* 2004; this study). This is why they display both up-dip and down-dip termination. The seaward terminations probably represent relatively long-lived palaeoshorelines created during higher-order highstands that occurred during the episodic fall in sea level (Figs 2 and 5). Broad, terraced morphology and lateral accretion characterize these channels (Fig. 6). The updip terminations probably result, at least in part, from transgressive ravinement, the depth of which varied as the rate of transgression varied (Siringan & Anderson 1994).

In reality, there are too many fluvial channels on the western Louisiana shelf to be linked to the known fluvial valleys onshore, specifically the Calcasieu and Sabine valleys (Anderson *et al.* 1991; Nicol *et al.* 1994; Fig. 7). To better understand this problem, we conducted a detailed analysis of seismic data and cores from the inner shelf aimed at extending the onshore valleys onto the continental shelf using offshore data. The results are shown in Figure 8. The map shows that onshore there are a number of dendritic channels, but there are only two deep valleys that can be traced across the shelf to where they merge with the previously mapped Trinity River incised valley (Thomas & Anderson 1994; Fig. 4) and that have surfaces at their bases with regional extent. Most of the other channels on the shelf lie below the sequence boundary and above the maximum flooding surface at different levels (Fig. 6).

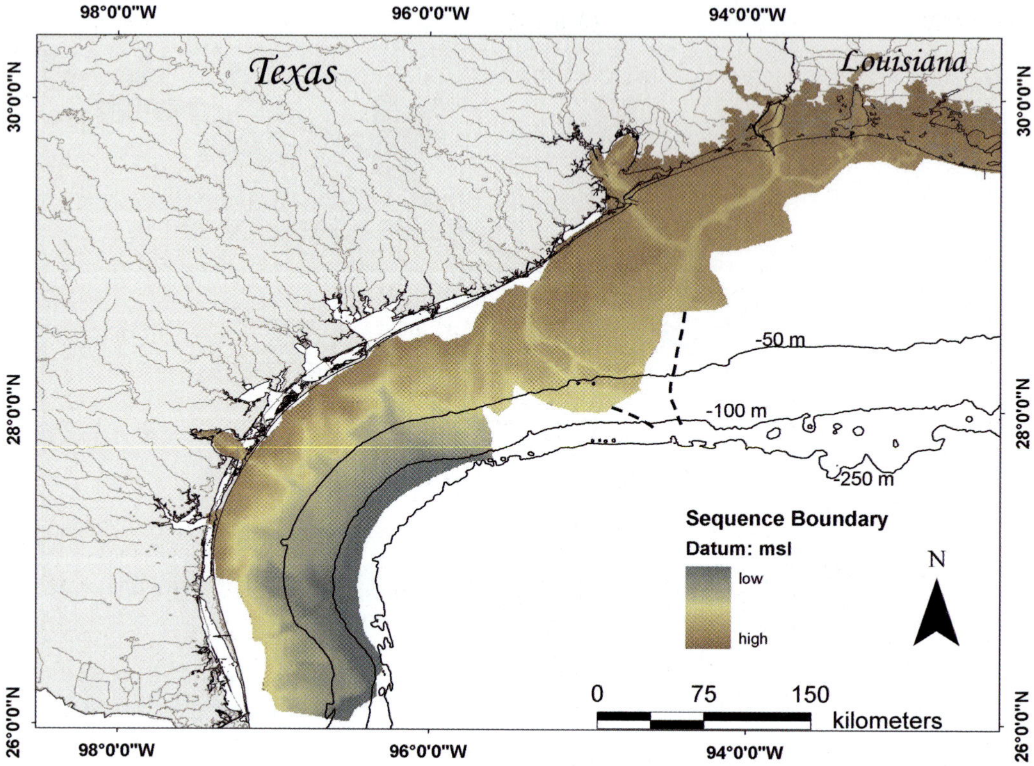

Fig. 4. Structure map of OIS2 sequence boundary showing locations of lowstand incised valleys. Dashed lines show the approximate locations of the valleys, based on the work of Abdulah *et al.* (2004) and Wellner *et al.* (2004).

The Calcasieu and Sabine River valleys merge and extend roughly parallel to the coast for tens of kilometres on the inner shelf (Fig. 8). This shore-parallel orientation of these valleys was influenced by presence of an older delta, the Western Louisiana Delta, which lies above the OIS5e maximum flooding surface but below the OIS2 sequence boundary (Anderson *et al.* 1996; Wellner *et al.* 2004), on the outer shelf. This demonstrates the importance of antecedent topography in controlling the flow directions of rivers. In addition, the palaeotopographic map for the Calcasieu and Sabine valleys (Fig. 8) shows several small tributaries. Abundant tributaries is one criterion for recognizing lowstand incised valleys (Posamentier 2001).

East Texas shelf

The Trinity valley and Sabine valley merge on the inner shelf (Fig. 8). Detailed mapping of the Trinity incised valley showed that it extends from the present Galveston Bay estuary, where the valley is about 40 m deep, to the shelf break where the valley is less discernable (Smyth *et al.* 1988; Thomas & Anderson 1994; Wellner *et al.* 2004).

Thomas & Anderson (1994) recognized fluvial terraces along the flanks of the offshore Trinity valley. They mapped these terraces up-dip to Galveston Bay where they correlated the submerged terraces to onshore terraces that formed during the initial fall of sea level (Fig. 9). Thus, the river began to down-cut its valley during the initial fall in sea level, but maximum incision cross-cuts these terraces to form the u-shaped incision shown in Figure 9. Radiocarbon dates from bayhead delta deposits above the fluvial section are as old as 9.6 ka (Thomas & Anderson 1994). No dates were obtained above the fluvial section on the inner shelf.

On the outer shelf, the valley becomes wider and shallower and is difficult to resolve in seismic data. Oil company platform borings were used to constrain its location on the outer shelf, where it merges with the Brazos valley (Fig. 4). On the outer shelf, seaward of where these valleys merge, platform borings sampled an average of 30 m of sand within the valley (Wellner *et al.* 2004). At the shelf break, the valley is linked to an incision in the upper slope that is situated within a salt-withdrawal minibasin (Anderson *et al.* 1996).

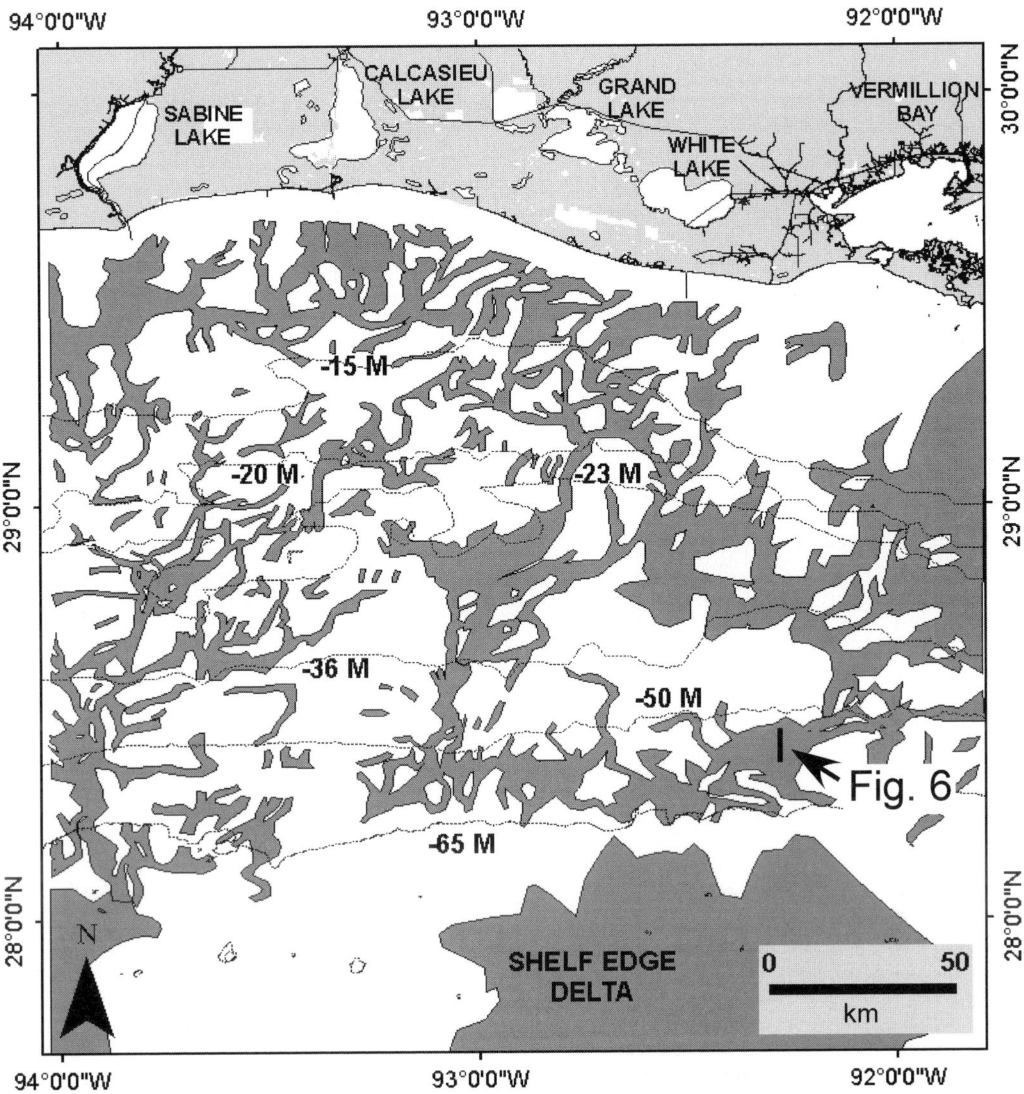

Fig. 5. Berryhill *et al.*'s (1986) map of fluvial channels on the western Louisiana shelf. The dashed lines are bathymetric contours in metres.

On the inner shelf, the Brazos valley has a terraced morphology, which indicates that it was initially cut during the falling limb of sea level (Anderson *et al.* 1996; Abdulah *et al.* 1997, 2004). During this fall in sea level, the Brazos River nourished a large delta that advanced across the shelf (Abdulah *et al.* 2004). The river also incised as sea level fell. However, near the shoreline the grade of the valley remained somewhat low due to the rapid progradation of the delta (Abdulah *et al.* 2004). When sea level neared its lowest position, the river shifted its course to the east and flowed around the margin of its former delta, ultimately merging with the Trinity/Sabine valley on the outer shelf (Fig. 4).

During the falling limb of sea level, the Colorado River also formed a large delta on the shelf (Abdulah *et al.* 2004). Three deep valleys occur offshore of the modern Colorado River mouth (Fig. 10). The eastern valley is linked to a prominent OIS3 delta on the outer shelf (Abdulah *et al.* 2004). The central and western valleys merge on the outer shelf and are linked to a shelf-margin delta, which is linked via small canyons to two

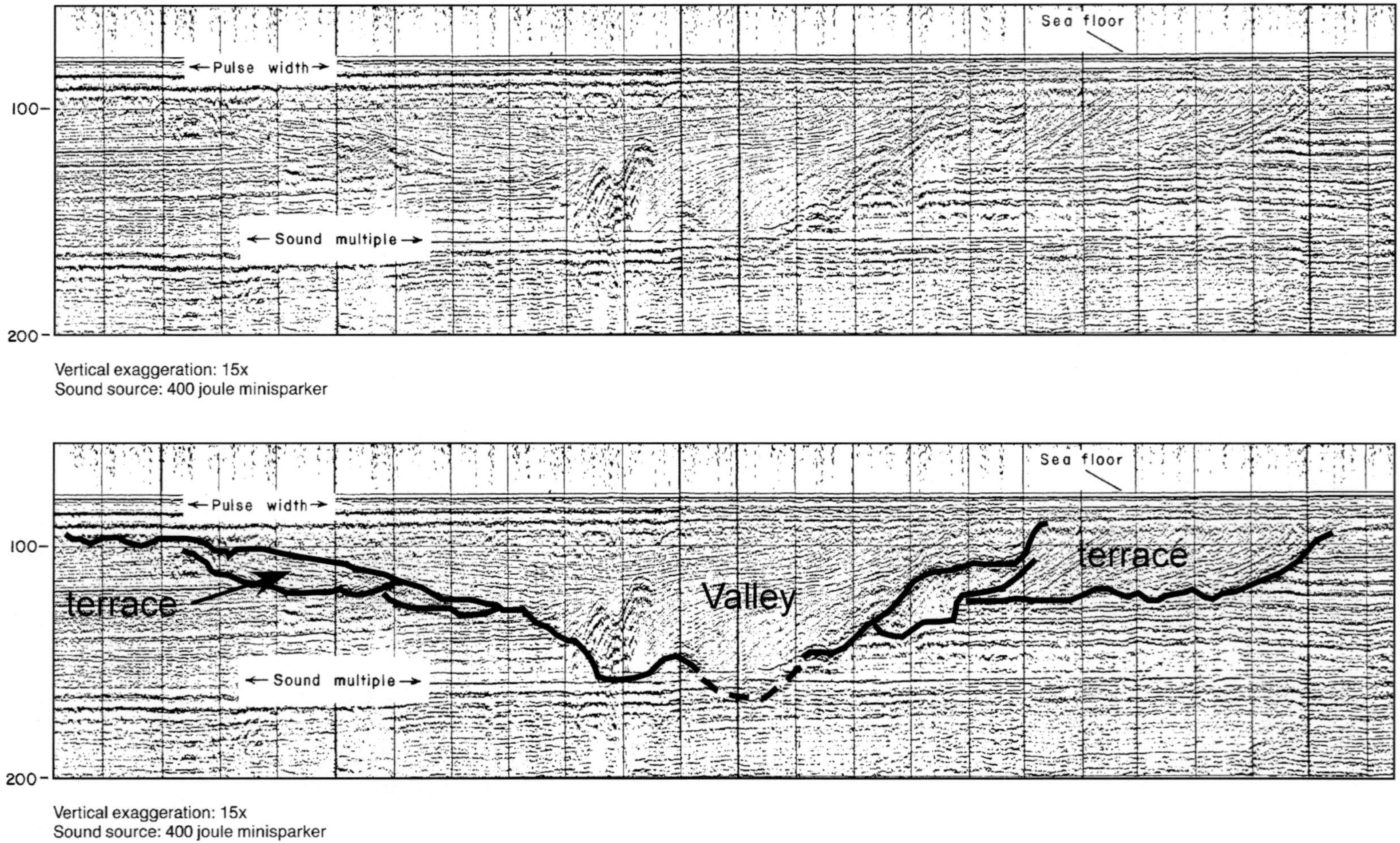

Fig. 6. Seismic line showing fluvial valley that formed during the fall in sea level, but not associated with the sequence boundary, with terraced morphology and lateral accretion (modified from Berryhill *et al.* 1986 AAPG©1986 reprinted by permission of the AAPG whose permission is required for further use).

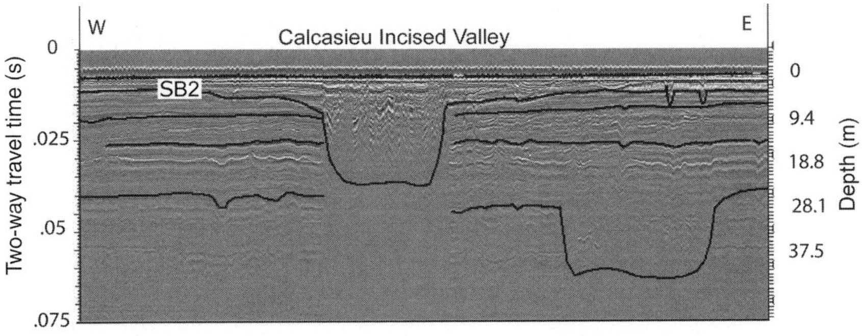

Fig. 7. Seismic line across the lowstand Calcasieu valley. Depth conversion based on a velocity of 1500 m s^{-1}. See Figure 8 for profile location.

upper-slope turbidite systems and the East Breaks Slides (Lehner 1969; Tatum 1977; Woodbury *et al.* 1978; Rothwell *et al.* 1991). No terraces were noted along the flanks of the valley on the shelf.

Central Texas shelf

On the central Texas shelf, the Lavaca, Nueces and other small rivers extended their valleys across the shelf as sea level fell. The ancestral Lavaca River occupied a deep valley where Matagorda Bay is now located (Fig. 11). The valley is incised to a depth of approximately 40 m at the modern coast. On the outer shelf the valley is less prominent and does not appear to extend to the shelf-break.

The San Antonio, Aransas, Nueces and Baffin Bay valleys can be mapped only to the mid-shelf, approximately −60 m water depth (Fig. 11). The valleys terminate where they encountered a decrease in shelf gradient seaward of the former OIS3 shoreline (Simms *et al.* 2006*b*; Fig. 12). Seaward of this inflection point in shelf gradient, valley incision was shallow enough that transgressive ravinement has removed them. Even though fluvial valleys do not occur on the outer shelf, the sequence boundary occurs as a prominent erosional surface that is draped by mud less than 7.5 ka old (Fig. 12).

On the inner shelf and within Corpus Christi Bay, prominent terraces flank the valley below the sequence boundary (Fig. 13). Simms *et al.* (2006*b*) correlated these to onshore terraces formed during the falling limb of sea level (Durbin *et al.* 1997). Durbin *et al.* (1997) obtained thermoluminescence dates ranging between 91 and 31 ka for these deposits.

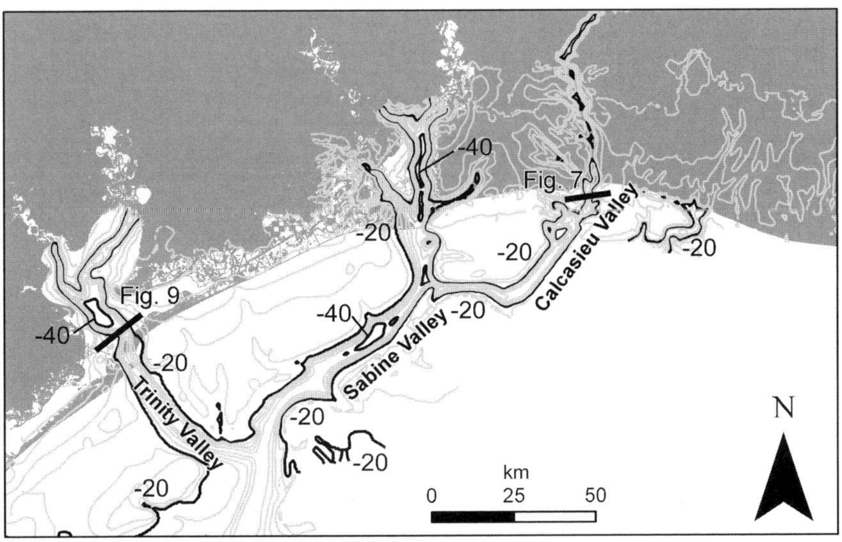

Fig. 8. Structure map of the OIS2 sequence boundary on the western Louisiana inner shelf.

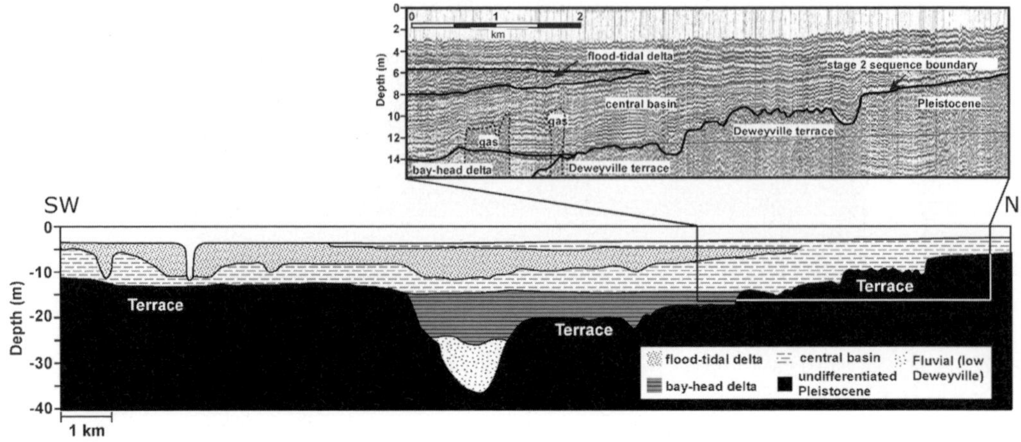

Fig. 9. Interpreted seismic profile from Galveston Bay crossing the Trinity incised valley. Inset shows segment of the line to illustrate the terraced morphology and deep incision of Stage 2 valley. See Figure 8 for line location.

South Texas shelf

The ancestral Rio Grande formed a prominent delta that prograded across the shelf throughout the period of falling sea level from OIS5 through OIS3 (Banfield & Anderson 2004). The deltaic deposits of the early fall in sea level on the shelf are wave-dominated, but were fluvial-dominated by the late fall, when sediment supply was greater (Banfield & Anderson 2004; Anderson 2005). The Rio Grande remained relatively fixed in its location on the shelf throughout the fall of sea level. During the maximum lowstand in sea level, the river shifted its course to a location south of the older deltas. Unfortunately, our data stop at the U.S. and Mexican border, so only the northern half of the valley was surveyed (Fig. 14). The data shows that the valley is linked to a shelf-margin delta and turbidite system and the valley experienced headward erosion, resulting in offshore deepening of the valley. This is similar to valleys on the Alabama shelf, as mapped by Bartek *et al.* (2004).

Chronostratigraphic results

It is not easy to date an unconformity. We know that the surface shown in Figure 4 cuts into deltas and shelf mud that were deposited prior to the maximum lowstand in sea level, and that the age of these deposits decreases in an offshore direction (Anderson *et al.* 2004). This observation simply reflects the incision into offlapping succession of strata deposited during the falling limb of sea level. The age of shelf-margin deltas also varies across the shelf (Abdulah *et al.* 2004; Wellner *et al.* 2004;

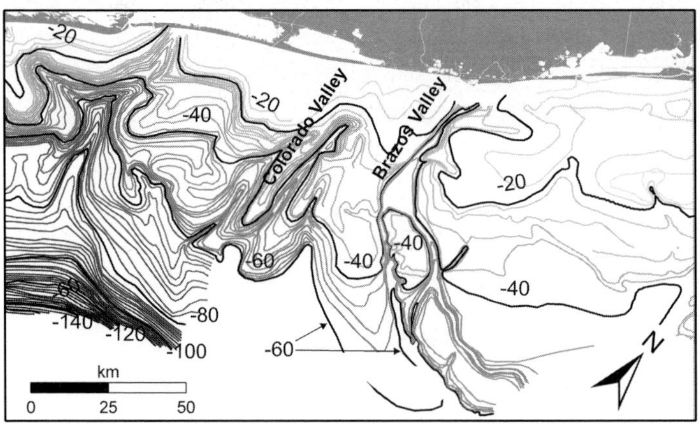

Fig. 10. Structure map of the OIS2 sequence boundary on the east Texas continental shelf.

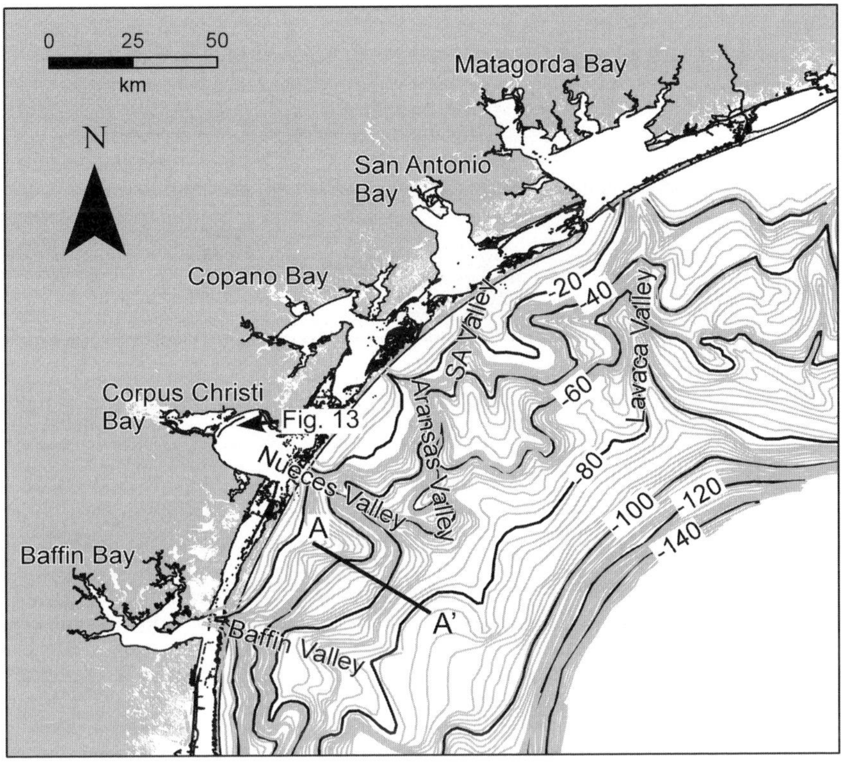

Fig. 11. Structure map of the OIS2 sequence boundary on the central Texas continental shelf. Contours in meters. SA, San Antonio.

Anderson 2005). Thus, the stratigraphic position of the conformable surface associated with the sequence boundary varies with relationship to shelf-margin deltas (Anderson 2005; Fig. 15). Without radiocarbon-age constraints, this variability in the position of the sequence boundary would be difficult to determine, forcing one to randomly place the sequence boundary either above or below the shelf-margin delta, creating a problem when prospecting for linked lowstand fans (Anderson 2005).

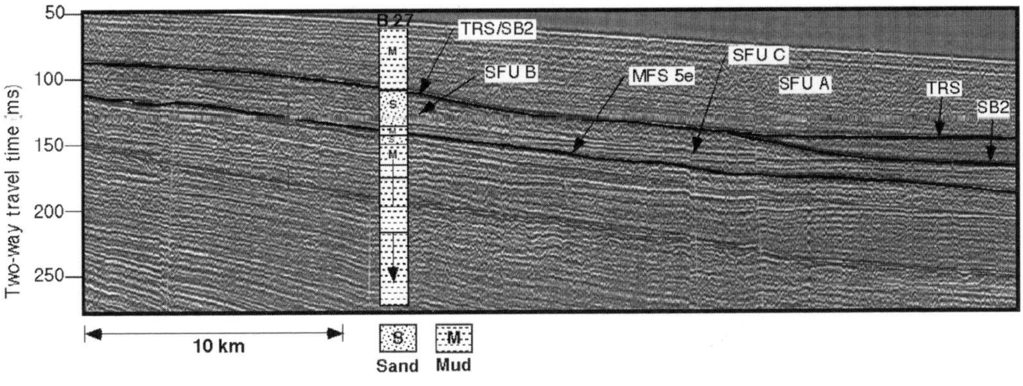

Fig. 12. Seismic profile across the central Texas shelf showing the change in shelf gradient associated with the OIS3 palaeoshoreline. Landward of this location there is deep fluvial incision. Seaward of this location, the sequence boundary is discernable on high-resolution seismic data, but may not be evident in core or outcrop. See Figure 11 for line location (modified from Eckles *et al.* 2004).

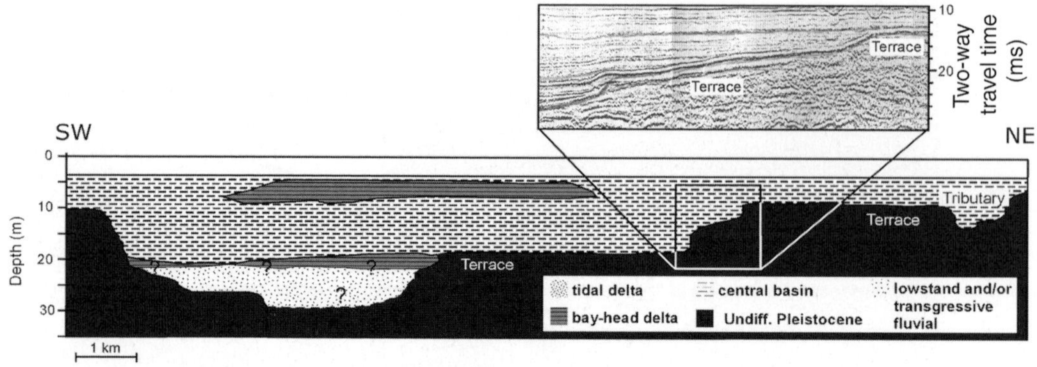

Fig. 13. Seismic profile across the Nueces valley within Corpus Christi Bay showing highstand fluvial terraces. See Figure 11 for location.

The radiocarbon-age data have been presented in prior publications (Rodriguez *et al.* 2000, 2004; Abdulah *et al.* 2004; Anderson *et al.* 2004; Banfield & Anderson 2004; Eckles *et al.* 2004; Wellner *et al.* 2004; Simms *et al.* 2006*a*) and are summarized in Figure 16. Again, the surfaces above and below the dates shown in Figure 16 has been tied using high-resolution seismic data (Fig. 1). At most core locations, a lack of datable material prevented acquisition of dates directly above the sequence boundary. We did not acquire radiocarbon dates below the sequence boundary in areas where existing chronostratigraphic data indicated that the deposits below the surface were older than 40 000 ka, the age limit of radiocarbon dating.

In general, the dates taken close to the sequence boundary show a crude pattern in which the ages above the sequence boundary decrease landward. The time interval represented by this surface increases landward, except for fluvial terraces deposits flanking the incised valleys (Thomas & Anderson 1994; Durbin *et al.* 1997). This pattern of dates is a result of the surface cutting an off-lapping succession of strata whose age decreases in a seaward direction. Sites on the outer shelf did, in fact, bracket the age of the sequence boundary to between 26 and 17 ka, the timing of the last maximum lowstand in sea level (Lambeck *et al.* 2002). What is significant is that there are no ages older than 18 ka, the accepted age for the Last Glacial Maximum, above the sequence boundary and none younger than 18 ka below the sequence boundary.

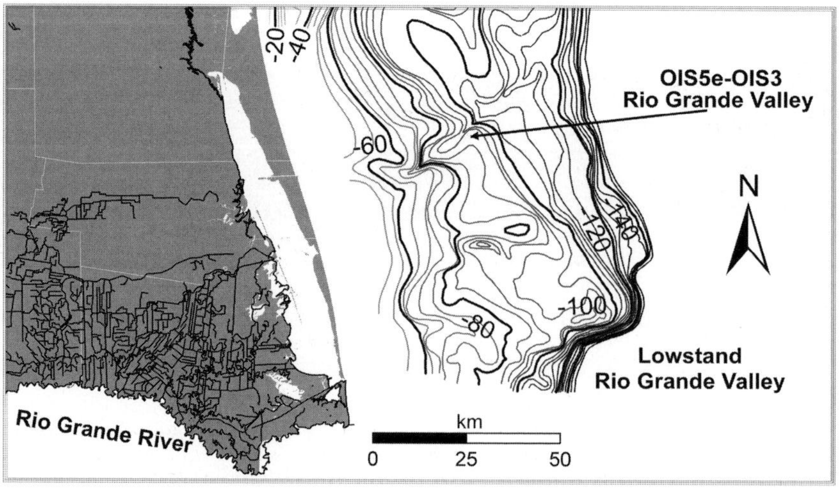

Fig. 14. Structure map of the Stage 2 sequence boundary on the south Texas shelf. OIS5e-OIS3 Rio Grande valley is the valley that formed during the falling limb of sea level.

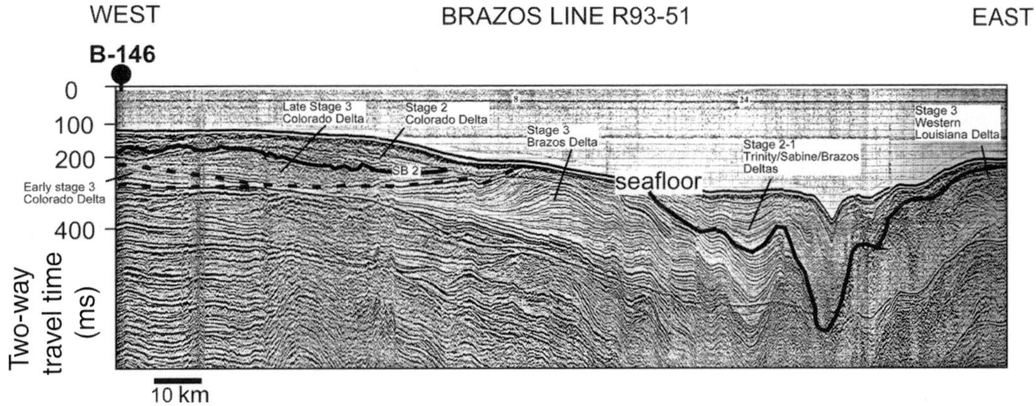

Fig. 15. Seismic line R93-51 from the outer shelf used to illustrate the relationship between shelf margin deltas and the OIS2 sequence boundary. Notice that shelf margin deltas exist both above and below the surface.

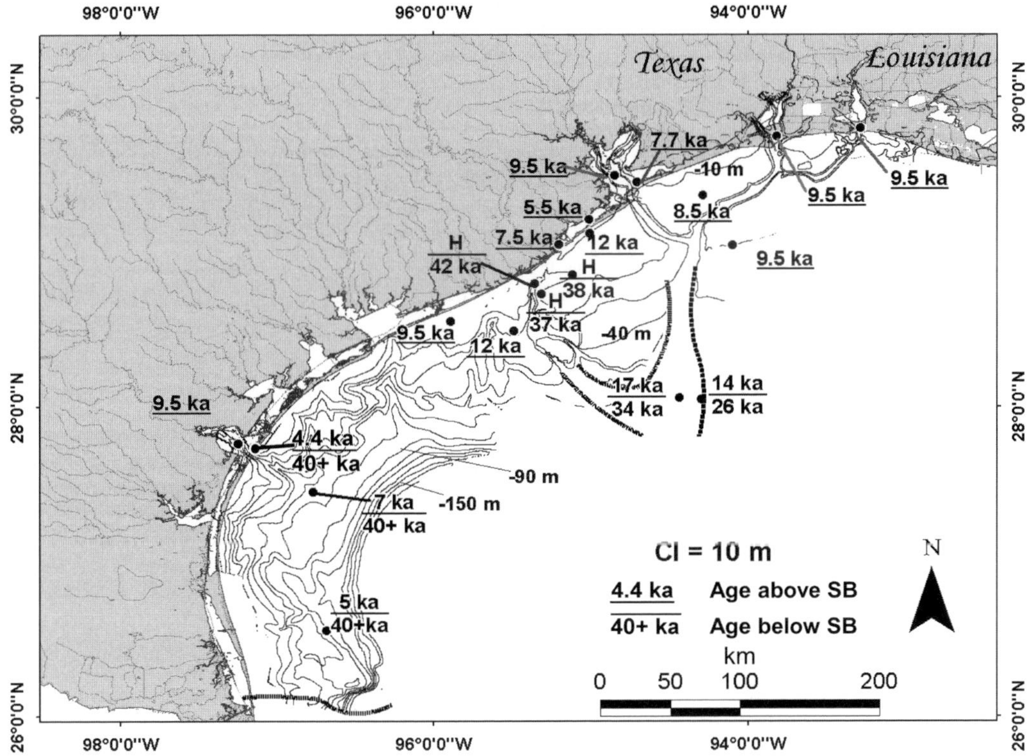

Fig. 16. Map showing the locations of cores used to obtain radiocarbon ages that help to bracket the OIS2 sequence boundary. H, Holocene age. The line represents relationship to the sequence boundary. If solid line above date, then date lies below the sequence boundary. If solid line lies below date, then date lies above the sequence boundary. The ages from above the boundary vary in part because datable material was not always available immediately above the surface. Still, a pattern of decreasing age in an onshore direction is discernable. Ages below the boundary are also shown and ages of greater than 40 000 years are considered radiocarbon dead.

Discussion

Geomorphology

Without exception, the rivers of the Texas and western Louisiana continental shelves extended their valleys onto the continental shelf as sea level fell during the last eustatic fall. Our results agree with those who have argued that incision occurs throughout the fall in sea level (i.e. Posamentier & Allen 1999; Tornqvist *et al.* 2003, etc.). However, some of the valleys we studied were abandoned before the maximum lowstand in sea level. Most of the valleys that formed during the fall in sea level have terraced morphologies and many terminate seaward on the shelf in deltas or palaeoshorelines other than the lowstand shoreline (Abdulah *et al.* 2004; Banfield & Anderson 2004; Wellner *et al.* 2004). The ancestral valleys of the Brazos River, western Louisiana rivers, and the Rio Grande provide good examples (Abdulah *et al.* 2004; Banfield & Anderson 2004; Wellner *et al.* 2004). The valleys that terminate on the shelf in deltas or palaeoshorelines other than the lowstand shoreline occur at different stratigraphic levels, thus they do not form a regional surface of relief. Only the valleys occupied during the maximum (OIS2) lowstand in sea level, representing the maximum incision for each river, have a surface of regional extent at their bases.

Rivers are prone to avulsion during transgression. This is especially true for the rivers with high sediment supplies because they are inclined to fill their valleys during transgression (Blum & Price 1998; Simms *et al.* 2006*b*). When this happens, rivers cut their new valleys to adjust to base level at the time avulsion occurs. Here again, these valleys do not extend to the shelf break, but rather terminate down dip at deltas or palaeoshorelines on the inner to mid shelf. These deltas and palaeoshorelines, as well as the river valleys, may be partially or entirely removed by transgressive ravinement, which in the study area eroded to a depth of 6–15 m (Rodriguez *et al.* 2001). The ancestral Rio Grande (Banfield & Anderson 2004), Colorado, and Brazos valleys (Abdulah *et al.* 2004) are good examples of rivers that avulsed and cut deep channels during the transgression of sea level. The valleys that formed during the transgression highlight another problem in identifying the sequence boundary, the fact that not all valleys are associated with sequence boundaries (Schumm & Etheridge 1994).

In contrast to the larger rivers, smaller rivers, like the ancestral Trinity River, generally remained fixed in their location on the shelf throughout the fall and into the lowstand as well as during the transgression (Thomas & Anderson 1994; Eckles *et al.* 2004; Simms *et al.* 2006*b*). Together, the larger and smaller valleys that were occupied during the maximum lowstand in sea level and the tributaries that flowed into them formed a surface of relief that is quite distinct and can be mapped on a regional scale (Fig. 4). The true incised valleys associated with the sequence boundary are located several tens of kilometres apart.

Some valleys deepen seaward; others become shallower seaward. The orientation of these valleys also varies and antecedent topography exerted a clear influence on fluvial drainage, thus geomorphology of the sequence boundary. Rivers that flowed across low-gradient shelves formed lowstand valleys that extend to the shelf break. However, headword erosion created linked incised valley and canyon systems only in the case of the Rio Grande and possibly Colorado rivers. These are the more bedload-dominated systems and they were more likely subject to erosion by sandy turbidity currents.

It is noteworthy that sea level did not fall below the shelf break during the last lowstand. Canyon incision would probably have been more widespread had sea level fallen below the shelf break. There is evidence that during the previous (OIS6) lowstand sea level did fall below the shelf break and that fluvial incision of the shelf margin was more pronounced at that time (Fraticelli & Anderson 2003).

Posamentier & Allen (1999), Posamentier (2001) and Wellner & Bartek (2003) have shown that incised valleys do not always extend to the shelf break. Rivers can respond to base-level fall in a variety of ways including adjustments to meander width, bed roughness, and competence (Schumm 1977). Tallings (1998) pointed out that the shelf must be concave down in order for incision to take place. Similarly, Posamentier *et al.* (1992) and Koss *et al.* (1994), using modern analogues and flume models, showed that incised valleys form at breaks in slope.

On the ramp-like central Texas shelf (Fig. 1), the depth of fluvial incision decreases abruptly at the mid-shelf where the shelf gradient changes abruptly. The topographic break corresponds to the OIS3 palaeoshoreline. Seaward of this location on the shelf, the valleys were so shallow that any trace of them was removed as a result of transgressive erosion that exceeded valley depth. Still, the sequence boundary is a prominent surface in seismic profiles marked by truncation below and onlap above (Fig. 12). In cores and outcrops, however, the sequence boundary would be difficult to discern on the outer shelf where marine mud overlies marine mud, as lithologic logs of platform borings provided by Fugro–McClellan show no changes at the sequence boundary, even though a distinct surface is seismically imaged.

Chronology

During the course of our investigation, we acquired radiocarbon dates from cores throughout the study area that bracket the sequence boundary. Dates from below the valley are generally radiocarbon dead, meaning that they indicate deposition prior to the lowstand of sea level. The exceptions are dates that were acquired from cores taken on the outer shelf, where incision was less pronounced and deltas were being formed during the late stages of sea-level fall. There, we acquired ages in the range of 25–34 ka below the sequence boundary and ages younger than 17 ka above the sequence boundary. In all cases the hiatus encompasses the time in which the shelf was exposed during OIS2 and deposition resumed shortly after the shelf was flooded or, within valleys, during the lowstand.

In order to provide the most time-significant surface, the sequence boundary should be marked by the surface representing the maximum lowstand in sea level as defined by Hunt & Tucker (1992). The diachronous nature of the sequence boundary is often a function of grouping the surfaces at the base of valleys from several higher-order sequences into one surface, such as the surface marking the base of the fluvial terrace deposits flanking the incised valleys. Thus, when dating the sequence boundary it will appear diachronous. The sequence boundary cuts into an off lapping succession whose age decreases in an offshore direction.

Conclusions

1. On the northwestern Gulf of Mexico continental shelf, the sequence boundary, maximum flooding surface, and transgressive surface all have their value in sequence stratigraphy. The 'Exxonian' sequence boundary, defined by an erosional surface and its correlative conformity, proved to be the most prominent surface and most useful for correlation of late Quaternary strata on the continental shelf. However, the correlative conformity on the shelf margin and upper slope is more difficult to identify and occurs at different stratigraphic levels within shelf-margin deltas. Arbitrary selection of this surface can result in erroneous interpretations about timing of delta formation and the occurrence of linked fan systems.
2. Maximum fluvial incision occurred during the maximum lowstand around 18 ka. Although the different rivers of the study area all formed incised valleys during the last sea-level fall, the geomorphic character of these valleys varies widely. The most important factors regulating valley morphology are the interval of the sea-level fall during which the valleys are occupied and the river gradient relative to the shelf gradient.
3. Not all valleys exhibiting significant truncation are true incised valleys associated with a sequence boundary and a subsequent lowstand systems tract. Regional mapping of a surface associated with a valley exhibiting great erosional relief will reveal if the valley in question is a true incised valley associated with a sequence boundary by correlation with other valleys. Distances of tens of kilometres separate the incised valleys of the study area. Thus, these valleys would be seen only in the best of rock exposures or in areas of extensive well control.
4. Within the study area, when the sequence boundary is chosen as the unconformity that formed during the maximum lowstand in sea level, the sequence boundary has time significance. All deposits below the sequence boundary are older than all deposits above the sequence boundary.

References

ABDULAH, K. C., ANDERSON, J. B., COPERUDE, S. & CANNING, A. 1997. Temporal and spatial sampling constraints in high-resolution 3-D seismic surveys. *Marine Geophysical Researches*, **19**, 97–113.

ABDULAH, K. C., ANDERSON, J. B., SNOW, J. N. & HOLDFORD-JACK, L. 2004. The late Quaternary Brazos and Colorado deltas, offshore Texas – their evolution and the factors that controlled their deposition. *In*: ANDERSON, J. B. & FILLON, R. H. (eds) *Late Quaternary Stratigraphic Evolution of the Northern Gulf of Mexico Margin.* Society for Sedimentary Geology, Special Publications, **79**, 237–269.

ANDERSON, J. B. 2005. Diachronous development of late Quaternary shelf-margin deltas in the northwestern Gulf of Mexico: implications for sequence stratigraphy and deep-water reservoir occurrence. *In*: BHATTACHARYA, J. P. *ET AL.* (eds) *River Deltas – Concepts, Models and Examples.* Society for Sedimentary Geology, Special Publications, **83**, 257–276.

ANDERSON, J. B., SIRINGAN, F. P., SMYTH, W. C. & THOMAS, M. A. 1991. Episodic nature of Holocene sea-level rise and the evolution of Galveston Bay. *In*: *Gulf Coast Section of the Society for Economic Paleontologists and Mineralogists 12th Annual Research Conference Program and Abstracts*, 8–14.

ANDERSON, J. B., ABDULAH, K., SARZALEJO, S., SIRINGAN, F. & THOMAS, M. A. 1996. *Late Quaternary Sedimentation and High-Resolution Sequence Stratigraphy of the East Texas, Shelf.* Geological Society, London, Special Publications, **117**, 95–124.

ANDERSON, J. B., RODRIGUEZ, A., ABDULAH, K. C., FILLON, R. H., BANFIELD, L. A., MCKEOWN, H. A. & WELLNER, J. S. 2004. Late Quaternary stratigraphic evolution of the northern Gulf of Mexico margin: a synthesis. *In:* ANDERSON, J. B. & FILLON, R. H. (eds) *Late Quaternary Stratigraphic Evolution of the Northern Gulf of Mexico Margin.* Society for Sedimentary Geology, Special Publications, **79**, 1–23.

BANFIELD, L. A. & ANDERSON, J. B. 2004. Late Quaternary evolution of the Rio Grande Delta:

Complex response to eustasy and climate change. *In*: ANDERSON, J. B. & FILLON, R. H. (eds) *Late Quaternary Stratigraphic Evolution of the Northern Gulf of Mexico Margin*. Society for Sedimentary Geology, Special Publications, **79**, 289–306.

BARD, E., HAMELIN, B. & FAIRBANKS, R. G. 1990. U–Th ages obtained by mass spectrometry in coral from Barbados: sea level during the past 130,000 years. *Nature*, **346**, 456–458.

BARTEK, L. R., CABOTE, B. S., YOUNG, T. & SCHROEDER, W. 2004. Sequence stratigraphy of a continental margin subjected to low-energy and low sediment supply environmental boundary conditions: late Pleistocene–Holocene deposition offshore Alabama. *In*: ANDERSON, J. B. & FILLON, R. H. (eds) *Late Quaternary Stratigraphic Evolution of the Northern Gulf of Mexico Basin*. Society for Sedimentary Geology, Special Publications, 85–109.

BERRYHILL, H. L., SUTER, J. R. & HARDIN, N. S. 1986. *Quaternary Facies and Structure, Northern Gulf of Mexico: Interpretations from Seismic Data*. American Association of Petroleum Geologists Studies in Geology, **23**, 289.

BLUM, M. D. & PRICE, D. M. 1998. Quaternary alluvial plain construction in response to glacio-eustatic and climatic controls, Texas Gulf Coastal Plain. *In*: STANLEY, K. W. & MCCABE, P. J. (eds) *Relative Role of Eustasy, Climate, and Tectonism in Continental Rocks*. Society for Sedimentary Geology, Special Publications, 31–48.

CATUNEANU, O., WILLIS, A. J. & MIALL, A. J. 1998. Temporal significance of sequence boundaries. *Sedimentary Geology*, **121**, 157–178.

CHAPPEL, J., OMURA, A., ESAT, T., MCCULLOCH, M., PANDILFI, J., OTA, Y. & PILLANS, B. 1996. Reconciliation of late Quaternary sea levels derived from coral terraces at Huon Peninsula with deep sea oxygen isotope records. *Earth and Planetary Science Letters*, **141**, 227–236.

CHRISTIE-BLICK, N. 1991. Onlap, offlap, and the origin of unconformity-bounded depositional sequences. *Marine Geology*, **97**, 35–56.

CHRISTIE-BLICK, N., MOUNTAIN, G. S. & MILLER, K. G. 1988. Sea level history: Technical comment. *Science*, **241**, 596–601.

CURRAY, J. R. 1960. Sediments and history of Holocene transgression, continental shelf, Northwestern Gulf of Mexico. *In*: SHEPHERD, F. P., PHLEGER, F. B. & VAN ANDEL, T. H. (eds) *Recent Sediments, Northwest Gulf of Mexico*. American Association of Petroleum Geologists, Tulsa, OK, 221–266.

DURBIN, J. M., BLUM, M. & PRICE, D. M. 1997. Late Pleistocene stratigraphy of the Lower Nueces River, Corpus Christi, Texas: Glacio-eustatic influences on valley-fill architecture. *Gulf Coast Association of Geological Societies Transactions*, **47**, 119–129.

ECKLES, B. J., FASSELL, M. L. & ANDERSON, J. B. 2004. Late Quaternary evolution of the wave-storm-dominated Central Texas shelf. *In*: ANDERSON, J. B. & FILLON, R. H. (eds) *Late Quaternary Stratigraphic Evolution of the Northern Gulf of Mexico Margin*. Society for Sedimentary Geology, Special Publications, **79**, 271–287.

EWING, T. E. 1991. Structural framework. *In*: SALVADOR, A. (ed.) *The Gulf of Mexico Basin*. The Geology of North America Volume J, Geological Society of America, 31–52.

FILLON, R. H., KOHL, B. & ROBERTS, H. H. 2004. Late Quaternary deposition and paleobathymetry the shelf-slope transition, ancestral Mobile River delta complex, northeastern Gulf of Mexico. *In*: ANDERSON, J. B. & FILLON, R. H. (eds) *Late Quaternary Stratigraphic Evolution of the Northern Gulf of Mexico Margin*. Society for Sedimentary Geology, Special Publications, Tulsa, OK, **79**, 111–142.

FRATICELLI, C. M. & ANDERSON, J. B. 2003. The impact of Brazos deltaic system on upper slope stratigraphic sequence evolution. *In*: *Shelf Margin Deltas and Linked Down-Slope Petroleum Systems: Global Significance and Future Exploration Potential. Gulf Coast Section of Society of Sedimentary Research Foundation 20th Bob Perkins Research Conference*, 325–354.

GALLOWAY, W. E. 1989. Genetic stratigraphic sequences in basin analysis I: architecture and genesis of flooding surface bounded depositional units. American Association of Petroleum Geologists Bulletin, **73**, 125–142.

HAQ, B. U., HARDENBOL, J. & VAIL, P. R. 1987. Chronology of fluctuating sea levels since the Triassic. *Science*, **235**, 1156–1167.

HUNT, D. & TUCKER, M. E. 1992. Stranded parasequences and the forced regressive wedge systems tract: deposition during base-level fall. *Sedimentary Geology*, **81**, 1–9.

JERVEY, M. T. 1988. Quantitative geological modeling of siliciclastic rock sequences and their seismic expression. *In*: WILGUS C. K., HASTINGS B. S., KENDALL, C. G. S., POSAMENTIER, H. W., ROSS, C. A. and VAN WAGONER, J. C. (eds) *Sea-level Changes: an Integrated Approach*. Society of Economic Paleontologists and Mineralogists, Special Publications, **42**, 47–69.

JORDAN, T. E. & FLEMINGS, P. B. 1991. Large-scale stratigraphic architecture, eustatic variation, and unsteady tectonism: a theoretical evaluation. *Journal of Geophysical Research*, **96**, 6681–6699.

KOHL, B., FILLON, R. H. & ROBERTS, H. H. 2004. Foraminiferal biostratigraphy and paleoenvironments of the Pleistocene Lagniappe delta complex, northeastern Gulf of Mexico. *In*: ANDERSON, J. B. & FILLON, R. H. (eds) *Late Quaternary Stratigraphic Evolution of the Northern Gulf of Mexico Margin*. Society for Sedimentary Geology, Special Publications, Tulsa, OK, **79**, 189–216.

KOSS, J. E., ETHRIDGE, F. D. & SCHUMM, S. A. 1994. An experimental study of the effects of base-level change on fluvial, coastal plain and shelf systems. *Journal of Sedimentary Research*, **B64**, 90–98.

LABEYRIE, L. D., DUPLESSY, J. C. & BLANC, P. L. 1987. Variations in mode of formation and temperature of oceanic deep waters over the past 125,000 years. *Nature*, **327**, 477–481.

LAMBECK, K. & CHAPPELL, J. 2001. Sea level change through the last glacial cycle. *Science*, **292**, 679–686.

LAMBECK, K., YOKOYAMA, Y. & PURCELL, T. 2002. Into and out of the last glacial maximum: sea-level change during oxygen isotope stages 3 and 2. *Quaternary Science Reviews*, **21**, 171–181.

LEHNER, P. 1969. Salt tectonics and Pleistocene stratigraphy on continental slope of northern Gulf of Mexico. *American Association of Petroleum Geologists Bulletin*, **53**, 2431–2479.

MIALL, A. D. 1997. *The Geology of Stratigraphic Sequences*. Springer, New York.

MITCHUM, R. M., VAIL, P. R. & SANGREE, J. B. 1977. Seismic stratigraphy and global changes in sea level, Part 6: Stratigraphic interpretation of seismic reflection patterns in depositional sequences. *In*: PAYTON, C. E. (ed.) *Seismic Stratigraphy – Applications to Hydrocarbon Exploration*. American Association of Petroleum Geologists, Memoirs, **26**, 117–133.

NICOL, S. L., BOYD, R. & PENLAND, S. 1994. Stratigraphic response of wave-dominated estuaries to different relative sea level and sediment supply histories: Quaternary case studies from Nova Scotia, Louisiana and Eastern Australia. *In*: DALRYMPLE, R., BOYD, R. & ZAITLIN, B. A. (eds) *Incised-Valley Systems: Origin and Sedimentary Sequences*. Society of Economic Paleontologists and Mineralogists, Special Publications, **51**, 265–283.

NYSTUEN, J. P. 1998. History and development of sequence stratigraphy. *In*: GRADSTEIN, F. M., SANDVIK, K. O. & MILTON, N. J. (eds) *Sequence Stratigraphy – Concepts and Applications*. Elsevier, Amsterdam, 31–116.

PAINE, J. G. 1993. Subsidence of the Texas coast: inferences from historical and late Pleistocene sea levels. *Tectonophysics*, **222**, 445–458.

PEKAR, S. F., CHRISTIE-BLICK, N., MILLER, K. & KOMINZ, M. A. 2003. Quantitative constraints on the origin of stratigraphic architecture at passive continental margins: Oligocene sedimentation in New Jersey, U.S.A. *Journal of Sedimentary Research*, **73**, 227–245.

POSAMENTIER, H. W. 2001. Lowstand alluvial bypass systems: incised vs. unincised. *American Association of Petroleum Geologists Bulletins*, **85**, 1771–1793.

POSAMENTIER, H. W. and ALLEN, G. P. 1999. *Siliciclastic sequence stratigraphy – concepts and applications*. Society for Sedimentary Geology, Concepts in Sedimentology and Paleontology, **7**, 210.

POSAMENTIER, H. W. & VAIL, P. R. 1988. Eustatic controls on clastic deposition II – sequence and systems tract models. *In*: WILGUS, C. K., HASTINGS, B. S., ROSS, C. A., POSAMENTIER, H. W., VAN WAGONER, J. & KENDALL, C. G. ST. C. (eds) *Sea-level Changes: An Integrated Approach*. Society of Economic Paleontologists and Mineralogists Special Publications, **42**, 125–154.

POSAMENTIER, H. W., ALLEN, G. P. & JAMES, D. P. 1992. High resolution sequence stratigraphy – the East Coulee Delta, Alberta, *Journal of Sedimentary Petrology*, **62**, 310–317.

POSAMENTIER, H. W., JERVEY, M. T. & VAIL, P. R. 1988. Eustatic controls on clastic deposition I – conceptual framework. *In*: WILGUS, C. K., HASTINGS, B. S., ROSS, C. A., POSAMENTIER, H. W., VAN WAGONER, J. & KENDALL, C. G. ST. C. (eds) *Sea-level Changes: An Integrated Approach*. Society of Economic Paleontologists and Mineralogists Special Publications, **42**, 110–124.

POTTER, E. & LAMBECK, K. 2004. Reconciliation of sea-level observations in the western North Atlantic during the Last Glacial Cycle. *Earth and Planetrary Science Letters*, **217**, 171–181.

ROBERTS, H. H., FILLON, R. H., KOHL, B., ROBALIN, J. & SYDOW, J. 2004. Depositional architecture of the Lagniappe Delta: sediment characteristics, timing of depositional events, and temporal relationships with adjacent shelf-edge deltas. *In*: ANDERSON, J. B. & FILLON, R. H. (eds) *Late Quaternary Stratigraphic Evolution of the Northern Gulf of Mexico Margin*. Society for Sedimentary Geology, Special Publications, **79**, 141–186.

RODRIGUEZ, A. B., ANDERSON, J. B., BANFIELD, L. A., TAVIANI, M., ABDULAH, K. & SNOW, J. N. 2000. Identification of a −15 m middle Wisconsin shoreline on the Texas inner continental shelf. *Palaeogeography, Palaeoclimatology, Palaeoecology*, **159**, 25–43.

RODRIGUEZ, A. B., FASSELL, M. L. & ANDERSON, J. B. 2001. Variations in shoreface progradation and ravinement along the Texas coast, Gulf of Mexico. *Sedimentology*, **48**, 837–853.

RODRIGUEZ, A. B., ANDERSON, J. B., SIRINGAN, F. P. & TAVIANI, M. 2004. Holocene evolution of the east Texas coast and inner continental shelf: along-strike variability in coastal retreat rates. *Journal of Sedimentary Research*, **74**, 406–422.

ROTHWELL, R. G., KENYON, N. H. & MCGREGOR, B. A. 1991. Sedimentary features of the South Texas continental slope as revealed by side-scan sonar and high-resolution seismic data. *American Association of Petroleum Geologists Bulletin*, **75**, 298–312.

SCHUMM, S. A. 1977. *The Fluvial System*. Wiley, New York.

SCHUMM, S. A. & ETHERIDGE, F. G. 1994. Origin, evolution, and morphology of fluvial valleys. *In*: DALRYMPLE, R. M., BOYD, R. & ZAILTIN (eds) *Incised-valley Systems: Origin and Sedimentary Sequences*. Society for Sedimentary Geology, Special Publications, **51**, 12–27.

SHACKLETON, N. J. 1987. Oxygen isotopes, ice volume, and sea level. *Quaternary Science Reviews*, **6**, 183–190.

SIMMS, A. R. 2005. *Late Quaternary/Holocene evolution of the Nueces incised valley, central Texas*. PhD dissertation, Rice University.

SIMMS, A. R., ANDERSON, J. B. & BLUM, M. D. 2006*a*. Barrier-island aggradation via inlet migration, Mustang Island, Texas. *Sedimentary Geology*, **187**, 105–126.

SIMMS, A. R., ANDERSON, J. B., TAHA, P. Z. & RODRIGUEZ, A. B. 2006*b*. Over-filled versus Under-filled Incised Valleys: Examples from the Quaternary Gulf of Mexico. *In*: DALRYMPLE, R. M., LECKIE, D. & TILLMAN, R. (eds) *Incised Valleys in Time and Space*. Society for Sedimentary Geology, Special Publications, **85**, 117–139.

SIRINGAN, F. P. & ANDERSON, J. B. 1994. Modern shoreface and inner-shelf storm deposits off the east Texas Coast, Gulf of Mexico. *Journal of Sedimentary Research*, **B64**(2), 99–110.

SMYTH, W. C., ANDERSON, J. B. & THOMAS, M. A. 1988. Seismic facies analysis of entrenched valley-fill: a case study in the Galveston Bay area, Texas. *Gulf Coast Association of Geological Societies Transactions*, **38**, 385–394.

STECKLER, M. S., REYNOLDS, D. J., COAKLEY, B. J., SWIFT, B. A. & JARRAD, R. D. 1993. Modeling passive margin sequence stratigraphy. *In*: POSAMENTIER, H. W. SUMMERHAYES, C. P., HAQ, B. U. & ALLEN, G. P. (eds) *Sequence Stratigraphy and Facies Associations*. International Association of Sedimentologists, Special Publications, **18**, 19–41.

TALLINGS, P. J. 1998. How and where do incised valleys form if sea level remains above the shelf edge? *Geology*, **26**, 87–90.

TATUM, T. E. 1977. *Shallow geological features of the upper continental slope, northwestern Gulf of Mexico*. MS thesis, Texas A&M University.

THOMAS, M. A. & ANDERSON, J. B. 1994. Sea-level controls on the facies architecture of the Trinity/Sabine incised-valley system, Texas continental shelf. *In*: DALRYMPLE, R. W., BOYD, R. & ZAITLIN, B. A. (eds) *Incised-valley Systems: Origin and Sedimentary Sequences*. Society for Sedimentary Geology, Special Publications, **51**, 63–82.

TORNQVIST, T. E., WALLINGA, J. & BUSSCHERS, F. S. 2003. Timing of the last sequence boundary in a fluvial setting near the highstand shoreline-insights from optical dating. *Geology*, **31**, 279–282.

TRAMP, K. L. & ANDERSON, J. B. 2004. Depositional architecture for a coastal river in a sequence stratigraphic framework, Sabine Pass, Texas and Louisiana. *AAPG Annual Convention Abstracts*, A139.

VAIL, P. R., MITCHUM, R. M. JR & THOMPSON, S., III. 1977*a*. Seismic stratigraphy and global changes of sea level, part 3: relative changes of sea level from coastal onlap. *In*: PAYTON, C. E. (ed.) *Seismic Stratigraphy – Applications to Hydrocarbon Exploration*. The American Association of Petroleum Geologists Memoirs, **26**, 63–81.

VAIL, P. R., MITCHUM, R. M. JR & THOMPSON, S. III. 1977*b*. Seismic stratigraphy and global changes of sea level, part 4: global cycles of relative changes of sea level. *In*: PAYTON, C. E. (ed.) *Seismic Stratigraphy – Applications to Hydrocarbon Exploration*. The American Association of Petroleum Geologists Memoirs, 83–97.

VAIL, P. R., MITCHUM, R. M., TODD, R. G., *ET AL*., 1977*c*. Seismic stratigraphy and global changes in sea level. *In*: PAYTON, C. E. (ed.) *Seismic Stratigraphy - Applications to Hydrocarbon Exploration*. The American Association of Petroleum Geologists Memoirs, **26**, 49–212.

VAN WAGONER, J. C., POSAMENTIER, H. W., MITCHUM, R. M., VAIL, P. R., SARG, J. F., LOUTIT, T. S. & HARDENBOL, J. 1988. An overview of the fundamentals of sequence stratigraphy and key definitions. *In*: WILGUS, C. K., HASTINGS, B. S., KENDEALL, C. G. S. C., POSAMENTIER, H. W., ROSS C. A. & VAN WAGONER J. C. (eds) *Sea-level Changes: an Integrated Approach*. Society of Economic Paleontologists and Mineralogists, Tulsa, OK, 39–45.

VAN WAGONER, J., MITCHUM, R. M., CAMPION, K. M. & RAHMANIAN, V. D. 1990. *Siliciclastic Sequence Stratigraphy in Well Logs, Core, and Outcrops: Concepts for High-resolution Correlation of Time and Facies*. American Association of Petroleum Geologists, Methods in Exploration Series, **7**, 55.

WELLNER, J. S., SARZALEJO, S., LAGOE, M. & ANDERSON, J. B. 2004. Late Quaternary stratigraphic evolution of the west Louisiana/east Texas continental shelf. *In*: ANDERSON, J. B. & FILLON, R. H. (eds) *Late Quaternary Stratigraphic Evolution of the Northern Gulf of Mexico Margin*. Society for Sedimentary Geology, Special Publications, **79**, 217–236.

WELLNER, R. W. & BARTEK, L. R. 2003. The effect of sea level, climate, and shelf physiography on the development of incised-valley complexes; a modern example from the East China Sea. *Journal of Sedimentary Research*, **73**, 926–940.

WOODBURY, H. O., SPOTTS, J. H. & AKERS, W. H. 1978. Gulf of Mexico continental-slope sediments and sedimentation. *In*: BOUMA, A. H., MOORE, G. T. & COLEMAN, J. M. (eds) *Framework, Facies and Oil-trapping Characteristics of the Upper Continental Margin*. American Association of Petroleum Geologists, Studies in Geology, **7**, 117–137.

Seismic stratigraphic and geomorphic analysis of deep-marine deposition along the West African continental margin

F. HADLER-JACOBSEN[1], M. H. GARDNER[2] & J. M. BORER[3]*

[1]*Statoil ASA N-7005 Rotvoll, Trondheim, Norway (e-mail: fhj@statoil.com)*

[2]*Montana State University, PO Box 173480, Bozeman, MT 59717-3480, USA*

[3]*Colorado School of Mines, 1516 Illinois St., Golden, CO 80401-1887, USA*

**Present address: El Paso Oil and Gas, 1225 17th St., Suite 19, Denver CO 80202, USA*

Abstract: The West African continental margin evolution is preserved in a small source-distant setting (20 × 30 km area) by changes in lobe-channel–levee seismic geomorphological elements within a threefold seismic stratigraphic hierarchy. The *c.* 32 Ma depositional record of rift, drift and depositional outbuilding of the margin by gravity-driven adjustment, deformation and deposition produced a hierarchy of second- through fourth-order stratigraphic cycles bounded by laterally continuous fine-grained drapes inferred to record prolonged periods of sediment starvation. The margin outbuilding phase, the focus of this contribution, consists of three second-order adjustment bounded cycles (ABCs) that record major adjustment and/or modification of the deep-marine depositional system. Seven third-order cycles also show changes in depositional trend and seismic facies architecture. Ten fourth-order cycles, best resolved within the upper part of the succession, consist of multiple, wedge-shaped and compensating, lobe–channel–levee complexes up to 20 km wide. These complexes show an upward increase in channel–levee and decrease in lobe proportion. They also show an upward change from lobes incised by sinuous channels to channels deflected to lobe flanks. Outcrop and shallow core-calibrated analogues from the Permian Brushy Canyon Formation, and modern Amazon and Zaire Fans help constrain these patterns. Changes in the sediment composition and volume of subaqueous flows at their point of origin, and subsequent gravitational deformation, syn-sedimentary mass-wasting and large-scale fan avulsion punctuating deep-marine sedimentation, adjust deep-marine depositional pattern during basin margin outbuilding. Lobe-channel-levee distributions in this sediment source-distant setting record a progressive increase in local topographic relief and gradient related to the basinward migration of deformation during depositional outbuilding of the continental margin. Two important conclusions derived from this record include (1) the importance of local seabed topography and gradient on producing changes in depositional pattern, and (2) that repeated and cyclic changes in these patterns reflect adjustment/deformation within, and probably restricted to, the deep-marine record. Integrated seismic stratigraphic and geomorphic analysis delineates multiple scales of these adjustment-bounded cycles. The evolving map patterns record adjustment by shifts in geomorphic pattern and orientation. These local geomorphic changes can be used to predict longer-term and larger-scale changes in the depositional record of the continental margin evolution. This analytical approach should have general utility along high shelf-to-basin relief margins with similar gravity-driven deformation.

Linking deep-water depositional system and continental margin evolution

There is a direct link between tectonics and sedimentation along continental margins. Thermomechanical processes that act along a continental margin produce a well-known tectonic evolution from rifting, crustal extension and faulting, and subsequent sea-floor spreading (McKenzie 1978). The final, open ocean basin phase records the outbuilding of a sediment prism along the continental margin (Pratson & Laine 1989). These prisms are composed of shallow- and deep-marine deposits, leaving a record of subaqueous flow deposition and gravitational deformation directed basinward into the open ocean basin. This study uses seismic stratigraphic and geomorphic analyses to document long-term trends in the proportion and style of lobe, channel and levee deposits during the outbuilding of the South Atlantic continental margin, west of Africa. The high-resolution three-dimensional seismic dataset provides an opportunity to characterize and understand the basinward translation of complex topography, subaqueous flow focus and

From: DAVIES, R. J., POSAMENTIER, H. W., WOOD, L. J. & CARTWRIGHT, J. A. (eds) *Seismic Geomorphology: Applications to Hydrocarbon Exploration and Production*. Geological Society, London, Special Publications, **277**, 47–84.
0305-8719/07/$15.00

local gradients as a response to gravitational deformation that continues to the present.

A seismic stratigraphic framework is established from regional correlation of the most continuous seismic reflections calibrated to dated condensed sections. These relatively conformable horizons are interpreted to represent zones of fine-grained, pelagic and hemipelagic drapes that record significant time. This framework allows for the construction of consistent time series maps showing changing seismic geomorphological seascapes. Seascape evolution depicted in these maps record normal deep-marine sedimentation punctuated by periods of gravitational deformation and adjustment during continental margin outbuilding. From this context, the evolving seafloor topography resulting from deep-marine deformation and deposition can be related to changes in subaqueous flow pathways, erosion, bypass and the distribution and proportion of resulting depositional elements (Nardin *et al.* 1979; Kneller *et al.* 1999; Posamentier & Kolla 2003).

We adopt the turbidite system classification proposed by Mutti, (1985) and Mutti & Normark (1991). In this scheme the down-current change from channelform to lobeform sedimentary bodies records the change from confined to unconfined flow conditions going from high to low gradient. Commonly, this decrease in depositional gradient is combined with a progressive decrease levee height (Piper & Normark 2001). Though originally conceived to describe down-profile changes in a deep-water depositional system, this channel-lobe classification scheme fundamentally links flow processes to gradient that can vary in an irregular fashion down the depositional profile. It is in this more fundamental sense of linking gradient to resulting sedimentary bodies and depositional pattern that we apply the model in this study.

Long-term evolution of gravity-driven slope and basin physiography

Continental margin evolution has been viewed primarily in terms of tectonic evolution related to thermo-mechanical processes of rifting and sea floor spreading (Vink 1982; Unternehr *et al.* 1988; Karner & Driscoll 1999). These observations were important to general acceptance of the plate tectonic theory (Menard 1972). Though deep-water depositional models were tied to this tectonic evolution by Shanmugam & Moiola (1988), the link was broad and general, with four deep-water systems related to (1) immature passive margin, (2) mature passive margin, (3) mixed margin and (4) active margin types. The evolution of lobe, channel and levee deposits is a higher resolution characterization than this basin classification study.

The West African continental margin is an 'Atlantic type' passive margin with a shelf, slope and rise in Miall's (1984) classification. It straddles continental and oceanic crust and overlies an earlier rift system. Mature ocean basin margins are prone to deep bathymetry because the margin is pinned by rift-related faulting and differential thermal subsidence of the oceanic crust. Gravity-driven subaqueous flow deposition and deformation record continual adjustment to the inherited structure and deep bathymetry along these ocean basin margins (Ross *et al.* 1994). This pattern of deposition and deformation is enhanced where major, long-lived, continental drainage systems promote significant depositional outbuilding of the margin. Gravity-driven, up-dip, extension and growth faulting is balanced by downdip compression, and thrust and fold-belt development (Fig. 1; Calassou & Moretti 2003; Brun & Fort 2004). This larger-scale pattern of gravitational deformation is recognized on other margins such as the Gulf of Mexico (Wu *et al.* 1990) and the Brazilian Atlantic margin (Cobbold & Szatmari 1991; Cobbold *et al.* 1995; Mohriak *et al.* 1995). All of the post-rift deformation is gravity-driven, as documented to the south in Angola (Duval *et al.* 1992; Lundin 1992; Spathopoulos 1996).

Along the West African margin, relatively long-lived, deep-water systems in Nigeria and Angola produced significant depositional outbuilding and slope adjustment and deformation. Well-developed, extensional and compressional deformation belts characterize these continental margin successions (Fig. 1; Duval *et al.* 1992; Calassou & Moretti 2003; Brun & Fort 2004). This paper documents the long-term (*c.* 32 Ma) cyclic record of deep-water deposition and deformation as expressed in a position more distant from the continental margin and dominated by channel and lobe elements. Published submarine fan depositional models (e.g. Mutti 1977; Normark 1978; Walker 1978; Reading & Richards 1994) do not account for the long-term topographic evolution of ocean basins due to slope adjustment, deformation and deposition during the outbuilding of a continental margin. Nonetheless the fundamental impact of seabed topography has been shown to be an important control on subaqueous flow transport and deposition (Kneller 1995; Kneller & McCaffrey 1999; Kneller *et al.* 1999; McCaffrey & Kneller 2001; Prather 2003). This paper integrates seismic stratigraphy with seismic geomorphic analysis to evaluate smaller-scale depositional patterns with the longer term evolution of progressively more complex margin structure and seabed topography through time.

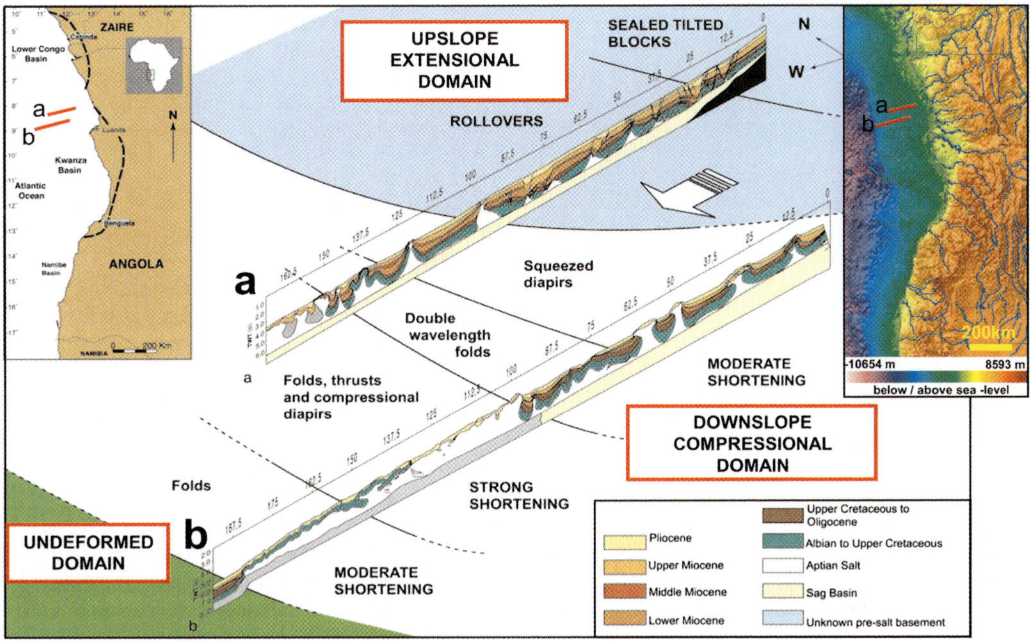

Fig. 1. Example of extensional and compressional domains of deformation along South Atlantic Margin, West Africa. The diagram illustrates structural zonation of post-salt sediments west of Angola. Slope-parallel cross section (**a**) highlights the upslope extensional domain dominated by growth faults and the downslope compressional domain dominated by salt diapirs. Cross-section (**b**) shows the compressional domain and the outer abyssal un-deformed domain. Section locations are shown on the map inserts. At the frontal part of section (b) (between 137 and 187.5), the upper Cretaceous to upper Miocene is too thin to be represented. Modified from Brun and Fort (2004). Reprinted from *Tectonophysics*, **382**, 129–150, copyright 2004, with permission from Elsevier. The combined topographic/bathymetric map shown in upper right corner is an Etopo2, 2 min surface grid from National Geophysical Data Center.

Brief outline of paper

The following sections of this paper describe the long-term (*c.* 32 Ma) depositional outbuilding of the West African continental margin based on: (1) temporal evolution of lobe–channel–levee complexes (LCLCs) from high-resolution seismic data; (2) dynamic internal architecture of LCLCs; (3) use of outcrop and shallow seismic analogues; (4) the relationship between changes in architectural elements and local gradient; and (5) the relationship between these attributes and basin margin deformation and slope adjustment. This leads to the recognition of adjustment bounded cycles (ABCs) that record changes in the morphology and trend/orientation of the deep-marine depositional system in this source-distant setting. Seafloor topography reflects both sediment deposition as well as structural deformation. This resultant and evolving topography controls successive subaqueous flow behaviour, trend and position with respect to erosion, transport and deposition (Fig. 2).

Data and methods

Study area and database

The study area is along the South Atlantic continental margin of West Africa (Fig. 3). The depositional setting is classified as a 'high shelf-to-basin relief (deep bathymetry: Steffens *et al.* 2003), sediment under-filled basin' in the scheme proposed by Hadler-Jacobsen *et al.* (2005; Fig. 4). Analysis of the long-term continental margin evolution is aided by *c.* 2.6 s vertical section (*c.* 2600 m thick, assuming 2 km/s velocity) of relatively 'complete' (conformable) and undeformed stratigraphy, providing an adequate vertical record of the margin outbuilding. However, the spatial area of the high-resolution seismic volume used in this study is limited (20 × 30 km map-view coverage) relative to the overall margin and depositional system dimensions (hundreds of kilometres by thousands of kilometres). The local high-resolution seismic data is interpreted within a regional, borehole-calibrated, seismic framework

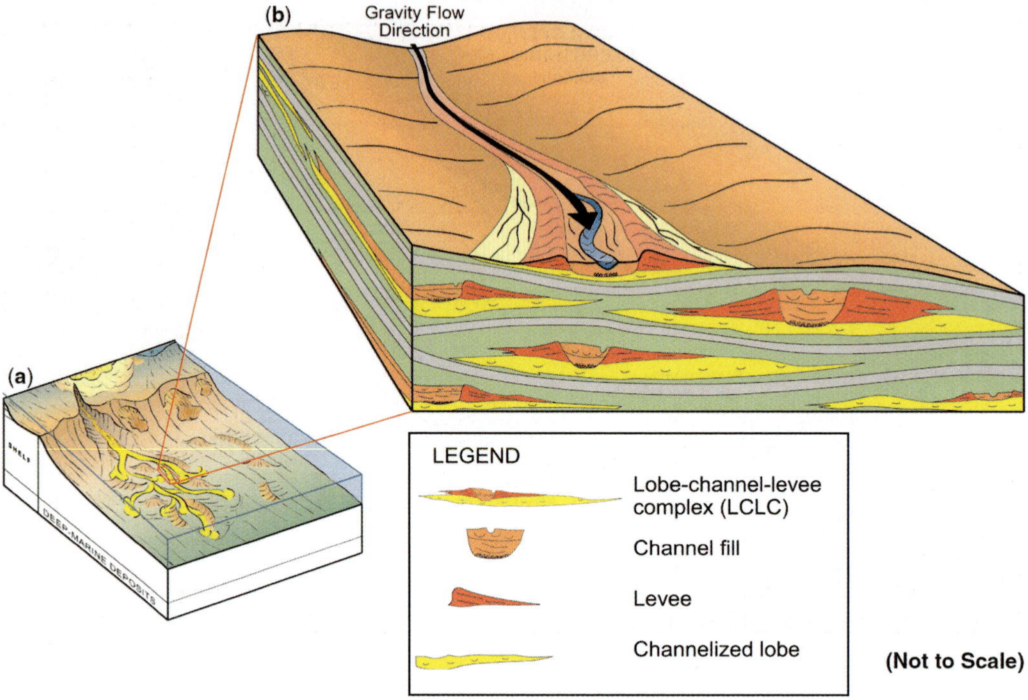

Fig. 2. Schematic box diagram illustrating how local gradient and topography control the distribution of sediment gravity flow deposits. Gravity flows follow lows at all scales, here illustrated at semi-regional scale (inherited structurally deformed seabed topography; 10's to 100+ kilometers) and local scale (compensating cycles of lobe–channel–levee complexes; 1 to 10's kilometers Mutti & Sonnino 1981). The degree of erosional and/or topographic confinement controls the proportion of channel versus lobe deposits. Modified from Hadler-Jacobsen *et al.* (2005).

that supports the conclusions on the long-term basin margin evolution presented here.

Seismic interpretation methods

Integrated seismic stratigraphic and seismic geomorphic analysis. Improved imaging of extensive three-dimensional (3D) seismic data from ocean basins in the last decade has significantly contributed to, and to a large degree altered, our perception of deep-water depositional systems (Beauboeuf & Friedmann 2000; Booth *et al.* 2000; Mayall & Stewart 2000; Pirmez *et al.* 2000; Posamentier *et al.* 2000; Abreu *et al.* 2003; Fonnesu 2003; Posamentier 2003). Advanced 3D seismic attribute techniques can now characterize subsurface seascape evolution, map thickness distributions, and in particular display compelling views of deep-water geomorphology unseen a decade ago. These methods are used to analyse high-resolution 3D seismic data from the West African margin. Seismic-based palaeogeographic maps are systematically examined within a hierarchy of stratigraphic cycles to determine the optimum stratigraphic window for constructing seismic maps that minimize the display of unrelated depositional elements. Seismic cross sections relate geomorphic lobe, channel and levee elements, displayed in map view, to the stratigraphic framework. This approach defines depositional cycle and element age relations based on cross-cutting relationships, termination geometries, condensed section identification and seismic facies analysis (Vail *et al.* 1977; Vail 1987; Mitchum 1985). Interpreting seismic geomorphology maps within a stratigraphic framework is critical for defining genetically related stratal packages for comparative analysis. The maps document the evolving distribution and variability of lobe, channel and levee depositional elements. The difference in the proportions of these elements is used to infer changing gradient conditions that are related to depositional outbuilding and deformation of the continental margin. Regional seismic and borehole data, not presented here, are utilized to calibrate these local subsurface maps and sections to the overall continental margin evolution. Architectural patterns are compared and calibrated to other modern and ancient analogous systems: (1) the Permain,

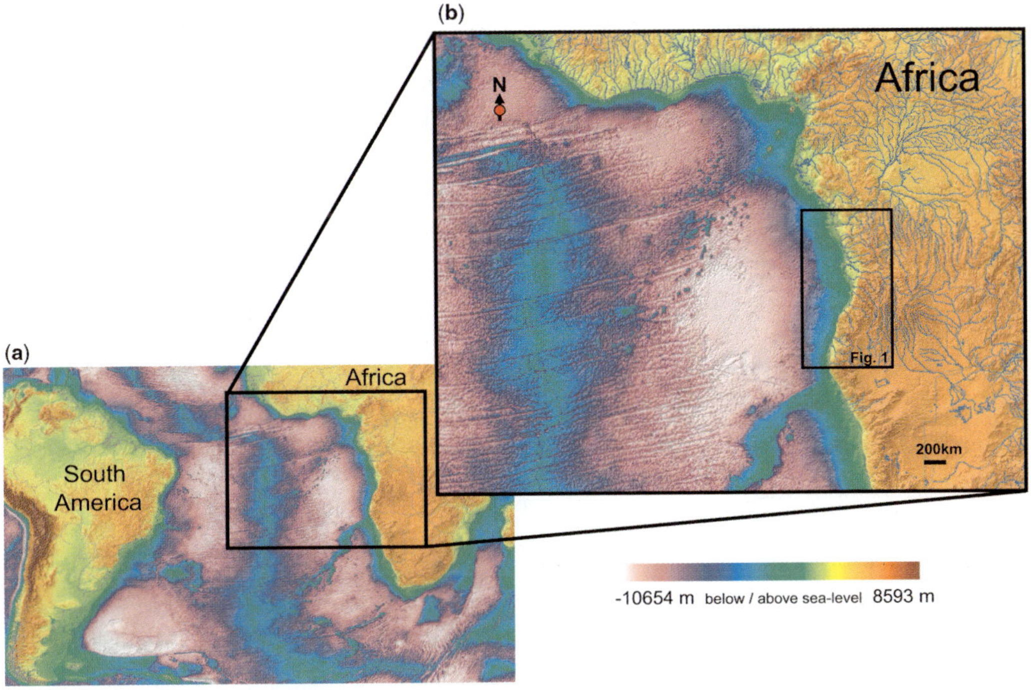

Fig. 3. Map of study area showing combined topography/bathymetry of South Atlantic region and detail of South Atlantic Margin, offshore West Africa. The dataset analysed in this study represents an unspecified part of this margin. The inset frame indicates the location of Figure 1, describing the deformation style along the Angolan Margin. The maps are derived from the Etopo2, 2 min surface grid from National Geophysical Data Center.

Brushy Canyon Formation (Zelt & Rossen 1995; Gardner & Sonnenfeld 1996; Beaubouef *et al.* 1999; Blikeng & Fugelli 2000; Carr & Gardner 2000; Gardner & Borer 2000; Gardner *et al.* 2003), and (2) modern core calibrated (ODP) Amazon Fan (Manley & Flood 1988; Flood *et al.* 1991; Pirmez & Flood 1995; Hiscott *et al.* 1997; Pirmez *et al.* 1997; Piper & Normark 2001).

Amplitude extraction. In this study, seismic amplitude extractions capture successive map-view expressions of stratigraphic amplitude populations. By means of the Landmark/SeisWorks/Post Stack PAL functionality, 'Maximum Trough Amplitudes' were extracted from stratigraphic intervals of interest. For each map a constant seismic two way travel time (ms) interval away from interpreted condensed section horizons (H0–H6) was set. In order to capture appropriate portions of the stratigraphy, the user-defined extraction interval ('thickness') was varied somewhat through the section. The typical amplitude extraction window for this study is 100 (±50) ms. The typical 100 ms-thick window captures multiple LCLCs with internal channelized lobe elements and channel-fill elements within one fourth-order cycle (Fig. 5). Strongly compensating stratigraphy leads to seismic geomorphology map extraction representing information from two or more LCLCs (Fig. 5). These seismic geomorphology maps therefore include depositional elements (i.e. sedimentary bodies), which in fact were deposited at different times. In this study the grouping of multiple LCLCs and thus map-extraction of diachronous depositional elements within the seascape depicted is the norm. However, as long as the diachronous regions of a geomorphology maps are understood through corresponding seismic cross-section analysis, this kind of composite geomorphic map can be used in 3D seismic stratigraphic analysis, and contributes significantly to facies/reservoir/seal prediction confidence.

Integrated seismic stratigraphic/geomorphic analysis

Chronostratigraphic framework

The north–south oriented seismic type-section displays the geology of a *c.* 3 s (two-way travel time)

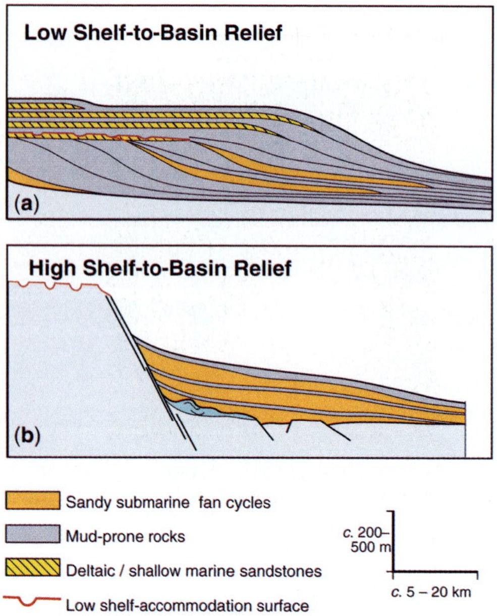

Fig. 4. Conceptual diagrams illustrating (**a**) low shelf-to-basin relief and (**b**) high shelf-to-basin relief margins (Hadler-Jacobsen *et al.* 2005). The diagram highlights the contrast in dip-oriented stacking pattern between the two end-members. Low relief margins tend to show the basin-ward directed shelf break migration along with successive basin-ward stepping/laterally compensation submarine fan cycles. High relief margins show a more stationary to aggradational shelf break and corresponding aggradational/compensational submarine fan stacking patterns. The South Atlantic margin is a high-relief margin that experienced increasing topographic slope and basin roughness through time due to gravity-driven deformation and slope adjustment. The adjustment/deformation processes include deep-water extensional growth faults, fold/thrust belts and diapirs associated with a mobile ductile substratum.

vertical section from the 20 × 30 km study area (approximately 3000 m thick, assuming 2 km/s average velocity; Fig. 6a). The chronostratigraphic ages assigned to this succession are based on seismic correlation to maximum flooding surfaces/condensed section (MFS/CS) mudstones identified and dated from wells located outside the study area. Complex structural deformation related to emplacement of salt canopies and salt-cored structures between these wells and the study area complicates correlation and adds uncertainty to this chronostratigraphy. Consequently, the proposed age dates for six seismic horizons are of moderate confidence and must be considered tentative (Fig. 6b). Within the study area, six of eight semi-regional seismic horizons are correlated to MFS/CS zones as follows (dating based on Berggren *et al.* 1995): H0 *c.* 32.4 Ma; H2 *c.* 23.5 Ma, H3 *c.* 16.8 Ma; H5 *c.* 14.3 Ma; H6 *c.* 11.2 Ma; H7 *c.* 5.4 Ma. The H1 and H4 seismic horizons lack chronostratigraphic calibration.

Stratigraphic cycle hierarchy

We define the long-term cycles (second-order, third-order) by their chronostratigraphic age, with the higher frequency cycles (fourth-order cycles and cyclic LCLCs) placed within a relative hierarchy based on physical stratigraphic correlations. The higher frequency cycles are not assigned a temporal duration, nor is their placement within the hierarchy based on an assumption of temporal equivalence. Furthermore, the time value of strata within a cycle, at all scales, is not assumed to be uniform, with the mudstone intervals apparently recording an order of magnitude more time than the sand-prone strata they bracket. Given the moderate chronostratigraphic confidence (no direct borehole calibration), the proposed cycle hierarchy is first and foremost a relative hierarchy based primarily on physical stratigraphic correlations calibrated to, but not constrained by, chronostratigraphic ages (Embry 1995).

First-order cycle. Prior to depositional outbuilding, ocean basins that flank passive continental margins generally record pre-syn- and post-rift tectonic phases that are not completely resolved in this seismic dataset. The lower portion of the type-section shows reflection patterns with highly irregular structural relief related to rifts (Brun & Fort 2004; Fig. 6a and b). The overlying 'post-rift' strata concordantly drape and 'heal' this underlying rift-topography. Highly uniform, parallel concordant seismic facies continue this in-filling pattern, and are interpreted to record a relatively quiescent 'drift' period related to continued sea-floor spreading and thermal subsidence of oceanic crust. In this study the first-order cycle comprising this margin's pre-syn- and post-rift evolution is not characterized beyond the fact that the post-rift margin outbuilding succession of interest (described below) represents a late phase of the overall first-order South Atlantic margin basin fill cycle. The onset of Cycle 1.1 corresponds in time (*c.* 32.4 Ma) to documented combined eustasy change and early Oligocene uplift of the African hinterland (Lunde *et al.* 1992), which suggests this initial second-order margin adjustment primarily is driven by re-arrangement of shelf sediment delivery systems. A threefold stratigraphic hierarchy of cycles (second-order through fourth-order) is established for the Oligocene (*c.* 32 Ma) to present margin outbuilding sedimentary record (H0 to Seabed) (Fig. 6b).

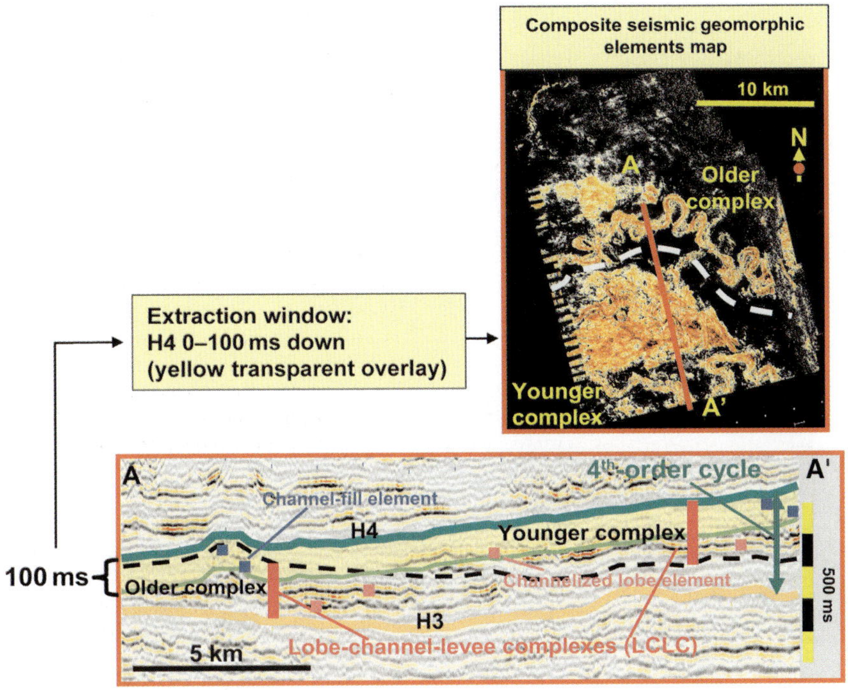

Fig. 5. Cross section and map illustrating the composite nature of amplitude extraction maps used in this study. Cross section analysis shows that the geomorphic elements represented on the map are two distinct lobe–channel–levee complexes, offset in time and space through topographic compensation. The constant 100 ms extraction interval down from H4 displays parts of both the older and the younger complex. Consequently, the map is a composite since it displays depositional elements that were never present at the same geologic instant in time. The high-resolution data allow for individual geobodies (e.g. channel fill, lobe element) to be delineated on the map by coherent amplitude patterns that render geomorphic detail. The seismic attribute type extracted from the 3D volume represents 'maximum trough amplitude' throughout this study.

The basic physical criterion for the definition of the proposed three-fold hierarchy is based on cross-sectional (type-section) expression of seismic amplitude. At all the three hierarchical scales relatively long correlation length seismic reflection patterns (seismic drape facies) define cycle boundary position, corresponding to periods of sediment starvation/condensation. Furthermore, at all scales successive contrasts in internal seismic facies expression corroborate the existence of families of cycles (longer term), and cycles within cycles (shorter term). Explicit cycle characterizations including applied cycle recognition criteria at each hierachical level are described below and in Table 1.

Second-order cycles. Within the long-term (*c.* 32 Ma) H0 to seabed succession three packages with prominently contrasting seismic expressions appear in the seismic type-section (Fig. 6a, b): (1) a low-amplitude reflection record showing variable but predominantly long lateral continuity internal facies characterized by subtle lobeform and local channelform (lower *c.* 1100 ms (*c.* m) interval); (2) a high-amplitude reflection record showing distinct lobeform and channel-leveeform internal seismic facies (middle *c.* 800 ms (*c.* m) thick interval); and (3) sub-parallel, concordant, very long correlation length reflection record with external wedgeform geometry showing internal chaotic facies where the cycle is thickest (upper *c.* 620–800 ms (*c.* m) interval). The upward increase in amplitude strength and resolution, in part, records the attenuation of the high frequency content and decreased seismic energy with depth. However, the upward contrasts in cycle pattern and seismic facies are primarily reflecting geological changes to evolving basin fill cycles. The three above described long-term seismic stratigraphic cycles are defined as second-order cycles C1, C2 and C3 in this study.

The base of the lowermost second-order cycle (C1) is defined at H0 where amplitudes change from uniform parallel concordant reflection pattern to slightly irregular seimic facies with subtle channelforms. This seismic facies contrast reflects initial

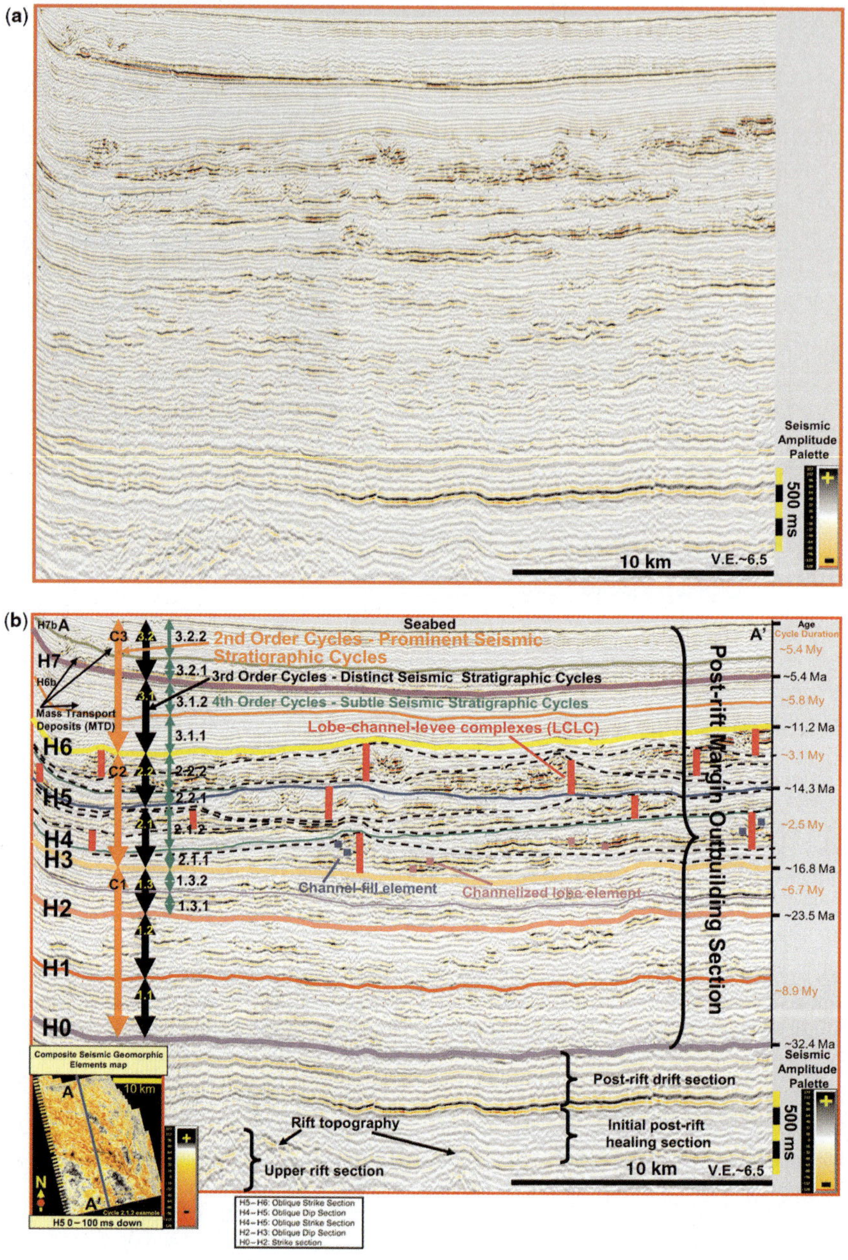

Fig. 6(a). Uninterpreted seismic type-section through study area. Grey/black colours represent hard acoustic responses. Yellow/red colours represent soft acoustic responses. **(b)** Interpreted seismic type-section through the study area, including condensed sections that define the seismic stratigraphic framework, H0–H7. The section shows pre-existing rift-topography at the deepest level, the 'upper rift section', an 'initial post-rift healing section' draping the early rift topography, and a regular sub-parallel 'post-rift drift section'. Chronostratigraphic data correlated from updip boreholes and corresponding approximate cycle durations are given at the right-hand margin. A threefold cycle hierarchy (second-order to fourth-order) is illustrated with arrows. Internal subdivision (dashed black lines and vertical bold red lines) delineate lobe–channel–levee complexes that are genetically linked sets of architectural elements. Individual channel-fill (blue boxes) or lobe (purple boxes) elements are the highest resolution features that can be detected. The index map is an extracted amplitude map (100 ms window below H5) that indicates the location of the section relative to the seismic cube. The inclined seismic horizons at the left-hand side of the section represent the lower flank of a significant anticline situated just outside the structurally quiet study area.

Table 1. *Cycle recognition criteria and characterization of threefold hierarchy of cycles and LCLCs (see also corresponding Fig. 6b)*

	Second-order cycles	Third-order cycles	Fourth-order cycles	LCLCs
Number of cycles	3 C1, C2 and C3	7 within 3 2nd-order cycles	10 within 5 3rd-order cycles	12 within 2nd-order C2
Cycle boundary recognition criteria	Top C3: seabed C2/C3: prominent seismic drape facies C1/C2: prominent seismic drape facies Base C1: onset irregular lobeform and channelform reflection pattern	Draping seismic reflectors between packages with distinctly contrasting seismic facies expression	Draping seismic reflectors between packages with contrasting LCLC style	Draping seismic reflectors between packages representin globe–channel –levee complexes
Cycle thickness (ms ≈ m)	Range: 620–1100 C3: *c.* 620–800 C2: *c.* 800 C1: *c.* 1100	Range: 280–500	Range 30–350	Range 30–300
Lateral extension (width)	Beyond the dataset	Beyond the dataset	Usually beyond dataset, but two 4th-order cycles pinch out to very thin record (<30 ms) within the seismic type-section	From 8 km (youngest complex within C2) to 25 km (oldest complex within C1)
Correlation to moderate confidence chronostratigraphic ages?	Yes C3: *c.* 11.2 Ma C2: *c.* 5.6 Ma C1: *c.* 15.6 Ma	Partly, from C1/1.3 and younger: Cycle 3.2: *c.* 5.4 Ma Cycle 3.1: *c.* 5.8 Ma Cycle 2.2: *c.* 3.1 Ma Cycle 2.1: *c.* 2.5 Ma Cycle 1.3: *c.* 6.7 Ma	No	No
Vertical seismic stratigraphic and/or seismic facies change between cycles or complexes of same order	Prominent contrasts in seismic facies expression between cycles. C3: sub-parallel seismic drapefacies with pockets of chaotic facies	Distinct contrasts in seismic facies expression between cycles: 2.2: more focussed/narrower lobe–channel–levee complexes	Trend from 1.3 through 2.2: Upward increase in channel–leveeform to lobeform proportion	Defined in 3.1, 2.1 and 2.2 only. Upward increase in channel–leveeform to lobeform proportion

(Continued)

Table 1. *Continued*

	Second-order cycles	Third-order cycles	Fourth-order cycles	LCLCs
	C2: distinct strong amplitude lobe–channel–levee forms C1: Irregular, isolated weak/moderate amplitude lobe- and channelforms	2.1: Larger and more clearly expressed lobe–channel –levee complexes. 1.3 (relative to 1.1 and 1.2): increasing lobeform proportion 1.1 and 1.2: upward increase in channe–leveeform to lobeform proportion.		Upward decrease in compensation cycle width reflecting successive upward increase in lobe–channel–levee focus
Typical vertical seismic facies change within cycles of same order	C1 and C2: upward increase in mounding geometries.	1.1–2.2: upward increase in channel– leveeform to lobeform proportion	Upward increase in channel–leveeform to lobeform proportion	Upward change from lobeform to channel–leveeform

impact of more active basin sedimentation as a response to combined eustasy change and early Oligocene uplift of African hinterland, and thus onset of more effective sediment delivery to slope and basin from the palaeo Congo drainage system (Lunde *et al.* 1992). The top of C1/base of C2 is defined at H3, representing a distinct *c.* 100 ms (*c.* m) thick seismic drape-facies between underlying moderate/low amplitude reflection succession and overlying strong amplitude reflection succession. The top of C2/base of C3 is defined at H6 where the high amplitude reflection record abruptly transforms to prominent sub-parallel concordant seismic drape facies. The top of C3 is defined at the seabed. Together, these second-order cycles (early Oligocene to present) represent the sedimentary record of depositional and deformational evolution of the West African continental margin.

Third-order cycles. Within the three above described second-order cycles, seven third-order cycles are defined by seismic correlation of laterally continuous drape-facies representing cycle boundaries (Fig. 6a, b). The key third-order cycle recognition criterion is related to significant vertical shifts in seismic facies pattern. The seismic horizons H0, H1, H2, H3, H5 H6, H7 and seabed represent third-order cycle boundaries that bracket strata with distinctly contrasting seismic stratigraphic geometries and facies (Fig. 6b). Except for H1, all third-order boundaries represent seismically interpreted horizons which are correlated to dated condensed sections, or fan abandonment zones in wells outside the study area. Estimated duration of these third-order cycles are: Cycle 1.1 + Cycle 1.2 (H0–H2), 8.9 Ma (individual cycle duration not estimated); Cycle 1.3 (H2–H3), 6.7 Ma; Cycle 2.1 (H3–H5), 2.5 Ma; Cycle 2.2 (H5–H6), 3.1 Ma; Cycle 3.1 (H6–H7), 5.8 Ma; and Cycle 3.2 (H7–seabed), 5.4 Ma (Fig. 6b). Within the seismic type-section Cycles 1.1, 1.2 and 1.3 show uniform thicknesses of *c.* 400 ms (*c.* m), *c.* 370 ms (*c.* m) and 300 ms (*c.* m), respectively. Cycle 2.1 shows distinct mound geometry varying in thickness from *c.* 350 to 500 ms (*c.* m). Cycle 2.2 thickness variability across the type-section is gentle, and in the order of *c.* 280–360 ms (*c.* m). Cycle 3.1 shows distinct external wedgeform, varying in thickness from 300 to 500 ms (*c.* m). Cycle 3.2 displays a rather uniform thickness around *c.* 350 ms (*c.* m).

Cycle 1.1 is dominated by irregular low amplitude seismic facies locally displaying subtle lobeforms and channelforms. Cycle 1.2 shows increasing amplitude reflections with lobeform and channelform geometry. Cycle 1.3 shows an abundance of laterally extensive (up to 12 km wide) sheet/lobeforms and an upward increase in channelforms. Seismic resolution within Cycle 2.1

significantly increases relative to the underlying Cycle 1.1, Cycle 1.2 and Cycle 1.3 third-order cycles. Cycle2.1 is characterized by internal seismic drape-facies subdividing the cycle into two compensation cycles with wide and distinct internal lobe-channel-leveeform. Cycle 2.2 show depositional thinning over mounded geometry in Cycle 2.1, with significantly narrower internal lobe-channel-leveeforms relative to Cycle 2.1. The externally wedge-shaped Cycle 3.1 succession is dominated by laterally extensive monotonous sub-parallel concordant seismic facies with an internal wedgeform package of chaotic seismic facies where the cycle is thickest. Cycle 3.2, representing the youngest third-order cycle within the long-term basin fill record is defined between the seismic horizon H7 and the seabed. Internal seismic facies resembles the facies observed in Cycle 3.1 both with respect to laterally extensive sub-parallel facies and pockets of chaotic facies. Cycle 3.2 show two stacked packages of chaotic seimic facies, resting at the structural flank displayed in northern part of the dataset (Fig. 6b).

Third-order Cycles 1.1, 1.2 and 1.3 together with third-order Cycles 2.1 and Cycle 2.1, which in total show distinct increasing clarity in seismic lobeform and channelform, represent the main margin outbuilding in this area. Third-order Cycle 3.1 and Cycle 3.2 show abrupt transition to a continuous sub-parallel reflection pattern, which is interpreted as a period of passive sedimentation representing a phase of fairway retreat in the study area. However, time equivalent to this phase of local margin retreat, active lobe-channel-levee deposition and preservation is observed north of study area, indicating northward avulsion of deep-marine fairway activity in this late phase of the overall margin outbuilding.

Fourth-order cycles. None of the fourth-order cycles described below are correlated to chronostratigraphically dated reflectors at both their top and base so cycle durations are unknown. The lower seismic resolution and poor chronostratigraphic control in the lower part of the seismic volume (second-order C1) precludes definition of fourth-order cycles within third-order Cycles 1.1 and 1.2 (Fig. 6a, b). Two gently compensating fourth-order cycles (1.3.1 and 1.3.2) are recognized within the relatively thin tabular lobeform-dominated Cycle 1.3, but no higher-order cycles are defined at this depth. The fourth-order cycle boundary is put at a somewhat subtle seismic drape facies within Cycle 1.3. The improved data resolution within third-order Cycles 2.1 and 2.2 (within second-order C2) permits recognition of four fourth-order cycles subdivided by seismic drape-facies (Fig. 6a, b). Drape facies bracket two wedge-shaped fourth-order cycles (2.1.1 and 2.1.2) that show large scale compensation patterns within the *c.* 300–400 m (*c.* 300–400 ms) thick succession between the H3–H5 seismic horizons. Internally these fourth-order compensation cycles display distinct lobe-channel-leveeform packages. Cycle 2.1.1 almost pinches out within the type-section. The two over lying fourth-order cycles (2.2.1 and 2.2.2) display lower confidence compensation patterns given incomplete coverage of cycle 2.2.1 (Fig. 6b).

The chaotic seismic facies unit within Cycle 3.1 serves as the criterion for splitting this third-order cycle into two fourth-order cycles 3.1.1 and 3.1.2. The fourth-order cycle boundary (H6b) is put at the base of the chaotic facies, which is interpreted as a local mass transport deposit (MTD) derived from the rising structure located north of, and outside the study area. Similarly, Cycle 3.2 is sub-divided into two fourth-order cycles 3.2.1 and 3.2.2 based on a prominent seismic drape facies (H7b) upon which chaotic seismic facies, interpreted as an MTD, rests.

Architectural elements

Large-scale architectural elements: LCLCs. Fourth-order cycles within the higher frequency portion of the seismic data (H2–H6 succession) show distinct lobe–channel–leveeform seismic facies defined by geometry and correlation of thin seismic drape facies (Fig. 6a, b). These architectural elements (Jackson 1975; Miall 1985; Miall 1989; Shanmugam & Moiola 1990) are characterized by distinctive high-amplitude wedge-shaped geometries and local thickness anomalies that promote lateral offset of overlying lobe–channel–leveeform packages. These seismic features are interpreted as genetically linked LCLCs. They represent the optimum extraction unit for deriving seismic geomorphological maps that minimize the mixing of unrelated depositional elements. The LCLCs represent the essential ‘source’ from which geomorphic elements are derived. However, given the pronounced stratigraphic compensation in the study area, the seismic extraction maps still typically capture elements from multiple LCLCs (Fig. 6a, b, Table 1).

Small-scale architectural elements: channelized lobe elements and channel-fill elements. This higher resolution seismic dataset also permits recognition of internal units within individual LCLCs, representing single reflector seismic cycles (Fig. 6b). Where seismic resolution is best (third-order Cycles 2.1 and 2.2), internal reflection patterns within the LCLCs show offset stacking of single seismic cycle reflectors. These locally distributed moderate/strong amplitude seismic reflectors, which are seen both in lobeform and in the channelform seismic features are interpreted as individual channelized lobe elements and channel fill elements respectively (Figs 5, 6, 12, and 13).

Seismic geomorphology within a stratigraphic hierarchy

Compilation of successive seismic geomorphologies through second-order C1, C2 and C3 succession documents evolving patterns of LCLCs (Figs 7–10). There are two scales critical to understanding the stratigraphy in this study area, the second-order margin adjustment scale, and the architectural element scale (LCLC scale; Fig. 6b). As seen from the seismic type-section, prominent offsets in seismic stratigraphic styles occur across the second-order cycle boundaries and are therefore essential to identify and understand. Seismic geomorphic extraction and analysis of displayed elements in composite maps requires understanding of the amplitude 'source' displayed in seismic sections. The 12 composite seismic geomorphology maps are therefore placed in stratigraphic order by numbers: 1–12 on a reference cross-section shown in Figures 7–10. Maps 1–12 document the seascape evolution of the main margin outbuilding phase. Interpretation confidence of lobeform and channelform elements is very high; strong acoustic contrast between seismic geomorphic lobeform/channelform elements and background strata enhance the portrayal of geomorphic patterns. Table 2 summarizes the following properties of the 12 seismic maps; these include: (1) channel trend/orientation; (2) seismic geomorphic style; (3) margin adjustment impact on deep-water fairway positioning, shape and trend.

LCLC architectural element analysis

High-amplitude continuous (*c*. 10–15 km wide) reflectors (HARPs), overlain by *c*. 2–3 km wide high-amplitude discontinuous reflectors (HARs) and flanked by wedge-shaped, low-amplitude reflections, form a distinctive and repetitive pattern of multiple LCLCs in the very high-resolution second-order C2 succession (Fig. 11). Similar features, but less clearly imaged in the seismic section, are also seen in underlying Cycle 3.1.

The fundamental deep-water depositional elements documented for analysis include: (1) channelized lobes (predominately lobeform in cross section); (2) channel-levees (cross-sectional channelform with wedge-shaped flanks; Fig. 11a, b). The channelized lobes correspond to 'frontal splays or distributary channel complexes', in the scheme of Posamentier & Kolla (2003). The analysis focuses on channel incision/aggradation/levee height in the upper part of Cycle C1 (1.3) and C2 (2.1–2.2). These lobe-channel-levee distributions including quantitative measurements are summarized in Table 3. Broad multilateral channelforms (orange) are flanked by subordinate and low-relief channel flank elements (Fig. 11b, green; Cycle 1.3). This architecture gives way to more erosive (red line) multilateral channelforms that show aggradation of discrete accretionary channel elements flanked by better developed wedge-shaped (levee) elements (Cycle 2.1). Channelforms in Cycle 2.2 show the most incision and aggradation, with flanking wedge-shaped (levee) elements best developed in this interval (Fig. 11b). These patterns produce a systematic upward increase in the proportion of channel–levee relative to lobeform elements (Table 3).

Dynamic genesis of lobe–channel–levee complexes

The observed lobe–channel–levee complex motif is documented in more detail in Cycle 2.1.1 between seimic horizons H3 (orange) and H4 (green; Figs 12 and 13). Both strike- and dip-oriented cross-sections and their corresponding high-resolution seismic maps show a close genetic relationship between lobeform and channelform depositional elements. The older LCLC within Cycle 2.1.1 records three depositional phases: (1) confined to unconfined subaqueous flow deposition of a frontal splay in a low-gradient setting; (2) advance of a confined highly sinuous feeder-channel that incises the preceding frontal splay it sourced; and (3) channel avulsion driven by mounded topography. Channel avulsion away from the lobe axis to flank establishes deposition in a new topographic low, which produces a deflected sediment-bypassing channel at the lobe/frontal splay flank.

Outcrop analogue: Brushy Canyon formation

Outcrop analogues are a useful resource for calibration of known sedimentary body architecture and lithology information to even the highest resolution seismic data, where this scale information is generally below data resolution. In addition, concepts, methods and models derived from more data-rich and detailed outcrop studies can help guide correlation and interpretation of sedimentation patterns indirectly imaged in subsurface datasets that may lack actual rock calibration, and/or contain sparse and widely spaced well control. We utilize the sand-rich, Permian Brushy Canyon Formation outcrop analogue from west Texas primarily for conceptual purposes, but also relate lobe–channel depositional patterns to seismic maps and cross sections (Gardner *et al.* 2003). This outcrop was selected because it is a simple

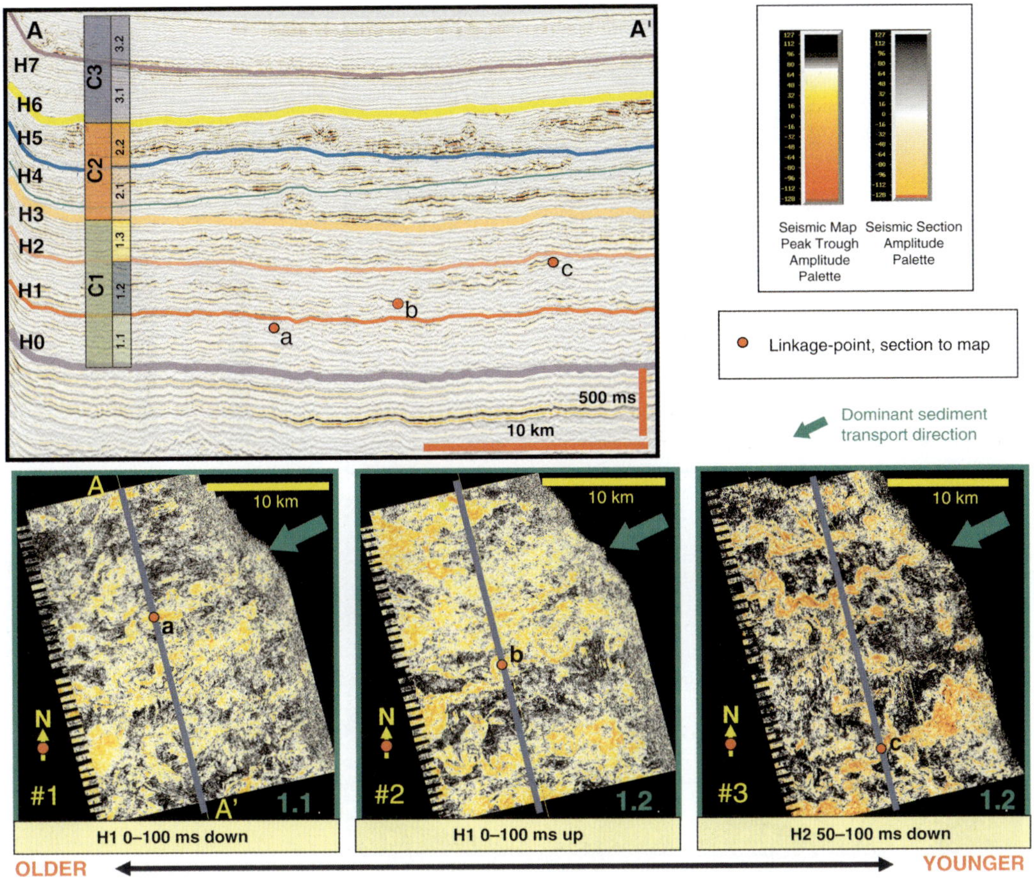

Fig. 7. Relationship between seismic cross section and geomorphic maps within third-order Cycles 1.1 and 1.2 (green family of maps). The diagram shows three successive amplitude maps extracted from seismic volumes defined below/above the H1 seismic horizon and below H2 seismic horizon. The relationship between seismic geomorphology and seismic stratigraphy is indicated with linkage points (**a**), (**b**) and (**c**). A description of each geomorphic map is given in Table 2. Note the west-southwest trend of low- to moderate-sinuosity channels and subtle increase in lobe elements on west (downdip) side of maps.

(structurally undeformed), but complete deep-water depositional system documented from its basin margin onlap to its depositional limit on the basin floor (Fig. 14). Seismic scale strata, in continuous shelf-to-basin exposures, are repeated on multiple fault blocks that dissect the 255 km^2 outcrop belt, with fault walls and canyon cliffs providing serial slices of deep-water architecture that mimic seismic cross sections. Sedimentation patterns common to both deep-marine successions include (1) the upward change from lobeform to channelform sedimentary bodies, at multiple scales, (2) the signature of depositional cycles superimposed within a stratigraphic hierarchy, (3) the role of seafloor topography on subaqueous flow behaviour, trend and position on the basin floor, (4) the planform relationship between channels and lobes within a branching channel network (Fig. 15), (5) the correlation and time value of organic-rich fine-grained intervals that form cycle boundaries and condensed sections, and (6) the relationship between basin-margin adjustment recorded by deformation near the margin and its correlation to changes in trend, focus, lithology, thickness and architecture on the basin floor to produce adjustment-bounded stratigraphic cycles. Recognition of the link between deformation and adjustment, in the outbuilding of the basin margin and deep-marine deposition infilling the basin proved to be a valuable concept that provided important

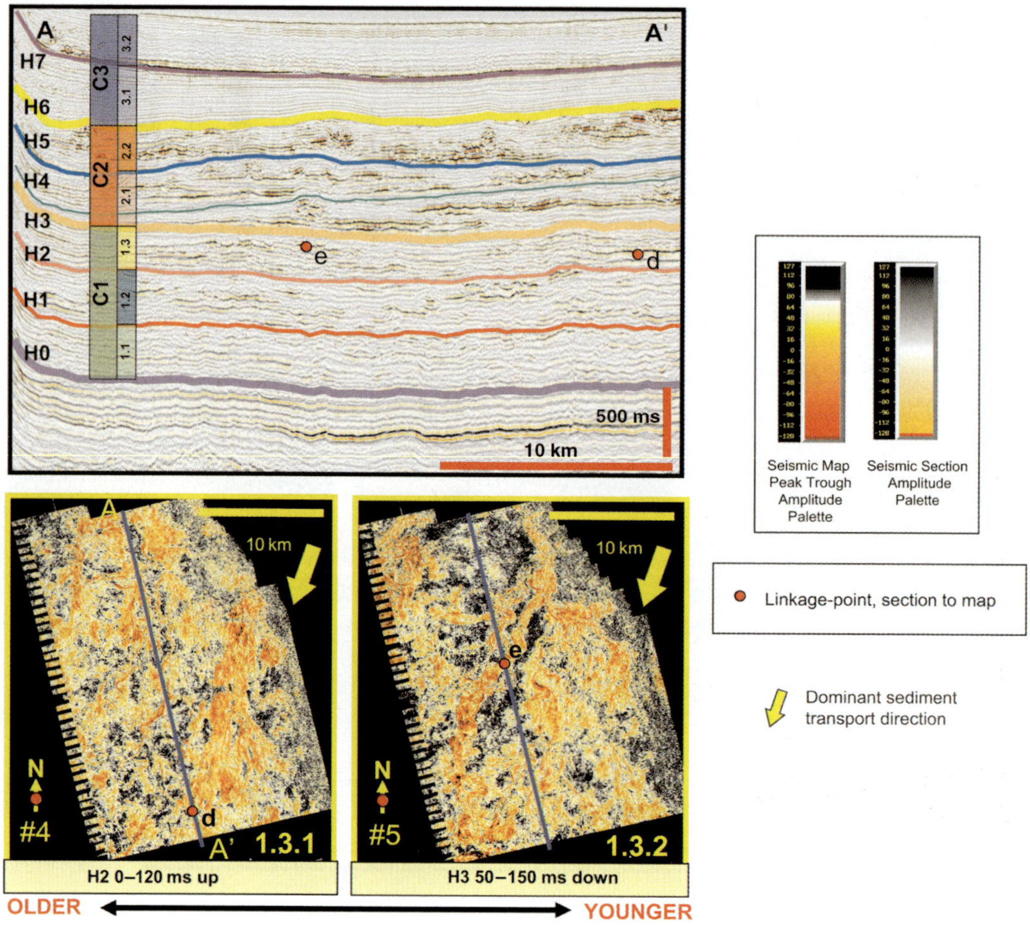

Fig. 8. Relationship between seismic cross section and geomorphic maps within third-order Cycle 1.3 (yellow family of maps). The diagram shows two successive amplitude maps extracted from seismic volumes defined above H2 seismic horizon and below H3 seismic horizon. The relationship between seismic geomorphology and stratigraphy is indicated with linkage points (**d**) and (**e**). Note the south–southwest channel trends, with increased branching to the south. Further descriptions of corresponding stratigraphic and geomorphic features are given in Tables 2 and 3.

context for understanding patterns documented from the aerially limited West African seismic volume.

Although the temporal and spatial scale of cycles is not comparable, several important concepts derived from the Brushy Canyon stratigraphic hierarchy are relevant to seismic correlations in this study. Most important are the predictive trends that can be extracted from a cycle hierarchy, where step-wise changes in depositional elements reflect their position within the hierarchy. For example, the continuity and organic richness of cycle-bounding condensed sections is greater in lower order cycles. Similarly, the largest and most channels within higher order cycles generally occur in those positioned in the middle to upper part of a lower order cycle (Hodgson *et al.* 2006). Similar step-wise changes in channelform and lobeform elements are apparent in the West Africa seismic data, where individual lobe–channel–levee complexes show incremental changes in channel morphology and lobe–channel proportions related to their stratigraphic position within six fourth-order cycles within three third-order cycles (Cycles 1.3, 2.1 and 2.2).

Two outcrop patterns used to document sub-seismic sedimentary body and facies arrangements in seismic maps and cross sections include: (1) sedimentary body hierarchy of channel and lobes; and (2) gradient control on sedimentary body type, arrangement and lithology. The outcrops illustrate how confined and unconfined, channelform and lobeform, sedimentary bodies are arranged in a scalar hierarchy (Fig. 15).

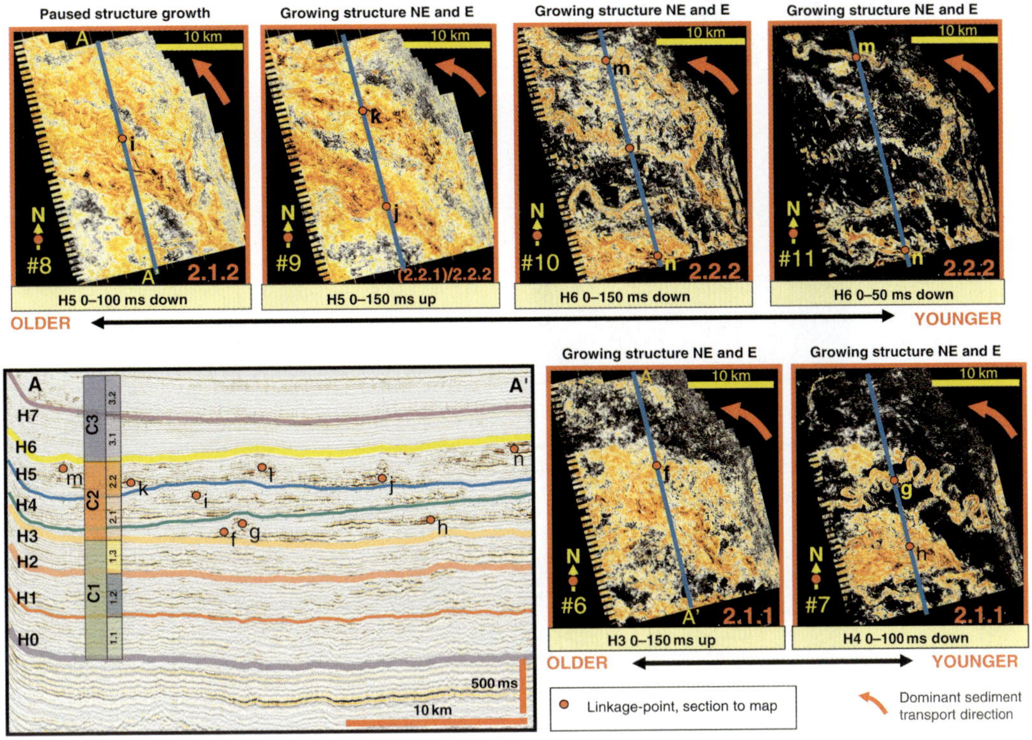

Fig. 9. Relationship between seismic section and geomorphic maps within fourth-order cycles Cycles 2.1.1, 2.1.2, 2.2.1 and 2.2.2 (red family of maps; see Fig. 6b for detailed cycle hierarchy). The diagram shows six successive amplitude maps extracted from seismic volumes related to H3, H4, H5 and H6 seismic horizons as indicated in text at base of each map. The relationship between seismic geomorphology and stratigraphy is indicated with linkage points (f)–(n). Note that channel trends have shifted *c.* 120–140° to the northwest; broad fan-shaped bodies alternate with high-sinuosity channels, and a high degree of depositional topography results in well-developed compensation patterns. Further descriptions of stratigraphic and geomorphic features are given in Tables 2 and 3.

Recognition of differences in lithology, channel–levee development, tectonic setting, geologic age, and the temporal and spatial scale of these deep-water systems established the limits to which this Permian outcrop analogue was applied in this study (Table 4, Fig. 17). However, the outcrop data were considered inappropriate for direct temporal correlation and scalar calibration of sedimentary body type, dimension, and lithology. This led to consideration of the recent Amazon and Zaire fan systems as additional analogue sources.

Modern analogues: Amazon fan and Zaire fan

Comparison of the well-documented Amazon Fan with the West African example from this study is relevant despite the fact that the data type & coverage from each area differ (Fig. 16; Manley & Flood 1988; Flood *et al.* 1991; Pirmez & Flood 1995; Hiscott *et al.* 1997; Pirmez *et al.* 1997; Piper & Normark 2001). The Amazon Fan shows longitudinal (downdip) changes in levee systems deposited in a very short interval of time (<10 Ka) but relatively undeformed sea-floor. By contrast, the West African succession records longer term (in total *c.* 32 Ma; refined analysis over *c.* 12.3 Ma) deformation and depositional outbuilding documented from a small (20 × 30 km), source-distant window.

The two areas exhibit similar lobe–channel–leveeform architectural styles that are comparable in geometry and spatial scale (Fig. 17). West African LCLCs typically are 10–20 km wide and 100–300 ms (*c.* m) thick, whereas individual channel-levee-system documented from the Amazon Fan typically are 20–40 km wide and 200–500 ms thick (175–440 m, assuming an interval velocity of 1750 m s^{-1}; e.g. Flood *et al.* 1991). Individual Amazon channel–levee systems tend to stack laterally and/or vertically to form leveecomplexes (Flood *et al.* 1991; Piper &

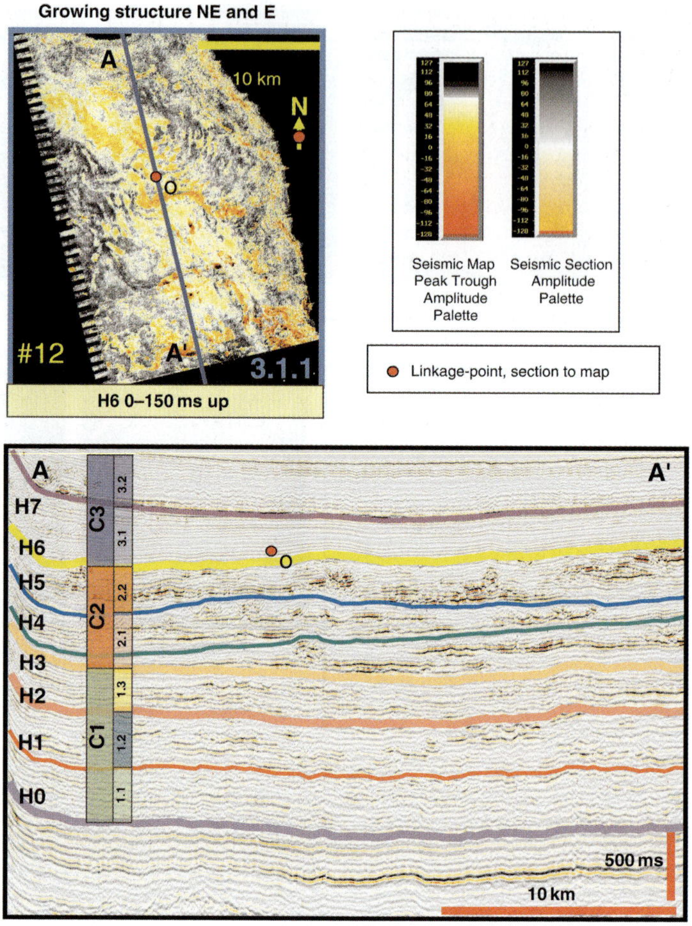

Fig. 10. Seismic stratigraphic/geomorphic relationships within fourth-order Cycle 3.1.1 (blue map). The amplitude map is extracted from a seismic volume 0–150 ms above horizon H6. The relationship between seismic geomorphology and stratigraphy is indicated with linkage point (o). The map shows curved deformation bands sub-parallel to margin strike orientation and 'ghosts' of channel geomorphology elements sitting below H6. A further description is given in Table 2.

Normark 2001). In the Amazon Fan 'high amplitude reflection packages' (HARP), representing lateral or terminal sheet-sand bodies, correspond to channelised lobe facies in this study. Our channel fill facies is comparable to 'high amplitude reflections' (HAR) in the Amazon. Both datasets show pronounced levee development expressed as low-amplitude wedgeform seismic facies flanking HAR facies.

Channel–levee systems within the youngest Amazon levee complex show a systematic decrease in levee height and increase in HARP proportion, from upper-fan to lower fan position (Fig. 16; Piper & Normark 2001). Amazon Fan cores are calibrated to seismic facies architecture, and therefore bridge seismic patterns and lithofaces (Figs 16 and 17). With no direct borehole calibration available for the West African study, lithology and sedimentary facies are inferred from observed seismic facies patterns and comparison to the outcrop and modern Amazon analogues (Fig. 17). This analogy suggests that the LCLCs within the West African subsurface dataset represent three key facies associations: (1) sandy channelized lobe facies (HARP); (2) sandy to heterolithic channel-fill facies (HAR); and (3) muddy to heterolithic levee facies. The consistent presence of seismic drape facies between LCLCs is interpreted as mud-prone background sedimentation from very low concentration gravity flows and hemipelagic deposition.

The Quaternary Zaire Fan (Congo-Angola margin SE Atlantic) exhibit channel–levee systems with seismic facies and external configurations very similar to the Amazon Fan and the LCLCs seen in this study. In particular, high-amplitude reflection units with high reservoir potential are recognized almost systematically as a basal component of channel–levee systems (Droz *et al.* 1996, 2003). Seventy per cent of the channel–levee systems of the Zaire Fan overlie a basal unit of high-amplitude, low frequency, and relatively continuous reflectors that, in the most typical cases, infill intersystem lows. (Droz *et al.* 2003). The consistent lobe–channel–leveeform seismic facies architecture observed within Amazon Fan, Zaire Fan and in the West African subsurface dataset reflects a common and generally valid relationship between unconfined and confined gravity flow deposition, which helps predict corresponding reservoir and non-reservoir rock distribution.

Gradient control on LCLC architectural style; longitudinal v. temporal (stratigraphic) trends

The two most important conclusions derived from this integrated seismic stratigraphic/geomorphic analysis include (1) local gradient control on architectural element genesis and (2) recognition of the predictive relationship between evolving margin basin adjustment/deformation and deep-water fairway deposition.

Local gradient control on architectural element genesis

Observed longitudinal (spatial) changes in Amazon Fan lobe–channel–levee architecture (upper, middle and lower fan style) resemble vertical (temporal) changes through the high-resolution West African stratigraphic record (upper C1 and C2 successions; Fig. 17a, b). Both show an upward increase in channel–levee relative to lobe proportion. It is proposed that these architectural trends, which are directly derivable from seismic data, are related to increased gradient and topographic focus. This fundamental conceptual relationship between architectural style and corresponding gradient is in concurrence with documented down-current change from channelform to lobeform sedimentary bodies (change from confined to unconfined flow conditions) and an associated change from high to low gradient, commonly combined with progressively decreased levee height down system (Mutti 1985; Mutti & Normark 1991; Piper & Normark 2001). Though originally conceived to describe down-profile changes in a deep-water depositional system, this channel–lobe classification scheme fundamentally links flow processes to gradient that can vary in an irregular fashion down the depositional profile, particularly relevant along severely deformed margins with mobile substratum.

In this study, increased gradient is related to local topographic changes produced by gravity-driven margin adjustment, deformation and folding processes, rather than a significant basinward extension of the submarine channel to a mid-fan position. Because of the significant deformation of the West African margin, it is appropriate to characterize the local topography and gradient, rather than assign upper fan, middle fan, lower fan and abyssal plain geomorphic divisions appropriate for the shorter duration of the Amazon fan analogue. The latter geomorphic terms imply and require application of Walther's Law, which is inappropriate to continental margins dominated by significant gravity driven deformation and basin margin adjustment.

Recognition of margin basin adjustment phases from integrated seismic stratigraphic/geomorphic analysis

Observed offsets in seismic stratigraphic and geomorphic style and orientation of lobe–channel–levee patterns (second-order cycles C1 and C2), or other seismic facies offsets (C3) within the margin outbuilding section suggest significant shifts in (1) subaqueous flow pathways and delivery mechanisms, and/or (2) physiography of the longitudinal profile (seascape gradient and topographic focus). All deep-water stratigraphic successions record some combination of these fundamental conditions. They control subaqueous flow pathways, trends and focus and contribute to the state of flow dynamics controlling sediment erosion, bypass and deposition. Basin margins bypassing significant volumes of sediment to the basin may or may not trigger significant mobilization of the basin substratum (e.g. salt or mud). Nonetheless, gravity-driven deformation impacts basin physiography. If significant mobilization of the substratum develops, the evolving seabed topography will control the position and style of deep-marine sediment pathways (e.g. Prather 2003). Continental slope located downdip of large deltas is prone to adjustment in the form of mass wasting, slumping and sediment creep. These processes, which do not require the presence of mobile substratum, also impact evolving seabed topography, and therefore subsequent subaqueous flow erosion, bypass and deposition dynamics. Changes in shelf width, magnitude and rate of relative sea-level change, position and number of shelf depocenters combine with gravitational deformation, syn-sedimentary mass-wasting on the slope and large-scale fan avulsion to produce a signature of adjustment in the deep-marine record.

The long-term stratigraphic and geomorphic evolution documented in this study (Figs 6 and 19) reflects an evolving deep-marine seascape. Where the deep-water system records a significant change in stratigraphic attributes, adjustment of the basin margin is inferred. Table 5 lists nine different stratigraphic recognition criteria related to margin adjustment. In this study, recognition criteria (2)–(8) (Table 5) are considered relevant, with changes in shelf delivery to the basin (SRC1) not analyzed. No significant submarine disconformities (SRC9) are recognized within the study area.

Margin adjustment bounded cycles

The combined effects of shifting staging area, gravitational deformation, syn-sedimentary mass-wasting and large-scale fan avulsion processes are defined as basin margin adjustment. Along continental margins there is a well-documented pattern of updip, gravity-driven extension and growth faulting which is balanced by downdip compression and thrust and fold-belt development. This has been documented in Gulf of Mexico, offshore Nigeria and offshore Angola (e.g. Brun & Fort 2004). The evolving stratigraphic and geomorphic patterns documented in this study are located in a structurally quiescent, source-distant position within an overall highly deformed continental margin sequence. The strong seismic facies contrast between the three successive second-order basin-fill records (C1, C2 and C3) suggests major margin adjustment between each of these cycles. Therefore these second-order cycles are termed Margin Adjustment Bounded Cycles, ABC1, ABC2 and ABC3 respectively (Fig 19, Table 6).

Second-order ABCs, which record prominent adjustment in the margin outbuilding evolution, are primarily defined using cross-sectional seismic stratigraphic criteria (Fig. 6, Table 6). The integrated seismic stratigraphic and geomorphic analysis permitted visualization of long-term seismic geomorphologic evolution. The most significant change in LCLC style and orientation or abrupt absence of lobe-channel-leveeform seismic facies occurs across second-order cycle boundaries (H0, H3 and H6 seismic horizons), and to a less extent across third-order boundaries. Third-order scale seismic stratigraphic and geomorphic data show more subtle margin adjustment and deformation (Figs 6 and 19 and Table 6).

Linkage between local stratigraphic evolution and regional margin deformation

The integrated analysis of evolving seismic stratigraphic and geomorphic patterns in this study permits analysis of relationships between margin deposition and gravity-driven deformation. In order to better understand these potential relationships, changes and offsets in LCLC architectural styles within the established threefold hierarchy of cycles (second-order through fourth-order) were systemized through the *c.* 32 Ma period of margin outbuilding (Tables 2 and 6).

ABC1: simple undeformed and initial extension and compression phase (green/yellow family of cycles/composite geomorphology maps) Cycles 1.1 and 1.2 show WSW trends of isolated, but freely migrating channel complexes that overlie local lobeforms. This depositional pattern indicates sediment transport across relatively smooth and stable slope/basin topography of the inherited ocean basin. Cycle 1.3 (yellow) records increased rates of sedimentation associated with the southward-directed widening of moderate/occasional high-amplitude lobeforms overlain by more distinctive distributary channelform elements. The almost *c.* 70° shift in lobe–channel orientation (from WSW- to S-trending channels in Cycles 1.1/1.2 and 1.3, respectively) suggest re-arrangement of updip sediment pathways to this position in the basin. Relative to the underlying Cycles 1.1 and 1.2 (green), there is a distinct shift in geomorphic style and trend, which suggests an increase in updip topographic focus leading to more confined sediment delivery to this position in the basin. The observed geomorphic transition from focused to distributary channel morphologies reflects a transitional confined to unconfined flow setting, inferred to represent a relatively low gradient setting during this period. There is a higher proportion of channelized lobes/splays/distributary channels relative to channel–levee within this succession. This suggests a significant change in basin physiography, combined with increased sediment delivery to the basin from a more efficient shelf sediment delivery system. The change in slope and basin physiography is interpreted to reflect gravity sliding and extension (upper regional slope) and initial fold-belt compression (medial regional slope), which produced a relatively thin succession of unconfined, lobe-dominated 'outer-fan-style' deposits within Cycle 1.3. Consequently the transition from Cycle 1.2 to 1.3 (green to yellow cycle/map family) is interpreted as an initial phase of margin extension and compression (Fig. 19).

ABC2: Basinward migration fold-belt phase (red family of cycles/composite geomorphology maps). The ABC2 succession records the highest sedimentation rates within the outbuilding succession in the study area (*c.* 800 m sediment deposited over a period of *c.* 5.6 Ma). The presence of successive prominent LCLCs records significant sediment deposition and bypass in study area during this period of margin outbuilding. Except for the H4–H5 section, all cycle strata thin toward the present-day structural flank north of study area (Fig. 5b, left section margin). This suggests structural growth initiated close to the study area, during this period, evolving over the H3–H4 and H5–H6 intervals, with a pause during H4–H5 interval. The NW–WNW trend of channels within the ABC2 cycle (almost *c.* 120–140° shift in lobe–channel orientation relative to underlying south directed trend) and northward thinning of higher frequency cycles document significant adjustment of updip deep-water physiography including initial growth of an anticline structure immediately north of study area. Somewhat less distinct 'deflection' of channel trends indicates cessation of structural movement during H4–H5 time. This sedimentation pattern suggests partial 'healing' of the structural topography. Compressional deformation controlled the position and trend of sediment pathways within ABC2 strata and indicates that deposition was contemporaneous with initial basinward migration of syn-sedimentary folding in the outer fold belt of this continental margin (Fig. 19). The observed 'middle-Amazon-fan-style' channel–levee architecture that characterizes ABC2 deposition suggests the development of increased local gradients, relative to depositional patterns within underlying ABC1/Cycle (1.3). Increased gradient is also reflected in the upward increase in the proportion of channel–levee relative to lobe elements. ABC2 records a phase of margin outbuilding characterized by the highest local sediment rates and gradients within the outbuilding succession. ABC2 strata were deposited during a period when initial and subtle compressional deformation occurred in the immediate proximity of the study area. This reflects basinward migration of an evolving outer fold-belt, which increased local sediment delivery, gradient and topographic focus (Fig. 19).

ABC3: outer fold-belt phase (blue family of cycles/composite geomorphology map) ABC3 deep marine sedimentation is dominated by the uniform sub-parallel seismic facies and wedge-shaped chaotic seismic facies. The latter seismic facies are interpreted as mass transport deposits that record deformation of emergent structures recording local margin adjustment. The sub-parallel seismic facies dominate the ABC3 succession and lack lobe–channel–levee elements within the study area. It is inferred that seafloor topography produced by emergence of the outer fold-belt near the study area created a topographic high and 'shadow zone' that shielded the study area from significant subaqueous flow deposition (Fig. 19). Regional analysis of the lower portion of ABC3 strata (Cycle 3.1), however, show significant lobe-channel-levee deposition away from (north of) this structural 'shadow zone', reflecting active deep-water sedimentation elsewhere along this margin at this time.

Discussion

To ensure the results from this study have broader applicability to other deep-marine systems, and to aid with the calibration necessary for this more general usage, this discussion focuses on issues related to (1) erecting a stratigraphic hierarchy, (2) differentiating between internal and external controls, (3) integrating outcrop and subsurface analogs in interpretation and (4) make fairway and reservoir predictions.

Cycle hierarchy

To be predictive, stratigraphic frameworks must account for different scales of cyclicity superimposed within the stratal succession. Two fundamentally different criteria, time or space, however, can be used to define a stratigraphic hierarchy (Jackson 1975; Miall 1985, 1995; Mutti & Normark 1987; Michum & Van Wagoner 1991; Gardner & Sonnenfeld 1996; Prather *et al.* 2000). The practice of mixing these criteria to build a cycle hierarchy is common because short-term cycles rarely are constrained by absolute dates. Consequently, the age of short-term cycles is estimated by dividing the number of short-term cycles by the duration of the longer-term chronostratigraphic cycle they comprise. These shorter duration cycles (e.g. fourth-order cycles in this study), however, are actually defined by physical stratigraphic criteria. Furthermore, a small spatial area, such as in this study, precludes corroboration of the longer-term cycles (second and third order) by physical stratigraphic criteria like bed termination geometry, cross cutting and continuity relationships used to define the short-term cycles. This approach produces a stratigraphic hierarchy where all the cycles within the hierarchy are defined by different criteria.

A problem arises when temporal and spatial criteria produce conflicting results, and one criterion has to be selected over another in building a stratigraphic hierarchy. For example, in this paper Cycle 1.3 is classified as a third-order cycle based on physical criteria, that is similarity in thickness and number of higher order cycles to the other six third-order cycles comprising the succession. This third-order cycle, however, spans more time (6.7 Ma) than the overlying second-order cycle C2 (5.6 Ma). This has the unfortunate consequence of making a third-order cycle longer in duration than a

Table 2. *Summary table of evolving seismic geomorphology patterns*

2nd/3rd-order cycle name Map no. Max trough amplitude extraction interval	Dominant downdip channel trend (orientation)	Dominant seismic geomorphologic style	Margin adjustment/deformation driven impact on fairway positioning and channelform/lobeform style/shape/orientation
Cycle 3.1.1, map 12 H6 0–120 ms up	—	Curved deformation bands sub-parallel to margin strike orientation.	Abrupt shut-down of lobe–channel–levee genesis in this position of the basin. Explained as consequence of significant 2nd-order margin adjustment and deformation effect on seabed topography. Study area was brought into local 'shadow zone', relative to active fairway which shifted (avulsed) northward, beyond study area.
Cycle 2.2.2, map 11 H6 0–50 ms down	NW bending WNW	Well-defined moderate sinuosity CCs. Hints of individual late phase internal low amplitude channel fill features.	Growing structure NE of dataset steers fairway from NW to WNW direction
Cycle 2.2.2, map 10 H6 0–150 ms down	NW bending WNW	Well-defined moderate sinuosity CCs compensationally oriented (deflected) relative to preceding tongues.	Growing structure NE of dataset steers fairway from NW to WNW direction
Cycle (2.2.1)/2.2.2, map 9 H5 0–150 ms up	NW bending WNW	Two tongue formed CLC with subtle channelization only. Initial lobe-attracted moderate sinuosity CCs erodes into southern CLC tongue.	Growing structure NE of dataset steers fairway from NW to WNW direction
Cycle 2.1.2, map 8 H5 0–100 ms down	NW	Fan shaped feature with two feeder CCs (high and low sinuous respectively). CLC overridden by prominent central mod/high sinuous CC	Gently NW steering of fairway only, indicating partial 'healing' of growing structure in NE.
Cycle 2.1.1, map 7 H4 0–100 ms down	NW bending W	Lobe-formed CL with prominent (overriding) high-sinuous channel feature Genetically linked moderate sinuous CC to the south positioned at flank to underlying CLC. Higher sinuous CC north of CL is genetically linked to CLC in map 6.	Distinctly steering of fairway from NW to W direction associated with growing structure NE of dataset.
Cycle 2.1.1, map 6 H3 0–150 ms up	NW bending W	CLC with subtle channel networks	Significant fairway orientation shift relative to underlying, explained as response to 2nd-order margin adjustment (*c.* 120–140° shift in orientation). Distinctly steering of fairway from NWto W direction indicating growing structure NE of dataset.

(Continued)

Table 2. *Summary table of evolving seismic geomorphology patterns (Continued)*

2nd/3rd-order cycle name Map no. Max trough amplitude extraction interval	Dominant downdip channel trend (orientation)	Dominant seismic geomorphologic style	Margin adjustment/deformation driven impact on fairway positioning and channelform/lobeform style/shape/orientation
Cycle 1.3.2, map 5 H3 50–150 ms down	SSW	Two CCs and associated CLCs with freely distributed branching channel networks, widening in S direction	No obvious impact relative to underlying cycle
Cycle 1.3.1, map 4 H2 0–120 ms up	SSW	CLC with freely distributed branching channel networks, widening in S direction	Significant fairway orientation shift relative to underlying, explained as response to 3rd-order margin adjustment (*c.* 70° shift in orientation).
Cycle 1.2, map 3 H2 50–100 ms down	WSW	Freely distributed, low/moderately sinuous CCs in distinct contrast to low amplitude background facies	No obvious impact relative to underlying cycle
Cycle 1.2, map 2 H1 0–100 ms up	WSW	Subtle, freely distributed, low/mod sinuous CCs occasionally transforming down-dip to lobeforms	No obvious impact relative to underlying cycle
Cycle 1.1, map 1 H1 0–100 ms down	WSW	Subtle, freely distributed, low sinuous CCs occasionally transforming downdip to lobeforms	Base of this cycle represents onset of active lobe–channel–levee deposition in the study area, explained as response to 2nd-order margin adjustment, and correlating in time to hinterland uplift (Early Eocene *c.* 32 Ma)

CC, channel complex representing upper/younger portion of each LCLC.
CLC, channelized lobe complex representing lower/older portion of each LCLC.

second-order cycle. The overlap in the duration of published second- and third-order cycles is well known (see Schlager 2004), and has been attributed to different datasets and classification schemes, but not to changes within the same hierarchy.

Temporal comparison of all cycle durations within this hierarchy demonstrates that the 1.1 million year difference in age between the second- and third-order cycles is less than the 3.6 million year range in age for all seven third-order cycles. Furthermore, all but the youngest of the three second-order cycles is shorter in duration than the 10–100 Ma age proposed for second-order cycles (Miall 1995). Strict application of this temporal criterion requires all but the youngest second-order cycle and all the third-order cycles that comprise the 32 Ma outbuilding phase be classified as third-and fourth-order cycles, respectively. This shift resolves the duration for two of the three third-order cycles, that is <10 Ma, but places the youngest third-order cycle (10.8 Ma) outside the 1–10 Ma age for third-order cycles. This creates a more serious problem for the duration of fourth-order cycles, which in this new scheme are an order of magnitude older than the 0.2–0.5 Ma age reported for fourth-order cycles (Miall 1995). This problem can be partly reconciled if different physical criteria are used, that is degree of channel trend and architecture change, to classify Cycle 1.3 strata as a second-order cycle, but the ages of the four second-order cycles in this scheme are still less than the published duration for second-order cycles. This new scheme does not significantly reduce the range of overlap in the age of four, second- and six, third-order cycles.

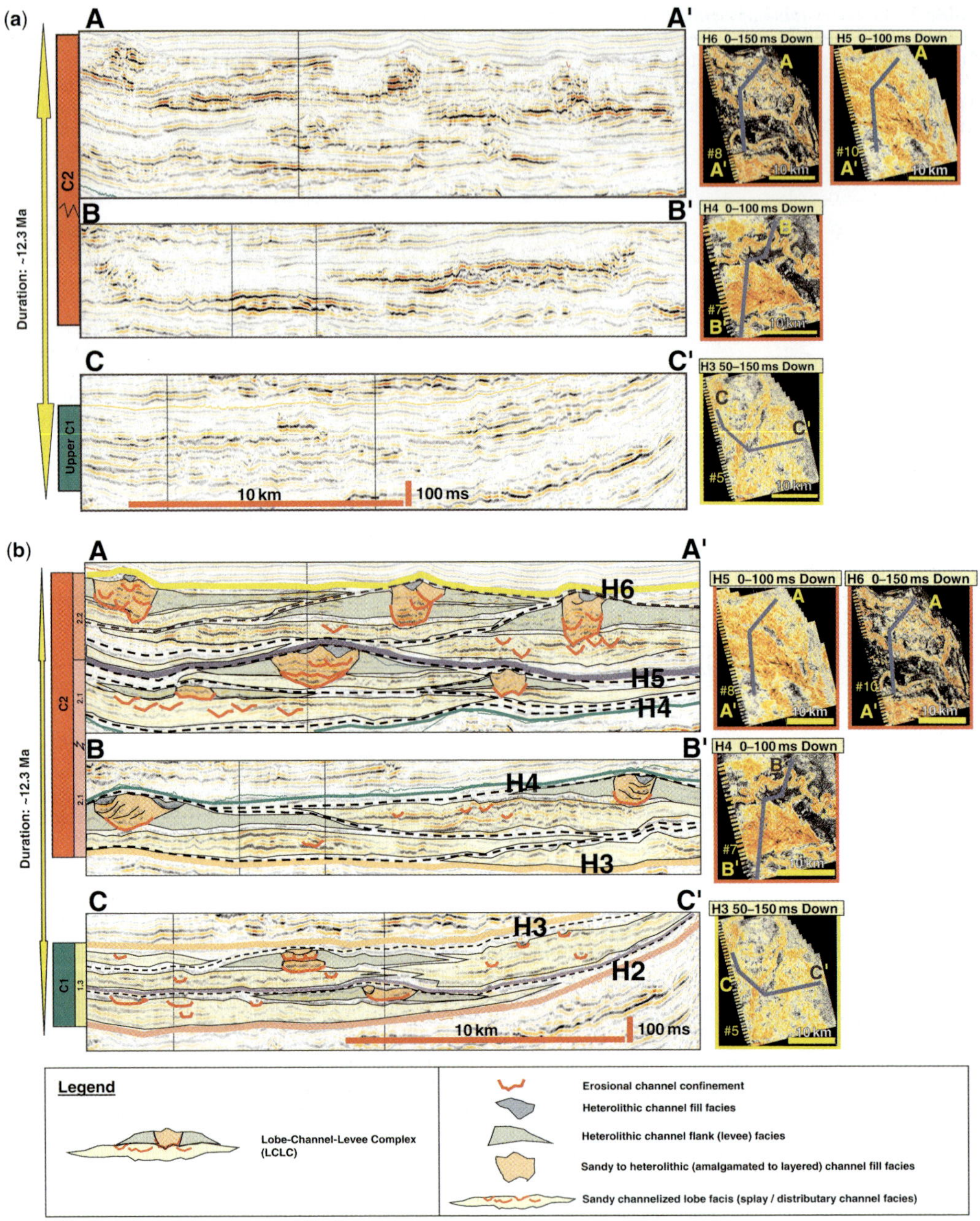

Fig. 11. Uninterpreted seismic strike sections showing stratigraphic evolution of lobeform and channelform seismic facies. Corresponding amplitude maps show the positions of strike sections (blue lines). Interpreted strike sections showing stratigraphic and geomorphic evolution of lobe–channel–levee complexes. An upward increase in channel and decrease in lobe proportion is evident through upper C1 (Cycle 1.3) and C2 (Cycles 2.1 and 2.2). Seismic facies interpreted as sets of channelized lobes are highlighted with yellow overlays. Interpreted channel fills are highlighted with orange overlays. Particularly low amplitude channel fills (typically found at the top of a channel fill succession) are shown with bluish overlay. Seismic facies interpreted as channel flank–levee complexes are highlighted with green overlays.

Table 3. *Evolving LCLC styles within Cycles 1.3, 2.1, 2.2*

Cycle name Horizon interval Map no.	Overall channel: lobe ratio	Compensation style between LCLCs	LCLC style CC upper/younger portion of each LCLC CLC (lower/older portion of each LCLC)
Cycle 2.2 (H5–H6) Maps 9–11	High	Distinct *c.* 5–10 km lateral offset compensation	CC: erosionally confined to highly aggradational levee-confined CCs deflected and un-deflected from underlying CLC depocentra Estimated depth of erosion (from seismic section): *c.* 80 ms (*c.* m) Maximum levee height: (from seismic section) *c.* 100 ms (*c.* m) CLC: tongueformed complexes with subtle loosely confined to freely migrating distributary channels attracted to associated lobes.
Cycle 2.1.2 (H4–H5) Map 8	Moderate-high	Distinct *c.* 10 km lateral offset compensation	CC: erosionally confined to aggradational levee-confined CCs deflected from underlying CLC depocentra Estimated depth of erosion (from seismic section): *c.* 55 ms (*c.* m) Maximum levee height (from seismic section): *c.* 85 ms (*c.* m) CLC: lobeformed complex with loosely confined to freely migrating distributary channels attracted to associated lobes.
Cycle 2.1.1 (H3–H4) Maps 6 and 7	Moderate	Distinct *c.* 15 km lateral offset compensation	CC: erosionally confined to aggradational levee-confined CCs deflected from underlying CLC depocentra. LAPS in deflected CCs Estimated depth of erosion (from seismic section): *c.* 40 ms (*c.* m) Maximum levee height (from seismic section): *c.* 85 ms (*c.* m) CLC: lobeformed complex with loosely confined to freely migrating distributary channels attracted to associated lobes.
Cycle 1.3 (H2–H3) (H2–H3) Maps 4 and 5	Low	Subtle wide *c.* 20 km lateral offset compensation	CC: loosely levee-confined CCs deflected from underlying CLC depocentra Estimated depth of erosion (from seismic section): *c.* 30 ms (*c.* m) Maximum levee height (from seismic section): *c.* 40 ms (*c.* m) CLC: lobeformed complex with loosely confined to freely migrating distributary channels attracted to associated lobes.

ABC, margin adjustment bounded cycle.
LAPS, laterally accreting packages (Abreu *et al.* 2003).

In this example, the selection of temporal or spatial criteria produces alternative cycle hierarchy schemes. The temporal discrepancies outlined above could reflect an error with the condensed section dates and/or their correlation from structurally complex regions outside the study area. It may reflect the problem with assuming that cycle thickness is related to time, which is compounded by the increased uncertainty in the time-depth conversion of a seismic volume lacking well control. Furthermore, the six absolute dates are derived from widely spaced, condensed intervals. These fine-grained drapes record sedimentation starvation and the majority of time within relatively conformable successions characterized by aggradational and compensational patterns. These active depositional patterns in the intervening dep osits record much higher sedimentation rates from

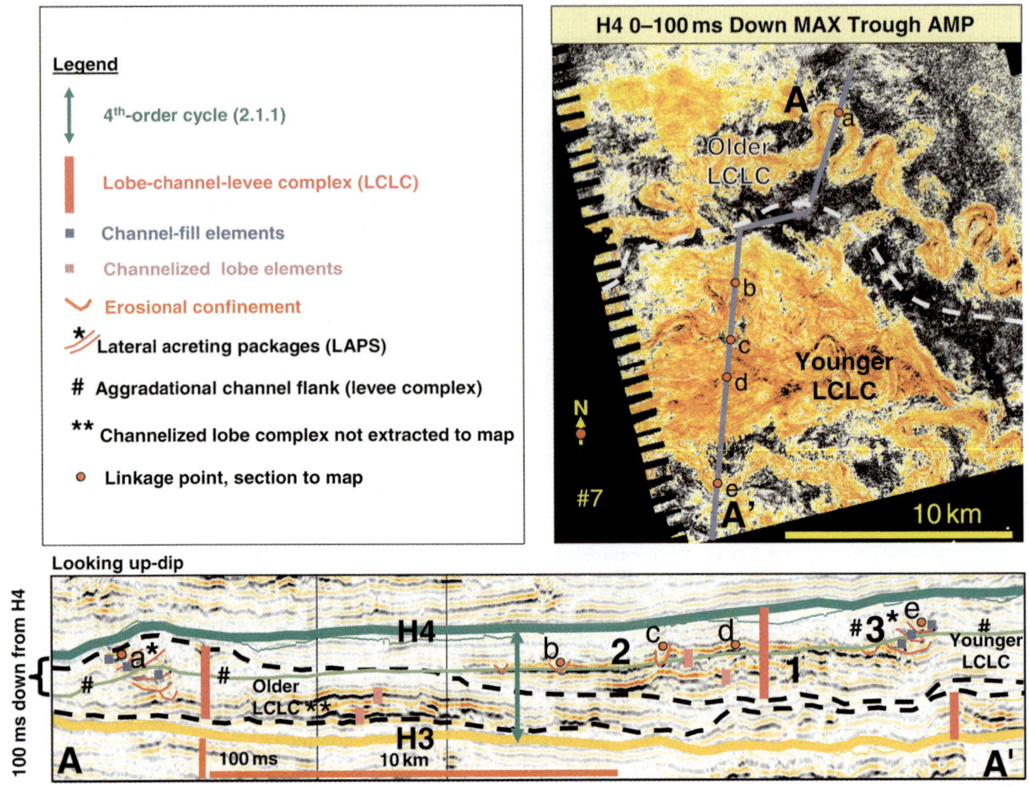

Fig. 12. Detailed seismic stratigraphic and geomorphic expression within fourth-order Cycle 2.1.1 (between H3 and H4). The optimized strike-section is dominated by two distinctly contrasting high-amplitude (sandy) seismic facies. The first seismic facies includes lobeform (1) and channelized lobeform (2) elements with strong amplitude. The second seismic facies is characterized by moderately strong amplitudes that form laterally accreting packages (LAPS, Abreu *et al.* 2003) marked by an asterisk. Corresponding amplitude maps delineate two sinuous channelforms, the older marked by point (a) and the younger marked by point (e). There is also an intermediate age fan shaped feature with an internal branching channel network related to an early distributary phase of the younger lobe–channel–levee complex. Point (b) marks a well-developed highly sinuous channel that overrides and erodes the lobeform (lobe-attracted channel). The amplitude extraction captures geomorphic features from parts of two distinct lobe–channel–levee complexes. Note that underlying lobeform of the younger northern complex is not captured in this particular map. Within the younger southern complex, the mounded channelized lobe deflects the overlying (e) sinuous channel complex (lobe-deflected channel-complex). Both the older and younger lobe–channel–levee complexes show asymmetrical cross-section relationship between the channelized lobe complex and genetically linked channel complexes. Clear LAPS (*) migrating towards outer bend are evident in positions where the cross-section cuts perpendicular to sinuous bends. Significant aggradational wedge shaped low amplitude channel margin facies (interpreted as levees) are marked with a hash sign. The abrupt (northwest to west-southwest) change in channel orientation observed in the northern part of the map is caused by structural growth north of the study area (see Fig. 6). The two lobe–channel–levee complexes show distinct compensation relative to each other (*c.* 15 km offset in strike direction). Each complex shows 2D mounding geometry in cross-section, as do individual lobe-elements.

subaqueous flows than indicated by the linear interpolation of time between the sparse and widely spaced dates from condensed sections. In fact, the most closely spaced dates for the highest-resolution interval span 2.5 Ma. These potential sources of error are compounded by the final calibration of cycle order to other published temporal hierarchy durations and/or datasets that show a range in values greater than the duration of the cycles comprising the hierarchy.

Despite the temporal ambiguity, the relative cycle hierarchy approach used in this paper serves two important purposes. It captures the inherent additive and dampening effects that result from superimposing different order stratigraphic cycles. Because the contribution from higher order cycles is embedded

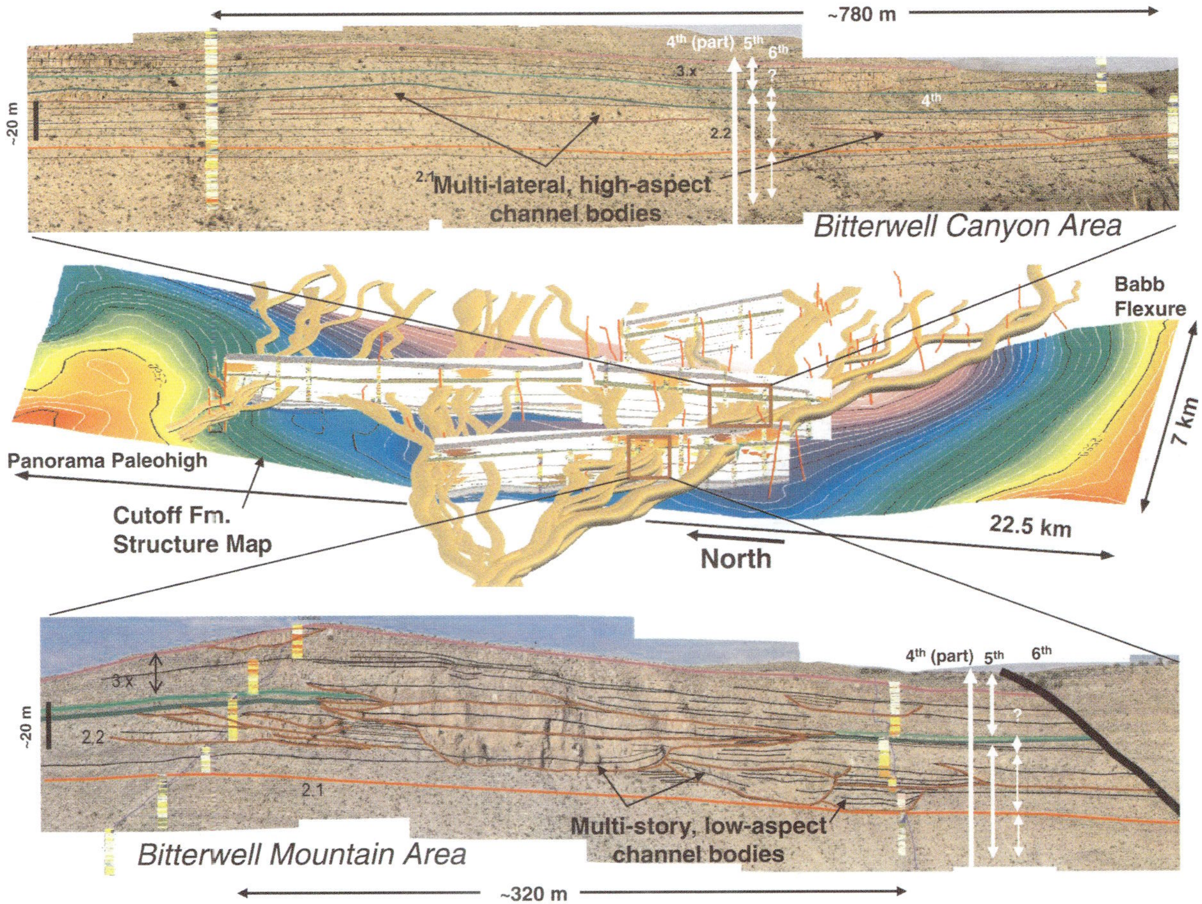

Fig. 15. Map and outcrop panels showing branching submarine channel network in Bitterwell area of the Brushy Canyon outcrop belt. This example illustrates the relationship between channel branching and architectural style. Updip exposures in the Bitterwell Mountain area (bottom photo) exhibit tightly-clustered (multistory) channel bodies with low aspect (width to thickness) ratios. The same channel complex crops out in the Bitterwell Canyon area (top photo) as high-aspect, multi-lateral channel bodies. The down-dip change in architectural style occurs over 2.9 km. The 3D perspective channel map is only for sixth-order Cycle 2.2. The channel network occupies a structural low at top of the Cutoff Formation that is bounded by the Panorama Point palaeo-high to the north and the Babb flexure to the south.

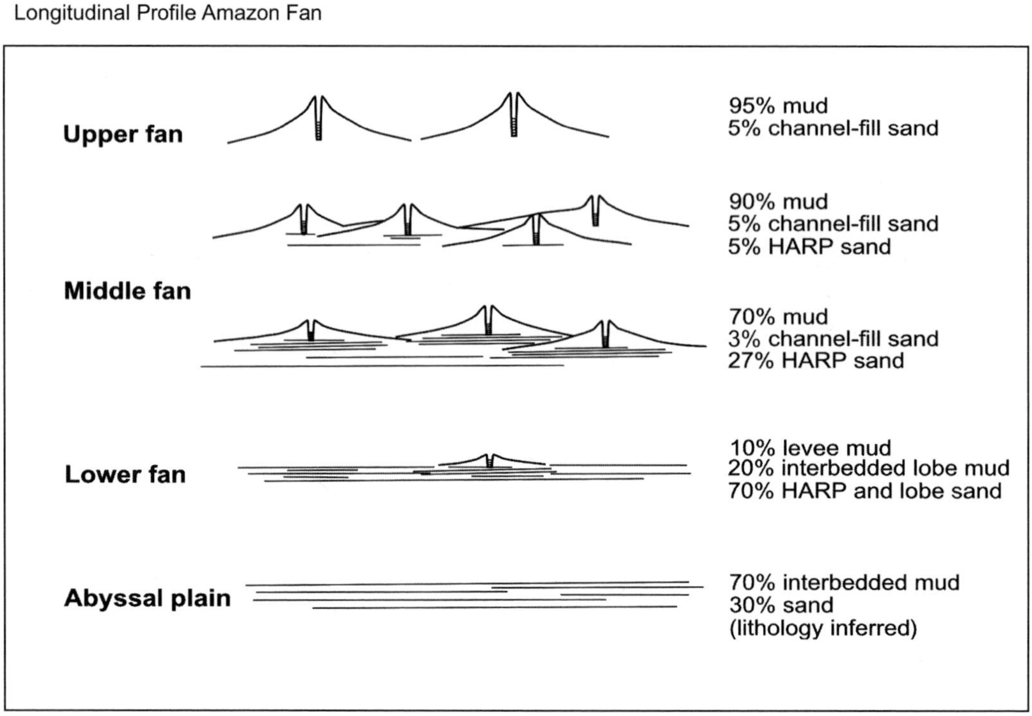

Fig. 16. Schematic strike oriented cross-sections of the youngest (last glacial cycle) levee complex (sets of channel–levee systems), Amazon Fan. The diagram illustrates downdip (longitudinal) changes in architectural style including core/log calibrated proportions of muddy and sandy sediment from high gradient to low gradient fan positions. (From Piper & Normark 2001.)

aggradational and compensational patterns, and similar seismic facies, require sedimentation rates consistent with deposition of the Amazon and Zaire Fan channel–levee systems. This conclusion assumes a similar lithology with few thin condensed intervals within the LCLCs.

The Upper Levee Complex of the Amazon Fan spans *c.* 22 to *c.* 36 ka and is thought to record deposition during a single glacial sea-level lowstand (Flood *et al.* 1991). The Upper Levee Complex contains at least seven channel–levee systems, and division by these bracketing dates leads to a 3–5 ky duration for each channel–levee system (Flood *et al.* 1991). Similarly, the upper Quaternary succession of the Zaire Fan (*c.* 780 ka to present) contains at least 80 channel–levee systems (Droz *et al.* 2003). Simple division of the number of channel–levee systems by this longer duration interval indicates <10 ky duration for each of these channel–levee systems. The factor of three age difference in these two recent channel–levee systems may reflect differences in fan-scale sedimentation rate. However, it more likely reflects an error related to the method of dividing the number of elements by the bracketing condensed section age, an error that increases with the interval age, and presumably reflecting the increased time value of condensed sections within older successions, as illustrated in this dataset. Regardless of the source of the age difference in the modern channel–levee systems, there is a three-order of magnitude difference in age between the modern, and the comparable, features documented in this study.

The age differences between the different modern analogues, and between the modern and ancient channel–levee deposits, produces a conundrum with using time as a calibration tool. Differences in the time value of fine and coarse sediment in deep-marine systems are significant, reflecting different rates and types of depositional processes. The infrequent and event-driven subaqueous flow deposition may enhance this difference, particularly when compared with other clastic depositional systems. Because fine-grained sediment is a common

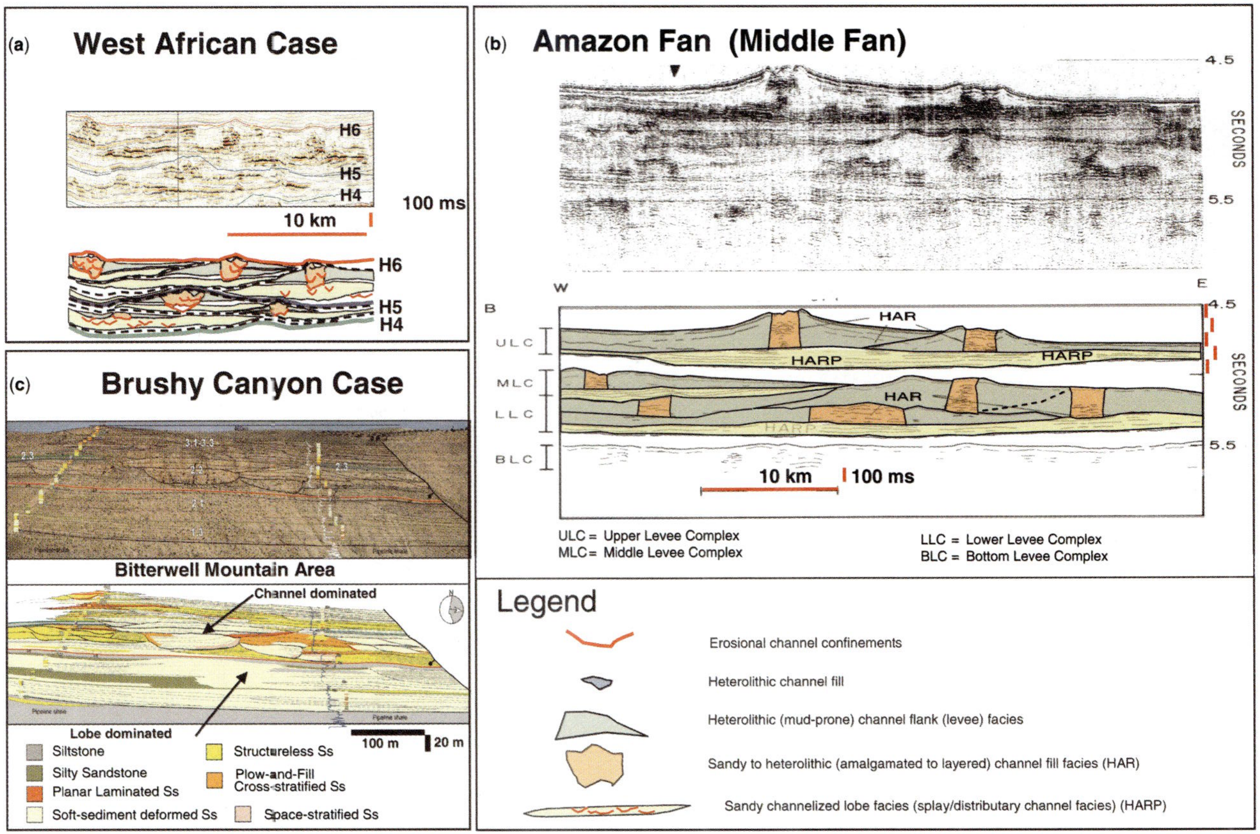

Fig. 17. Comparison of strike oriented seismic cross-sections through lobe–channel–levee architectural elements (displayed at similar scale): (**a**) West African case (this study); (**b**) Amazon Fan (middle fan). The two datasets show similar architectural style and spatial scale. The relationship between seismic facies architecture, and sedimentary facies/lithology as derived from Amazon is as follows: (1) wedgeform channel-flank seismic facies (often transparent) represent mud-prone (heterolithic) levees; (2) HARs represent sand-prone (sandy to heterolithic) channel-fill facies; (3) HARPs represent sand-prone, frontal or lateral splays/lobes originating from leveed channels (Flood *et al.* 1991; Piper & Normark 2001). Given the close geometric and spatial resemblance between seismic channel–levee forms, the basic seismic facies relationships and lithology trends calibrated for the Amazon Fan are also considered valid for the West African deep-marine succession. The lower Brushy Canyon outcrop example from Bitterwell Mountain (**c**) is displayed at significantly smaller scale; however, this stratigraphic record also shows a vertical motif of channelforms overlying lobeforms. Significant constructive levee facies are not developed in the lower Brushy Canyon Fm. This is primarily explained as an effect of mud shortage associated with gravity flows entering the deepwater system at this time. Notice the hierarchical nature of architectural elements within the channel complex. Interpreted panel is courtesy of Diah Hanggoro.

Table 4. *Similarities and contrasts between West African case and Brushy Canyon formation*

	West African Case	Delaware Mountain Gp/Brushy Canyon Formation (Gardner *et al.* 2003)
Age	Early Oligocene to present	Middle Permian
Tectonic setting /margin type	South Atlantic post-rift passive margin with mobile substratum	Intra cratonic passive margin Siliciclastic system associated with inherited carbonate platform
Lithology	Siliciclastic sandy system High grain-size variability	BCF: Siliciclastic, very sandy system Low grain-size variability (fine grade dominant) Influenced by carbonate clasts
Shelf-to-basin relief	High (1000 m)	BCF: high (400–600 m)
Temporal window	Early Oligocene to present *c.* 32 Ma (H0–seabed)	DMG: *c.* 15 Ma BCF: *c.* 3–4 Ma
Channel architecture	Aggradational channel–levee complexes common	BCF: Aggradational channel–levee complexes rare
Spatial window (map dimension)	20 × 30 km ('incomplete')	160 × 100 km ('complete')
Number of 2nd-order cycles*	3 within post-rift margin outbuilding phase C1, C2 and C3 (H0–H3; H3–H6; H6–seabed)	DMG: 1
Number of 3rd-order cycles*	7 within 3 2nd-order cycles	DMG: 3 (BCF; Cherry Canyon Fm and Bell Canyon Fm)
Number of 4th-order cycles*	10 within 5 3rd-order cycles (H2–seabed)	BCF: 3 within 1 3rd-order cycles (Lower, Middle, Upper BCF)
Margin adjustment/deformation expression	Significantly shifting lobe–channel–levee complex style and trend. Increasing gravity driven up-dip extensional and down-dip compressional deformation associated with mobile substratum (Brun & Fort 2004)	Gravity driven mass transport complexes (MTCs) below and above BCF

*Cycles within a relative hierarchy based primarily on physical seismic stratigraphic correlations, partly calibrated to, but not constrained by chronostratigraphic ages.

constituent of subaqueous flows, these mudstones must be differentiated from those recording suspension fallout and sediment starvation. Distinguishing between these different mudstone types requires additional facies and architectural criteria like bed continuity, sedimentary body-type and/or associated facies (Jackson 1975; Miall 1985). Estimating the age of a deep-marine succession using dates from the condensed sections that brackets them further contributes to this time discrepancy. The linear interpolation of time produces several orders of magnitude difference in the duration for similar scale depositional patterns. For example, the ancient channel–levee deposits in this study are (*c.* 1–1.5 Ma) longer in duration than their modern counterparts. Another factor may be the source-distant setting of this study area, where even more time is recorded by the condensed mudstones. These relatively distal mudstones would record even more sediment starvation than their counterparts from the more proximal setting of the modern analogs.

There is a conspicuously low proportion of chaotic seismic facies, corresponding to MTDs, in this study, compared with Amazon Fan levee complexes that are commonly separated by MTDs

Table 5. *Stratigraphic recognition criteria related to deep-marine margin adjustment*

Stratigraphic recognition criterion no.	Stratigraphic recognition criterion description
SRC1	Shifts in shelf-margin delivery to basin (inferred volume; sand:mud ratio; entry point position and type/character)
SRC2	Shift in position of slope and basin (submarine fan) depocentra
SRC3	Shift in trend of flow pathways (lobe–channel–levee trend/orientation)
SRC4	Offset in seismic facies and seismic geomorphic (depositional element) type, style and proportion
SRC5	Gradient change (estimated from temporal change in lobe–channel–levee proportion)
SRC6	Change in flow dispersion and/or focus (sediment fairway clustering v. 'shadow-zones')
SRC7	Major lithology change (e.g. transition from carbonates to siliciclastics or sand-prone to mud-prone record)
SRC8	Presence of mass transport deposits (MTDs)
SRC9	Presence of submarine unconformities and/or slump scars

(e.g. Flood *et al.* 1991). The two MTDs present in the uppermost part of this succession flank a locally growing anticlinal structure just north of study area.

Local (internal) controls v. regional (external) controls

Differentiating between external and internal controls on deep-marine sedimentation is important, but difficult to establish without complete information from the entire deep-marine depositional system. External forcing of deep-marine systems is generally related to tectonic, climatic, sea level and source area, i.e. drainage basin and sediment composition, controls. Varying dependency among these primary variables, coupled with intrinsic thresholds and feedback responses operating within the basin, combine to produce non-unique depositional responses that are poorly understood. Consequently, external controls are often analysed independently (see Reading & Richards 1994; Shanmugam & Moiola 1998 as examples of this approach), or they are combined and simply described as relative sea level, or staging area controls (e.g. Garfield *et al.* 1998; Posamentier & Kolla 2003). Combined with the incomplete stratal record often preserved in proximal shallow water and continental settings, these simplifications significantly reduce the ability to uniquely link external controls to a deep-marine sedimentation pattern. Within the basin, local gradient and topography also contributes to changes in the subaqueous flow composition and volume, which further complicates isolating the contribution of internal and external controls on deep-marine sedimentation.

External controls can operate in concert with, or independent of, local internal controls (e.g. cycle compensation and channel avulsion, fairway trend and number of avulsion nodes, mass transport distributions, and the local generation of seafloor topography). Internal controls on the adjustment of channel style and trend that change from cycle to cycle in a source-distant setting, therefore, cannot be easily differentiated from external controls operating significant distances from the depositional site. The record of depositional outbuilding of the continental margin shows significant slope adjustment and deformation restricted to the basin. Local structural movements near the depositional site that may have controlled the complete abandonment of lobe–channel–levee deposition in the youngest deposits are consistent with this deformation record.

Despite uncertainty in determining the exact contribution from internal and external controls, local changes in the deep-marine system are necessary to sufficiently explain the high frequency depositional patterns documented in this source-distant setting. The conspicuous upward increase in levee proportions in cycle C1 (ABC1) to C2 (ABC2) strata are consistent with locally increased gradient and seafloor focus at the depositional site, which also corresponds to the basinward migration of deformation in the outer foldbelt (Fig. 18). Also consistent with this long-term secular trend is the higher frequency LCLCs that show a progressive decrease in sand:mud ratio related to local levee development. Alternatively, the decreased sand:mud ratio has been interpreted in other datasets to reflect long-term external changes in sediment composition in the shelf staging area (Posamentier & Kolla 2003).

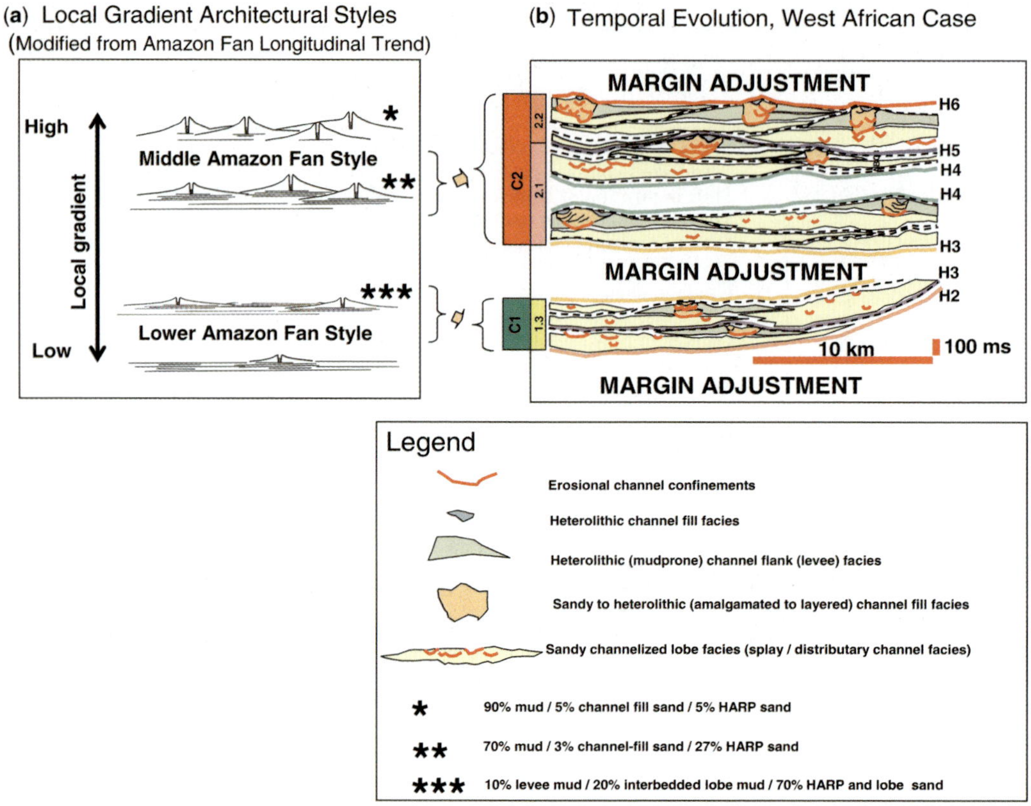

Fig. 18. Illustration of the relationship between middle and lower Amazon Fan architectural styles and the observed West African lobe–channel–levee architecture. A modified version of the schematic Amazon Fan cross-sections (modified from Piper & Normark 2001) emphasizes the general relationship between lobe–channel–levee architectural style and local gradient: high levee construction and a high proportion of channel–levee relative to HARP/lobe proportion correlates to high local gradient. Low levee construction and a low proportion of channel–levee relative to HARP/lobe proportion correlates to low local gradient. This conceptual relationship suggests there is an upward (temporal) increase in local gradient through the studied West African stratigraphy (C1/1.3 through C2 record).

This study and the modern Amazon and the Zaire analogues show a common depositional pattern referred to, in this paper, as a LCLC and characterized by a basal, lobate depositional element (usually HARP) overlain by 'gull-wing' channel–levee deposits (usually HAR-transparent wedgeform). This common motif can be produced by channel extension, with the frontal splay attached and incised by the advancing channel. In this case the depositional topography created by the frontal lobe deposit is insufficient to cause the channel to shift laterally. In other examples, frontal lobe deposition creates topography sufficient to deflect the channel away from a constructional seafloor mound, which produces an attached lateral splay pattern. In this study both styles of frontal lobe development are observed. Frontal lobes that construct sea-floor topography, termed 'terminal lobes,' are also described from the Zaire Fan. They are characterized by discontinuous and contorted seismic facies and are grouped into lens-shaped units. They are considered to be genetically distinct from more continuous 'basal high-amplitude units' related to either lateral splay and/or channel deflected lobe processes (Droz *et al.* 2003). Alternatively, the LCLC pattern can record lateral sedimentation related to channel avulsion and infilling of topographic lows flanking constructional channel–levee systems. This is a common pattern in the Amazon Fan where the increased levee height creates high local gradients that produce lateral splays detached from the levee (HARPs). The detached lateral splays originate from a channel avulsion node and are not found updip of a channel bifurcation site (Flood *et al.* 1991; Piper & Normark 2001).

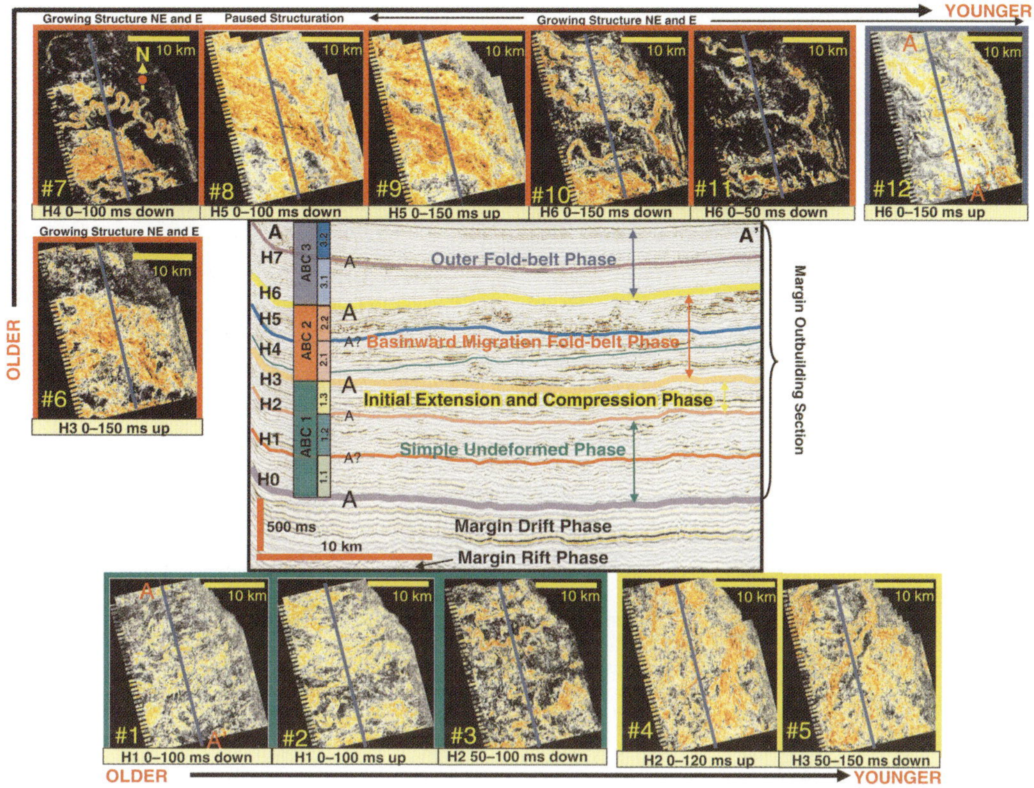

Fig. 19. Summary of West African margin evolution as expressed in seismic stratigraphic and geomorphic patterns though time. Diagram shows twelve successive seismic geomorphology displays (maps 1–12) derived from seismic volumes defined above or below interpreted seismic horizons (condensed sections, H1–H6). The stratigraphic record is subdivided into three, second-order adjustment-bounded cycles (ABC 1, green/yellow; ABC 2, red; and ABC-3, blue), based on significant offsets in seismic facies and shifts in channel/lobe style and orientation. The relationship between seismic geomorphology and seismic stratigraphy is indicated with corresponding colours. Primary (second-order) margin deformation and adjustment events are labelled 'A', whereas, less extreme (third-order) margin adjustment/deformation events are labelled 'A'.

In this study, there is a systematic and repetitive upward change in lobe shape and incision style of channels overlying frontal lobes (e.g. Fig. 9, Cycle 2.1.1, map 7). The lobes change upward from lobate to elongate tongue-shaped geometries (e.g. Fig. 9, Cycle 2.2.2, map 9) that reflect high frequency and local changes in gradient related to lobe depositional topography. In this scheme, the lobate frontal lobes (frontal/terminal HARPS) are deposited under lower local gradient conditions (e.g. Cycle 2.1, map 7). Depositional topography created by these frontal lobes plugs the channel updip of this site, which causes the channel to shift laterally. The convexity of the preceding lobe topography locally increases the gradient for the laterally shifted channel, which deposited a more elongate and possibly detached lobe axis at the extended channel mouth, and lateral to the older frontal lobe (e.g. Cycles 2.1.2 and 2.2.2, maps 8 and 9). The increased channel/lobe proportion, levee height and lobe elongation reflect the locally increased gradient. The increased levee height also reflects the constructive effect of increased flow confinement, height and gradient during this phase of deposition (Fig. 18).

This repetitive LCLC pattern is summarized in a three-stage model characterized by (1) short-lived and lower sediment volume unconfined lobe (frontal splay) deposition, (2) lobe incision by sinuous channels bypassing the most coarse sediment and building depositional topography, followed by (3) channels plugged and deflected to the flank of the

Table 6. *Recognized margin adjustment boundaries defining ABCs*

2nd-order adjustment boundaries	Stratigraphic adjustment recognition criterion no. (see also Figs 6b and 19 and Table 5)
ABC2/ABC3 (H6)	SRC4, SRC7 ABC3: Figs 6 and 10; map 12 relative to underlying maps 6–11 (ABC2) Change from strong amplitude distinct lobeform and channel–leveeform facies to uniform sub-parallel seismic drape facies and wedge-shaped chaotic seismic facies
ABC1/ABC2 (H3)	SRC2, SRC3, SRC4, SRC5, SRC6 ABC2: Figs 6 and 9; maps 6–11 relative to underlying maps 1–5 (ABC1) Change from weak-moderate amplitude lobeform dominated facies to strong amplitude distinct lobeform and channel–leveeform facies *c.* 150–180° shift in LCLC orientation
Base ABC1 (H0)	SRC1; SRC4; SRC6, ABC1: Figs 6, 7 and 8; maps 1–5 relative to underlying drift section Change in depositional style from passive post-rift drift-phase (seismic drape facies) to active depositional outbuilding (subtle lobe and channelform facies)
3rd-order adjustment boundaries	
3.1 /3.2 (H7)	SRC8 (Fig. 6; mass transport deposit
2.1/2.2	No significant margin adjustment recognition
1.2/1.3 (H2)	SRC2, SRC3, SRC4, SRC6 Maps 4–5 relative to underlying maps 2–3 Change in lobe-channel style and orientation *c.* 70° shift in orientation
1.1/1.2 (H1)	No significant margin adjustment indicators
*4th-order adjustment boundaries**	
3.2.1/3.2.2 (H7b)	SRC8 (MTD)
3.1.1/3.1.2 (H6b)	SRC8 (MTD)

SRC, stratigraphic recognition criterion indicating margin adjustment.
*No significant internal margin adjustment recognition except local MTDs within ABC3.

mounded lobe, and the incision and development of constructive levees that promote sediment bypass that constructs detached and more elongate lobes that begin the next cycle of LCLC deposition (Figs 12 and 13). This model describes depositional responses to internal controls produced by local depositional topography and gradient changes operating at the mouth of a submarine channel (Adeogba 2005). Though the sediment volume and composition may be modulated by external changes, these internal controls are important because lobe deflection and incision enhances levee construction and confinement, which provides a local mechanism for increasing the gradient and extending confined flows in channels farther into the basin.

Implications to reservoir and play fairway analysis

This stratigraphic hierarchy illustrates a variety of deep-marine depositional patterns that can be used for sand fairway and play analysis (e.g. channel-lobe positioning, style, trend and proportion relative to structural and stratigraphic traps). This multi-scalar approach integrates local high-resolution 3D seismic volumes with sparse regional seismic 2D datasets. Seismic geomorphology maps placed within the stratigraphic hierarchy, in a structurally quiescent area of the basin, are used to establish the long-term basin margin evolution. Lithology prediction is achieved by using this long-term record to determine if there is a consistent relationship between the record of gravity-driven adjustment and deformation and subaqueous flow trends and depositional patterns. Progressive changes in local topographic relief, focus and gradient are related to the basinward migration of deformation produced by depositional outbuilding. The long-term margin evolution translates more complex topography basinward through time, which promotes subaqueous flow focus and higher local gradients in source-distant settings that continue to the present. The inherent linkage of dynamic margin deformation and subaqueous flow pathways controlling reservoir fairway distributions provides a holistic and multi-scalar approach for play-, trap- and prospect analysis.

Conclusions

Seismic geomorphologic and stratigraphic analysis of a local, high-fidelity deep-marine succession documents a *c.* 32 Ma record of gravity-driven margin adjustment and deposition related to outbuilding of the West African continental margin. A threefold hierarchy of stratigraphic cycles (second- through fourth-order) constrains seismic geomorphologic maps that show systematic and repetitive changes in LCLCs. Depositional outbuilding is recorded by an upward change from lobe-dominated successions capped by margin perpendicular oriented channels, to lobe–channel–levee complexes showing multiple channel trends oriented oblique to the margin. Within this interval the levees increase and lobes decrease in proportion upward through the succession. Discrete LCLCs change upward from frontal splays incised by sinuous channels that back-fill and shift laterally to lobe flanks to form lobe-deflected sinuous channels with thicker levees. These LCLCs form 10 km-wide, wedge-shaped geometries that are offset laterally, with the compensational stacking indicating aggradational conditions. By contrast, a significant reduction in sedimentation in the upper third of this succession is indicated by a change to thick successions of parallel continuous reflectors interpreted to represent a mudstone-dominated succession bracketed by mass transport deposits. These local deformational features are the culmination of a progressive increase in gradient and development of contractional structures that deflect sedimentation away from this site and record basinward migration of deformation related to depositional outbuilding of the continental margin.

Three predicted facies associations characterize the LCLC evolution; these include (1) sandy frontal splay deposits organized as distributary channel complexes (HARP); (2) sandy to heterolithic channel fill facies (HAR); (3) heterolithic (mud-prone) levee facies. These lithology predictions are supported by similar core-calibrated, seismic architecture from the modern Amazon and Zaire Fans as well as Brushy Canyon Fm outcrop data.

Two important conclusions derived from this study are (1) the importance of local gradient on depositional patterns, and (2) the linkage between basin margin adjustment/deformation and deep-marine fairway deposition. Local gradient changes controlling the documented architectural styles were produced by gravity-driven margin adjustment/deformation/folding processes. Long-term depositional outbuilding of the continental margin deformation produced local topography and gradient changes rather than upper fan, middle fan, lower fan and abyssal plain divisions that more appropriately describe the geomorphology of modern fan systems. The latter category of geomorphic terms implies basinward translation of facies tracts described by Walther's law, which is inappropriate to fixed continental margins that record outbuilding and basinward migration of gravity-driven deformation and basin margin adjustment and not translation of water-depth and/or profile dependent facies tracts.

Seismic stratigraphic patterns of second-order ABCs document three phases of gravitational adjustment, deformation and deposition. Correlation to more proximal borehole calibrated data indicates that the ABCs are semi-regional events. The long-term stratigraphic evolution reflects basinward translation of progressively more complex topography, stronger sediment-gravity flow focus and higher local gradients that continue to the present.

The authors would like to thank Statoil colleagues B. Blikeng, E. P. Johannessen and Ø. Steen for providing essential regional geological context and correlation basis for this study. I. Prince is acknowledged for his biostratigraphic input to the study. Many thanks to S. Goldsmith of the Colorado School of Mines Slope and Basin Consortium, as well as E. Storsten and L. Reistad, Statoil, for their drafting assistance. Furthermore, the authors would like to acknowledge Statoil Exploration Management, Statoil Technology Management and Colorado School of Mines Faculty for the opportunity to conduct this research during a Colorado School of Mines sabbatical. Thanks are also due to WesternGeco for permission to use the seismic data presented in this paper. Finally many thanks to H. Posamentier and E. Duerichen, whose reviews significantly improved the structure and content of this paper.

References

Abreu, A., Sullivan, M., Pirmez, C. & Mohrig, D. 2003. Lateral Accretion Packages (LAPs): an important reservoir element in deep water sinuous channels. *Marine and Petroleum Geology*, **20**, 631–648.

Adeogba, A. A., McHargue, T. R. & Graham, S. A. 2005. Transient fan architecture and depositional controls from near-surface 3-D seismic data, Niger Delta continental slope. *AAPG Bulletin*, **89**, 627–643.

Beaubouef, R. T. & Friedman, S. J. 2000. High-resolution seismic/sequence stratigraphic framework for the evolution of Pleistocene intra slope basins, Western Gulf of Mexico: depositional models and reservoir analogs. *In*: Weimer, P., Slatt, R. M., Coleman, J., Rosen, N. C., Nelson, H.,

BOUMA, A. H., STYZEN, M. J. & LAWRENCE, D. T. (eds) *Deep-Water Reservoirs of the World: Gulf Coast Society of the Society of Economic Paleontologists and Mineralogists Foundation, 20th Annual Research Conference*, 40–60.

BEAUBOUEF, R. T., ROSSEN, C. R., *ET AL.* 1999. *Deep-water sandstones, Brushy Canyon Formation, West Texas.* AAPG Continuing Education Course Note Series, 40, Field guide for AAPG Hedberg Field Research Conference, 15–20 April 1999.

BERGGREN, W. A., KENT, D. V., SWISHER, C. C., III. & AUBRY, M.-P. 1995. A revised Cenozoic geochronology and chronostratigraphy. *In*: BERGGREN, W. A., KENT, D. V., AUBREY, M.-P., & HARDENBOL, J. (eds.) *Geochronology, Time Scales and Global Stratigraphic Correlation.* SEPM (Society for Sedimentary Geology) Special Publications, **54**, 129–212.

BLIKENG, B. & FUGELLI, E. 2000. Application of results from outcrops of the deep water Brushy Canyon formation, Delaware Basin, as analogs for the deep water exploration targets on the Norwegian Shelf. *In*: WEIMER, P., SLATT, R. M., COLEMAN, J., ROSEN, N. C., NELSON, H., BOUMA, A. H., STYZEN, M. J. & LAWRENCE, D. T. (eds) *Deep-Water Reservoirs of the World: Gulf Coast Society of the Society of Economic Paleontologists and Mineralogists Foundation, 20th Annual Research Conference*, 61–81.

BOOTH, J. R., DUVERNAY, A. E. III, PFEIFFER, D. S., & STYZEN, M. J. 2000. Sequence stratigraphic framework, depositional models, and stacking patterns of ponded and slope fan systems in the Auger Basin: Central Gulf of Mexico Slope. *In*: WEIMER, P., SLATT, R. M., COLEMAN, J., ROSEN, N. C., NELSON, H., BOUMA, A. H., STYZEN, M. J. & LAWRENCE, D. T. (eds) *Deep-Water Reservoirs of the World: Gulf Coast Society of the Society of Economic Paleontologists and Mineralogists Foundation, 20th Annual Research Conference*, 82–103.

BRUN, J. P. & FORT, X. 2004. Compressional salt tectonics (Angolan margin). *Tectonophysics*, **382**, 129–150.

CALASSOU, S. & MORETTI, I. 2003. Sedimentary flattening and multi-extensional deformation along the West African margin. *Marine and Petroleum Geology*, **20**, 71–82.

CARR, M. & GARDNER, M. H. 2000. Portrait of a basin-floor fan for sandy deep-water systems, Brushy Formation, West Texas. *In*: BOUMA, A. H. & STONE, C. G. (eds) *Fine-Grained Turbidite Systems.* AAPG Memoirs, 72/SEPM Special Publications, **68**, 1020–1036.

COBBOLD, P. R. & SZATMARI, P. 1991. Radial gravitational gliding on passive margins. *Tectonophysics*, **188**, 249–289.

COBBOLD, P., SZATMARI, P., DEMERCIAN, L., COELHO, D. & ROSSELLO, E. A. 1995. *In*: JACKSON, M. P., ROBERTS, D. & SNELSON, S. (eds) *Seismic and Experimental Evidence for Thin-skinned Horizontal Shortening by Convergent Radial Gliding on Evaporites, Deep-water Santos Basin, Brazil Salt Tectonics: a Global Perspective.* AAPG Memoirs, **65**, 305–321.

DROZ, L., RIGAUT, F., COCHONAT, P. & TOFANI, R. 1996. Morphology and recent evolution of the Zaire turbidite system (Gulf of Guinea). *GSA Bulletin*, **108**, 253–269.

DROZ, L., MARSSET, T., ONDREAS, H., LOPEZ, M., SAVOYE, B. & SPY-ANDERSON, F.-L. 2003. Architecture of an active mud-rich turbidite system: The Zaire Fan (Congo-Angola margin southeast Atlantic): results from ZaiAngo 1 and 2 cruises. *AAPG Bulletin*, **87**, 1145–1168.

DUVAL, B., CRAMEZ, C. & JACKSON, M. P. 1992. Raft tectonics in Kwanza Basin. Angola. *Marine and Petroleum Geology*, **9**, 389–404.

EMBRY, A. F. 1995. Sequence boundaries, sequence hierarchies: Problems and proposals. *In*: STEEL, R. J., FELT, V. L., JOHANNESSEN, E. P. & MATHIEU, C. (eds) *Sequence Stratigraphy on the Northwest European Margin.* Norwegian Petroleum Society (NPF), Special Publications, 5. Elsevier, Amsterdam, 1–11.

FLOOD, R. D., MANLEY, P. L., KOWSMANN, R. O., APPI, C. J. & PIRMEZ, C. 1991. Seismic Facies and late Quaternary growth of the Amazon submarine fan. *In*: WEIMER, P. & LINK, M. H. (eds) *Seismic Facies and Sedimentary Processes of Modern and Ancient Submarine Fans.* Springer, New York, 415–433.

GARDNER, M. H. & BORER, J. M. 2000. Submarine channel architecture along a slope to basin profile, Permian Brushy Canyon Formation, West Texas. *In:* BOUMA A. H. & STONE, C. G. (eds), *Fine-Grained Turbidite Systems.* AAPG Memoirs, 72/SEPM Special Publications, **68**, 195–215.

GARDNER, M. H. & SONNENFELD, M. D. 1996. *Stratigraphic Changes in Facies Architecture of the Permian Brushy Canyon Formation in Guadalupe Mountain National Park, West Texas.* Permian Basin Section, SEPM Publications, **96–38**, 17–40.

GARDNER, M. H., BORER, J. M., MELICK, J. J., MAVILLA, N., DECHESNE, M. & WAGERLE, R. M. 2003. Stratigraphic process-response model for submarine channels and related features from studies of Permian Brushy Canyon outcrops, West Texas. *Marine and Petroleum Geology*, **20**, 757–787.

GARFIELD, T. R., JENNETTE, D. C., GOULDING, F. J. & SICKAFOOSE, D. K. 1998. An integrated approach to deep-water reservoir prediction. Extended abstracts volume *In*: MELLO, M. R. & YILMAZ, P. O. (eds) *ABGP/AAPG International Conference & Exhibition*, 8–11 November, Rio de Janeiro, Brazil. AAPG Bulletin, **82**, 1883–1984.

HADLER-JACOBSEN, F., JOHANNESSEN, E. P., ASHTON, N., HENRIKSEN, S., JOHNSON, S. D. & KRISTENSEN, J. B. 2005. Submarine fan morphology and lithology distribution, a predictable function of sediment delivery, gross shelf to basin relief, slope gradient and basin topography. *In*: DORÉ, A. G. & VINING, B. A. (eds) *Petroleum Geology: North-West Europe and Global Perspectives – Proceedings of the sixth-Petroleum Geology Conference.* Geological Society, London, 1121–1145.

HISCOTT, R. N., PIRMEZ, C. & FLOOD, R. D. 1997. Amazon submarine fan drilling: a big step forward for deep-sea fan models. *Geoscience Canada*, **24**, 13–24.

HODGSON, D. M., Flint, S. S., HODGETTS, D., DRINKWATER, N. J. & JOHANNESSEN, E. P. 2006. Stratigraphic evolution of fine-grained submarine fan systems, Tanqua Depocenter, Karoo Basin, South Africa. *Journal of Sedimentary Research*, 2006, **76**, 20–40.

JACKSON, R. G. 1975. Hierarchical attributes and a unifying model of bed forms composed of cohesionless material and produced by shearing flow. *Geological Society of America Bulletin*, **86**, 1523–1533.

KARNER, G. D. & DRISCOLL, N. W. 1999. Tectonic and stratigraphic development of the West African and East Brazilian margins: insights from quantitative basin modeling. *In*: CAMERON, N. R., BATE, R. H., CLURE, V. S. & BENTON, J. (eds) *Oil and Gas Habitats of the South Atlantic*. Geological Society of London, Special Publications, **153**, 11–40.

KNELLER, B. 1995. Beyond the turbidite paradigm: physical models for deposition of turbidites and their implications for reservoir prediction. *In*: HARTLEY, A. J. & PROSSER, D. J. (eds), *Characteristics of Deep Marine Clastic Systems*. Geological Soceity Special Publications, **94**, 31–49.

KNELLER, B. & MCCAFFREY, W. D. 1999. Depositional effects of flow nonuniformity and stratification within turbidity currents approaching a bounding slope; deflection, reflection, and facies variation. *Journal of Sedimentary Research*, **69**, 980–991.

KNELLER, B. C., BENNETT, S. J. & MCCAFFREY, W. D. 1999. Velocity structure, turbulence and fluid stresses in experimental gravity currents. *Journal of Geophysical Research*, **104**, 5381–5391.

LUNDE, G., AUBERT, K., LAURITZEN, O. & LORANGE, E. 1992. Tertiary uplift of the Kwanza Basin in Angola. *In*: CURNELLE, R. (ed.), *Geologie Africaine. Bulletin des Centres de Recherches*. Exploration-Production Elf-Aquitaine, Pau, France, 99–117.

LUNDIN, E. 1992. Thin-skinned extensional tectonics on a salt detachment, northern Kwanza basin, Angola. *Marine and Petroleum Geology*, **9**, 405–411.

MANLEY R. L. & FLOOD, R. D. 1988. Cyclic sediment deposition within Amazon deep-sea fan. *AAPG Bulletin*, **72**, 912–925.

MAYALL, M. & STEWART, I. 2000. The architecture of turbidite slope channels. *In*: WEIMER, P., SLATT, R. M., COLEMAN, J., ROSEN, N. C., NELSON, H., BOUMA, A. H., STYZEN, M. J. & LAWRENCE, D. T. (eds), *Deep-water Reservoirs of the World: Gulf Coast Society of the Society of Economic Paleontologists and Mineralogists Foundation, 20th Annual Research Conference*, 578–586.

MCCAFFREY, W. & KNELLER, B. 2001. Process controls on the development of stratigraphic trap potential on the margin of confined turbidite systems and aids to reservoir evaluations. *AAPG Bulletin*, **85**, 971–988.

MCKENZIE, D. P. 1978. Active tectonics of the Alpine-Himalayan belt: the Aegean Sea and surroundings regions. Geophys. *Journal of the Royal Astronomical Society*, **55**, 217–254.

MENARD, H. W. 1972. Citations in a scientific revolution. *Geological Society of America Memoir*, **132**, 1–5.

MIALL, A. D. 1984. *Principles of Sedimentary Basin Analysis*. Springer, New York, 503.

MIALL, A.D. 1985. Architectural-element analysis – a new method of facies analysis applied to fluvial deposits. *Earth-Science Reviews*, **22**, 261–308.

MIALL, A. D. 1989. Architectural elements and bounding surfaces in channelized clastic deposits; notes on comparisons between fluvial and turbidite systems. *In*: TAIRA, A. & MASUDA, F. (eds) *Sedimentary Facies in the Active Plate Margin; Advances in Earth and Planetary Sciences*: Terra Scientific Tokyo, 3–15.

MIALL, A. D. 1995. Whither stratigraphy? *Sedimentary Geology*, **100**, 5–20.

MITCHUM, R. M., 1985. Seismic stratigraphic expression of submarine fans. *AAPG Memoirs,* **39**, 117–136.

MITCHUM, R. M. JR. & VANWAGONER, J.C. 1991. High-frequency sequences and their stacking patterns: sequence-stratigraphic evidence of high-frequency eustaic cycles. *Sedimentary Geology*, **70**, 131–160.

MOHRIAK, W., MACEDO, J. & CASTELLANI, R. 1995. *In*: M. P. JACKSON, D. ROBERTS, & S. SNELSON (eds), *Salt Tectonics and Structural Styles in the Deep-water Province of the Cabo Frio Region, Rio de Janeiro, Brazil. Salt Tectonics: a Global Perspective*. AAPG Memoirs, **65**, 273–304.

MUTTI, E. 1977. Distinctive thin-bedded turbidite facies and related depositional environments in the Eocene Heco Group, south central Pyrenees, Spain. *Sedimentology*, **25**, 107–131.

MUTTI, E. 1985. Turbidite systems and their relationships to depositional sequences. *In*: ZUFFA, G. G. (ed.) *Provenance of Arenites*. Reidel, Dordrecht, 65–93.

MUTTI, E. & NORMARK, W. R. 1987. Comparing examples of modern and ancient turbidite systems; problems and concepts. *In*: LEGGERT, J. K. & ZUFFA, G. G. (eds) *Marine Clastic Sedimentology; Concepts and Case Studies*. Graham and Trotman, London, 1–38.

MUTTI, E. & NORMARK, W. R. 1991. An integrated approach to the study of turbidite systems. *In*: WEIMER, P. & LINK, M. H. (eds) *Seismic Facies and Sedimentary Processes of Submarine Fans and Turbidite Systems*. Springer, New York, 75–106.

MUTTI, E. & SONNINO, M. 1981. Compensation cycles – a diagnostic feature of turbidite sandstone lobes. *In*: *International Assocciation of Sedimentologists Second European Regional Meeting*, Bologna, Italy, Abstracts Volume, 120–123.

NARDIN, T. R., HEIN, F. J., GORSLINE, D. S. & EDWARDS, B. D. 1979. A review of mass movement processes, sediment and acoustic characteristics, and contrasts in slope and base-of-slope systems versus canyon-fan-basin-floor systems. *In*: DOYLE, L. J. & PILKEY, O. H. (eds) *Geology of Continental Slopes*. Society of Economic Paleontologists and Mineralogists, Special Publications, **27**, 61–73.

NORMARK, W.R. 1978. Fan valleys, channels and depositional lobes on modern submarine fans – character for recognition of sandy turbidite environments: *AAPG Bulletin*, **62**, 912–931.

PIPER, D. J. W. & NORMARK, W. R. 2001. Sandy fans – from Amazon to Huemene and beyond. *AAPG Bulletin*, **85**, 1407–1438.

PIRMEZ, C. & FLOOD, R. D. 1995. Morphology and structure of Amazon Channel. *In*: FLOOD, R. D., PIPER, D. J. W. KLAUS, A. & PETERSON, L. C. (eds) *Proceedings of the Ocean Drilling Program: Initial Reports*. College Station, TX (Ocean Drilling Program), **155**, 23–45.

PIRMEZ, C., HISCOTT, R. N. & KRONEN J. D. JR 1997. Sandy turbidite successions at the base of channel–levee systems of the Amazon Fan revealed by FMS and cores: unravelling the facies architecture of large submarine fans. *In*: FLOOD, R. D., PIPER, D. J. W., KLAUS, A. & PETERSON, L. C. (eds) *Proceedings of the Ocean Drilling Program: Scientific Results*. College Station, Texas (Ocean Drilling Program), **155**, 7–33.

PIRMEZ, C., BEAUBOUEF, R. T., FRIEDMANN, S. J. & MOHRIG, D. C. 2000. Equilibrium profile and base level in submarine channels: examples from Late Pleistocene systems and implications for the architecture of deep-water reservoirs. *In*: WEIMER, P., SLATT, R. M., COLEMAN, J., ROSEN, N. C., NELSON, H., BOUMA, A. H., STYZEN, M. J. & LAWRENCE, D. T. (eds) *Deep-Water Reservoirs of the World: Gulf Coast Society of the Society of Economic Paleontologists and Mineralogists Foundation, 20th Annual Research Conference*, 782–805.

POSAMENTIER, H. W. 2003. Depositional elements associated with a basin floor channel-levee system: case study from the Gulf of Mexico. *Marine and Petroleum Geology*, **20**, 677–690.

POSAMENTIER, H. W. & KOLLA, V. 2003. Seismic geomorphology and stratigraphy of depositional elements in deep-water settings. *Journal of Sedimentary Research*, **73**, 367–388.

POSAMENTIER, H. W., MEIZARWIN, WISMAN, P. S. & PLAWMAN, T. 2000. Deep water depositional systems – ultra deep Makassar Strait, Indonesia. *In*: WEIMER, P., SLATT, R. M., COLEMAN, J., ROSEN, N. C., NELSON, H., BOUMA, A. H., STYZEN, M. J. & LAWRENCE, D. T. (eds) *Deep-Water Reservoirs of the World: Gulf Coast Society of the Society of Economic Paleontologists and Mineralogists Foundation, 20th Annual Research Conference*, 806–816.

PRATHER, B. E. 2003. Controls on reservoir distribution, architecture and stratigraphic trapping in slope settings. *Marine and Petroleum Geology*, **20**, 529–545.

PRATHER, B. E., KELLER, F. B. & CHAPIN, M. A. 2000. Hierarchy of deep-water architectural elements with reference to seismic resolution: implications for reservoir prediction and modeling. *In*: WEIMER, P., SLATT, R. M., COLEMAN, J., ROSEN, N. C., NELSON, H., BOUMA, A. H., STYZEN, M. J. & LAWRENCE, D. T. (eds) *Deep-water Reservoirs of the World: Gulf Coast Society of the Society of Economic Paleontologists and Mineralogists Foundation, 20th Annual Research Conference*, 817–835.

PRATSON, L. F. & LAINE, E. P. 1989. The relative importance of gravity-induced versus current-controlled sedimentation during the quarternary along the mid-east U. S. outer continental margin revealed by 3.5 kH echop character. *Marine Geology*, **89**, 87–126.

READING, H. G. & RICHARDS, M. 1994, Turbidite systems in deep-water basin margins classified by grain size and feeder system. *AAPG Bulletin*, **78**, 792–822.

ROSS, W, HALIWELL, B. A., MAY, J. A., WATTS, D. E. & SYWITSKI, J. P. M. 1994. Slope readjustment – a new model for the development of submarine fans and aprons. *Geology*, **22**, 511–514.

SCHLAGER, W. 2004. Fractal nature of stratigraphic sequences. *Geology*, **32**, 185–188.

SHANMUGAM, G. & MOIOLA, R. J. 1988. Submarine Fans: characteristics, models, classification and reservoir potential. *Earth-Science Reviews*, **24**, 383–428.

SHANMUGAM, G. & MOIOLA, R. J. 1990. Submarine fan 'Lobe' models: implications for reservoir properties. *In*: *Gulf Coast Coast Association of Geological Societies and Gulf Coast Section SEPM, 40th Annual Meeting, Transactions*. Gulf Coast Association of Geological Societies, **40**, 777–791.

SPATHOPOULOS, F. 1996. An insight on salt tectonics in the Angola basin, South Atlantic. *In*: ALSOP, G. I., BLUNDEL, D. J. and DAVISON, I. (eds) Salt Tectonics. Geological Society of London Special Publications, 153–174.

STEFFENS, G. S., BIEGERT, E. G., SUMNER, H. S. & BIRD, D. 2003. Quantitative bathymetric analyses of selected deep-water siliciclastic margins: receiving basin configurations for deep-water fan systems. *Marine and Petroleum Geology*, **20**, 547–561.

UNTERNEHR, P., CURIE, D., OLIVET, J. L., GOSLIN, J. & BEUZART, P. 1988. South Atlantic fits and intra plate boundaries in Africa and South America. *Tectonophysics*, **153**, 169–179.

VAIL, P. R. 1987. Seismic-stratigraphy interpretation using sequence stratigraphy. Part. 1: seismic stratigraphy interpretation procedures: *In*: BALLY, A. W. (ed.), *Atlas of Seismic Stratigraphy.* American Association of Petroleum Geologists Studies in Geology, **27**, 1–10.

VAIL, P. R., MITCHUM, R. M. JR, TODD, R. G., WIDMIER, J. M., THOMPSON, S., III, SANGREE, J. B., Bubb, J. N. & HATLELID, W. G. 1977. Seismic stratigraphy and global changes of sea level. *In*: C. E. PAYTON (ed.) *Seismic Stratigraphy – Applications to Hydrocarbon Exploration*. AAPG Memoirs, **26**, 49–212.

VINK, G. E. 1982. Continental rifting and the implications for plate tectonic reconstructions. *Journal of Geophysical Research*, **87**, 10677–10688.

WALKER, R. G. 1978. Deep water sandstone facies and ancient submarine fans: models for exploration and stratigraphic traps. *AAPG Bulletin*, **62**, 932–966.

WU, S., BALLY, A. W. & CRAMEZ, C. 1990. Allochthonous salt, structure and stratigraphy of the north-eastern Gulf of Mexico. *Marine and Petroleum Geology*, **7**, 318–370.

ZELT, F. B. & ROSSEN, C. 1995. Geometry and continuity of deep-water sandstones and siltstones, Brushy Canyon Formation (Permian) Delaware Mountains, Texas. *In*: PICKERING, K. T., HISCOTT, R. N., KENYON, N. H., RICCI-LUCHI, F. & SMITH, R. D. A. (eds) *Atlas of Deep Water Environments – Architectural Style in Turbidite Systems*. Chapman and Hall, London, 167–183.

The use of two- and three-dimensional seismic to understand sediment transfer from fluvial to deepwater via sinuous channels: example from the Mahakam shelf and comparison with outcrop data (south central pyrenees)

P. CRUMEYROLLE, I. RENAUD & J. SUITER

TOTAL S.A., Centre Scientifique et Technique, Avenue Larribau, 64000 PAU, France

Abstract: While being very different in term of horizontal and vertical resolution, the three-dimensional seismic and the two-dimensional lines straddling the Mahakam shelf complement each other to refine the classical two-dimensional seismic-based sequence stratigraphy models and the associated depositional environments interpretation by establishing closer three-dimensional genetic relations between fluvial, deltaic and turbiditic sediments using palaeolandscape morphologies. The Mahakam lowstand seismic geomorphology study can be compared with ancient sinuous fluvial to slope and deep marine channels outcropping in the cenozoic of the Spanish Pyrenees. This comparison emphasizes the interaction of various processes such as sea-level cycles, shelf-edge failures and the behaviour of flood/fluvial flows shaping the fluvial conduits that fed turbiditic systems during the lowstand period. Fluvial-derived flows by-passing the depositional shoreline break or the shelf break appear to be one of the main mechanisms for sediment transfer from shelf to deep basin both in the Mahakam and the Sobrarbe delta despite the obvious difference in tectonic setting, slope dip and flow efficiency.

During the last 20 years, sedimentary imagery derived from three-dimensional (3D) seismic has become one of the most important techniques for exploring, developing and characterizing hydrocarbon reservoirs. Outside the prospective sections of sedimentary basins, the shallow portions of the conventional three-dimensional seismic can also be used as a tool for research for sedimentologists and stratigraphers. These data provide continuous images of the sediment architecture and morphology with an incomparable spatial resolution compared with outcrop, core studies, or well-log correlations.

Three-dimensional seismic surveys are also reaching higher resolution than was previously possible. Using these data with advanced visualization tools, spectacular palaeolandscapes and morphologies are revealed in sedimentary successions providing, in particular, new insights into sediment transfer mechanisms from the basin margins to the slope and deep basin. This allows the interpreter to have a more visual understanding of what a typical depositional environment looks like to use as an analogy for other areas where less data is available.

In intensively explored areas, like the cenozoic of the Mahakam Delta (Indonesia), 3D seismic surveys are sufficiently numerous and cover large areas, providing a detailed picture of a whole margin relatively undisturbed by syn-sedimentary tectonics. It is thus possible to refine the classical 2D seismic-based sequence stratigraphy models and the associated depositional environments by establishing closer 3D genetic relations between fluvial, deltaic shelfal and turbiditic sediments using palaeolandscape morphologies.

Nevertheless, a gap still remains between 3D imagery and the geological observations made with core, outcrop or well-log data as highlighted by Posamentier & Allen (1999). Ideally, the variety of depositional morphologies observed with 3D seismic should be interpreted using available geological calibration, especially when dealing with high-frequency cycles.

Complex morphologies like those revealed, for example, from the sinuous turbidite systems of divergent margins generate a need to return to the field outcrop analogues with particular questions in mind raised by the nature of the sedimentary processes generating these sinuous turbidite channels (Kolla *et al.* 2001; Abreu *et al.* 2003; Posamentier 2003).

At the same time the body of this research derived from 3D seismic has been accompanied by a renewal of interest in the processes associated with hyperpycnal flows at the mouth of rivers in modern and ancient settings and the way in which these flow regimes evolve into turbidites (Mulder & Syvitski 1995, 1996; Mutti *et al.* 1996, 2003).

Mutti *et al.* (1996) highlighted the importance of flood-dominated fluvio-deltaic systems in basin margin settings and discussed the associated spectrum of facies characterizing river mouth to slope regions. He presented the concept that many

From: Davies, R. J., Posamentier, H. W., Wood, L. J. & Cartwright, J. A. (eds) *Seismic Geomorphology: Applications to Hydrocarbon Exploration and Production*. Geological Society, London, Special Publications, **277**, 85–103.
0305-8719/07/$15.00

ancient fluvio-deltaic systems need to be viewed in terms of catastrophic processes and are thus difficult to compare and to perceive from what we know from the recent facies assemblages and modern landscape, which was demonstrated with numerous facies observations and inferred depositional environment processes.

In this paper we propose to illustrate the strong differentiation in processes and resulting morphologies that occur between the modern highstand Mahakam delta and the late lowstand Quaternary delta (a period of less than 20 000 years). The modern highstand tide-dominated deltas are morphologically different from the lowstand deltas and the associated incised valleys feeding the turbidite systems over the shelf break. These morphological differences are revealed by spectacular images from the shallow portions of 3D seismic over the Mahakam shelf.

We describe various sinuous morphologies in non-marine and delta-front settings of the lowstand Mahakam delta, extending into the deep marine. We then compare these morphologies with selected outcrops of the Sobrarbe delta system from the Tertiary of the Spanish Pyrenees.

Geological setting of the Mahakam Delta

The Mahakam Delta is located in the Kutei basin on the eastern side of Borneo. The Kutei basin is one of the largest cenozoic basins of Borneo. The basin occupies a foreland depression on the eastern edge of the Malaysian craton. It is bounded to the north by the Mangkalihat Peninsula, to the east by the deep Makassar Strait and to the south by the Paternoster platform. To the west, the high relief mountains of central Kalimantan form the drainage basin of the Mahakam Delta developed on the eastern coast of Kalimantan.

Since the mid-Miocene, the basin has accumulated over 5000 m of deltaic sediments on a rapidly subsiding shelf. The rapid eastward prograding evolution of the basin causes an 80 km shift of the shelf break toward the deep basin.

The modern Mahakam Delta

The modern Mahakam Delta is a meso-tidal range delta affected by low wave energy and relatively constant fluvial discharge exhibiting common features of tide to fluvial dominated environments (Allen *et al.*

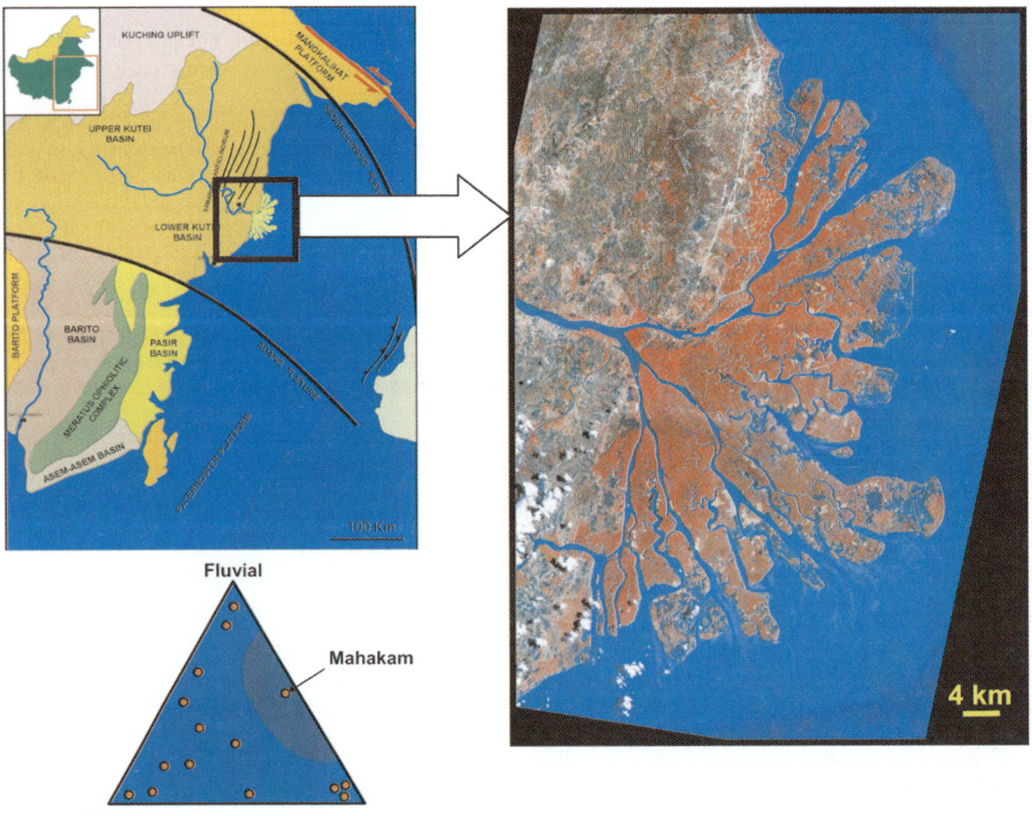

Fig. 1. Location and physiography of the modern Mahakam Delta.

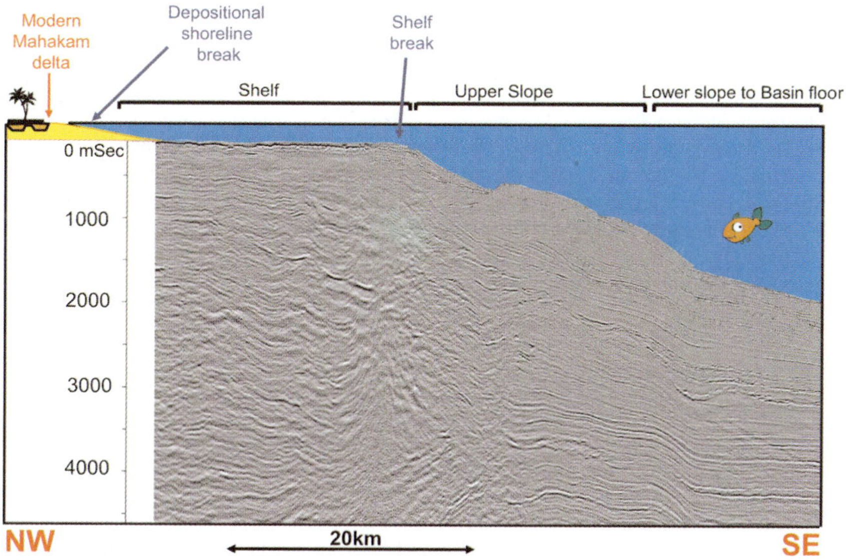

Fig. 2. The modern Mahakam delta and its adjacent shelf profile as revealed by a seismic profile. The shelf break is at −110 m water depth.

1979; Gastaldo and Huc 1992; Allen & Chambers 1998).

The sediment supply to the modern Mahakam delta has been described as moderate with a 3000 m^3 s^{-1} discharge with a tropical monsoonal climate associated with a moderate tidal range and a very low wave energy (Figs 1 and 2).

A detailed description of the modern depositional environment was given by Allen *et al.* (1979) in previous studies of the subaerial delta plain and the shallow water delta front of the Mahakam. The Holocene facies were studied and described using aerial photos, shallow coring and sea bed samples. The sedimentology and the stratigraphy of the Pleistocene section were studied using the data collected with the Misedor boring (Carbonel & Moyes 1987; Caratini & Tissot 1987).

The modern Mahakam Delta has prograded 20–40 km during the last 8000 years over the Holocene maximum flooding surface. The subaerial delta has a lobate symmetrical geometry with a large intertidal delta plain composed of organic clay incised by distributary and tidal channels (Fig. 1).

Modern distributaries are rectilinear and sand-filled, while the tidal channels are sinuous and contain organic-rich mud as they are unconnected to fluvial feeders. In the distal region, the distributaries are more tide-dominated and grade progressively to sandy mud.

Contrary to the deep Miocene section cored in exploration wells, the modern delta lacks alluvial levees and fluvial overbank deposits, which is perceived as the more likely effect of tidal currents associated with a constant fluvial discharge.

The late Pleistocene Mahakam lowstand delta

The late Pleistocene Mahakam lowstand delta is located seaward from the modern delta on the shelf and down to the shelf break (Fig. 2). This lowstand delta was analysed using the shallow portions of several 3D seismic volumes combined with several 2D high-resolution profiles in order to study the depositional geometries, and the morphological changes occurring throughout the relative sea level variations during the Quaternary glacial cycles.

These 100 ka duration cycles played a key role on the ultimate stratigraphy and also the distribution of potential reservoirs on the Mahakam shelf and also on the equivalent deepwater deposits. These deltaic cycles and their associated turbidite deposits were first recognized and dated with a limited amount of 3D data by Suiter (1996) on the shelf and also more recently described in great detail by Posamentier *et al.* (2000) and Saller *et al.* (2004) over a limited area covering the northern part of the Mahakam shelf.

The late Pleistocene lowstand deposits and the modern Mahakam Delta deposits record an evolution from an initial sea level rise followed by a period of sea level fall ending with a rapid transgression and the modern highstand delta deposition as inferred from late Pleistocene global oxygen isotope curves (Bard *et al.* 1990).

The lowstand deltas are characterized on 2D profiles by prograding clinoforms and associated incised valleys reaching the shelf break and developed beyond the seaward limit of the highstand delta. The depositional geometries of these lowstand deposits vary laterally as a result of relative sea level changes and subtle fault tectonic activity with two separate main depocenters identified both with 3D seismic amplitude maps and more regionally constrained by the larger mapping derived from the 2D high resolution profiles.

As a result, the understanding gained from the morphologies can provide good analogies for fields or prospects occurring in the same physiographic position. This study was particularly focused on the understanding of the development of these lowstand deltaic deposits accumulated on the shelf and the associated incised valleys formed during the high order glacio-eustatic cycles.

Methodology and data summary

Methodology

The observations made with the combined 3D and 2D seismic data sets allows us to describe the depositional architecture and the sedimentary facies in 3D of the incised valleys and the lowstand deltaics related to the last glacio-eustatic cycles. The distribution of the depositional systems was mapped using more than 1500 km of 2D seismic profiles and several tens of square kilometres of 3D seismic.

It is well established that high-order glacio-eustatic cycles are playing a key role on the Pleistocene stratigraphy (Carbonel & Moyes 1987). During these 100 ka cycles on the Mahakam Delta shelf, the distributary channels developed during the previous highstand delta are incised and converted into deeper incised valleys with characteristic adjacent dendritic tributaries particularly well expressed in areas experiencing fault tectonic uplift. These areas are located in the central and northern part of the shelf while, to the South, rapid subsidence and the formation of a shelf-perched depocentre prevented distributaries from developing pronounced incisions.

Over and around the Mahakam blocks operated by Total, several adjacent 3D seismic surveys have been acquired, covering a large part of the shelf beyond the modern Mahakam Delta and also straddling the shelf break and the slope. For this study, several continuous horizons have been picked manually or tracked automatically in the shallow portions of these 3D seismic surveys. These horizons revealed spectacular palaeodepositional environment morphologies which are described in the following paragraphs. The 3D images, the associated palaeomorphologies and the observed sinuous shapes are compared and matched with the detailed internal sedimentary geometries seen with 2D high-resolution seismic lines.

Some comparisons are provided with the modern Mahakam Delta, which tends to be abusively used as an oversimplified analogue to the shelfal lowstand Miocene reservoirs in the Kutei basin. The variety of sinuous shapes observed on 3D seismic images on the Mahakam shelf are illustrated in this paper and should be looked at with caution and not interpreted solely on the base of morphologies with simple comparison with those observed in the modern Mahakam Delta.

Nonetheless, the stratigraphic model derived from the combined study 2D and 3D seismic data sets was calibrated using bottom samples, cores and data recovered from several platform site surveys borings shows that stratigraphic architecture of the Holocene/late Pleistocene Mahakam deltaics reflects the subtle interaction of eustacy, subsidence and synsedimentary tectonics which control alternating periods of deltaïc clastics progradation and carbonate build up deposition.

Study area and data summary

The study area is covered with several large 3D seismic surveys covering a surface of several tens of square kilometres and also with several 2D high-resolution seismic profiles located on the shelf offshore of the modern Mahakam Delta (see Figs 3 and 6).

Despite the lack of homogeneity and some misties between the different 3D seismic surveys it is

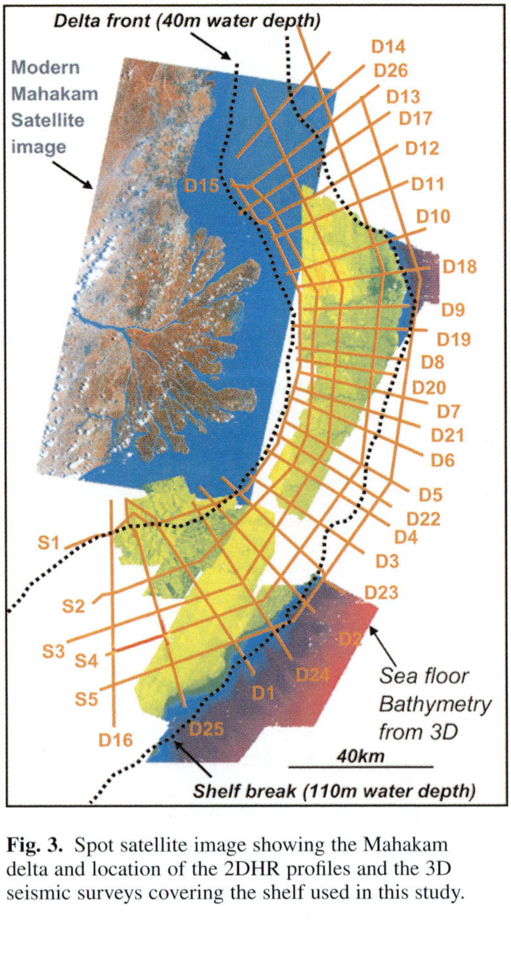

Fig. 3. Spot satellite image showing the Mahakam delta and location of the 2DHR profiles and the 3D seismic surveys covering the shelf used in this study.

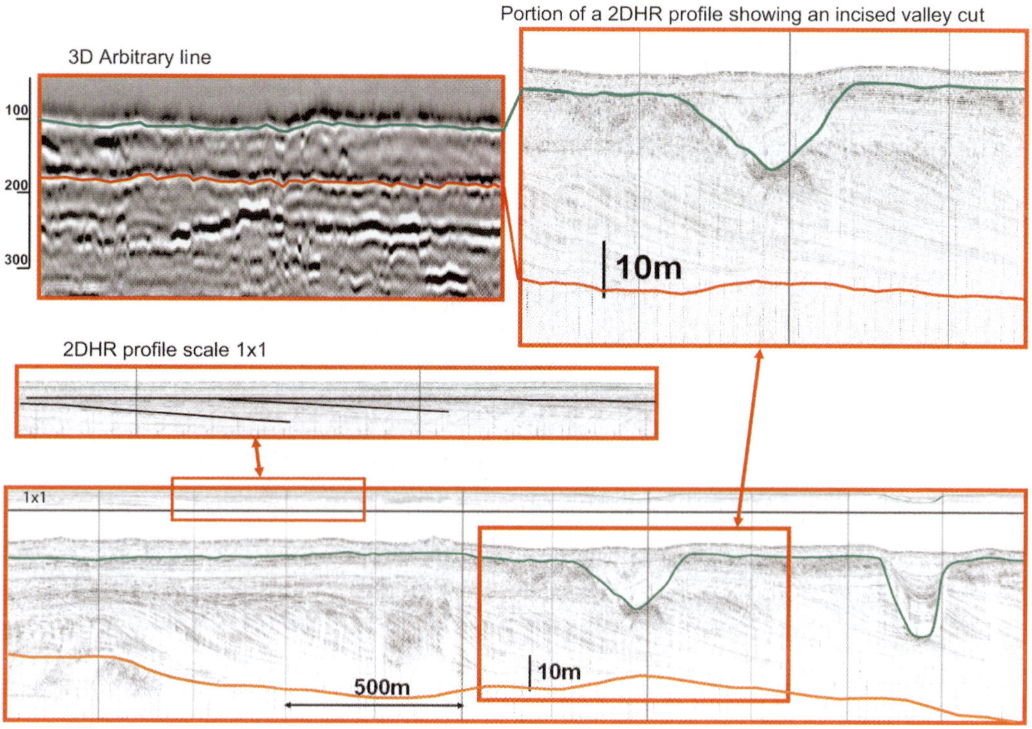

Fig. 4. Comparison of a random line and a 2D high-resolution profile (green horizon near sea bed imaging the 8000 years (TS1) lowstand topography and red horizon (DLP1) illustrating the 120 000 years lowstand topography.

possible to merge and pick the same main depositional surfaces on the 3D surveys and on the 2D high-resolution seismic profiles distributed on the shelf (Fig. 3). Figure 4 shows an example of a portion of a 3D arbitrary line compared with a 2D high-resolution profile over the same feature.

The 2D high-resolution seismic profiles were acquired by Louisiana State University in 1992–1993; the survey is widely spaced on a grid covering most of the offshore area between the modern delta and the modern shelf edge. These data, along with more recent 2D lines, were studied in great detail by Roberts & Sydow (2003).

The 2D seismic profile grid is spaced 5–10 km and was acquired using a Geopulse boomer and a Seismic System Water Gun. The Geopulse system produces a wide-band signal of 400–14,000 Hz with a resolution around 2 m and a penetration of 250 ms two-way time.

The 2D seismic profiles available in numerical format were analysed using the Total's in-house Sismage™ Research Software. These 2D HR lines were positioned on the 3D survey grid to display the portions of the sections which overlap the 3D seismic.

While being very different in terms of horizontal and vertical resolution, the 3D seismic and the 2D lines complement each other in many ways and their combined use allows a deeper understanding of the Makakam shelf stratigraphy.

The 2D lines are of variable quality due to acoustic wipeout problems but the vertical resolution is about 2 m compared with the theoretical resolution of the 3D seismic, which is around 20 m at these shallow depths. The difference in lateral resolution is also obvious between the 3D volume data, having a mid-point spacing of 12.5 m while the 2D is based on the overall grid spacing of 5–10 km.

The erosive cuts seen on 2D with great are interpreted as incised valleys. They appear as v-shaped channel features in the plane-view images from the 3D seismic. These data were gathered in order to understand the continuum of morphologies and facies from non-marine to deepwater environments developed during lowstand sea level periods (see Fig. 5).

The merged 3D seismic shallow portions of the entire Late Pleistocene Mahakam shelf were compared with new 2D high-resolution profiles and also with sea bed sonar images. The 2D HR data are used mainly as a tool for looking at the internal geometry of the depositional elements while the 3D data are used to display images of the external architecture and reconstruct the large-scale geometries.

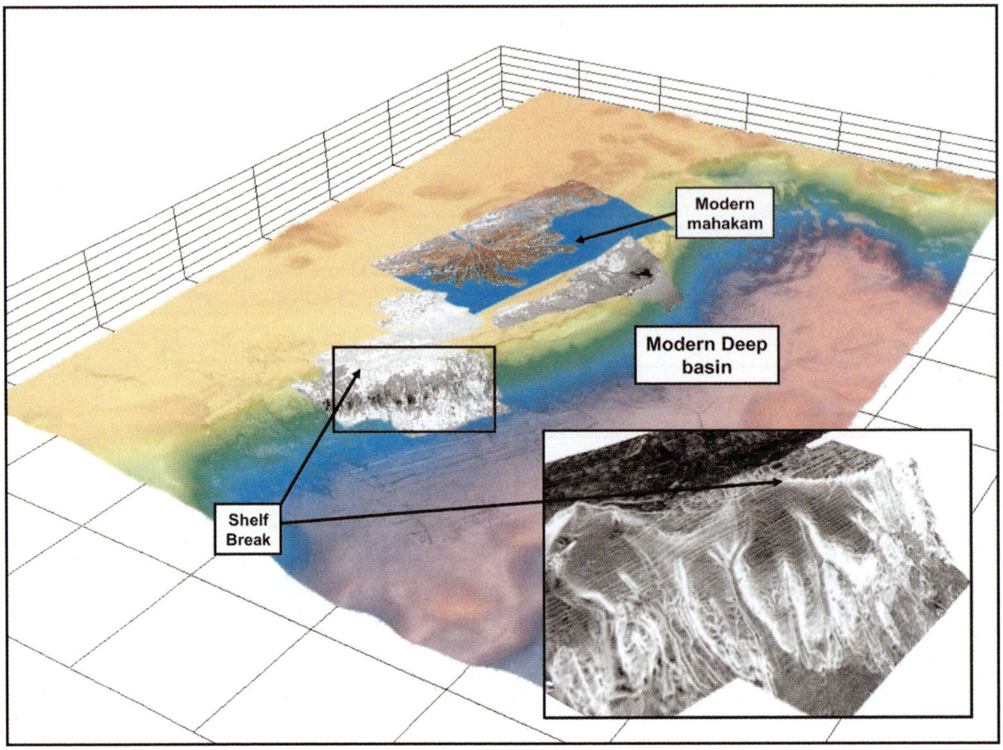

Fig. 5. Sea bed morphology illumination derived from various data illustrating the shelf break morphology and the main physiographic domain of the Mahakam delta area. The areas studied in greater detail are in grey. Bluish to pink areas represent the lower slope to basin plain domain.

The comparison of these data highlights (see Figs 6 and 8) the limits of stratigraphic and sedimentological interpretations derived from conventional 3D seismic slices. While being very different in terms of horizontal and vertical resolution, the 3D seismic and the 2D HR lines complement each other in many ways. The erosive cuts seen in great detail on 2D HR can be interpreted as incised valleys with greater confidence by the supporting 3D images. It is thought that these cuts were formed by rivers extending onto and cutting into the shelf break during late lowstand with the sediment load being discharged into the deep basin by long-lived turbidity currents.

The 2D data are thus used mainly as a tool for looking at the internal geometry of the depositional elements while the 3D data is used to display images of the external architecture (i.e. palaeolandscape).

Stratigraphic interpretation

Holocene/Pleistocene stratigraphy and mapped horizons

Data from the modern Mahakam highstand delta and seismic images from the late Pleistocene lowstand furnish a complete 3D picture of the non-coeval, laterally juxtaposed succession of depositional systems, which developed during a glacioeustatic depositional sequence during successive lowstand, transgressive and highstand periods of sea level changes of only 100 ka duration.

The Holocene section equivalent to the modern Mahakam Delta is a relatively thin, muddy blanket deposited above the erosional Pleistocene Holocene transgressive surface (TS1) clearly identified on the 2D high-resolution profiles. Radiocarbon dating indicates that these transgressive deposits are younger than 12 ka before present (Saller *et al.* 2004).

The late Pleistocene succession has been subdivided using continuous correlatable surfaces traceable both on 2D high-resolution profiles and on 3D seismic arbitrary lines. These surfaces were tracked semi-automatically to produce various time structure and seismic attribute (dip azimuth, amplitude, coherency, spectral decomposition) maps which provided a significant improvement of sedimentary imagery.

The most correlatable surfaces are the downlap surfaces (Figs 4 and 7), which have been labelled from top to bottom: seabed, TS1, DLP1, DLP2. The DLP1

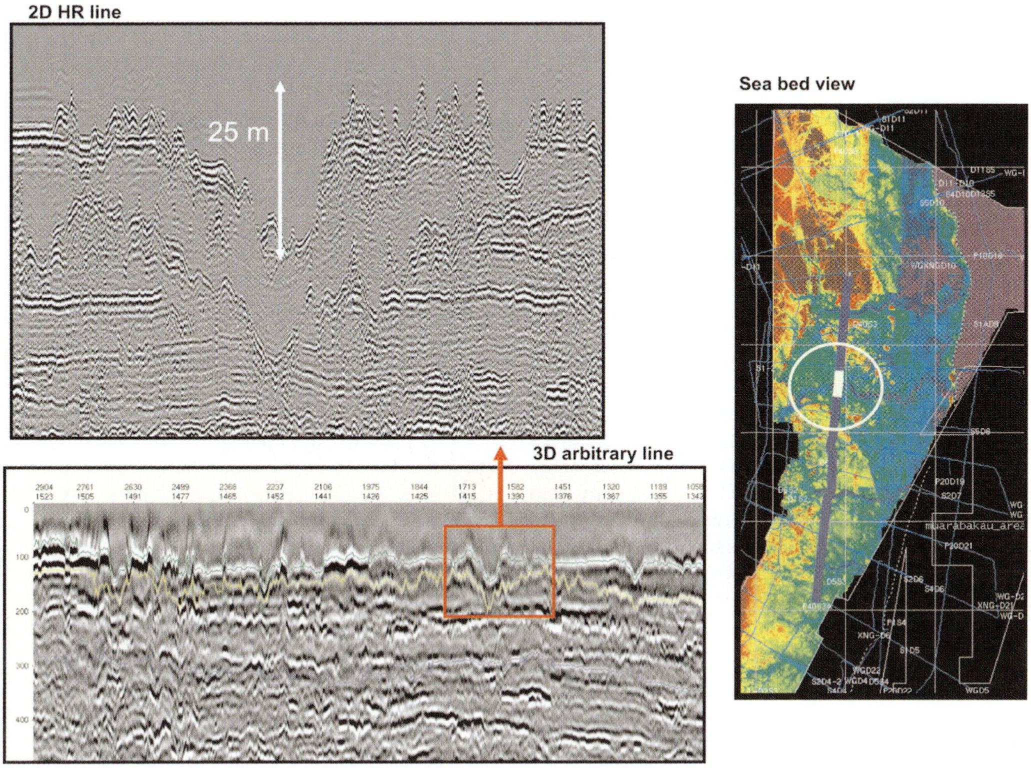

Fig. 6. Comparison of a 2D HR profile and a 3D random line illustrating the morphology of the erosive cut formed by an incised valley and the corresponding 3D image (DLP1). The white line indicates the portion of 2D and 3D random line displayed to the left of the picture.

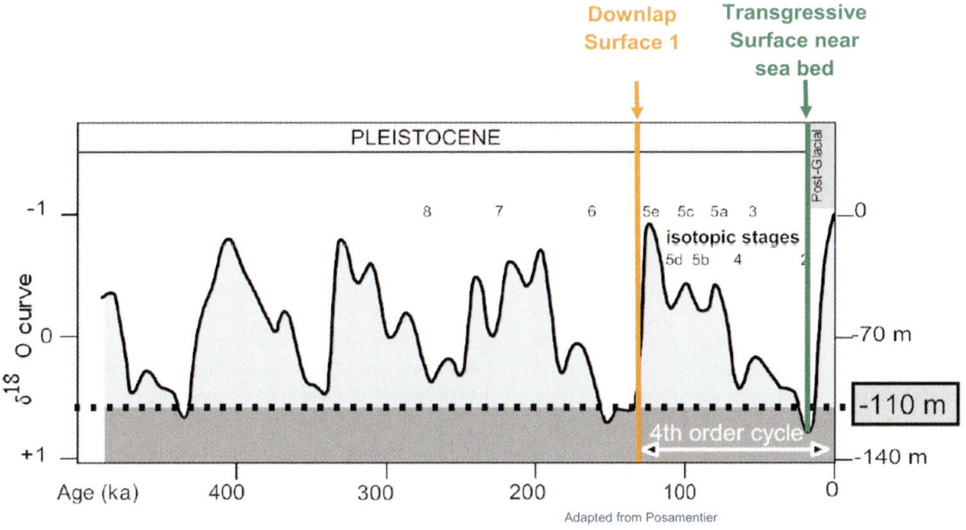

Fig. 7. Late pleistocene to Holocene sea level curves based on oxygen isotope data (Bard *et al.* 1990, drawing adapted from Posamentier 1999) and tentative calibration of the downlap surface 1 and transgressive surface corresponding to the surfaces traced on the seismic lines of Figure 4.

and DLP2 surfaces correspond to the downlapping base of the lowstand deltas clinoforms seen on the 2D seismic lines, which are also expressed as more subtle but continuous horizons on 3D seismic.

Similar to the modern seabed, the buried downlap surfaces are irregular surfaces formed by the carbonate mounded buildups being buried by the prograding modern Mahakam delta front deposits.

The TS1 is a horizon just beneath or concordant with the sea bottom over the major part of the 3D surveys where the Holocene sediments are thin. This horizon gives a clear image of the 18 ky lowstand morphology formed before present sea level transgression. The seabed surface has thus a morphology largely inherited from the last glacial lowstand period.

At least three main distinct lowstand deltaic wedges are seen on 2D profiles but, due to the quality of the data, only the last lowstand wedge located between the seabed horizon and the Downlap Surface 1 was mapped (Roberts & Sydow 2003) and compared with 2D lines. This latest cycle, called P1 in the work of Sydow and Roberts, comprises two main depocentres developed in the northern and southern parts of the shelf.

An attempt to calibrate the late Pleistocene events was made using the published glacial sea level cycles derived from the oxygen isotope data from deep-sea foraminifera. These record the history of global continental ice volume during the Quaternary time in relation with climate changes. During Quaternary time, climate is changing cyclically in relation with 400, 100, 40 and 20 ky duration cycles, which are known as Milankovitch cycles.

Published radiometric C14 dating indicates a more likely age of less than 8000 years for the Mahakam Delta. The transgressive sediments are dated 9000 years underneath the modern prodelta sediments; thus the top of the late lowstand could be around 18 000 years old. Below the Downlap Surface 1, the pronounced transgressive event could be equivalent to the 135 000 years sea level rise of the Lalou chart. Thus the duration of the first lowstand wedge cycle could be in the range of 100 000 years (see Fig. 7).

The seismic architecture of the late Pleistocene shelf perched delta is characterized by an alternance of packages of seaward prograding clinoforms downlapping on top of mounded bioherms. Clinoforms packages are capped by horizontally stratified beds with a network of nested channels (Fig. 10) or abruptly truncated by large erosional features: incised valleys (Figs 6 and 11) and finally by extensive mounded or isolated solitary pinnacles.

Clinoform geometry is interpreted to represent the depositional profile of a prograding lowstand comprising delta front and prodelta deposits. The youngest prograding clinoforms are presently buried by the modern Mahakam prodelta muds in water depths of 30–100 m. The individual superimposed deltaïc clinoforms do not occupy the entire shelf and tend to stack in a compensatory fashion. A detailed description of the deltaic lobe stacking pattern was given by Sydow (1996).

The horizontal-bedded clinoform tops were deposited in a 5–10 m bathymetry equivalent to the modern Mahakam delta plain and shallow delta front.

The isolated channels nested in parallel-bedded strata (see the southern part of the map in Fig. 12) are 7–10 m thick and show possible side-bars development on 2D lines. These are interpreted as delta plain to delta front distributary channels and occur preferentially in the Southern depocentre. These sinuous shapes are clearly imaged forming a network of erosive amalgamated shoestring channels less than 1 km wide with a contrasting 3D morphology compared with the incised valleys showing pronounced dentritic patterns adjacent to tributaries and only present in the central and northern part of the Mahakam shelf (Fig. 12). Similar dentritic tributaries were recognized on the Java shelf by Posamentier (2001).

The northern clinoforms present a clear toplap geometry on 2D lines truncated by incised valleys while the southern depocentre clinoforms show a progressive change from clinoforms with toplap geometry passing into parallel bedded strata truncated by delta plain channels. This coeval contrasting style of deposition is related to the variation in subsidence occurring during the last 100 000 year cycle. The southern part is subject to pronounced active subsidence while, to the north, uplift faulted panels enhanced fluvial incision and palaeosoil development in the interfluves (Figs 8 and 12); (Palaeosoil photography from Corindon oceanic campaign, Boichard *et al.* 1988).

Transition from shelf to slope. The lowstand deltaic wedges are developed near the shelf break during lowstand progradation. During transgression the shelf break is marked by the *Halimeda* algae buildups forming solitary pinnacles with abundant encrusting coralline algal nodules up to 40 m high (Fig. 11). They also may be seen forming irregular patches with thicknesses from 20 to 35 m.

Halimeda green algae are described as photosynthetic organisms living in the photic zone and requiring clear water conditions for their development. These conditions prevailed during transgression on the middle of the shelf and especially in the shallower part of the uplifted faulted blocks of the northern and central part of the shelf.

The *Halimeda* bioherms formed during the last transgressive period are now being buried by the prograding clastics of the modern highstand Mahakam delta. Modern *Halimeda* buildups are less active now as a result of the rapid sedimentation rates of the Mahakam delta, which brings a lot of turbid sediment to the shelf. Cored sections recovered from the carbonate mounds encountered a thin drape of prodelta clastic mud on the top of the mounds.

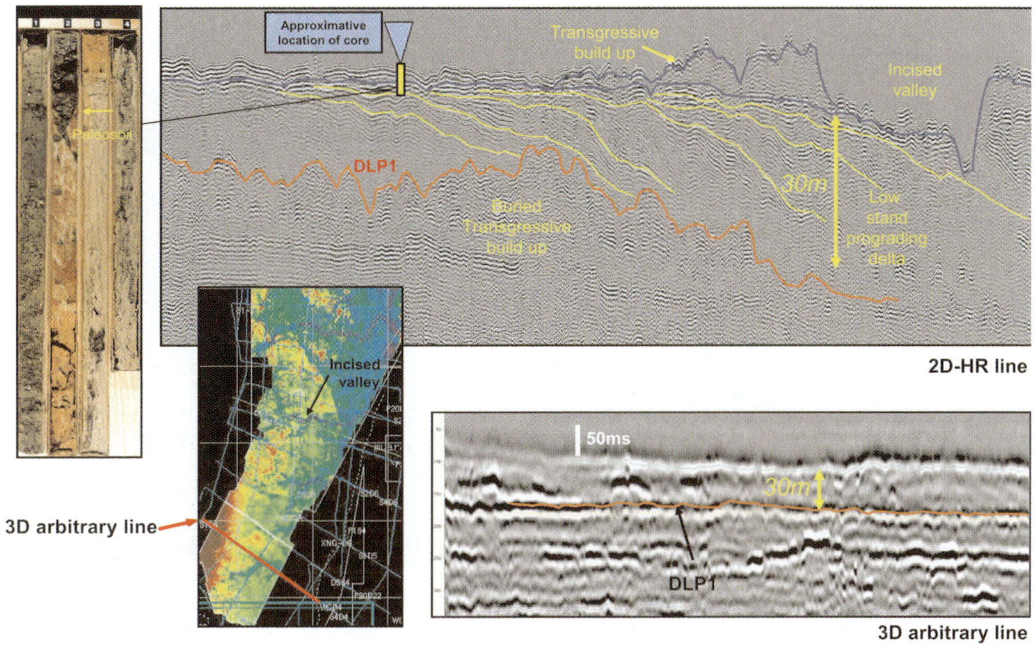

Fig. 8. Comparison of a 2D HR profile and a 3D random line illustrating the clinoforms associated with lowstand deltas progradation and comparison with 3D line. A cored section recovered in the interfluve of incised valleys is demonstrating the emersion of the Mahakam shelf.

The morphology of the shelf break is also controlled by minor faults as seen on southern part of the Mahakam shelf sea floor image and in the northern part of the external axis merged 3D. A large volume of sediments has moved down-slope as debris flow and turbidites as a result of the creation of shelf instability features such as slump scars. This is due to the steep nature of the shelf break and the syn-sedimentary movement of faults controlling the geometry of the shelf break.

Sea floor images and horizon maps also revealed that river incision occured near the shelf break during

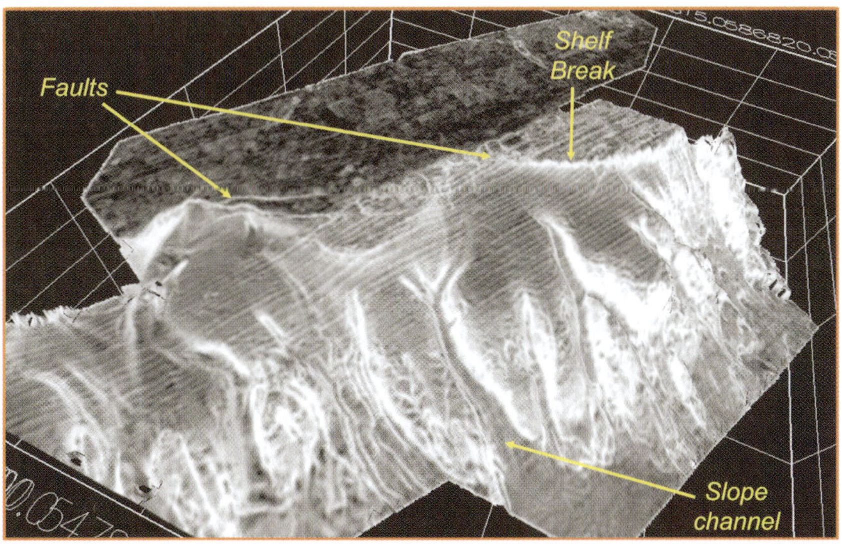

Fig. 9. Slope turbidite channels adjacent to the shelf break.

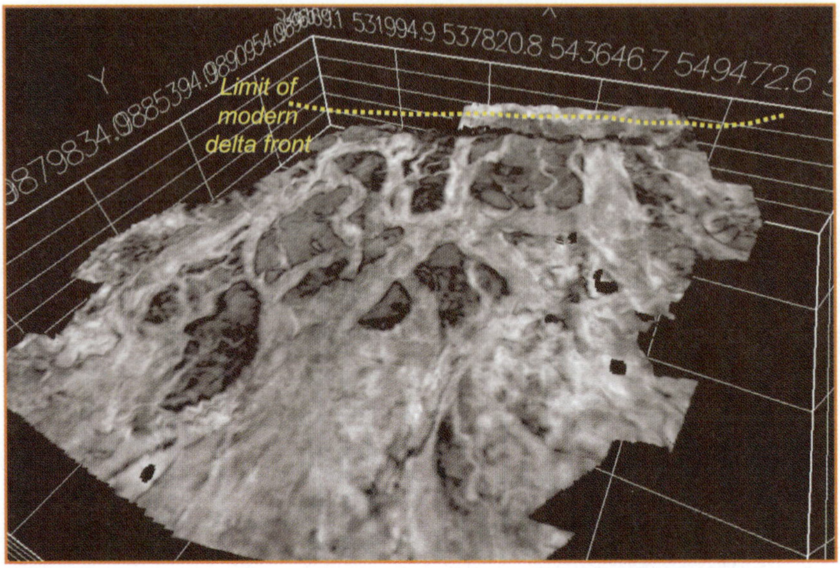

Fig. 10. Detail 3D view showing a network of erosive channels developed on the southern part of the Mahakam shelf located updip from the slope channels shown in Figure 9.

the late lowstand period. As rivers reached the shelf break they delivered their sediment discharge directly into the deep basin as tur bidite flows (Figs 9 and 11).

Systems tracts of the quaternary depositional sequences

The combined interpretation of 2D and 3D seismic data imaging part of the modern Mahakam highstand delta and a major part of the late Pleistocene lowstand furnish a complete 3D picture of the noncoeval laterally juxtaposed succession of depositional systems which developed during the last glacioeustatic cycle with successive lowstand, transgressive and highstand periods of sea level changes. The observation of this interpretation allows the following discussion around each of the distinct periods of deposition.

Highstand period. On the shelf, prograding deltaic sediments formed during early lowstand periods are

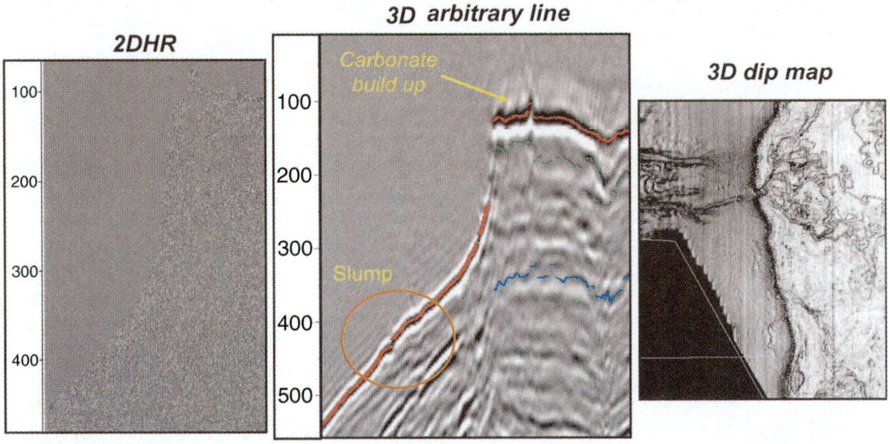

Fig. 11. Comparison of the slope morphology imaged with 3D and 2D profiles and a 3D image illustrating the connection between fluvial lowstand and turbidite channels. Note that in these sections east is to the left.

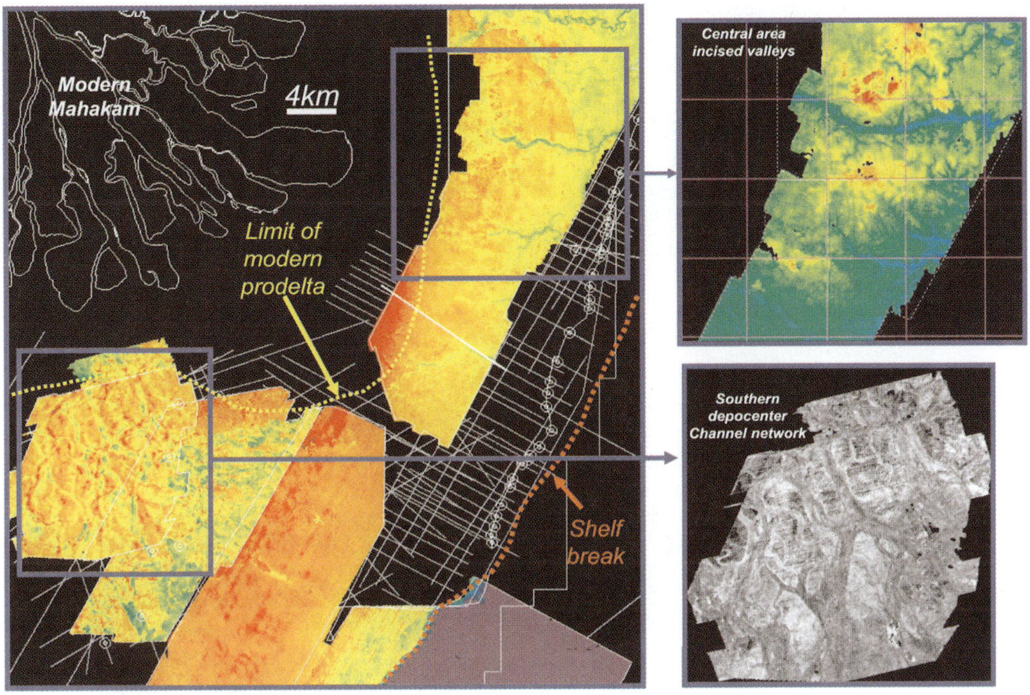

Fig. 12. Contrasting style between coeval incised valleys in the northern and central part of the shelf and the sinuous network of channels developed in the southern Mahakam shelf.

volumetrically the most significant component of the Quaternary Mahakam depositional cycles compared with highstand deposits. Highstand deltaic deposits are related to the early phase of progradation and are restricted to a wedge of sediments developed landward from the lowstand deltaïcs. These are partly cannibalized during the lowstand periods when distributaries formed during highstand are converted into incised valleys. These incised valleys represent by-pass conduits active during late lowstand period with fluvial-derived hyperpycnal currents flowing directly into the steep slope to feed turbidite systems. The interfluve of these incised valleys is subject to emersion and is interpreted to become the site of palaeosoil development.

Early lowstand period. The well-expressed regionally mappable surfaces which have been labelled downlap surfaces correspond to the maximum flooding surfaces draped during highstand by a thin horizon of shale and buried by downlaping clinoforms during early lowstand period. The progressive decreasing elevation of toplapping clinoforms is

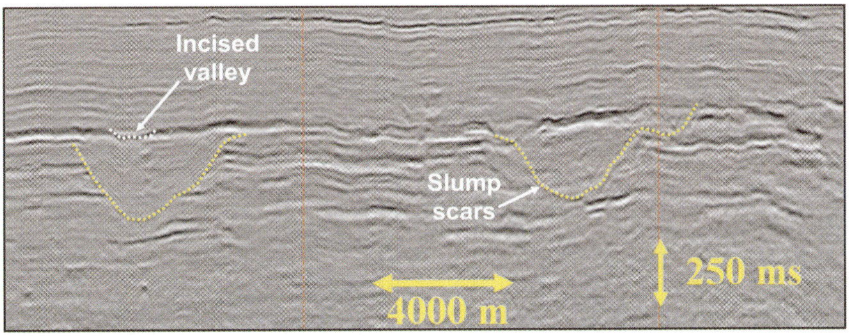

Fig. 13. Large slump scars and incised valleys.

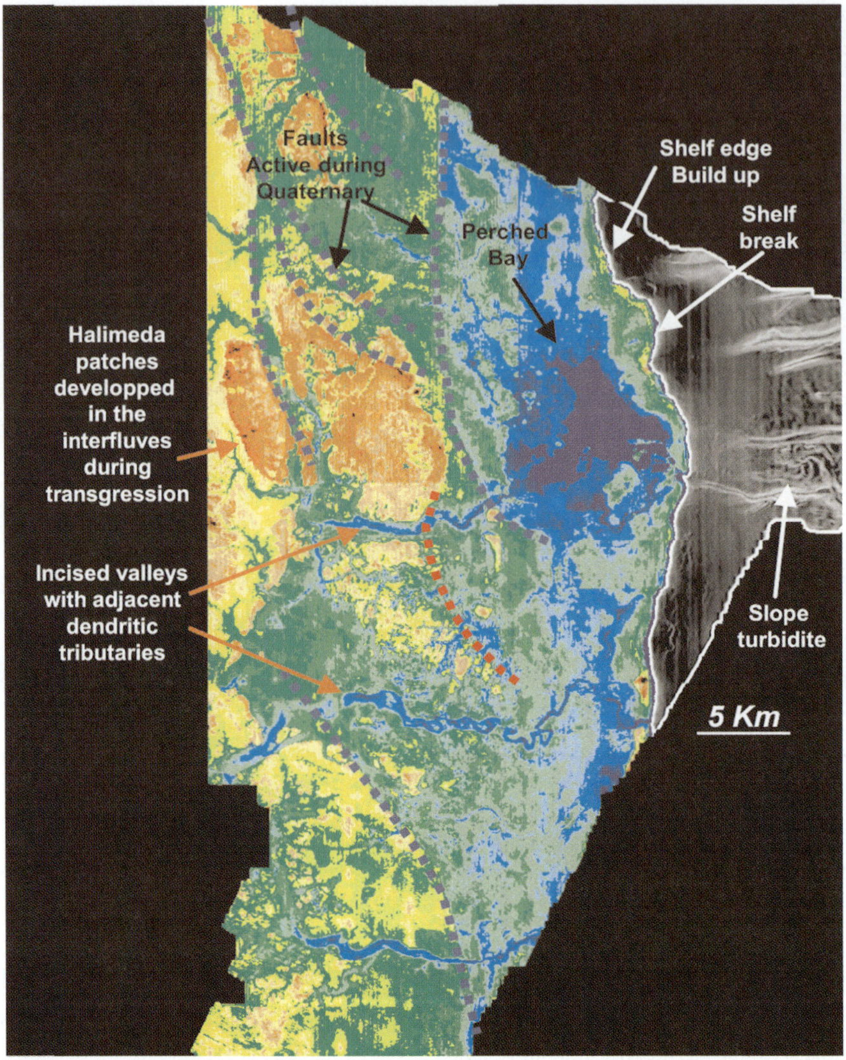

Fig. 14. Detail view of a perched bay embayment feeded by fluvial incised valleys developed on the northern part of the Mahakam shelf during lowstand period. Note the morphological expression of faults on the seabed.

obviously related to successive base level lowering modified by local structural deformation in relation to normal faulting (see fault traces on Fig. 4).

According to the classical sequence stratigraphy terminology, the sequence boundary by definition separates the highstand systems tract (HST) from the overlying lowstand systems tract (LST).

The progradational clinoform located landward and having higher elevation belong to the HST until a relative sea level fall occurs at the depositional shoreline break (Posamentier & Allen 1999).

Theoretically, the first deltaic clinoforms that shift seaward with a toplap surface forming at a lower elevation compared with highstand clinoforms should correspond to the early sealevel fall. These clinoforms are progressively overridden and incised by the advancing and downcutting fluvial system as the delta progrades during the later part of sea level fall. Thus, part of the earlier formed clinoforms could eventually be situated below the major sequence boundary. Contrarily to 3D sections, 2D HR seismic profiles clarify this question, indicating that the sequence boundary extends below these clinoforms and extend seawards from the downlap surface.

This conformable sequence boundary has been called the 'master' sequence boundary by Posamentier & Allen (1999). As already noted,

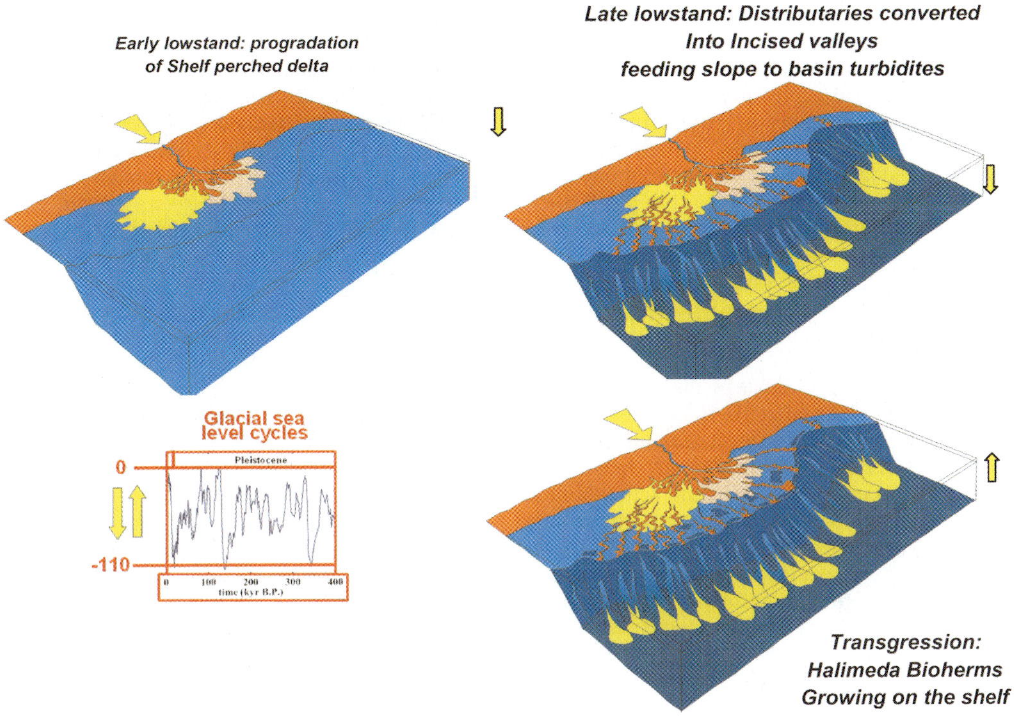

Fig. 15. 3D conceptual model of the Mahakam delta development during a glacial sea level cycle. The highstand delta is converted into perched lowstand deltas that are progressively incised by distributaries converted into valleys as the sea level drops to the shelf break, finally feeding directly turbidite fans.

there are no obvious criteria with low-resolution seismic data and when looking at a part of 3D arbitrary line that could be used to separate clinoforms formed prior to or post fluvial incision (see Fig. 8).

Over the Mahakam Delta area, progradational clinoform deposits on the inner to middle shelf, which potentially formed during the actual highstand in sea level, have a lower preservation potential compared with those formed during forced regressions. Shallow water depths on the inner and middle shelf (less accommodation) usually imply that the prograding delta deposits in the shelf are relatively thin compared with highstand deltas. Erosion by the incising alluvial valleys (including localized drainage on the subaerial exposure surface) and also by transgressive ravinement could remove a large portion of the landward-thinning clinoform wedge developed on the shelf.

This should be reinforced in basins with smaller subsidence rates compared with the Mahakam basin, and the lowstand deltaic will then appear as detached isolated sand packages on the shelf (see the discussion in Posamentier & Allen 1999).

In this study, transgressive and incised fluvial facies are relatively more easily identified than lowstand deltas on the shelf, particularly when dealing with low-resolution 3D seismic data.

Late lowstand period. The fluvial deposits associated with incised valleys on the shelf belong to the lowstand systems tract. Also incorporated in the late lowstand is a significant associated volume of turbidite deposits that are deposited when fluvial incision reached the shelf break.

There are also several erosional canyons identified on the upper slope in continuation with some of the fluvial valleys which incise the morphologic shelf break. Many of the canyon-like features developed during late lowstand seem to be connected up-slope to a fluvial incised valley feeder system and down-slope develop a fan shaped accumulation that is finally dissected by a network of sinuous channel features. These latter indicate a rapidly decreasing efficiency of the turbidite system concomitant with the starting of transgression on the shelf. The stratigraphically lower part of the basin-floor fan is formed by broad lobes with relatively continuous reflectors in section. The stratigraphically higher part is formed by a sinuous probably muddier channel–levee complex that accumulated over lower sandier fans. The stratigraphic evolution of an individual fan system was described with great detail by Saller *et al.* (2004) further north along the Mahakam shelf.

A shelf-perched depression tectonically induced with a bay-like geometry is visible in the northern part of the external axis (Fig. 14), which demonstrates syn-sedimentary fault-influenced deposition during lowstand on the shelf. This small shelf-perched embayment is fed by one of the identified sinuous incised valleys clearly identified on the external axis. Seawards, the tectonically induced embayment is bordered by a belt of transgressive *Halimeda* reefs, mainly built up during ensuing transgression. This pronounced effect of tectonic activity is particularly visible on the northern part of the Mahakam Delta affected by several faults in relation with a regional deep seated fault system.

To the south it seems that steady more active subsidence occurs at the same time with less faulting. This continuous subsidence activity seems to have prevented rivers from being strongly incised and the resulting geometry of channels differs also significantly from the northern incised valleys. Observed channelized features form smaller shoestring channels isolated into parallel bedded facies similar to distributary channels associated with delta plain.

These channels were formed during less prolonged period of emersion with less amount of incision. This is also demonstrated by the lack of dendritic erosive pattern, as is usually observed with incised valleys.

The southern more subsiding depocentre is seen as an aggrading delta plain recording a continuous deposition with less erosion and by-pass compared with the northern more tectonically induced area where incised valleys are seen.

The Quaternary sequence of the Mahakam Delta display a complete 3D scenario of a high sediment-supply fluvial system incising the morphologic shelf break with morphological lateral variations as suggested from the two classical end-members models (type 1 and type 2) of sequence stratigraphy.

The actual shelf-break depth in the Mahakam Delta (75–110 m), as well as in many other parts of the world, corresponds fairly closely to the sea level minimum estimates (−110 m) for the last glacial age.

Some large canyons features are also seen at the deeper horizons (Fig. 13). These are thought to be the result of massive slope failure and slumping in the structurally unstable shelf margin vicinity. They form large depression whose depth is greater than any sea level drop recorded in the Quaternary. It is thus suggested that sediment loading and oversteepening of the upper slope and outer shelf by the deltaïcs also occasionally initiate large-scale mass wasting and canyon formation.

Transgressive period. As sea level rises quickly, the rivers start filling the incised valleys, which were rapidly abandoned as transgression continued. *Halimeda* bioherms then grew on the transgressive ravinement surface formed by the interfluve of the late low stand incised valleys. Also, as transgression occurrs quickly, incised valleys remain underfilled and the troughs created by rivers remain empty, partly filled by muddy sediments. As transgression continues, some *Halimeda* keep growing, forming pinnacles with up to 40 m relief.

The Holocene transgression was particularly rapid and extremely important in terms of relative sea level rise (about 110 m). While looking at 2D lines, a thin drape of sediment is deposited in the moribund conduits left empty by incised valleys. It is clear that there was not enough sedimentation to fill the existing topography as rapid transgression occurred. This is demonstrated by shallow continuous coring performed for Total platform site surveys (not shown here). Only few metres of sand were encountered at the base of these incised valleys. The rest of the fill was constituted by marine shaly sediments. Thus the lowstand incised valley morphology was largely preserved and draped in a transgressive blanket of shale, and finally buried when highstand clastics prograded again. The transgressive patches of Halimeda reefs formed on the shelf are preferentially developed on the shelf in structurally controlled highs located in the interfluve of incised valleys along the uplifted part of faulted blocks and along the shelf break.

Summary of the Mahakam 3D morphological evolution during a glacial age depositional sequence

This combined study of 2D and 3D seismic data allows a precise reconstruction of the morphology, and the internal depositional geometries and the timing of events leading to incised valleys formation and lowstand deltas progradation over the shelf contemporaneously with the associated development of deepwater turbidite channels (Fig. 15). Tracking of continuous markers such as downlap surfaces or maximum flooding surfaces and imaging of seismic attribute maps permitted a three-dimensional reconstruction of the palaeoenvironments and palaeomorphologies of the lowstand system tract never obtained until now.

During relative sea level fall period, Suiter (1996) demonstrated with 3D images that distributary channels are converted into incised valleys with adjacent characteristic dendritic tributary channels. These features, previously interpreted as shelfal tidal channels and hard to map with the 2D seismic grid, clearly indicate that the northern part of the shelf was entirely exposed over a prolonged period of time and flooded during the following Holocene rapid sea level rise. Few main incised valleys reached the shelf break, bringing sediments beyond the shelf break.

During the rapid sea level rise, the incised valleys are rapidly flooded and remain largely underfilled with a thin drape of marine transgressive sediments while the interfluves are colonized by mounded Halimeda calcareous bioherms. Growth and fusion of Halimeda bioherms continue until the next high stand delta progradation. The distribution of the bioherms is also strongly influenced by the presence of small tilted faulted panels on top of which Halimeda bioherms developed.

The effect of tectonics is also marked with pronounced incision of the valleys while crossing upthrown block and with a more sinuous unconfined profile in the downthrown compartments. Also the southern depocentre mapped with 2D lines shows a different architecture with thicker prograding perched low stand deltaic wedge capped with a network of channels showing limited incision and less abundant Halimeda build-up. This demonstrates the effect of larger scale shelf physiography on the coeval development of a southern thick deltaic lowstand deltas while to the north incised valleys appear as tectonically induced features.

The proposed Quaternary 3D lowstand stratigraphic model provides good analogies for fields or prospects showing channelized features located near the shelf break. In the next paragraph, the Mahakam shelf channelized morphologies are discussed in light of some outcrop examples from the cenozoic of Spain (Sobrarbe Delta sediments) trapped at river mouth.

Comparison of the sinuous morphologies seen over the Mahakam shelf with outcrop data

Turbidite channel-fills showing lateral accretion surfaces indicating a sinuous morphology were recognized in ancient turbiditic succession as a marginal feature before the arrival of 3D seismic imagery in divergent margins. One of the first examples described in the literature was by Mutti in the Apennines and in the South Pyrenees and discussed in a paper highlighting the difficulty of comparing modern with ancient turbiditic succession (Mutti & Normark 1991).

For that reason, turbidite channelized reservoir analogues described in the literature have been generally considered as having a rectilinear, low-sinuosity geometry associated with terminal sandy lobes contrasting with the muddier sinuous channel-levee modern analogues described by oceanographers (Mutti 1992).

With deep offshore exploration, recent 3D seismic data sets have highlighted that sinuous channels with spectacular internally complex meandering belts are a common feature of many cenozoic deep marine sections of divergent margins. These sinuous channel belts are associated with large muddy levees and a variety of lateral to frontal splays or lobes.

The South Pyrenean Eocene basin comprises a wide range of fluvial, shallow marine and deep-water clastic sediments. In this basin, channelized sandbodies showing lateral accretions were also recognized in various depositional settings: the Morillo deep-water turbidite channels, the Sobrarbe delta-front channels and the fluvio-estuarine channels of the Montanana sequence are among the best examples.

Despite the different style of the associated fluvio-deltaic sedimentation and basin-margin physiography of the South Pyrenean Foreland, sinuous morphologies appear as an important morphological feature, particularly with the channels belonging to the Sobrarbe Delta Formation and the Upper Eocene Morillo Formation (Hecho Group, Spain). These outcrops raised questions on the nature of processes that are able to generate laterally accreted sinuous channels not only in the deep marine environment but also on slope to delta front with a similar architecture to fluvial meanders.

The sinuous, relatively deepwater channels recognized in the Morillo and lower part of the Sobrarbe Formations lie seaward of their time equivalent lowstand deltas. In the Sobrarbe, delta front laterally accreted channels are also recognized in shallow water settings and are more likely ponded in a tectonically induced topography caused by syn-sedimentary thrust folding.

The shallower delta front channels can be differentiated based on facies analysis and with their stratigraphic relationships with deltas from those developed in the relatively deeper marine setting (Morillo Formation). This differentiation will remain difficult, particularly if similar channels were seen with limited exposures or with limited data without palaeobathymetric indicators. Despite some differences in facies, the overall geometry of these delta front channelized systems show some strong similarities also with the typical meandering fluvial systems, such as the classical example of Montanana sequence (see the upper photograph of Fig. 16).

The Sobrarbe channel-fills developed on the delta front show lateral accretion surfaces formed by parallel-bedded or graded sandstone beds, with flute casts and Bouma-like sequences, which can be misleading to establish the delta front environment and the water depth of this sub-depositional system (Fig. 17).

Fig. 16. Examples of fluvial, slope and turbidite sinuous channels showing lateral accretion surfaces: fluvial example (i.e. point bar) in the Montanana Formation; slope example in the Sobrarbe Formation; turbidite example (i.e. deep-water) in the Morillo Formation).

These channelized systems are considered as being mixed depositional systems built up by gravity flows produced by hyperpycnal flows as described by Mutti *et al.* (2003). These flows were forced to deposit their sediment load in topographic depressions or low-gradient slope segments in front of a depositional shoreline break formed by delta front and before acquiring sufficient acceleration to become efficient bipartite turbidity currents.

The laterally accreted channels described here developed in shallow to deepwater channels are probably related to a short period of time with sediment supplied by quasi-continuous hyperpycnal currents issued from fluvial feeders and passing to the delta mouth into deepwater. These flow conditions were relatively similar in term of processes to what was prevailing during the late lowstand over the Mahakam shelf with fluvial channels feeding directly over the shelf break the turbidite channels.

In the Morillo Formation lateral accreted channelised units are also seen forming several channel complexes approximately 20–80 m deep and 800 m to 1 km wide, which resulted from the lateral and vertical stacking of several single small channels showing lateral accretion occurring above a basal mass transport complex.

When occurring as individual isolated sinuous channels, these form lenticular isolated bodies 300 m wide and 3–10 m thick inbedded in muddy sediments with tabular thin bedded sandy facies having spectacularly preserved sand dunes (F_6 in the Mutti classification of turbidite facies).

An important characteristic of the Morillo channel fill is the presence of laterally accreted packages of strata showing a broad similar internal geometry as the shingled seismic reflections observed in the sinuous channels of West Africa divergent margin. These accretion packages recognized with 3D seismic have been named lateral accretion packages (LAPs) (Pirmez *et al.* 2000; Abreu *et al.* 2003) or internally LOSC (i.e. laterally-offset channels; Kolla *et al.* 2001).

Despite their smaller dimensions, the Morillo laterally accreted channels can be seen as a good analogue for understanding lateral migration revealed by 3D seismic. Very few other examples have been described in the rock record in shelfal to slope setting.

The combined occurrence of meandering delta front to slope channels in the Sobrarbe Delta front associated with instability shelf edge failures (slump scars) suggests that flood-related flows issued from fluvial channel could explain the relatively continuous nature of flows needed to generate and maintain such sinuous channels over a short period of time.

These hyperpycnal flows could be seen as submarine 'rivers', i.e. fluidal turbulent flows with enough

Fig. 17. Examples of facies forming the infill of laterally accreted channels in the Sobrarbe Delta front (case 2 of Fig. 17) with highly concentrated flows and cross bedded facies indicating by-pass similar to the F_6 of Mutti turbidite facies classification (Mutti 1992; Mutti *et al.* 1999).

fine-grained sediment concentration to form density currents. Most of the sand deposited in these meandering belts, exhibiting spectacular laterally accreted bars, could be partly the 'bedload' of these submarine 'rivers', resulting from the erosion and resuspension of underlying coarser-grained sediment deposited by dense sandy flows.

In summary, both small (shallower) and large (deeper) channel-levee complexes of continental margins can be compared with the described examples developed in the prodeltaic wedge of the Sobrarbe delta, a relative small fluvio-deltaic system fed by relatively coarse grained fluvial rivers.

These channelized systems spreading over short distance from river mouths, followed by deposition in shelfal and slope regions, also remain relatively smaller compared with the sheet sands or lobes connected to large canyons recognised in older stratigraphic intervals of the same basin.

1. In the Morillo turbidite system, outcrops revealed channelized sinuous turbidite channels with lateral accretion surfaces associated with a range of facies showing analogies to sinuous migrating deep-water channels complexes recognized with 3D high resolution in the cenozoic of West Africa (Kolla *et al.* 2001).
2. In the shallow marine deltaic sediments of the Sobrarbe Formation, sinuous 'turbidite like' channels showing lateral accretion were also recognized and interpreted as a result of flood-related hyperpycnal currents entering in the delta front (less than 50 m water depth).
3. In the fluvio-estuarine sediments of the Montanana sequence, laterally accreted channels were classically described as point bars.

Despite the significant difference in water depth and depositional processes, the described examples of laterally accreted channels show stringing facies and architecture similarities.

Lateral accretion surfaces occuring in the Morillo turbidite system are the result of the deposition of successive turbidite currents forming individual isolated sandbodies or larger channel complexes, mostly made of F_5 and F_2 facies in bedded with thin-bedded overbank wedges.

In the Sobrarbe delta-front, sandbodies with lateral accretion occur in bedded in slope to delta front nummulite-rich mudstones and siltstones disrupted by numerous truncations related to slump scar fill. These channels can be traced physically into flood dominated delta-front proximal and fluvial depositional systems. The description of these laterally outcropping turbidite or turbidite-like channels, ranging from deep to shallow water depth, allow understanding of the interaction of shelf edge failures and hyperpycnal flood-related currents derived from fluvial channels as a mechanism for sediment transfer from shelf to deep basin. These outcrop observations are analogous to similarly observed with 3D images over the Mahakam, keeping in mind some obvious differences in tectonic setting and glacial sea level cycle effects.

Conclusion

While being very different in term of horizontal and vertical resolution, the 3D seismic and the 2D lines straddling the Mahakam shelf complement each other to refine the classical 2D seismic-based sequence stratigraphy models and the associated depositional environments interpretation by establishing closer 3D genetic relations between fluvial, deltaic and turbiditic sediments using palaeolandscape morphologies.

Mahakam lowstand seismic geomorphology study can be compared with ancient sinuous fluvial to slope and deep marine channels outcropping in the Tertiary of the Spanish Pyrenees. This comparison emphasizes the interaction of various processes such as sea-level cycles, shelf-edge failures and the behaviour of flood/fluvial flows shaping the fluvial conduits that fed turbiditic systems during lowstand period.

Fluvial-derived flows by-passing the depositional shoreline break or the shelf break appear to be one of the main mechanisms for sediment transfer from shelf to deep basin both in the Mahakam and the Sobrarbe Deltas, despite the obvious difference in tectonic setting, slope dip and flow efficiency.

The various sinuous channel outcropping examples ranging from subaerial to subaqueous delta-front channels recognized in the Sobrarbe Delta can be differentiated based on facies analysis and stratigraphic relationships from sinuous turbiditic channels developed in deep marine settings. The differentiation of purely fluvial channels or incised valleys from turbidite channels could remain difficult if these sinuous shapes are identified only with morphologies seen with 3D seismic images and with limited geological information.

With the Sobrarbe Delta example, delta-front sinuous subaqueous channels are interpreted as being mixed depositional systems built up by gravity flows produced by the combination of sediment failures and hyperpycnal flows imbedded in slope to delta-front mudstone. However, contrary to the Mahakam lowstand delta, strong evidence of emersion in the interfluve is missing with the Sobrarbe delta.

The different examples of sinuous channels recognised with 3D seismic over the Mahakam shelf or with outcrop ranging from fluvial to deep water settings can better explained by a continuum of depositional processes between non-marine and turbidity currents.

References

ABREU, V., SULLIVAN, M., PIRMEZ, C. & MOHRIG, D. 2003. Lateral accretion packages (LAPs): an important reservoir element in deep water sinuous channels. *Marine and Petroleum Geology*, **20**, 631–648.

ALLEN, G. P., LAURIER, D. & THOUVENIN, J. 1979. *Etude sédimentologique du delta de la Mahakam*. Notes et Memoires, **15**. Total, Paris.

ALLEN, G. P. & CHAMBERS, J. L. C. 1998. *Sedimentation in the modern and Miocene Mahakam Delta*. Indonesian Petroleum Association Guidebook.

BARD, G. P., HAMELIN, R. G. & FAIRBANKS, R. G. 1990. U–Th obtained by mass spectrometry in corals from Barbados: sea level during the past 130,000 years. *Nature*, **346**, 456–458.

BOICHARD, R., GRANIER, B. & SEGURA, F. 1988. *Corindon campagne océanographique: La sédimentation quaternaire de la plate-forme orientale de Kalimantan*. Internal Report Total.

CARATINI, C. & TISSOT, C. 1987. Paleogeographical evolution of the Mahakam delta in Kalimantan, Indonesia, during the Quaternary and late Pliocene. *Review of Palaeobotany and Palynology*, **55**, 217–228.

CARBONEL, P. & MOYES, J. 1987. Late Quaternary Palaeoenvironments of the Mahakam delta (Kalimantan, Indonesia). *Palaeogeography, Palaeoclimatology, Palaeoecology*, **61**, 265–284.

GASTALDO, R. A. & HUC, A. Y. 1992. Sediment facies, depositional environments, and distributions of phytoclasts in the recent Mahakam River delta, Kalimantan, Indonesia. *Palaios*, **2**, 229–240.

KOLLA, V., BOURGES, PH., URRUTY, J.-M. & SAFA, P. 2001. Evolution of deepwater Tertiary sinuous channels offshore Angola (west Africa) and implications for reservoir architecture. *AAPG Bulletin*, **85**, 1373–1405.

MULDER, T. & SYVITSKI, J. P. M. 1995. Turbidity currents generated at river mouths during exceptional discharges to the world oceans. *Journal of Geology*, **103**, 285–299.

MULDER, T. & SYVITSKI, J. P. M. 1996. Climatic and morphologic relationships of rivers: implications of sea level fluctuations on river loads. *Journal of Geology*, **104**, 285–299.

MUTTI, E. 1992. *Turbidite Sandstones*. Agip S.p.A.-Istituto di Geologia Universita di Parma, Special Publication.

MUTTI, E. & NORMARK, W. R. 1991. An integrated approach to the study of study of turbidite systems. *In*: WEIMER, P. & LINK, H. (eds) *Seismic Facies and Sedimentary Processes of Submarine Fans and Turbidite Systems*. Springer, Berlin, 75–106.

MUTTI, E., DAVOLI, G., TINTERRI, R. & ZAVALA, C. 1996. The importance of fluvio-deltaic systems dominated by catastrophic flooding in tectonically active basins. *Memorie di Scienze Geologiche*, **48**, 233–291.

MUTTI, E., TINTERRI, R., REMACHA, E., MAVILLA, N., ANGELLA, S. & FAVA, L. 1999. An introduction to the analysis of ancient turbidite basins from an outcrop perspective. *AAPG Course Notes*, **39**, 93.

MUTTI, E., TINTERRI, R., BENEVELLI, G., DI BIASE, D. & CAVANNA, G. 2003. Deltaic, mixed and turbidite sedimentation of ancient foreland basins. *Marine and Petroleum Geology*, **20**, 733–755.

NORMARK, N. R. 1978. Fan valleys, channels, and depositional lobes on modern submarine fans: characters for recognition of sandy turbidite environments. *AAPG Bulletin*, **62**, 912–931.

PIRMEZ, C., BEAUBOUEF, R. T., FRIEDMANN, S. J. & MOHRIG, D. C. 2000. Equilibrium profile and base level in submarine channels: Examples from late Pleistocene systems and implications for the architecture of deepwater reservoirs: GCSSEPM Foundation 20th Annual Research Conference, Deep-Water Reservoirs of the World, Dec, 3–6, p. 782–805.

POSAMENTIER, H. W. 2001. Lowstand alluvial bypass systems: Incised vs unincised. *AAPG Bulletin*, **85**(10), 1771–1793.

POSAMENTIER, H. W. 2003. Depositional elements associated with a basin floor channel-levee complex: case study from the Gulf of Mexico. *Marine and Petroleum Geology*, **20**(6), 677–690(14).

POSAMENTIER, H. W. & ALLEN, G. P. 1999. *Siliciclastic Sequence Stratigraphy – Concepts and Applications, No. 7*. SEPM (Society for Sedimentary Geology).

POSAMENTIER, H. W., MEIZARWIN, WISMAN, P. S. & PLAWMAN, T. 2000. Deep water depositional systems Ultra-deep Makassar Strait, Indonesia. *In*: WEIMER, P., SLATT, R. M., COLEMAN, J., ROSEN, N. C., NELSON, H., BOUMA, A. H., STYZEN, M. J. & LAWRENCE, D. T. (eds.) *Deep-water Reservoirs of the World: Gulf Coast Society of the Society of Economic Paleontologists Foundation and Mineralogists, 20th Annual Research Conference*, pp. 806–816.

ROBERTS, H. H. & SYDOW, J. 2003. Late Quaternary stratigraphy and sedimentology of the offshore Mahakam Delta, East Kalimantan (Indonesia). *In*: SIDI, H. & NUMMEDAL, D. (eds) *Tropical Deltas of the Southeast Asia—Sedimentology, Stratigraphy, and Petroleum Geology*. SEPM Special Publication, **76**, 125–145.

SALLER, A. H., NOAH, J. T., RUZUAR, A. P. & SCHNEIDER, R. 2004. Linked lowstand delta linked lowstand delta to basin-floor fan deposition, offshore Indonesia: an analog for deep-water reservoir systems. *AAPG Bulletin*, **88**, 21–46.

SUITER, J. S. 1996. Shallow 3-D seismic analysis of late Pleistocene lowstand deltas (Mahakam, Indonesia). *Proceedings, Indonesian Petroleum Association, 25th Silver Anniversary Convention*, October 1996, 347–351.

SYDOW, 1996. *Holocene to late Pleistocene stratigraphy of the Mahakam delta, Kalimantan, Indonesia*. PhD dissertation, Louisiana State University.

Using multi-attribute neural networks classification for seismic carbonate facies mapping: a workflow example from mid-Cretaceous Persian Gulf deposits

U. P. BAASKE[1], M. MUTTI[1], F. BAIONI[2], G. BERTOZZI[2] & M. A. NAINI[3]

[1]*Institut fuer Geowissenschaften, Universitaet Potsdam, PO Box 601553, D-14415 Potsdam, Germany (e-mail: ubaaske@geo.uni-potsdam.de)*

[2]*Edison S.p.A., Foro Buonaparte 31, 20121 Milan, Italy*

[3]*NIOC Exploration Directorate, Geophysics Department, 8th Central Building, Yaghma Alley, Jumhoori Avenue, Teheran, Iran*

Abstract: Seismic facies mapping in large seismic surveys can be time consuming, even if only a basic overview of the facies distribution is needed. Therefore this study outlines an approach for the use of volume-based seismic attributes from 2D surveys for automated seismic facies mapping within carbonate settings. The study area is located in the central Persian Gulf, offshore Iran. The interval of interest is the mid-Cretaceous Sarvak Formation, which is part of the extensive Cretaceous shallow-water carbonate platform of the eastern Arabian Plate. A set of nine volume-based seismic attributes, calculated from time, amplitude and frequency information of post-stacked 2D seismic data, was chosen to characterize geological information within the interval of interest. The volume-based attributes were supplemented by two grid-based attributes to highlight structural elements. The geological significance of each attribute was evaluated by comparing it with results of seismic sedimentological/geomorphological studies. Furthermore, statistical methods were applied to highlight direct relationships amongst the attributes. The results of these tests were then used to choose a limited set of attributes for neural network-based multi-attribute classifications. The results show that seismic attributes derived from 2D surveys can be used to map basic seismic facies types in carbonate settings and that the outlined, general approach might be applied in other studies.

Seismic attributes often highlight anomalies in seismic datasets. Depending on the attribute and extraction method, these anomalies might be linked to geological characteristics of the rock. Therefore seismic attribute studies are often used to defined seismic intervals for reservoir characterization purposes (e.g. Gastaldi *et al.* 1997; Skirius *et al.* 1999; Chambers & Yarus 2002). In order to do so the attributes have to be calibrated to wells, which is often the major drawback of the method insofar as wells are not always sufficiently available.

An alternative use of seismic attributes is to utilize them for automated seismic facies mapping. Seismic facies analysis is the description and geologic interpretation of seismic reflection parameters (Mitchum *et al.* 1977, p. 121) and is part of the seismic stratigraphic concept pioneered by Vail *et al.* (1977). By outlining groups of reflections with distinct parameters (e.g. seismic reflection continuity, frequency or amplitude), seismic facies analysis allows the creation of three-dimensional seismic facies units which can be interpreted in terms of environmental setting or depositional process (Mitchum *et al.* 1977, p. 122). Such units can be mapped and, utilizing the environmental and depositional interpretation, provide sedimentological and stratigraphic information on the studied sequence (Mitchum & Vail 1977; Macurda & Nelson 1988).

Because seismic facies mapping in large surveys can be very time-consuming, automated mapping methods are used to provide seismic interpreters with a fast analysis of the basic seismic facies distribution. This analysis can be used as a basis for further, spatially more detailed studies and offers geological characterization possibilities in areas with low well control. The advantage of such an approach is that the seismic attributes can be 'calibrated' on the existing seismic data and its geological interpretation.

This study integrates the results of a seismic sedimentological/geomorphological analysis, providing the geological background, with a neural network based multi-attribute classification. The main aim is not the use of seismic attributes for the characterization of a single, lithological feature like porosity, but to outline a basic approach and workflow for the use of volume-based seismic

From: DAVIES, R. J., POSAMENTIER, H. W., WOOD, L. J. & CARTWRIGHT, J. A. (eds) *Seismic Geomorphology: Applications to Hydrocarbon Exploration and Production*. Geological Society, London, Special Publications, **277**, 105–120.
0305-8719/07/$15.00

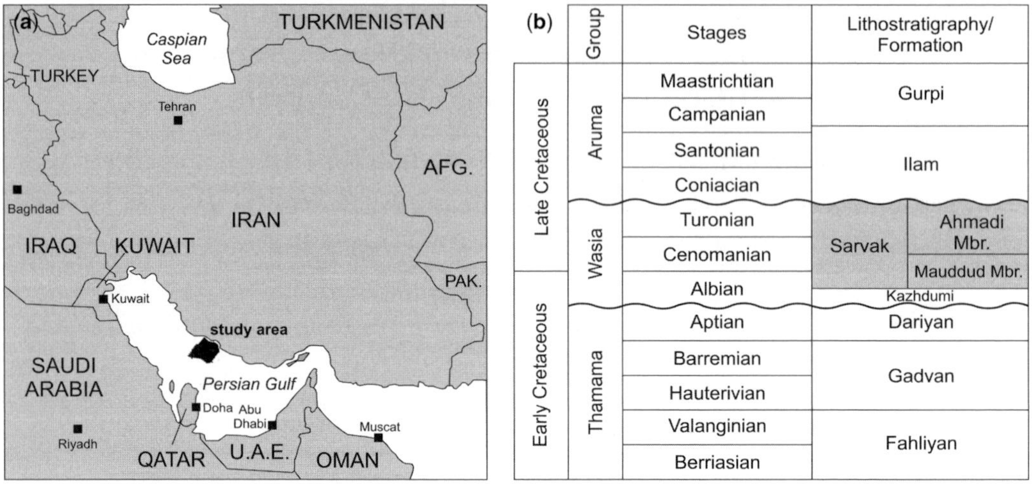

Fig. 1. (**a**) Geographic overview of the Persian/Arabian Gulf region showing the location of the study area. (**b**) Generalized stratigraphic overview of the Cretaceous of the coastal Fars area, Iran (modified after James & Wynd 1965; Christian 1997). The interval addressed by this study is shaded.

attributes for automated seismic facies mapping within a carbonate depositional setting. Moreover, the study tests if a 2D-based dataset can be used for such an approach.

The study area chosen to test this method is located in the central Persian/Arabian Gulf, offshore Iran (Fig. 1a). The interval of interest is the mid-Cretaceous Sarvak Formation (Fig. 1b), which is part of the extensive Cretaceous shallow-water carbonate platform of the eastern Arabian Plate (e.g. Alsharhan & Nairn 1997; Sharland *et al.* 2001).

Database

The dataset used consists of a 2D seismic survey covering approximately 7000 km^2 with about 6300 km of seismic lines in a 2 × 2 km grid. Only one well was available within the study area, but wells from adjacent areas were accessible. All wells include the complete interval of interest and had a spectrum of borehole logs available (e.g. gamma ray and sonic). Two seismic horizons, completely interpreted in all lines within the survey, define the top and the base of the Sarvak Formation (Fig. 2a) and the volume used to calculate the volume-based seismic attributes. The horizon/grid-based seismic attributes used were computed on the top horizon of the Sarvak Formation (Fig. 2b), a major regional unconformity often referred to as the Wasia-Aruma-Break (Harris *et al.* 1984). Using seismic sedimentological/geomorphological methods, this dataset was utilized to establish a general depositional background for the Sarvak Formation of the study area.

Seismic sedimentological background

Geological setting

Today the study area is part of the NE dipping Zagros foreland basin and the stratigraphic packaging shows clear influence of deep-seated salt tectonics (Fig. 2a), which had peak movements during the Jurassic/Cretaceous and Neogene (Ala 1974; Edgell 1996). During the middle Cretaceous the study area was situated on the eastern edge of the Arabian Plate and was part of the extensive shallow-water carbonate to mixed carbonate-siliciclastic shelf area of the Cretaceous southern Neo-Tethys (Setudehnia 1978). The palaeo-topography of the shelf during the mid-Cretaceous was flat, but still offered enough deep relief to generate intrashelf basin areas (Murris 1980). Examples of such intrashelf basins, in outcrop and in the subsurface, can be found in the mid-Cretaceous Natih Formation of Oman (van Buchem *et al.* 2002*b*; Droste & Van Steenwinkel 2004), but also in Iran (van Buchem *et al.* 2002*a*).

Seismic facies types

To describe the seismic characteristics of the Sarvak Formation and to provide insight into the sedimentological history, a basic set of typical seismic facies types was determined (Fig. 3) and mapped throughout the study area (Fig. 2c). There are two classes of seismic facies types present: (1) facies types 1–4, which are defined by the thickness and the stacking pattern of the reflections; and (2) facies types A–D, which are a characterization of the stratal geometries of the reflections.

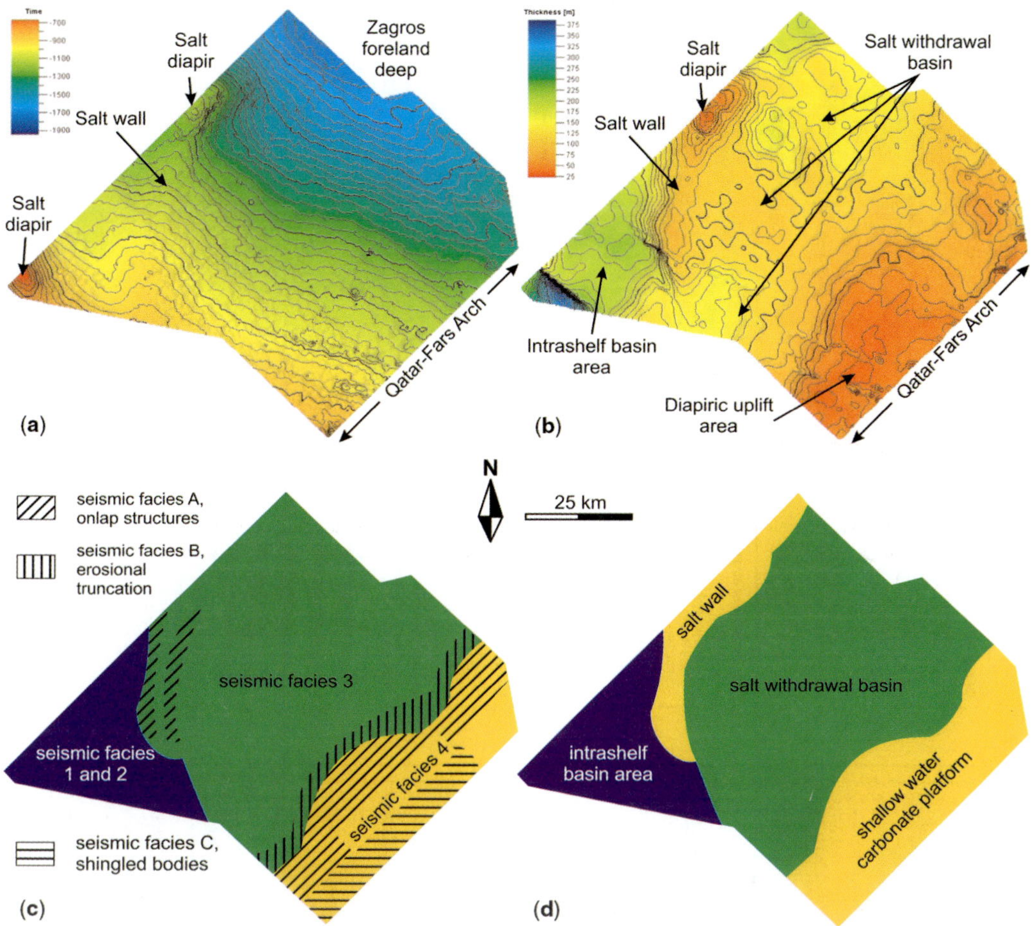

Fig. 2. (**a**) Time–structure map of the top Sarvak Formation indicating an overall NE dip of the study area due to the Zagros orogeny (scale: two-way-travel time). (**b**) Thickness map of the Sarvak Formation showing the main structural elements of the study area. The thickness values are the result of a depth-conversion process and are subject of process-related uncertainties. (**c**) General distribution of the seismic facies types. (**d**) Large-scale depositional domains based on seismic and structural interpretation.

Facies types 1 and 2, the stacked parallel reflections comprising seven or more seismic loops, typically occur in the central part of the study area and probably reflect thick deposits with relatively high lithologic diversity, i.e. in the form of large-scale carbonate-shale intercalations. Type 3, stacked parallel reflections comprising five to six loops, also occur in the centre and reflect mainly carbonate lithologies. Facies type 4 is present on top of topographic highs and is characterized by either a single cycle or stacked parallel reflections with up to four loops. Facies types A and B, onlap structures and erosional truncations, appear on the flanks of topographic highs and point to combined effects of eustatic changes and syn- to post-depositional salt tectonics. The shingled bodies of facies C can be found mainly in combination with facies 4 on top of high structures, therefore classified as 4C, but sometimes also within the top reflectors of facies types 1 and 2, resulting in 1C and 2C. The deformation structures of facies type D are mostly post-depositional and occur all over, overprinting the other seismic facies types.

Large-scale depositional setting

Three large-scale depositional domains could be identified based on the thickness distribution of the Sarvak Formation (Fig. 2b, thickness map based on depth-conversion of the seismic data), the spatial distribution

Facies No.	Seismic facies	Example
1	**Parallel reflections - 7+ loops** Reflection geometry: Sub-parallel continuous; 7 or more loops Amplitude characteristic: Medium to high amplitude reflections Spatial distribution: Occurs mainly in high thickness/intrashelf basin areas, not on palaeo-topographic highs	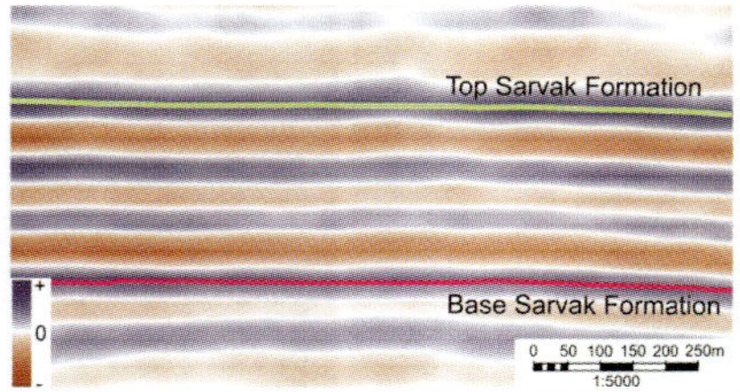
2	**Parallel reflections - mixed amplitude - 7+ loops** Reflection geometry: Sub-parallel continuous; 7 or more loops Amplitude characteristic: Medium to high and low to medium amplitude reflections vertically grouped and stacked Spatial distribution: Occurs in the thicker parts of the study area, not on palaeo-topographic highs	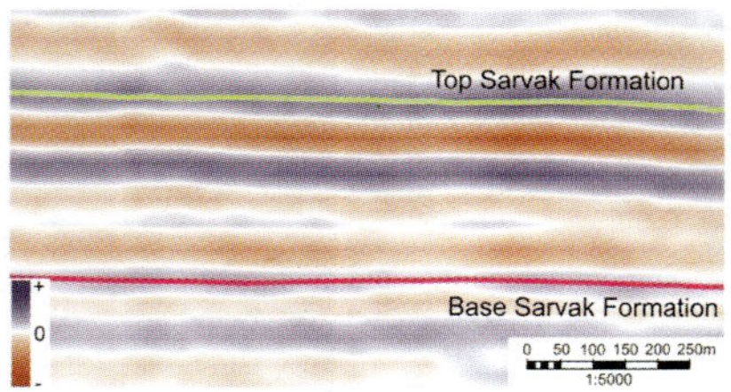
3	**Parallel reflections - low to medium amplitude - 5 to 6 loops** Reflection geometry: Sub-parallel continuous, 5 to 6 loops Amplitude characteristic: Low to medium amplitude reflections Spatial distribution: Occurs all over the study area, also on top of palaeo-high structures	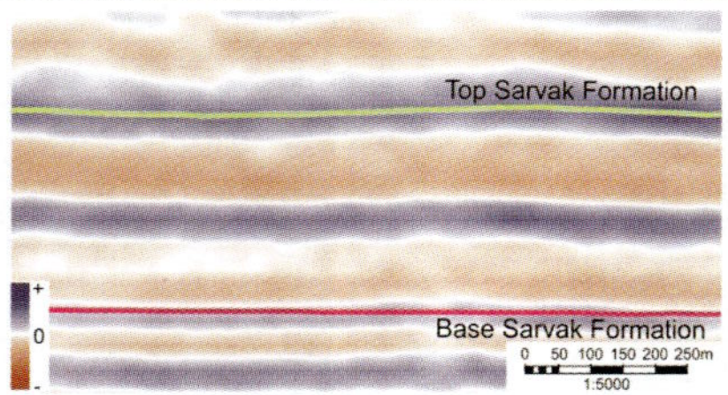
4	**Parallel reflections - medium amplitude - 1 to 4 loops** Reflection geometry: Sub-parallel continuous, 1 to 4 loops Amplitude characteristic: Low to medium amplitude reflections Spatial distribution: Occurs along the flanks and especially on top of palaeo-high structures	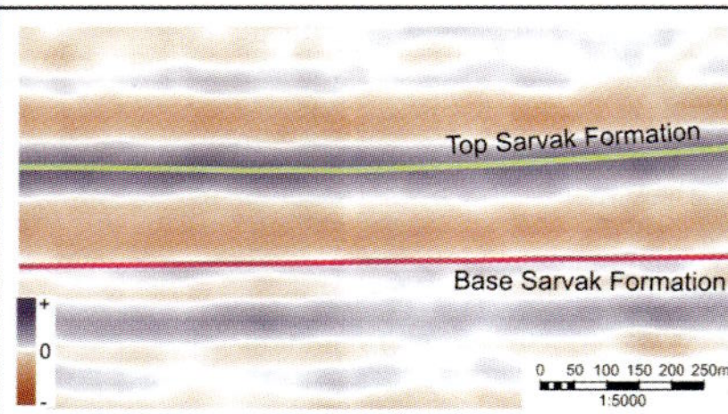
A	**Onlap structures** Reflection geometry: Sub-parallel stacked with 'enclosed' onlap/pinch out structures Amplitude characteristic: High to low amplitudes; sometimes chaotic appearance Spatial distribution: Along palaeo-topographic/structural highs; along palaeo-depressions	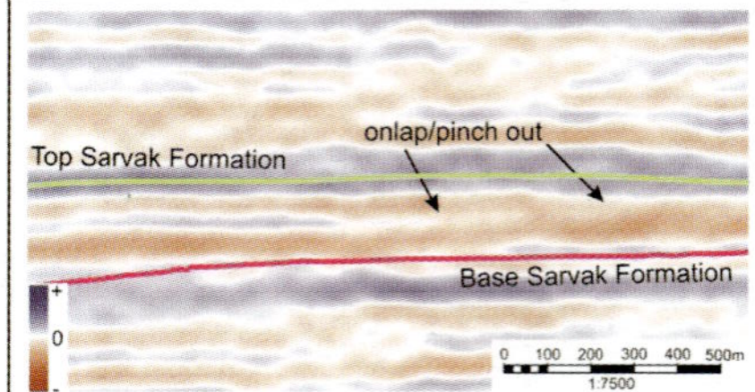
B	**Erosional truncations to onlap structures** Reflection geometry: Sub-parallel stacked with erosional truncation by following reflections; might furthermore enclose truncated onlap/pinch out structures Amplitude characteristic: High to low amplitudes; sometimes chaotic appearance Spatial distribution: Usually along palaeo-topographic/structural highs	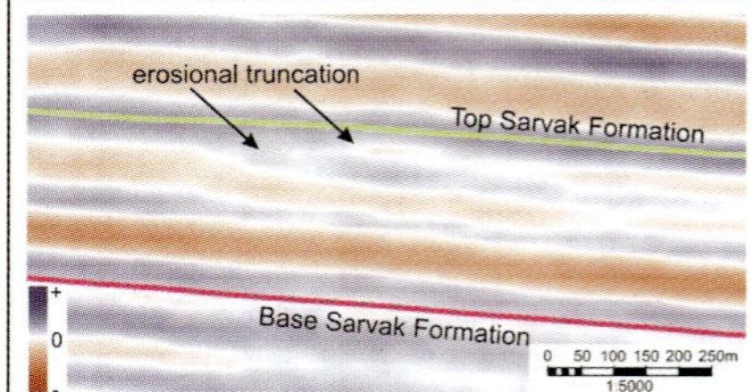
C	**Shingled/prograding bodies** Reflection geometry: Discontinuous; in parts chaotic; usually only very few stacked reflections Amplitude characteristic: Medium to high amplitude reflections Spatial distribution: Occurs on top of the QFA structure and occasionally also on top of reflection sets in 'basinal' areas	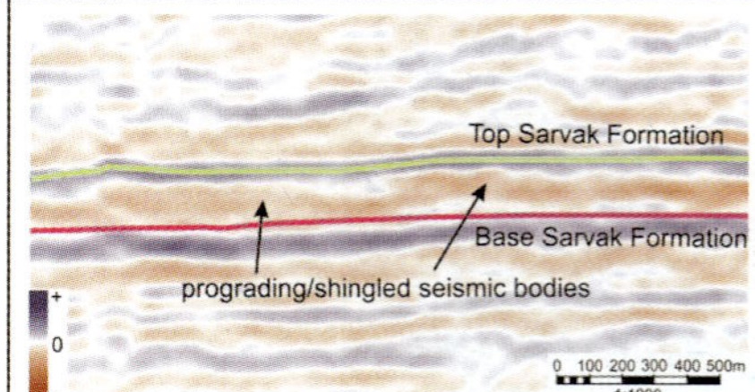
D	**Deformation structures** Reflection geometry: Sub-parallel to chaotic, often discontinuous; different range of reflection stacking (1 to 7+ loops) Amplitude characteristic: Low to high amplitude reflections Spatial distribution: Small scale structures (<1 km), bound to faults, occur all over the study area; large scale structures (>1 km), bound to salt withdrawal, occur only locally	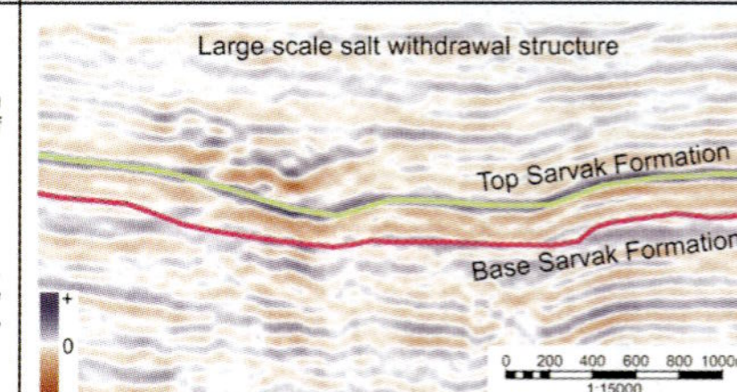

Fig. 3. Overview of the seismic facies types within the Sarvak Formation. Facies types 1–4 are based on the reflector stacking pattern while facies types A–D are based on the strata geometries. Several combinations of the two facies kinds, e.g. medium stacked reflector sets with onlap structures, classified as facies 3A, are present in the studied survey.

of the main seismic facies types (Fig. 2c) and the structural reconstruction of the area (Fig. 2d). The eastern part of the study area is dominated by a shallow water carbonate platform developed on top of the Qatar-Fars Arch. A nearby well confirms that this area is characterized by 40–50 m of carbonate depths.

The central part of the study area is interpreted as a salt withdrawal basin filled with up to 150 m of mixed carbonate and argillaceous lithologies. Thicknesses of 100 m and less on the western boundary of the salt-withdrawal basin indicate the presence of an north–south to NE–SW trending salt wall covered by clear carbonates. A high thickness area in the westernmost corner of the study area represents an intrashelf basin within the Sarvak Formation. There thicknesses of 250 m and more comprise large-scale carbonate-shale intercalations.

Multi-attribute classification with 2D-based seismic attributes for seismic carbonate facies mapping

The first step of a multi-attribute classification for seismic carbonate facies mapping is the seismic attribute creation. These attributes, derivates of basic seismic information, i.e. time/velocity, amplitude, frequency and attenuation (Brown 1996, 1999), form the basis for the classification and can, in the form of attribute maps, also be integrated in the geological model (Fig. 4).

Seismic attributes

For this study a set of nine volume-based seismic attributes was computed from post-stack 2D seismic data. To do this, the whole set of 2D lines was transformed into virtual 3D volumes, which then could be treated as a single 3D seismic volume for the seismic attribute generation. The computed seismic attributes were then visualized as attribute maps calibrated to a fixed colour spectrum for comparison (Fig. 5). Nevertheless one must be aware of the fact that the original dataset is made up by a 2 × 2 km grid. Therefore all attribute maps include interpolated areas in between the seismic lines, subject to uncertainty.

The volume-based seismic attributes were supplemented by the grid/horizon-based attributes dip and azimuth, calculated from the top horizon of the Sarvak Formation. This was done in order to highlight structural elements within the study area. The seismic attribute selection at this point of the study was guided by successful application of these attributes in previous studies. On the other hand it was constrained by the capabilities of the seismic interpretation software used to compute the seismic attributes.

Multi-attribute classification

The advantage of a multi-attribute classification is to integrate the information of numerous seismic attributes in a single classification process. The

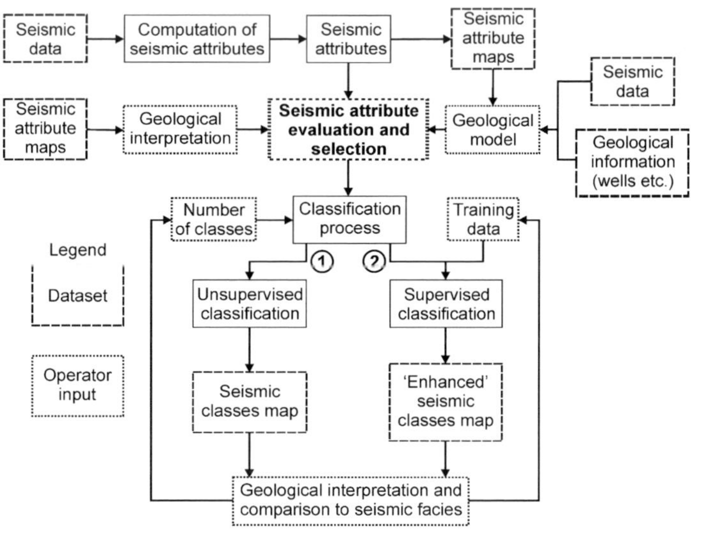

Fig. 4. The workflow of this study, summarizing the multi-attribute classification process. The 'seismic attribute evaluation and selection' is a central procedure of the workflow because of its obvious effect on the overall outcome of the classification. This step integrates geological and geophysical observations which are the foundation for the final geological interpretation of the results.

objective of this method is to provide an integrated result comprising better information than each single attributes can offer. This study utilized the multi-attribute classification to map the distribution of the different seismic facies.

In this study the commercial neural-network based, generalized inversion tool 'SeisClass' (Schlumberger) was used for the multi-attribute classification. The software package utilizes competitive learning algorithms to cluster seismic

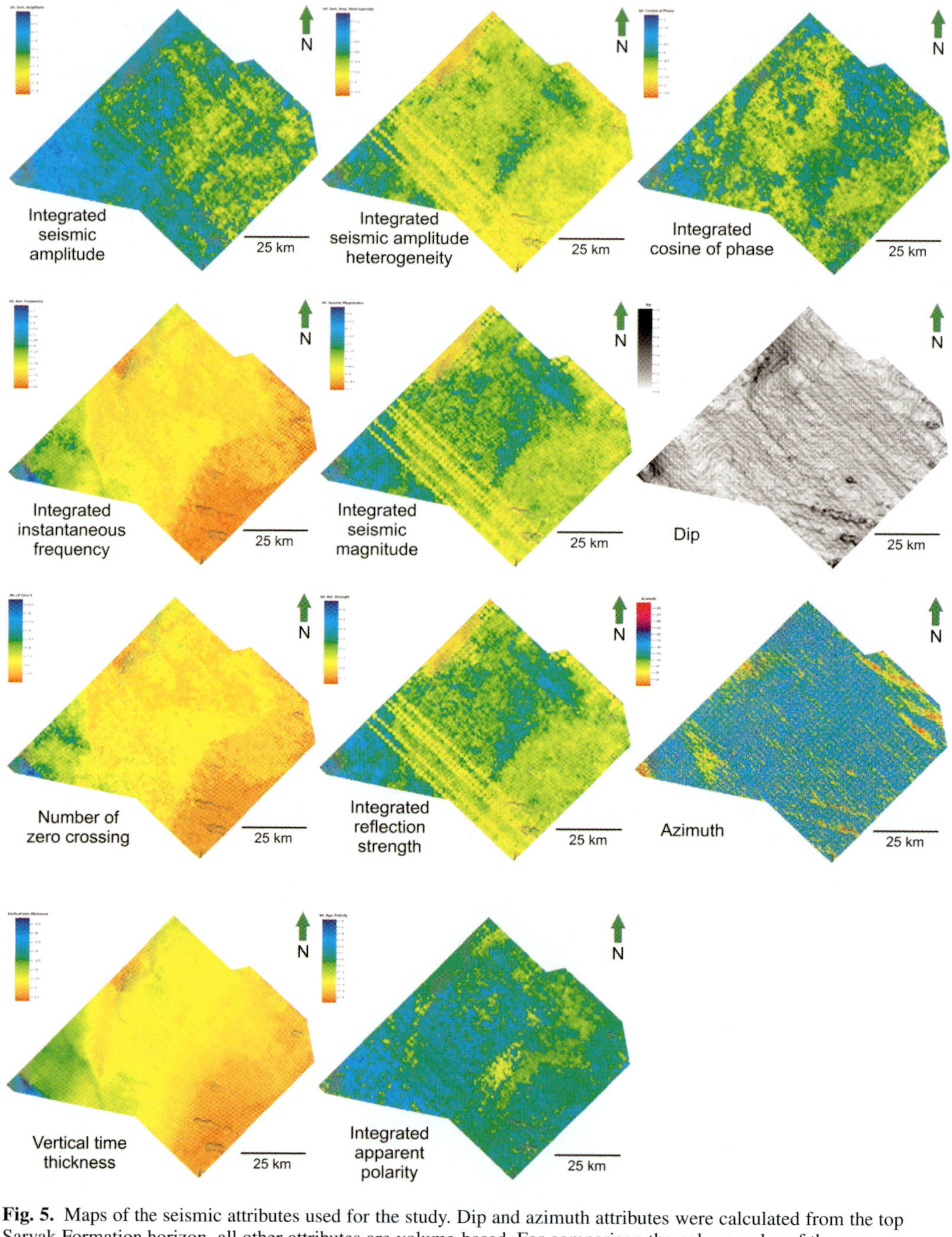

Fig. 5. Maps of the seismic attributes used for the study. Dip and azimuth attributes were calculated from the top Sarvak Formation horizon, all other attributes are volume-based. For comparison the colour scales of the maps, excluding dip, are modified to display the same colour range.

attribute-vectors in an n-dimensional attribute space. In this way seismic attribute combinations can be used to statistically catalogue different classes of distinct attribute combinations.

Classification workflow

A multi-attribute classification process for seismic carbonate facies mapping is only useful if the result is interpretable in a geological sense, i.e. if the result can be correlated with the seismic attribute data and its geological implications. To achieve this, it is useful not to use all seismic attributes available for the classification, but to apply only a few, selected ones to keep the number of geological uncertainties limited. Therefore attribute evaluation and selection processes have a central role in the classification workflow (Fig. 4).

Attribute evaluation

To reduce the number of geological uncertainties that have to be considered for the interpretation, a variable reduction was carried out. This approach should be considered in datasets with a large number of seismic attributes to reduce the potential risk of 'spurious (false) correlations' linked to large variable sets. Spurious correlations are correlations occurring by simple chance and without real, physically based correlation (Kalkomey 1997). The probability for spurious correlations will be proportional to the amount of seismic attributes used for the classification and inversely proportional to the number of data control points (Chambers & Yarus 2002).

Modifying the approach of Hart & Balch (2000), the attribute evaluation process applied in this study integrated several iterative and interactive approaches:

1. *Comparison of attributes* – first the seismic attributes maps were visually examined and compared with each other and to the geological model to find spatial and geological trends and correlations. Some attributes, like integrated instantaneous frequency, number of zero crossings and time thickness distribution, are related to each other and therefore correlate quite well (Fig. 5). Since the overall aim of the attribute evaluation is to find independent variables, such good correlations were marked to exclude some of the related attributes from the classification. In a parallel step correlation coefficients for the volume based attributes were calculated (Table 1). As a further statistical test an R-mode cluster analysis (Ward Method with squared Euclidean distance using the program SPSS 12.0.1 for Windows) was carried out, again to find close relationships within the volume-based dataset and to select independent variables (Fig. 6). All numerical methods carried out during this step are basically similar to the qualitatively visual comparison, but allow a numerical and therefore more quantitative statement. In order to keep the chance of spurious correlations to a minimum, a set of 105,211 points for each attribute was used for the statistical approach. The result of this data evaluation step proved the close relationships of several seismic attributes, demonstrating that some of them could be excluded from a multi-attribute classification.
2. *Assignment of geological meaning* – another important stage during this phase of the workflow was to determine the geological significance of the single seismic attributes. To do this, experience with seismic attributes from previous studies and also literature information (e.g. Brown 1996; Chen & Sidney 1997) was used to evaluate the geological significance for each seismic attribute (Table 2), which then was cross-checked with the seismic sedimentological study of the area to test its plausibility. After testing and grouping the seismic attributes into important and less important, e.g. redundant attributes due to high similarities or lacking geological significance, the final variable reduction was carried out.

Attribute selection

Following the evaluation step the attribute selection was accomplished by skipping ineffective attributes from the classification. To decide about the value of an attribute and to allow an objective selection process, a set of guidelines for the attribute selection was established:

- A limited set of attributes (Brown 1999 suggested three to five) should be chosen. This limits the number of seismic attributes in the classification, and limits the possibility of spurious correlations. This is especially important in studies with a high number of seismic attributes.
- The set of chosen seismic attributes represents the maximum available range of the seismic signal. This guideline was applied since the study did not focus on single rock properties for reservoir characterization, but aimed to classify seismic facies distribution on the basis of a spectrum of seismic information (i.e. time/velocity, amplitude and frequency). The intention of this guideline is therefore to put the different aspects of the seismic signal into effect. For a study focusing on special petrophysical properties, linked only to single

Table 1. *Correlation coefficients (after Pearson) of the volume based seismic attributes*

		1	2	3	4	5	6	7	8	9
1	Integrated reflection strength	—	1.0	0.980	0.724	0.762	0.747	0.421	0.223	−0.004
2	Integrated seismic magnitude	1.0	—	0.981	0.725	0.764	0.750	0.424	0.234	−0.005
3	Integrated seismic amplitude heterogeneity	0.980	0.981	—	0.765	0.776	0.731	0.396	0.286	0.030
4	Number of zero crossings	0.724	0.725	0.765	—	0.973	0.929	0.266	0.464	0.082
5	Integrated instantaneous frequency	0.762	0.764	0.776	0.973	—	0.970	0.445	0.299	0.061
6	Vertical time thickness	0.747	0.750	0.731	0.929	0.970	—	0.298	0.419	0.021
7	Integrated apparent polarity	0.421	0.424	0.396	0.266	0.299	0.298	—	0.177	−0.348
8	Integrated seismic amplitude	0.223	0.234	0.286	0.464	0.445	0.419	0.177	—	0.251
9	Integrated cosine of phase	−0.004	−0.005	0.030	0.082	0.061	0.021	−0.348	0.251	—

aspects of the seismic signal, the attribute set would have to be selected accordingly.

- Each single attribute has a 'good' geological significance. This was necessary to ensure the geological interpretability of the classification results, even though some attributes, e.g. seismic amplitude, are influenced by several geological factors (Table 2).
- The single attributes are statistical independent. In general using several statistically closely related attributes does not add information to a classification. Therefore only one attribute of the attribute-clusters in this study (Fig. 6 and Table 1) was used. For the selection of the attribute the previous guidelines were applied.

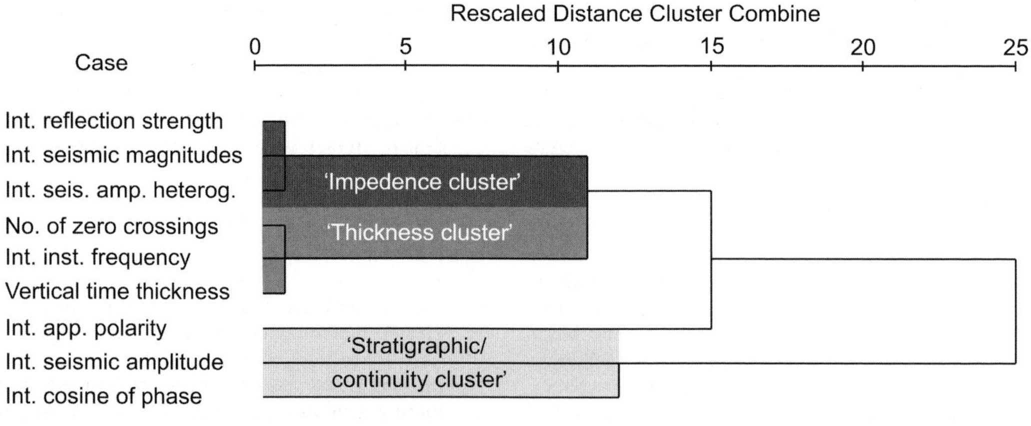

Fig. 6. An R-mode cluster analysis (Ward method with squared Euclidean distance using the program SPSS 12.0.1 for Windows; $n = 105{,}211$) of the nine volume-based seismic attributes to find close statistical relationships within the attributes.

Table 2. *The geological significance of the seismic attributes used for this study*

Seismic attribute	Bedding continuity	Lithology	Fluid content/ porosity	Structural changes	Thickness
Integrated seismic amplitude	✔	✔	✔		
Integrated instantaneous frequency		✔	✔		(✔)
Integrated apparent polarity	✔				
Integrated cosine of phase	✔				
Integrated seismic amplitude heterogeneity	✔	✔	✔		
Number of zero crossings	✔			(✔)	✔
Vertical time thickness					✔
Integrated seismic magnitude		✔		(✔)	(✔)
Integrated reflection strength	✔	✔	✔		
Dip				✔	
Azimuth				✔	

With the help of these selection guidelines, a combination of five volume-based and one grid-based seismic attributes was chosen for the classification process: (1) integrated seismic amplitude; (2) integrated instantaneous frequency; (3) integrated reflection strength; (4) integrated apparent polarity; (5) integrated cosine of phase; and (6) dip (grid-based). Table 3 summarizes the application of the selected attributes in this study and shows which basic geological information is linked to them.

Number of classes

Usually neural networks-based classification software allows the user to set the number of classes (NOC) for the classification result. This NOC-value is an important factor influencing the classification process (Fig. 4). A low NOC value will provide a very basic classification with a relatively high amount of background, possibly obscuring important information (Fig. 7a). With an increasing NOC value, the level of detail in the classification increases (Fig. 7b), though high NOC values can produce numerous (sub-) classes which are not necessarily representative and which might complicate the interpretation (Fig. 7c). For this reason the decision on the NOC value is a sensible, though not completely objective, step in the workflow and includes an interpretation measure which directly influences the overall classification process.

For the present study the number of classes was oriented on the number of principle seismic facies types to map. With four classes, representing the stacked parallel reflections (seismic facies 1 and 2 in dark blue, seismic facies 3 in green and seismic facies 4 in yellow) as well as a background class, the basic facies distribution was already met (Fig. 7a). A further four classes were added with the intent that they could represent the influence of the more structural/geometric oriented seismic facies classes (seismic facies A–D, Fig. 3). The result showed that eight classes provided a good correlation between the automatically mapped and geologically interpreted seismic facies (Fig. 7b). Increasing the NOC value to 12 (Fig. 7c), to allow more combinations of the stacking pattern facies types with the geometric facies types, did not bring the desired result and, moreover, decreased the geological interpretability.

Neural networks-based, multi-attribute classification: unsupervised v. supervised

The classification process can be operated either as unsupervised classification, where the neural network is looking for structure in the data, or as supervised classification where the clustering process is supported by training and validation data chosen by the operator (Fig. 4). Typically the training data is based on well information or on the results of a previous, possibly unsupervised classification and

Table 3. *Seismic attributes chosen for the presented classification*

Seismic attribute	Basic seismic information	Application in this study
Integrated seismic amplitude	Amplitude	Stratigraphic and structural interpretations
Integrated instantaneous frequency	Frequency	Stratigraphic interpretation; thickness and attenuation information
Integrated reflection strength	Amplitude (energy)	Stratigraphic interpretation and impedance link
Integrated apparent polarity	Amplitude (wave shape)	Bedding continuity and fluid content/porosity
Integrated cosine of phase	Time (phase)	Bedding continuity
Dip (grid-based)	Time	Structural changes

defines good examples of the classes to be mapped for the neural network to learn.

This study started with unsupervised classification runs, creating an unsupervised classification that was validated with the geological model of the study area. After that, good examples of each single seismic facies were selected from the final unsupervised classification to define training and validation areas for the supervised classification.

Discussion

Seismic facies mapping

Both the unsupervised and the supervised classifications reproduced the distribution of the stacking-pattern seismic facies types well (Fig. 8a–c). Classes 2 and 3 (blue colours) correlate well with the general distribution of the stacked reflection sets with seven or more loops represented by seismic facies types 1 and 2 (Table 4). The classes 5 and 7 (green colours) are directly associated with seismic facies 3, the stacked parallel reflections comprising five to six loops. Class 6 (dark yellow) is linked to the stacked parallel reflections with a single or up to four loops. This reflection configuration outlines seismic facies types 3 and 4 and is present on top of the Qatar-Fars Arch area.

With the geometry-based facies types A–D, the correlation was less pronounced but still acceptable. The onlap areas of seismic facies type A are included in classes 5, 6 and partly also in class 7. This implies that onlap structures in the study area occur mostly in areas with stacked reflections with four or five loops, which dominate the classification. Even with validation data the classification could not be trained to higher resolutions for this facies type. Seismic facies type B, the erosional truncation to onlap structures, were mapped very well by class 8 (light yellow). Especially in the supervised classification, the application of the training data improved the mapping result for this facies tremendously, removing imperfectly mapped areas, especially in the centre of the salt withdrawal basin. In Figures 8(b) and 9 the light yellow colours indicate areas where differential uplift/subsidence lead to erosional truncation of the basal Ahmadi Member deposits. The shingled bodies of seismic facies type C could not be mapped. This is mostly related to the fact that shingled bodies occur predominantly on top of the uplifted areas where stacked reflections comprising up to four loops were mapped as class 6. On the few locations where shingled bodies occur within stacked reflections comprising seven and more loops (seismic facies 1 and 2), the stacking pattern classification outweighs the geometry element again and these locations were mapped as class 2. The mapping of the deformation structures, seismic facies type D, was more successful. Based on abrupt changes in the grid-based dip attribute, most of the post-depositional major structural elements like faults and collapse structures were mapped as class 1 (red colour). Structural deformations occurring in seismic facies 1 and 2 were grouped into class 3 (light blue). Similar to the erosional truncation structures, the mapping of the deformation structures improved in the supervised classification. As with class 8, the locations of the mapped deformation structures match the previous interpretations and the geological model of the seismic survey.

Large-scale depositional domains

Both classification results (Fig. 8a, b) allow the recognition of the three major depositional domains of the initial geological model (Fig. 2d). The intrashelf basin area (Fig. 2d) is now identified by classes 2 and 3 (Fig. 8c), the salt withdrawal basin area is identified by class 7 and the shallow water carbonate platform on top of the Qatar-Fars Arch is identified by class 6. All three areas, as defined by the classification, correlate well with the geological interpretation of the seismic data and the surrounding

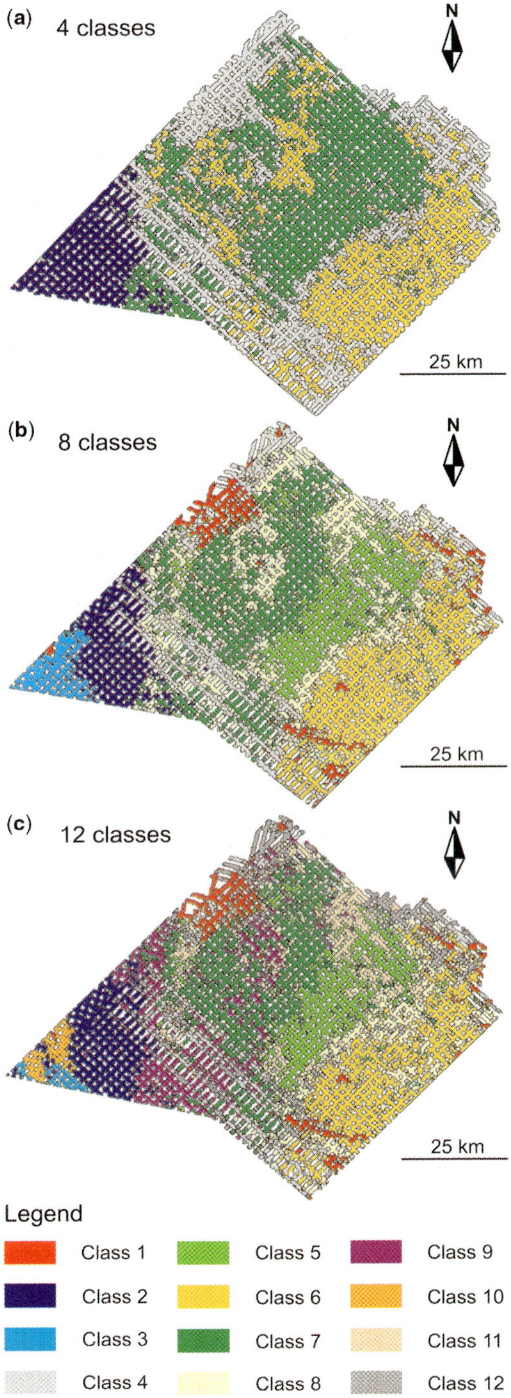

Fig. 7. Unsupervised classification results with the final set of seismic attributes shown in Table 4, for (**a**) four, (**b**) eight and (**c**) 12 classes. The comparison shows that the major depositional areas are mapped quite well in each run, whilst the amount of mapped detail in the centre of the study area increases with higher number of classes. With 12 classes the development of sub-classes, with no real geological meaning, is noticeable.

Table 4. *Summary and interpretation of the classification results*

Class	Dominant seismic attributes	Seismic facies type	Geological interpretation
1	High dip	D	Structural class controlled by strong change in dip value; not bound to lithology Post-depositional tectonics, e.g. faulting, collapse structures, salt inflation/deformation
2	High integrated seismic amplitude	1	High lithological diversity (carbonate–shale intercalations)
	High integrated instantaneous frequency	2	High thickness Intra-shelf basin facies comprising low energy deposits
	High integrated reflective strength		
3	High integrated seismic amplitude	1D	High lithological diversity (carbonate–shale intercalations)
	High integrated instantaneous frequency	2D	High thickness Influenced by post-depositional tectonics and salt deformation
	High integrated reflective strength		Intra-shelf basin facies
	High dip		
4	Background		
5	High integrated reflective strength	3	Medium lithological diversity (carbonate facies changes) Low to medium thickness
	High integrated cosine of phase	3A	Partly discontinuous 'Platform progradation front' on slopes; high to low energy facies
	Low apparent polarity		
6	Low integrated seismic amplitude	4	Low lithological diversity (carbonate facies changes)
	Low integrated instantaneous frequency	4A	Low thickness; partly discontinuous
	Low integrated reflective strength	4C	Mainly high energy carbonate facies with exposure surfaces
7	High integrated reflective strength	3	Medium lithological diversity (carbonate–shale intercalations and carbonate facies change)
	Low integrated cosine of phase	(3A)	Medium thickness Salt withdrawal basin facies; low to high energy deposits
	High apparent polarity		
8	High integrated seismic amplitude	B	Low lithological diversity (carbonate facies change) Low thickness
	High integrated cosine of phase	(3B)	Discontinuous with erosional truncation caused by diapiric uplift; high energy deposits

well logs. In addition to the three major depositional domains, new sub-classes were created by the classification. These new classes map transitional areas or domains with special structural influence and improve the geological model (Fig. 8c). Class 5 represents the transition from the shallow water carbonate platform (class 6) to the adjacent salt withdrawal basin (class 7). This class is characterized by onlap structures and stacked parallel reflections and is interpreted as a platform progradation front of the Mauddud Member (Fig. 9). To the west class 5 also outlines a structural high, an uplifted deep-seated salt wall. Besides class 1, indicating major structural elements like faults and collapse structures, structural influence is also mapped by class 8. This class occurs on the fringes of elevation areas and

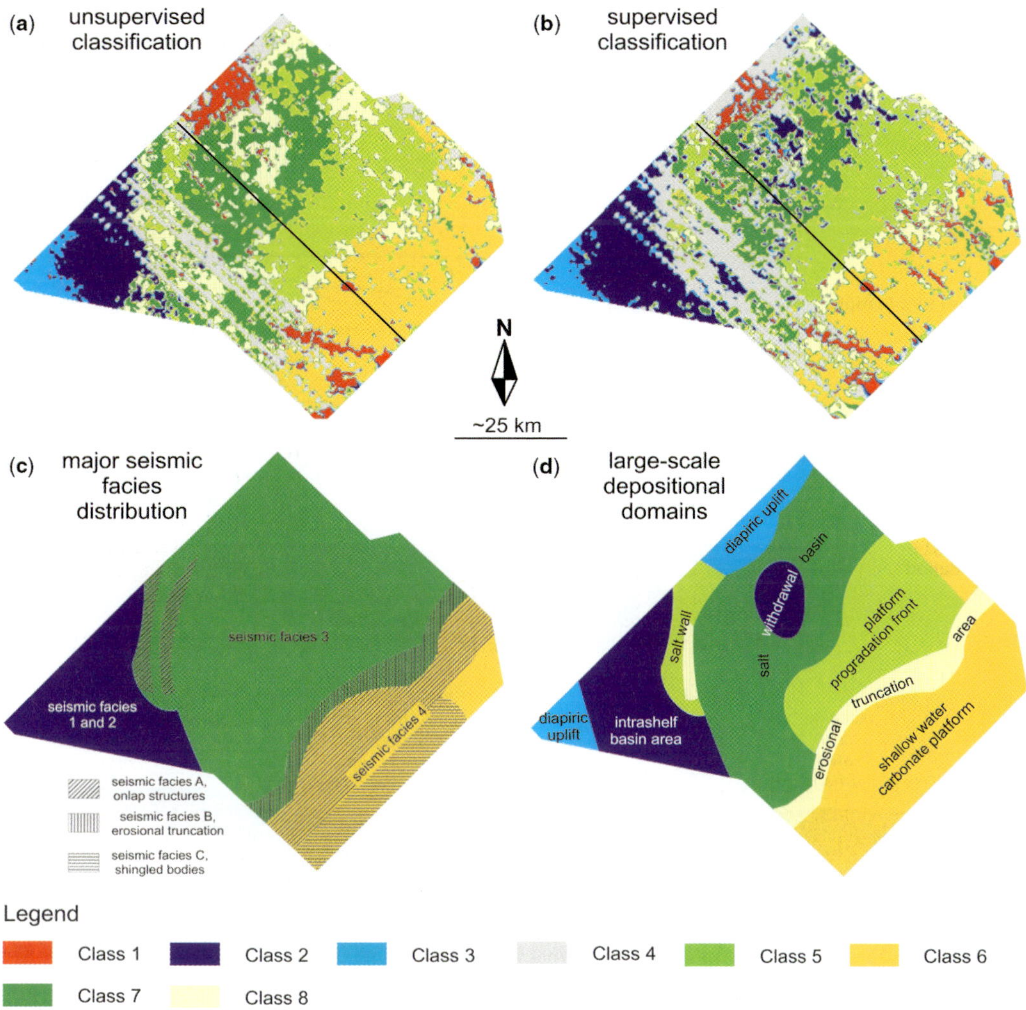

Fig. 8. The final results of the (**a**) unsupervised classification and (**b**) supervised classification runs. The supervised classification is improved by training and validation data chosen on the basis of the unsupervised results, indicated as lines in Figure 7(a). (**c**) Large-scale depositional domains based on the classifications. (**d**) General distribution of the seismic facies types. The black line in (a) and (b) indicates the position of the seismic lines shown in Figure 9.

represents regions where syn- and post-depositional diapiric uplift lead to deformation and erosional truncation of carbonates of the lower Ahmadi Member (Fig. 9). Taken as a whole, the classification results confirmed the geological interpretation and further improved it by adding new subclasses to the large-scale depositional domains.

Unsupervised v. supervised classification

The unsupervised classification offered a good overview of the seismic facies distribution, and provided a map of the reflection stacking-based facies types. Only small areas differ from the observation based on geological interpretation of the seismic lines. Since an unsupervised classification can be done quite rapidly, it should always be considered in the workflow before a supervised one, even when training data is already available.

The use of training data for the supervised classification enhanced the seismic facies mapping in this study. In particular, the mapping of the geometry-controlled features like the deformation structures (class 1, red) and the erosional truncation structures (class 8, light yellow) improved. Also the stacking

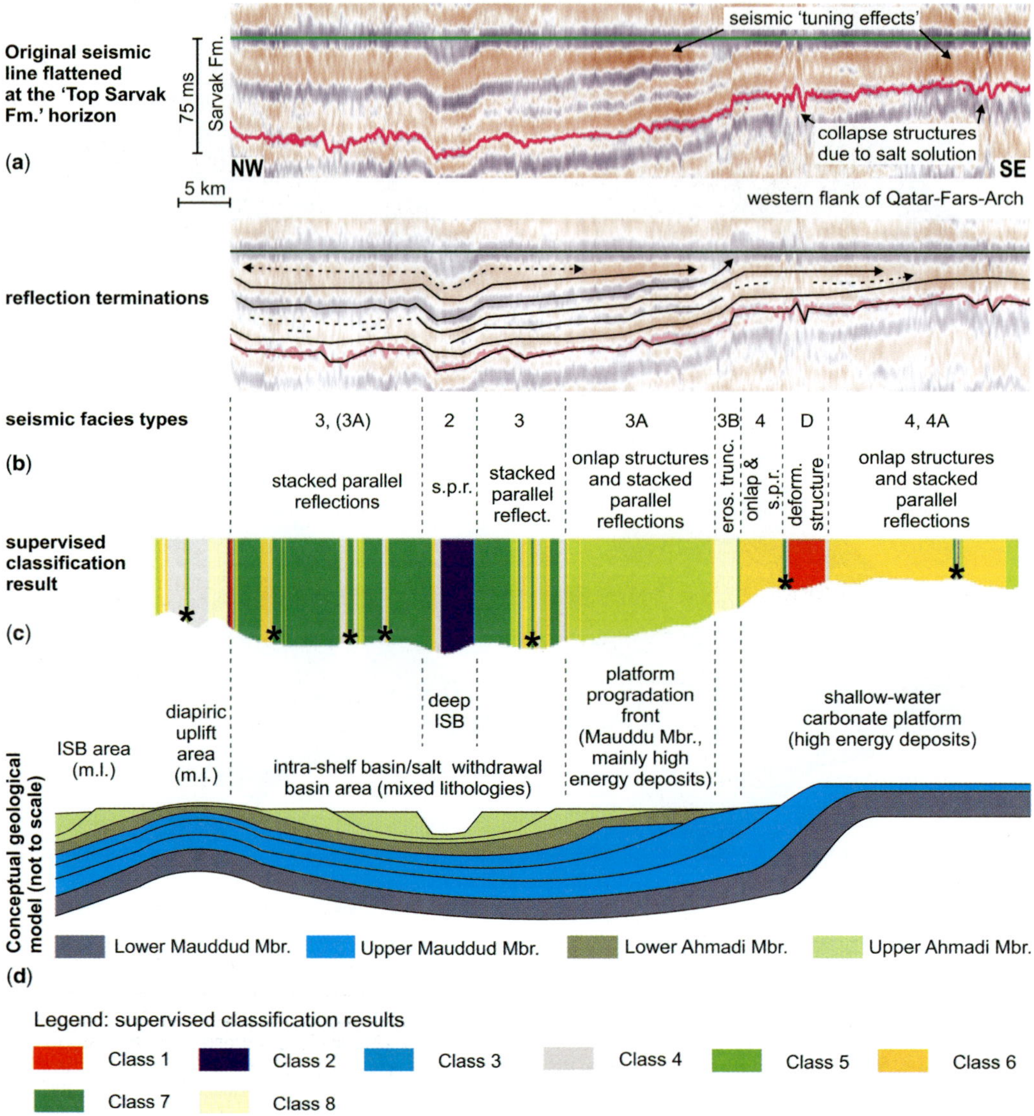

Fig. 9. Comparison of (**a**) a sample seismic line (the position of the line within the survey is shown in Fig. 8) with (**b**) seismic facies, (**c**) supervised classification result, and (**d**) conceptual geological model and interpretation. This comparison shows that the seismic facies types based on reflection stacking were well mapped, as well as most of the reflection terminations and structural features. The positions of the main depositional areas, i.e. shallow-water platform, platform progradation front and intrashelf/salt withdrawal basin, are replicated. By integrating these results with other methods, e.g. outcrop analogue studies, a more detailed characterization (e.g. facies, lithologies) of the seismic interval could be achieved. *Note*: asterisks indicate the position of 'classification artefacts', i.e. classification uncertainties produced by the gridding process during the 2D to 3D computation. s.p.r., stacked parallel reflections; ISB, intrashelf basin; m.l., mixed lithologies; Fm., Formation; Mbr., Member.

pattern bound mapping improved in a way that areas falsely mapped as class 8 (light yellow) are now mapped as class 2 (dark blue), which fits the observations in the corresponding seismic lines (Fig. 9).

These examples highlight the advantage of a supervised classification. Using validated, i.e. typical, examples of the classes as training data provides the classification software with a relatively sharp definition of each class, enhancing the mapping. The application of training areas is especially useful when classes are based on one dominating attribute and differ only in a few other, less important attributes. Classes 2 and 8, for example, are both dominated by high integrated seismic amplitude values and differ in the instantaneous frequency (high in class 2), reflection strength (high in class 2) and cosine of phase (high in class 8) values (see Table 4). In the unsupervised classification areas with high seismic amplitude but medium values in the other three attributes, were mainly mapped as class 8 (see Fig. 8a). In most cases this did not reflect the observations in the seismic lines. By providing training areas of each class, the supervised classification produced a more realistic map, since now the neural network was able to use the training data as well-defined class examples.

Conclusions

1. This study demonstrates that seismic attributes, based on 2D seismic surveys, can be used to map basic seismic facies types in carbonate depositional settings.
2. The paper outlines a new method for a fast seismic facies analysis and the mapping of depositional domains in large seismic surveys.
3. This first-order information provides the basis for more detailed and spatially focused studies.
4. The geological significance of the seismic attributes must be understood before the classification results can be interpreted in a regional perspective. The evaluation and reduction of seismic attributes before the classification is a very sensible step.
5. The outlined approach for automated classification of seismic carbonate facies can be applied in other carbonate settings. The success of this method depends on the spacing of the seismic lines and the scale of geological variability. However, the seismic attributes to use, the attribute selection and the number of classes have to be adjusted for each new study area to fit the specific regional geological, seismic facies framework and geological question of interest.

We are grateful to Giorgio Bolis (Edison S.p.A., Milan) for kindly providing the opportunity of this study and permission to publish the results. NIOC provided the data. We thank our colleagues in Milan and Potsdam for helpful discussions and especially Andrea Knoerich for the comments on the manuscript. The helpful and constructive reviews by Henry Posamentier and John Logel have helped to improve this paper.

References

ALA, M. A. 1974. Salt diapirism in Southern Iran. *AAPG Bulletin*, **58**, 1758–1770.

ALSHARHAN, A. S. & NAIRN, A. E. M. 1997. *Sedimentary Basins and Petroleum Geology of the Middle East*. Elsevier, Amsterdam.

BROWN, A. R. 1996. Seismic attributes and their classification. *The Leading Edge*, **15**, 1090.

BROWN, A. R. 1999. *Interpretation of Three-Dimensional Seismic Data*. The American Association of Petroleum Geologists and the Society of Exploration Geophysicists, Tulsa, OK.

CHAMBERS, R. L. & YARUS, J. M. 2002. Quantitative use of seismic attributes for reservoir characterization. *CSEG Recorder*, **27**, 14–25.

CHEN, Q. & SIDNEY, S. 1997. Seismic attribute technology for reservoir forecasting and monitoring. *The Leading Edge*, **16**, 445–456.

CHRISTIAN, L. 1997. Cretaceous subsurface geology of the Middle East Region. *GeoArabia*, **2**, 239–256.

DROSTE, H. & VAN STEENWINKEL, M. 2004. Stratal geometries and patterns of platform carbonates: the Cretaceous of Oman. *In*: EBERLI, G. P., MASAFERRO, J. L. & SARG, J. F. (eds) *Seismic Imaging of Carbonate Reservoirs and Systems*. The American Association of Petroleum Geologists, Tulsa, OK. AAPG Memoirs, **81**, 185–206.

EDGELL, H. S. 1996. Salt tectonism in the Persian Gulf Basin. *In*: ALSOP, G. I., BLUNDELL, D. J. & DAVISON, I. (eds) *Salt Tectonics*. The Geological Society, London. Geological Society, Special Publications, **100**, 129–151.

GASTALDI, C., BIGUENET, J.-P. ET AL. 1997. Reservoir characterization from seismic attributes: an example from the Peciko Field (Indonesia). *The Leading Edge*, **16**, 263–266.

HARRIS, P. M., FROST, S. H. ET AL. 1984. Regional unconformities and depositional cycles, cretaceous of the arabian Peninsula. *In*: SCHLEE, J. S. (ed.) *Interregional Unconformities and Hydrocarbon Accumulations*. AAPG, Tulsa, OK. AAPG Memoirs, **36**, 67–80.

HART, B. S. & BALCH, R. S. 2000. Approaches to define reservoir physical properties from 3-D seismic attributes with limited well control: an example from the Jurassic Smackover Formation, Alabama. *Geophysics*, **65**, 368–376.

JAMES, G. A. & WYND, J. G. 1965. Stratigraphic nomenclature of Iranian Oil Consortium Agreement Area. *AAPG Bulletin*, **49**, 2182–2245.

KALKOMEY, C. 1997. Potential risks when using seismic attributes as predictors of reservoir properties. *The Leading Edge*, **16**, 247–251.

MACURDA, D. B. JR & NELSON, H. R. JR. 1988. Interactive interpretation of a submarine fan, offshore Ireland: a case history. *The Leading Edge*, **7**, 28–34.

MITCHUM, R. M. J. & VAIL, P. R. 1977. Seismic stratigraphy and global changes of sea level, part 7: seismic stratigraphic interpretation procedure. *In*: PAYTON, C. E. (ed.) *Seismic Stratigraphy – Applications to Hydrocarbon Exploration*. The American Association of

Petroleum Geologists, Tulsa, OK. AAPG Memoirs, **26**, 135–143.

MITCHUM, R. M. J., VAIL, P. R. *ET AL.* 1977. Seismic stratigraphy and global changes of sea level, part 6: stratigraphic interpretation of seismic reflection patterns in depositional sequences. *In*: PAYTON, C. E. (ed.) *Seismic Stratigraphy – Applications to Hydrocarbon Exploration*. The American Association of Petroleum Geologists, Tulsa, OK. AAPG Memoirs, **26**, 117–133.

MURRIS, R. J. 1980. Middle East: stratigraphic evolution and oil habitat. *AAPG Bulletin*, **64**, 597–618.

SETUDEHNIA, A. 1978. The Mesozoic Sequence in south-west Iran and adjacent areas. *Journal of Petroleum Geology*, **1**, 3–42.

SHARLAND, P. R., ARCHER, R. *ET AL.* 2001. *Arabian Plate Sequence Stratigraphy*. Gulf PetroLink, Manama, Bahrain.

SKIRIUS, C., NISSEN, S. *ET AL.* 1999. 3-D seismic attributes applied to carbonates. *The Leading Edge*, **18**, 384–393.

VAIL, P. R., MITCHUM, R. M. J. *ET AL.* 1977. Seismic stratigraphy and global changes of sea level. *In*: PAYTON, C. E. (ed.) *Seismic Stratigraphy – Applications to Hydrocarbon Exploration*. The American Association of Petroleum Geologists, Tulsa, OK. AAPG Memoirs, **26**, 49–212.

VAN BUCHEM, F. S. P., GAUMET, F. *ET AL.* 2002*a*. Middle and Upper Cretaceous sedimentation patterns in the Dezful embayment, SW Iran. 5th Middle East Petroleum Geosciences Conference, GEO 2002, Abstracts. *GeoArabia*, **7**, 313–314.

VAN BUCHEM, F. S. P., RAZIN, P. *ET AL.* 2002*b*. Stratigraphic organization of carbonate ramps and organic-rich intrashelf basins: Natih Formation (middle Cretaceous) of northern Oman. *AAPG Bulletin*, **86**, 21–53.

Parameterization of meander-belt elements in high-resolution three-dimensional seismic data using the GeoTime cube and modern analogues

I. R. RABELO[1,2], S. M. LUTHI[1] & L. J. VAN VLIET[3]

[1]*Department of Geotechnology, Delft University of Technology, Mijnbouwstraat 120, 2628 RX Delft, The Netherlands*

[2]*Present address: Petrotechnical Data Systems, Lange Kleiweg 60-C, 2288 GK Rijswijk, The Netherlands (e-mail: israel.riverarabelo@shell.com)*

[3]*Quantitative Imaging Group, Delft University of Technology, Lorentzweg 1, 2628 CJ Delft, The Netherlands*

Abstract: The parametric quantification of geological bodies from high-resolution seismic data helps in understanding and predicting their occurrences, but is often hampered by layer distortions caused by post-depositional processes. A method called GeoTime cube is presented that overcomes this by creating a seismic volume between two near-isochronous geological markers in which the vertical axis corresponds approximately to relative geologic time. This volume is no longer affected by post-depositional deformations, a feature that greatly facilitates the extraction of sedimentary elements of interest. A case study of a fluvio-estuarine reservoir from Suriname demonstrates how fluvial point bars, channel fills and crevasse splays can be extracted from the GeoTime cube. Their geometries are quantified with the help of recent analogues. Meandering rivers are found to show relatively constant curvatures and a characteristic spacing of their meander loops. Cubic splines are suitable parametric descriptors of such river paths. Point bars are their main depositional product and can be approximated by two intersecting circle segments, representing the initial and the final position of the meander loop. The axis joining the circle centres corresponds to the direction of accretion, and the normals to these axes describe the drainage trend. Knowledge of these parameters from a limited area can be used to stochastically model the meander belt in the up- and downstream direction.

High-resolution three-dimensional seismic (HRS) data can provide unique insights into sedimentary systems through their structure (Bakker 2002). They will never exhibit the same amount of detail as seen in outcrops, but their true 3D nature adds an element that cannot be obtained from outcrops.[1] One can think of HRS as representing the envelope of sedimentary bodies at a scale that depends on the frequency of the survey. The very nature of HRS, however, causes the data set to have a considerable drawback for visualization: it is not possible to view the entire data set at once, simply because there is a data point at every x,y,z-position (a voxel) of the survey, and, in whatever direction one may look, there are always voxels that stand in the way of others behind them. A generally used method for overcoming this is to make part of the survey transparent, or opaque, so that another part of the volume becomes visible. In the simplest way of doing this, one just cuts out a sub-cube such that the rest becomes visible. Alternatively, one can extract certain parts of interest in the survey and dismiss the rest. This becomes particularly powerful if object extraction methods are used that isolate certain geological bodies from the rest of the data. In petroleum geology, obviously, the targets of such an extraction are commonly the reservoir layers. There is a large variety of algorithms designed to extract these objects, all adapted from image processing (e.g. Hassibi & Ershagi 1999). Once extracted, the shape and internal structure of these bodies can be analysed, often visually and with the help of 3D visualization centres. From analogy with sedimentary models, an interpretation is then made of their origin and, if possible, of their reservoir properties.

HRS data also offers the unique opportunity to quantify and parameterise the 3D geometry of these bodies. This is, strictly speaking, not possible from outcrops, although lower and upper bounds for their dimensions can be derived from cliff faces as well as

[1]There are no outcrops providing *full* three-dimensional information (so-called 'three-dimensional outcrops').

From: DAVIES, R. J., POSAMENTIER, H. W., WOOD, L. J. & CARTWRIGHT, J. A. (eds) *Seismic Geomorphology: Applications to Hydrocarbon Exploration and Production*. Geological Society, London, Special Publications, **277**, 121–137.
0305-8719/07/$15.00

from plan views, or from a combination of the two (Alexander 1992). The purpose of parameterizing the geometry of geological bodies is to obtain numerical values that can be used for reservoir modelling and prediction. Knowing, for example, the sizes and spacings of point bars in an ancient fluvial system, the volume outside the HRS data set can then be populated with improved confidence. The goal of this paper is to describe an approach developed for doing this using an HRS data set from a fluvio-estuarine setting and recent fluvial systems thought to be representative analogues.

Data set

The HRS data set used here is from the Tambaredjo oil field in Suriname (Guyana Coast, South America), operated by Staatsolie Maatschappij Suriname N.V. (Wong 1998). The survey was acquired in 2001 over about 80 km^2 with a bin size of 6.25 m in both lateral directions, and a depth resolution of 2.5–4 m. The reservoir is in the Palaeocene fluvio-estuarine Saramacca Formation at an average depth of 350 m and contains heavy oil. Data from more than 500 wells drilled at a spacing of 200 m were also available and proved crucial for validation. The HRS data set was pre-stack migrated and processed with a focus on obtaining optimal data at the depth of interest. An inversion to acoustic impedance was made using well data as control points. Furthermore, an interpretation was made by the operating company for the main horizons, which, in stratigraphically ascending order, are: (1) the top of the Cretaceous, which is a significant regional unconformity; (2) the T_1 and T_2 reservoir sands, both of which are laterally highly discontinuous; and (3) the 'hardebank' (Dutch for 'hard layer'), a laterally extensive limestone layer above the two reservoir sands that causes a very strong acoustic reflection because of its relatively high density compared with the surrounding rocks. The Palaeocene is essentially contained between the first and third of these reflectors and has on average a thickness of 40 m, indicating that this is a sequence with very low accommodation space (cf. Hardage *et al.* 1996).

Figure 1 shows a perspective view of the 'hardebank' reflector over the whole area with selected wells shown as vertical lines. The surface is shaded as a function of the reflected amplitude. It shows a general northward dip and subtle but distinct topographic features, mostly in the form of an east–west striking fault that offsets the northern block downward with respect to the southern block. Such faults are typical for the northern part of the Guyana shield (Wong 1976, 1998). The figure also illustrates the slightly undulating character of the surface, which is caused by secondary small-scale faulting, differential compaction and slight tectonic folding. In time slices, this deformation makes it very difficult to identify sedimentary bodies, as one continually changes into different stratigraphic levels. It has become clear, however, that the faulting is to a significant degree syn-sedimentary, and that it has considerably influenced the fluvial drainage system during Palaeocene times (Leeder 1993). While the upthrown (southern) blocks commonly display a channelized drainage pattern, in the downthrown blocks a much more dispersive pattern can be observed with east–west oriented valley-fills at the foot of the faults, such as the dark area in Figure 1. From these larger sedimentary units geological bodies were identified and extracted by processing the seismic images using a variety of 3D low-level

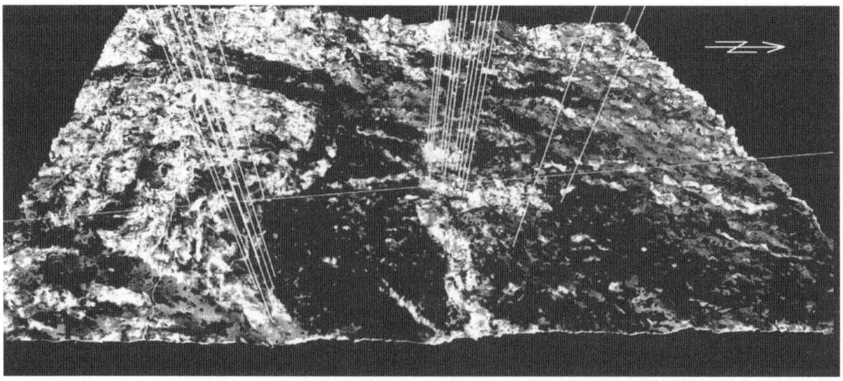

Fig. 1. Perspective view of the 'hardebank' surface and a few selected wells drilled into the underlying sandstone reservoirs. The area covers roughly 10 × 8 km^2. North is to the right (arrow). Notice the subtle northward tilt of the surface with a normal fault and associated changes in the seismic reflectivity, indicating fault-controlled sedimentation. Lighter shades represent higher acoustic reflectivities.

image processing techniques and high-level pattern recognition methods.

Construction of the GeoTime cube

Because of the irregular shape of the geological markers and the numerous small-scale faults, geological object extraction proved exceedingly difficult. It was therefore decided to create a new volume wherein the 'hardebank' marker and the Top Cretaceous marker represent the top and the bottom of the volume. These two markers are then flattened and the data in between is interpolated such that the volume contains an equal number of data points at each geographic location. If one assumes, to a rough approximation, that the two bounding markers of this limited volume are isochronous surfaces, then the vertical distance in between represents geologic time, albeit not in a linear fashion since the sedimentation rates certainly have varied throughout the Palaeocene. This method of creating a seismic volume in which the vertical axis is geologic time is a full 3D version of the 'stratal slicing' proposed by Zeng *et al.* (1998*a*, *b*) and of the 'proportional slicing' discussed by Posamentier *et al.* (1996). Stark (2004) also refers to a volume in which the vertical axis is 'relative geologic time' and which he recommends as a method for detailed sedimentological interpretations of seismic data.

The interpolation of the data between the two delimiters can be done using different methods such as nearest-neighbour interpolation, linear interpolation or spline interpolation. Because the interval to be interpolated is very small (ranging from 15 to *c*. 30 m), choosing the spline interpolation turned out to be very effective. The resulting volume is named GeoTime cube, and its thickness is a parameter that can be freely chosen. Possible choices for this are: the minimum time difference, the maximum time difference or any other value in between. In order to avoid aliasing in the resampling process, the maximum thickness is preferred. The procedure is thus nothing more than the horizon flattening commonly used in seismic interpretation, but here it is done on two seismic reflectors simultaneously (Fig. 2). The main purpose of this procedure is to facilitate the tracking of sedimentary elements, and to extract them subsequently. Ideally, any regional dip present during deposition is eliminated (i.e. set to zero), and all post-depositional deformations such as faulting or folding are undone by the flattening of the two delimiters. In practice, however, this is a very delicate operation both processing- and interpretation-wise. The method is applicable when prominent seismic markers are likely to have been deposited over relatively short geologic times, but it is to be avoided when prominent clinoforms or unconformities are present. Inaccurate horizon tracking of either delimiter can lead to unwanted and exaggerated distortions on the GeoTime cube. In general, the method is considered more applicable in areas where vertical aggradation is dominant and where the vertical interval is small. The procedure is reversible, i.e. once extracted, the vertical dimension of the objects can be restored into two-way travel time or depth. For proper volumetric calculations, this reverse step is a necessity but in order to complete it, a record of the mapping function from original seismic data to the GeoTime cube has to be kept.

Object extraction

Artificial neural networks (ANNs) have previously been used to cluster and classify seismic volumes (Aminzadeh *et al.* 1999; Poulton 2001). In experiments using unsupervised ANNs it became clear that a combination of seismic attributes from the Palaeocene interval can successfully be used to segment the entire data set into regions corresponding to different depositional settings, such as channelized fluvial belts or mud-prone floodplains. The principal goal, however, is to extract the essential building blocks of these depositional environments.

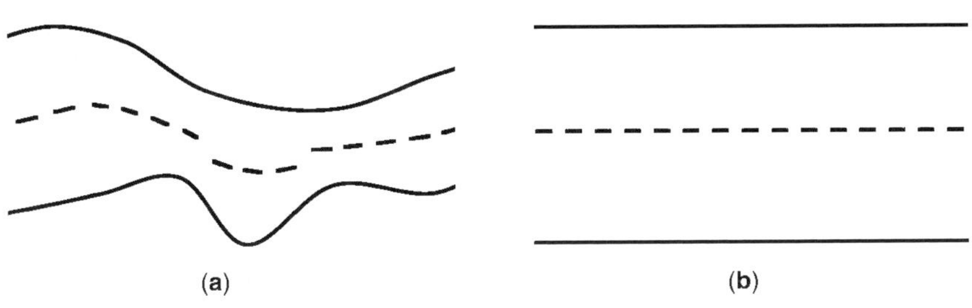

Fig. 2. Schematic sketch illustrating the construction of the GeoTime cube: (**a**) the original seismic data with two geological isochronous markers indicated by solid lines and an intercalated geological object shown by the dashed line; (**b**) the two markers are simultaneously flattened at their highest and lowest point respectively, and the data in between is interpolated. The geological object is now continuous.

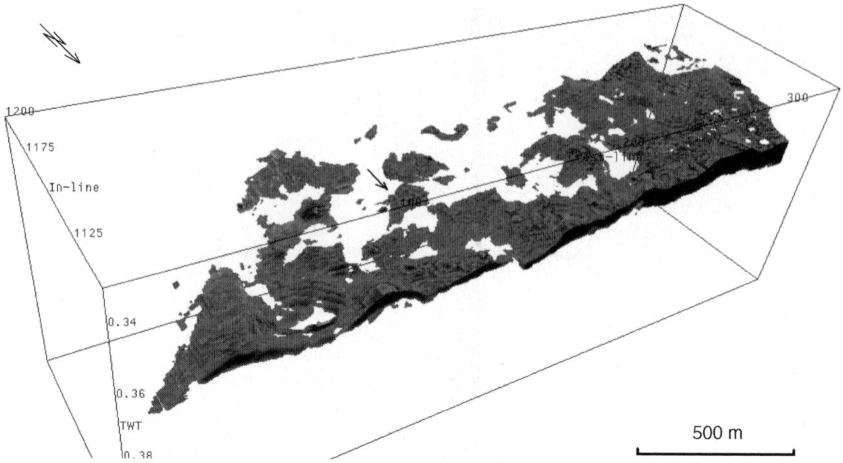

Fig. 3. Perspective view of the T_2 sands extracted with a supervised ANN over the test area. Arrow indicates point bar.

Two approaches were found to be particularly suited for this task: seismic classification using trained (or supervised) ANNs, and voxel-growing.

The supervised ANN used is a feed-forward back-propagation multi-layer perceptron which is trained using a set of control points. The latter is selected from well data where the lithologies can be accurately determined from core and log data. These control points are considered the ground truth, and hence the training targets for the ANN. Five lithologies are distinguished: limestones, lignites, shales and the two reservoir sands (T_1 and T_2). The training set for the neural network consisted of the known lithofacies at a number of well locations, and their corresponding instantaneous seismic attributes (Yilmaz 1987). Instead of the original seismic reflection data, the inverted acoustic impedance data was used because it contains information that is more directly related to lithological and petrophysical properties. Through trial and error the following seismic attributes were found to be best suited: seismic energy, Laplace edge enhancement, velocity fan, mathematical difference stack, amplitude average and a Laplace filtering on similarity (Aminzadeh *et al.* 1999). Typically, these attributes are computed using sliding windows covering $2 \times 2 \times 2$ or $3 \times 3 \times 3$ data points. Once trained, the ANN then determines for every voxel a probability for each of the five lithologies. Post-processing evaluates the most likely lithology, whereby neighbourhood criteria are taken into account in order to reduce the number of isolated voxels. The output of the neural network is a 3D probability cube for each lithology.

An HRS area of approximately 3 km^2 was selected to test the supervised ANN segmentation. Although small, the area has a high number of wells (27) with good wireline logs and it contains a sufficient variety of lithologies as well as synsedimentary faulting. Figure 3 shows the classification result for the T_2 sand. It illustrates that the sand is not a continuous sand body, but that it consists of many small patches, some of which have distinct shapes that can be geologically interpreted. For example, the sickle-shaped body indicated by an arrow is interpreted to be a point bar. There is a clear westward thickening trend of the sandstone with a concurrent welding, or amalgamation, of the sandstone bodies.

The second method used to extract geological objects is a 3D growing algorithm tool developed by Myers & Brinkley (1995) and implemented in the 3D visualization software[2] used here. A voxel belonging to the object to be segmented is defined as a seed point, and the range of allowed variation in both the vertical and the horizontal directions is defined for the growing process. The algorithm then searches in all directions and identifies spatially connected voxels within the specified range as belonging to the object. The growing can be limited by defining an area or surface beyond which the growing is not allowed to proceed. The method was applied to acoustic impedance as well as seismic reflection data. Figure 4 shows a channel identified with this method. It has two areally extensive appendices that are interpreted as crevasse splays.

In general, it was found that channel segments are often missing due to erosion or (to a lesser degree) data acquisition problems, such as in the topmost

[2]Inside Reality, a Mark of Schlumberger.

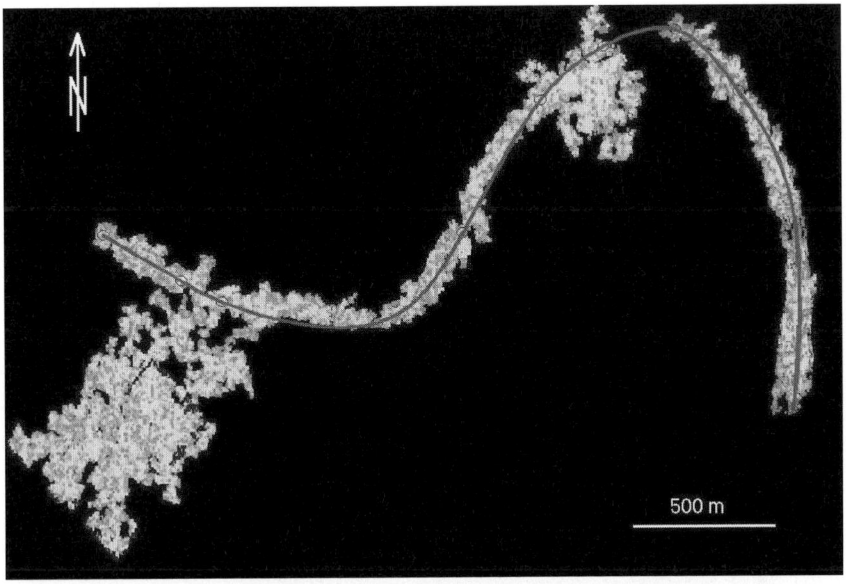

Fig. 4. Plan view of a channel with possible crevasse splays (left and top) extracted with the voxel-growing method.

part of the channel shown in Figure 4. In order to complete the channel path and thus be able to continue the search for its continuation in the data set, a cubic spline is fit to the channel axis. Points where the channel curvature is either maximum or minimum are selected, and a spline is fitted through these points, defining the channel axis also over the missing pieces (solid line in Fig. 4). As long as the missing gaps are not too long, this method is found to give visually acceptable results. If, however, an inflection point is missing, the fit can become relatively poor.

Parameterization of geological objects

The geological objects extracted from the data set very often had been partially eroded or otherwise affected by post-depositional processes such that they do not form shapes that can be quantitatively described in a simple way. Since parameterization is a principal objective of this work, it was decided to analyse recent systems that are much less affected by erosion and tectonics and that are thought to be suitable analogues. They are studied with the aid of aerial photographs and satellite images.

Recent analogues

Figure 5 shows a satellite image of the Cosewijne River, located a few kilometres south of the Tambaredjo oil field. It is one of many rivers that flow from the Guyana shield northward into the Atlantic Ocean. These rivers currently do not carry large sediment loads, mostly because the hinterland consists of basement rocks that erode very slowly (the Precambrian Roraima sandstone formation has been largely eroded in Suriname), but also because accommodation space is low under the present high sea level conditions. These rivers meander through mud-rich flood plains in their lower reaches, but they show accretion surfaces that indicate point bar developments, perhaps formed at times of greater accommodation space. The present situation is, therefore, to some degree comparable to the one during the Palaeocene, although the large amounts of mud transported by the Guyana current from the mouth of the Amazon contrast with the situation during the Palaeocene, when clastic supply from the Andes was not yet effective.

Other analogues are found in areas where accommodation space is greater and sand bodies could better develop, such as the Western Siberian Lowlands. In this vast region there are numerous meandering rivers that range from small to very large. Figure 6 shows a single, abandoned meander loop next to the currently active river in the vicinity of the town of Omsk. The snow highlights the accretion surfaces of the abandoned loop, indicating that the entire inner bend belongs to the point bar. This example can therefore be used to evaluate suitable parameterization methods for point bars, which are the principal sedimentary deposits of meandering rivers (Miall 1996). On another example, shown

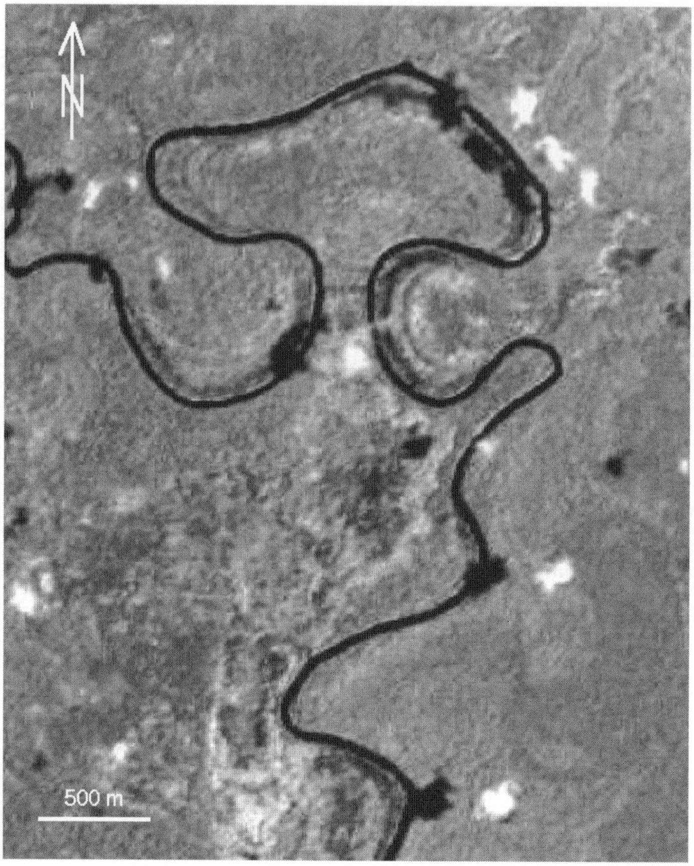

Fig. 5. Satellite image of the Cosewijne River in Suriname. Although the present sediment load in the rivers is low, there are clear indications of point bar development in some of the meander loops.

in Figure 7, the path of a meandering river can be followed over 10 meander loops, each of which shows point bar accretion, although not as clearly as in the previous figure. This example can be used to parameterize the channel as well as the point bars.

Parameterization of channels

Natural cubic splines (Lee 1989; Farin 1997; Kharab & Guenther 2002) are suitable descriptors for modelling the paths of sinuous rivers. The natural cubic spline is an interpolation method that connects two adjacent control points. In contrast to other curve fitting methods each curve segment has a unique equation, while still constraining the curve to fit the data properties at the control points. At these control points the spline is continuous and twice differentiable, a property that is needed for computing the curvature along the path of the spline. The average width of the river is easy to measure on the images and, for the sake of simplicity, is here assumed to be constant over the area analysed. A number of properties can be obtained from the spline fit, but the focus of attention is on the variations of the curvature along the channel path. Additional properties to be derived from the imagery include the main drainage trend of the meander belt and the sinuosity of the river (defined as the distance along the channel axis over a certain interval divided by the distance from the beginning to the end point of the interval).

Figure 8 shows two natural cubic spline fits to the channel of Figure 7. The fit at the top uses a large number of control points, while the fit at the bottom uses a much smaller number. In the second case, only those points with maximum curvature and the inflection points between the left and right turning loops are used. The second curve still offers a very good description of the river path and is represented by the spline coefficients and the coordinates of the control points only. As is observed, all relevant features are present in the second curve despite this sparse representation.

The parameterization of ancient rivers starts with the extraction of channel voxels in the HRS volume. They are identified using the GeoTime cube and

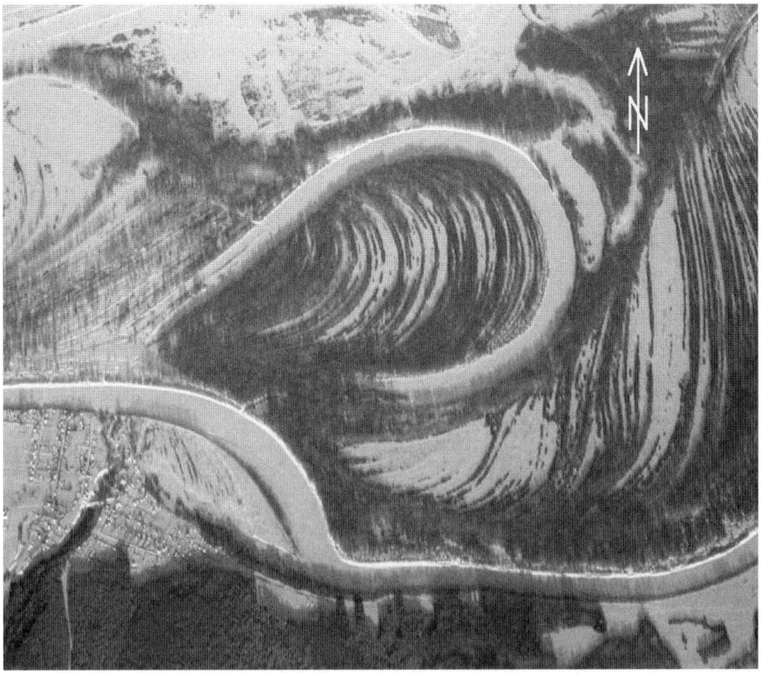

Fig. 6. A single abandoned meander loop next to the active channel of a river in the Western Siberian Lowland. Point bar accretion is highlighted by the difference in snow cover on flat and vegetated land. The photograph is taken from an airplane and is approximately corrected for parallax. The settlement in the lower left corner indicates approximate scale.

image filters that reduce noise and sharpen edges (Gonzalez *et al.* 1992; van Vliet 1993; Soillé 1999) such that the identification and extraction of geological bodies becomes easier. The coherency enhancing diffusion filter (Weickert 1999), when applied to a uniform filtered image, gave good results with the acoustic impedance cube. Next, the identified channel voxels are modelled as curvilinear objects with the central path described by natural cubic splines.

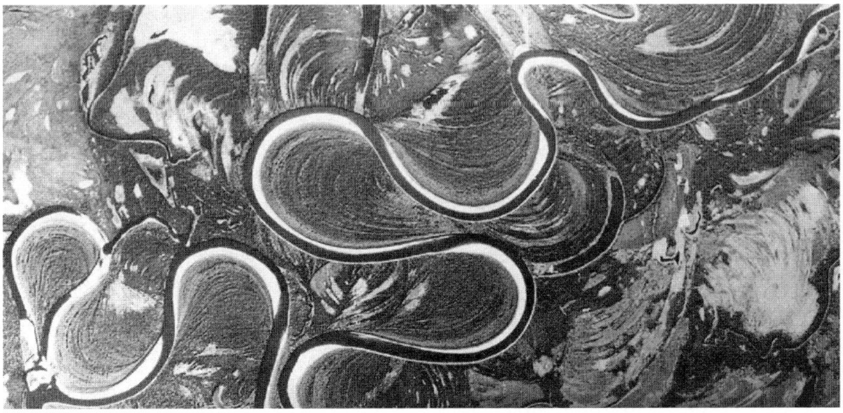

Fig. 7. Aerial photograph of a high-sinuosity river with prominent accretion in the meander loops leading to large point bars. Notice older meander systems away from the currently active channel. Orientation unknown.

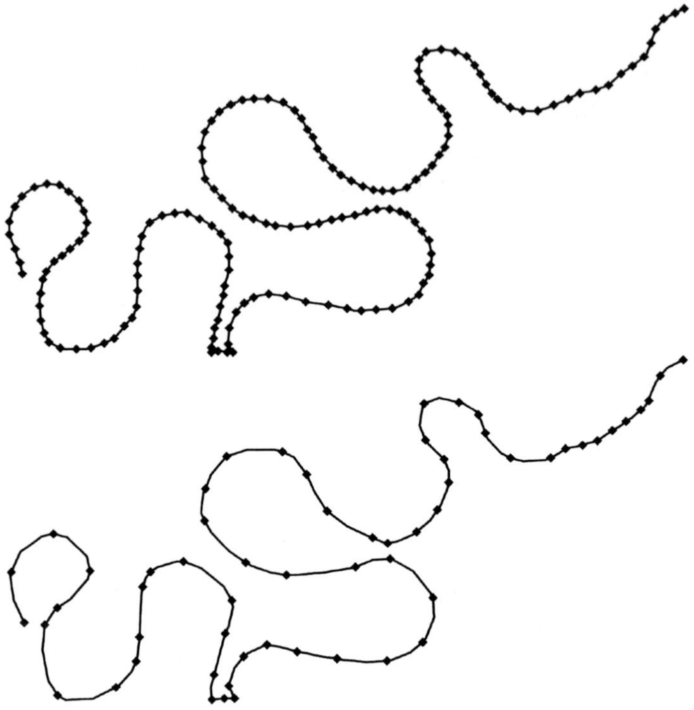

Fig. 8. Curve generated by the natural cubic spline interpolation to model the river channel of Figure 7 using a large number of control points (top) and a minimum set of points (bottom). Notice that one loop is truncated at the bottom of the figure.

The channel width is determined by analysing the voxel properties perpendicular to the channel axis. The shape of the channel has been extracted with one of the methods described earlier, and the procedure now consists of finding intervals within which the borders, determined by the voxel growing method do not change significantly. Plotting the amount of voxels grown for selected cross sections while varying the growing parameters yields an approximately constant interval in the resulting curve. The average across this interval is then taken as the local channel width. A possibility is to carry this calculation out at every control point used for the spline fitting, or alternatively to calculate it for the entire interval. In Figure 9, a channel is shown as an ensemble of voxels extracted from the seismic data set (top), and the channel representation with a minimum number of parameters from the cubic spline fit and a constant channel width (bottom). This average channel width has been determined by analysing the number of voxels that fall within the object for a given range allowed by the voxel-growing algorithm and determined by the user. By increasing the range, the channel volume will also increase. When plotting these two variables against each other, an interval can be seen where the volume hardly grows when increasing the range (Fig. 10), indicating that these are the natural boundaries of the channel. The average width within this range is then taken as the channel width along the entire interval. Figure 9 also shows that, despite this simple approximation, the comparison between the originally extracted channel and the parameterized model is reasonably good. However, important considerations are that the modelled channel is continuous, i.e. the missing segments have been adequately filled in, and that the model representation is several orders of magnitude smaller than the seismic data. In cases where the path of the channel cannot be modelled with a single spline curve, it is recommended to subdivide the path into small segments that can be modelled using a single spline. This subdivision can be complex but is important for the connectivity in case the channel fill is a reservoir.

Curvature along the river path

The considerable natural variation in sinuosity[3] of rivers is one of the bases for classifying fluvial

[3]Most rivers paths cannot be well approximated by sinusoids, but we maintain the term here.

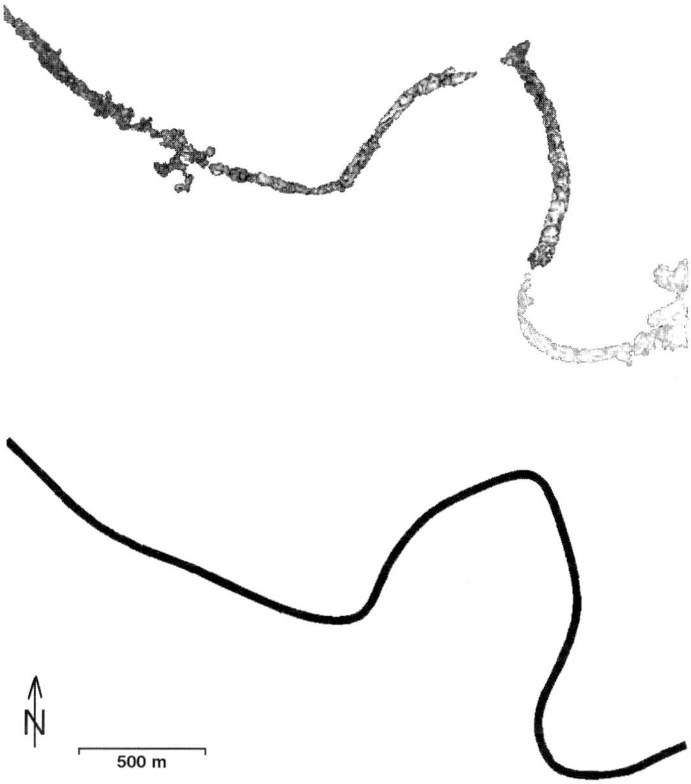

Fig. 9. Channel extracted from the seismic data (top) and modelled channel using cubic splines and an average constant width (bottom) determined with the aid of the graph in Figure 10.

systems (e.g. Reading 1986), although the causes for rivers developing sinuous courses are still not fully understood. Variation in sinuosity is equivalent to variation in the curvature of the curve that fits the river path. The curvature of a curve is mathematically defined as the rate of change per unit length of the direction of the curve (Weisstein 2002). The simplest form of curvature is an extrinsic curvature. In two dimensions, let a plane curve be given by Cartesian parametric equations $c: R \to R^2$, $c(t) = [x(t), y(t)]$. Then the curvature κ is defined by

$$\kappa(t) = \frac{d\phi}{ds} = \frac{\frac{d\phi}{dt}}{\frac{ds}{dt}} = \frac{\frac{d\phi}{dt}}{\sqrt{\left(\frac{dx}{dt}\right)^2 + \left(\frac{dy}{dt}\right)^2}} = \frac{\frac{d\phi}{dt}}{\sqrt{(x')^2 + (y')^2}} \quad (1)$$

where R is the set of real numbers; x, y are the Cartesian coordinates of the curve c; t is the parameter for curve interpolation (Lee 1989); ϕ is the tangential angle and s is the arc length. As can be seen from the definition, curvature has a unit of inverse length and the sign of the curvature allows discrimination between changes in clockwise and counterclockwise directions. For parametric curves as produced by our splines, the term $d\phi/dt$ is eliminated and the curvature expressed as a function of first and second derivatives of x and y with respect to t:

$$\kappa(t) = \frac{x'y'' - y'x''}{[(x')^2 + (y')^2]^{3/2}} \quad (2)$$

Special attention to the computation of the curvature $\kappa(t)$ was paid, not only because curvature is an indicator of the amount of bending of the curve at every position t along the river path, but also because one of the goals of this work was to study the sediment deposition at the inner bank of the river bends. If the curvature at every point of a river path segment is zero, then the path is a line segment

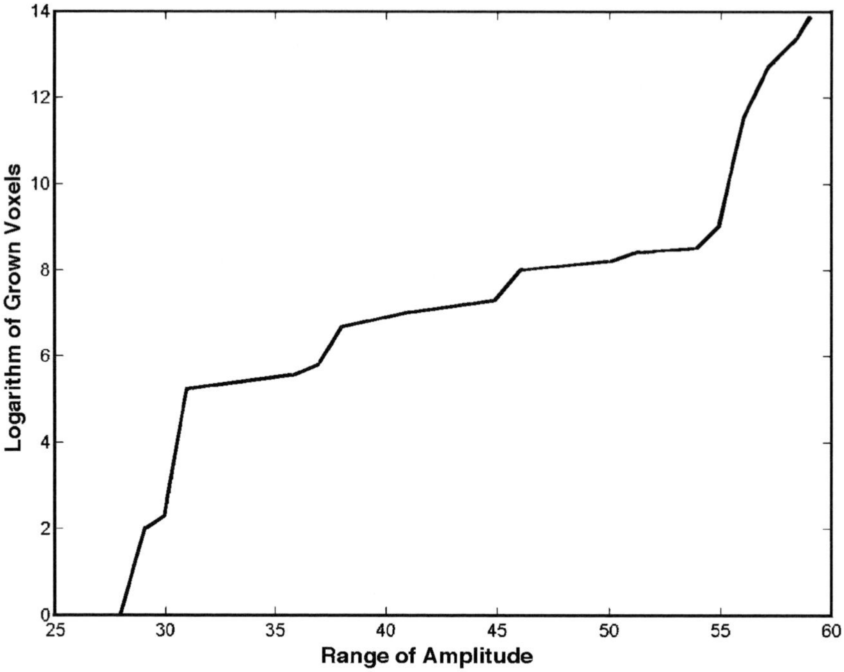

Fig. 10. Cross-plot of the amplitude range allowed in the growing algorithm vs. the total number of voxels in the extracted channel. The relatively flat interval in the middle of the curve is interpreted as indicating the optimum range of amplitudes for extracting the channel.

with no 'bending'. If the curvature at each point of a specific segment along the river path is constant, the river path follows a circular arc. The sediments deposited at the bends form point bars, which can form important hydrocarbon reservoirs.

The computation of the derivatives that constitute the curvature can be taken at various scales (Witkin 1983; Koenderink 1984). The derivative at scale σ is defined as

$$x'_\sigma = \frac{x_\sigma}{dt} \equiv \frac{x(t) * g(t;\sigma)}{dt} \quad (3)$$

where σ has the dimension length, * denotes the convolution operation and $g(t;\sigma)$ is a normalized Gaussian function:

$$g(t;\sigma) = \frac{1}{\sqrt{2\pi\sigma^2}} \exp\left(-\frac{t^2}{2\sigma^2}\right) \quad (4)$$

Using the derivatives at a particular scale allows selection of the proper curve interval for the calculation, i.e. focussing on the meandering loops rather than insignificant local variations caused by noise. This scale parameter will therefore be adapted to the inherent dominant size of the river's turns. This works as a feature selection process and yields a robust result that is also independent of the initial set of control points.

The radii of curvatures along the river path are plotted in Figures 11 and 12 for the two recent rivers, and for the ancient channel extracted from the seismic volume in Figure 13. The radius of curvature is defined by

$$R = \frac{1}{\kappa} \quad (5)$$

which preserves the sign of the computed curvature. At a given point on a curve, R is the radius of the osculating circle.

From Figures 11–13 it can be seen that for the scale parameter $\sigma = 5, 7, 10$ the curvature is equal or close to zero in the straight-line segments and inflection points between left and right turns. This means that the limit of the radius of curvature goes to infinity in those regions. Otherwise the radius of curvature has a finite value and is approximately constant. From this it is concluded that the meander loops can be approximated by circle segments whereas their connections can be approximated by straight-line segments. A manual fit of circles to

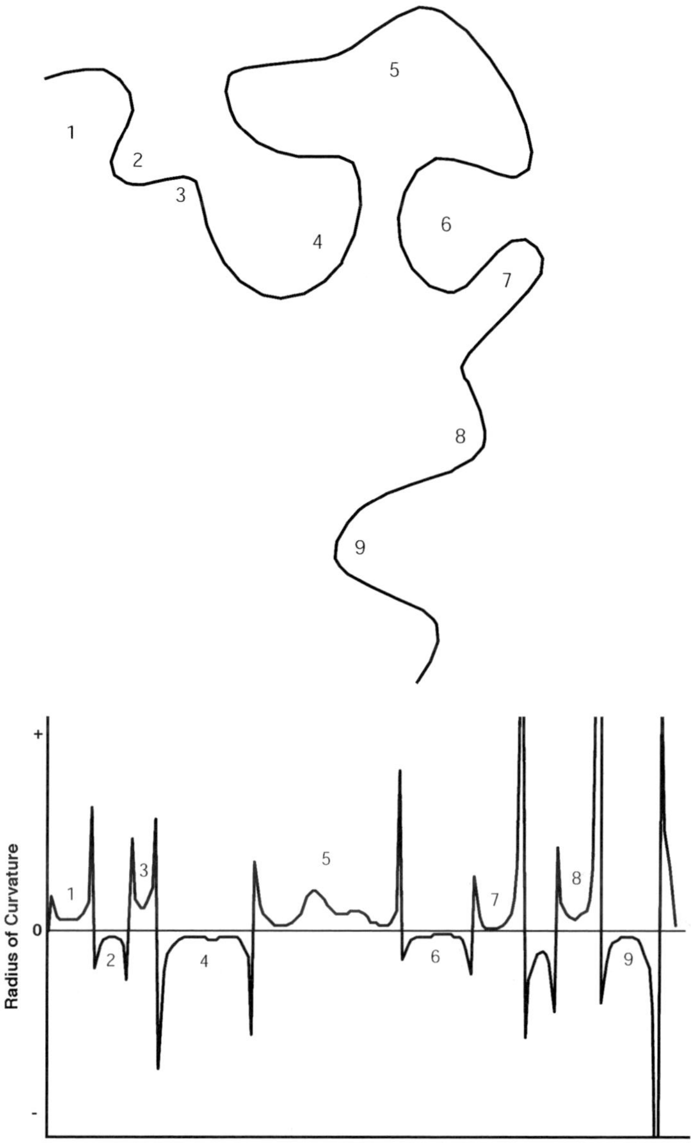

Fig. 11. Radius of curvature along the Cosewijne River (Fig. 5) computed with scale parameter $\sigma = 5$ as a function of the distance along river path (in arbitrary units). Numbering of meandering loops on map (top) is the same as on the cross-plot (bottom).

meander loops is shown in Figure 14. The sum of the lengths of the circular arcs fitted to all meander loops plus the straight-line segments determine the sinuosity, which here is found to be larger than $\pi/2$. This is very high, indicating that the two ends of a loop get fairly close together and channel cut-off may occur. The river will abandon the loop and find a new path, where new meandering will be initiated.

Parameterization of point bars

Point bar deposits are formed by sedimentary accretion on the inner bank of a meander loop. They have a corrugated accretionary topography, the scroll bar, which results from episodic lateral accretion (Schumm 1977; Reading 1986; see also Fig. 6). Accretion surfaces represent past positions of the

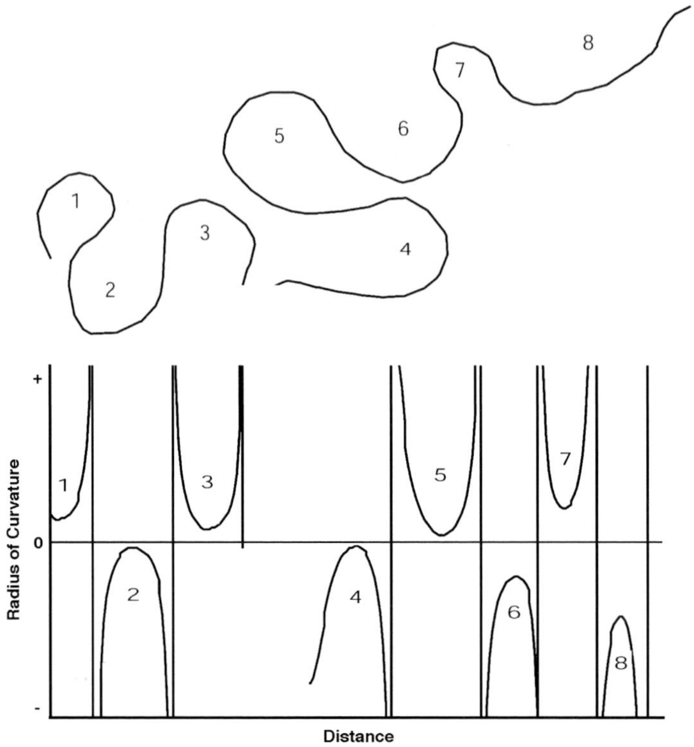

Fig. 12. Radius of curvature along the river shown in Figure 7 computed with scale parameter $\sigma = 7$ as a function of river path (in arbitrary units). Numbering of meandering loops on map (top) is the same as on the cross-plot (bottom).

river channel that evolved through time until the meander loop was abandoned. From the previous analysis it was concluded that meander loops can be modelled with a circular segment. If a point bar is defined as the sedimentary deposit between an initial and a final position of the meander loop, then its outline can be modelled with two intersecting circle segments, forming a crescent as shown in Figure 15. It is defined by four parameters: the two centres of the circles (indicated by c1 and c2); and their two radii. In practice, however, these parameters cannot be directly determined. Rather, the outline of the deposit has to be determined and a fit of two circles has to be made to this outline. Ideally, three points are needed to define the position of a circle, or six points to define two circles. If, however, two of these points are chosen at the end points of the crescent, then they are common to both circles and the outline of the point bar can be determined from four points alone (Rivera Rabelo *et al.* 2005). These four points are indicated as P1–P4 on Figure 15.

The approach can be tested with the point bar from Figure 6. For each distinct accretion line, a circle is fitted by choosing several points along it. The result is shown in Figure 16. Not all accretion surfaces are well fitted by the circles, because the depositional process varies over time, but the overall match is considered satisfactory. There are a few interesting observations to be made from Figure 16. First of all, the circle centres, indicated by numbers on the image, shift along an almost straight line. Secondly, the circle radii grow slightly as the meander loop evolves over time. There is also an area in the southwestern (lower left) area where the fit is less good because the accretion surfaces appear more elliptical than circular, but this area covers less than 10% of the total area of the point bar.

The sequence of circle centres in Figure 16 can be considered the axis of accretion, or the line along which the centre of the meander loop shifts. As a consequence of this migration, together with the observation that the radius of the meander loops increases with time (Fig. 17), the sinuosity of the river increases with time. The circular segments fitted to the loops increase for each successive accretion episode, meaning that the river erodes successively more of the older sediments. This is clearly visible in Figure 16, where considerably less than a semi-circle is preserved of the first accretion surface to which a

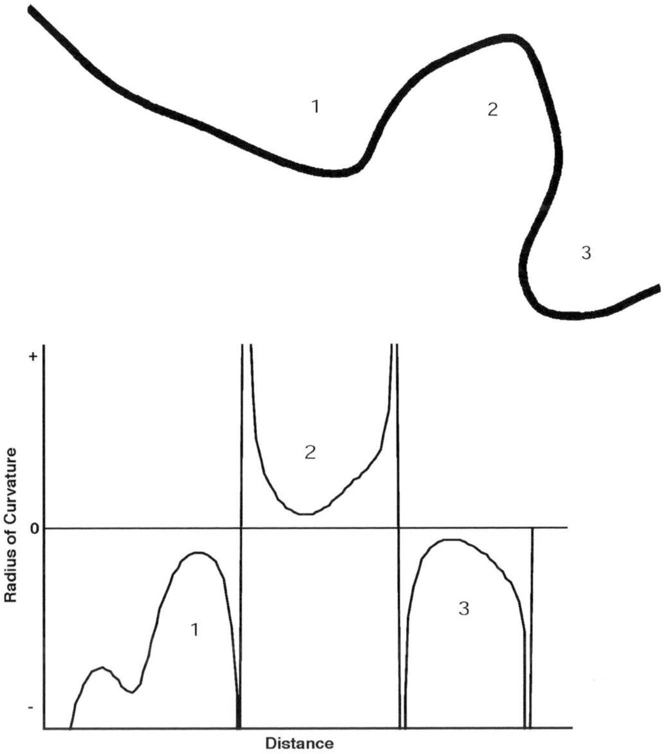

Fig. 13. Radius of curvature along the channel extracted from the seismic data (Fig. 9) computed with scale parameter $\sigma = 10$ as a function of river path (in arbitrary units). Numbering of meandering loops on map (top) is the same as on the cross-plot (bottom).

circle was fitted, while slightly more than a semicircle is currently being deposited, as shown by the last (rightmost) circle. These active depositional areas are visible as white sandbanks on Figure 7. A fit of successive circles to the meander loops of that river is shown in Figure 18. Again, most meander loops are seen to grow in size over time, and most circle centres follow a more or less rectilinear path.

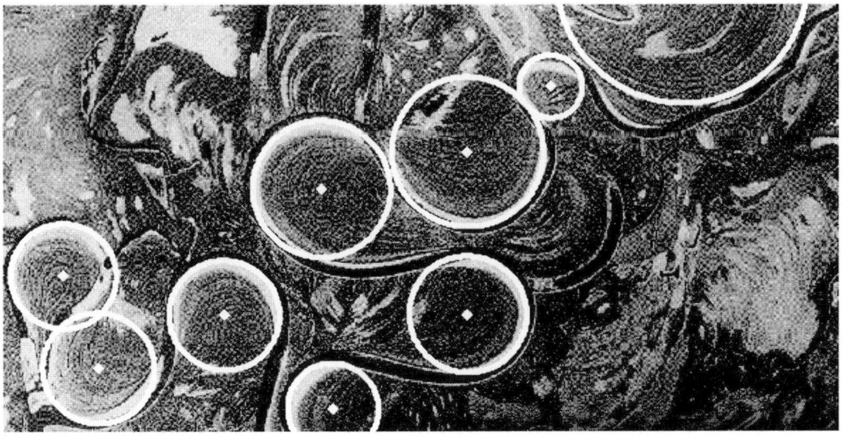

Fig. 14. Circles fitted to the meanders loops of the river in Figure 7. The fits are good over more than a semi-circle for most meander loops, indicating a very high river sinuosity. Notice that most connections between meander loops are short and can be approximated by straight lines.

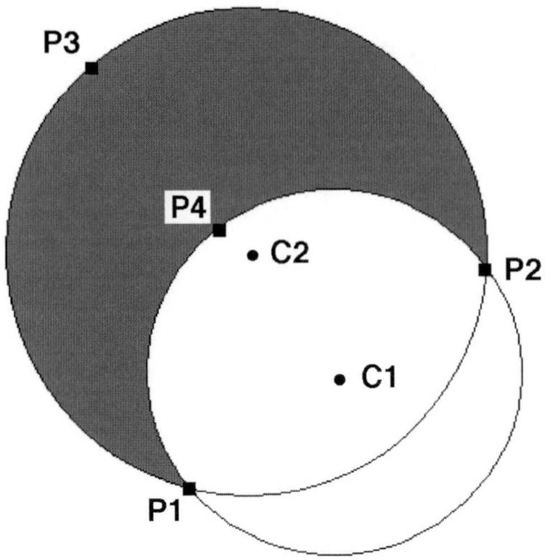

Fig. 15. Two-dimensional model of a point bar (grey).

The average local flow trend of the channel belt in a meander loop can be defined as being orthogonal to the axis of accretion. This line is parallel to the connection of the end point of the point bar crescent (points P1 and P2 in Fig. 15). Connecting these lines from a number of successive meander loops, or point bars, gives the average flow trend of the river. Another way of determining the general flow trend is by connecting the circle centres of the most recent accretion surfaces, shown in Figure 19 by thin white lines. The average of these individual directions over the entire area gives the regional flow or drainage trend and is indicated the thick white line.

The procedure can be applied to ancient point bars. From the HRS data set (specifically from ANNs outputs volumes classified as sand units), several point bars are identified. They are extracted from horizontal slices of the GeoTime cubes as features with crescent shapes. The lower sandstone interval, the T_1 reservoir, has a high degree of amalgamation and significant erosion that makes it poorly suited for this purpose. The higher T_2 layer shows more isolated channels and associated point bars. Four point bars extracted from it are shown in Figure 20. After extraction of these geological bodies, circles are fitted to the inner and outer limits of the crescent. It is obvious that not all point bars are complete crescents, but that some parts are missing. This happens commonly at either end point and is likely caused by erosion. It is, therefore, not possible to determine the end points of the crescent as suggested in Figure 15, and other points along the two circle segments must contribute to the best fit of the two circles. Therefore, points are chosen along the inner and outer boundaries of the crescent, and two circles are fitted as shown in Figure 20. The connections of the two circle centres are shown by straight lines and represent the lines of accretion. Under the assumption that the normal to their average direction indicates the general regional drainage

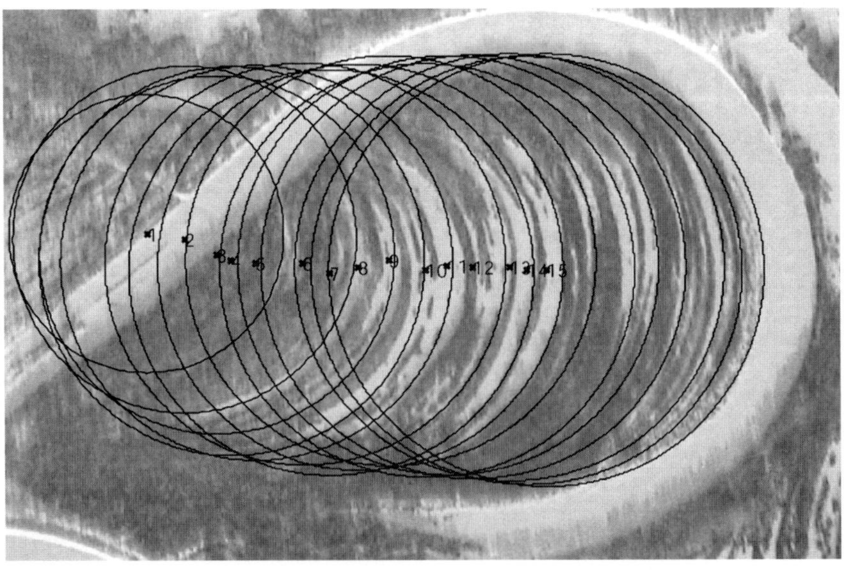

Fig. 16. Circles fitted to accretion lines on the point bar, with their centres indicated by dots and numbers. In total 15 circles have been fitted.

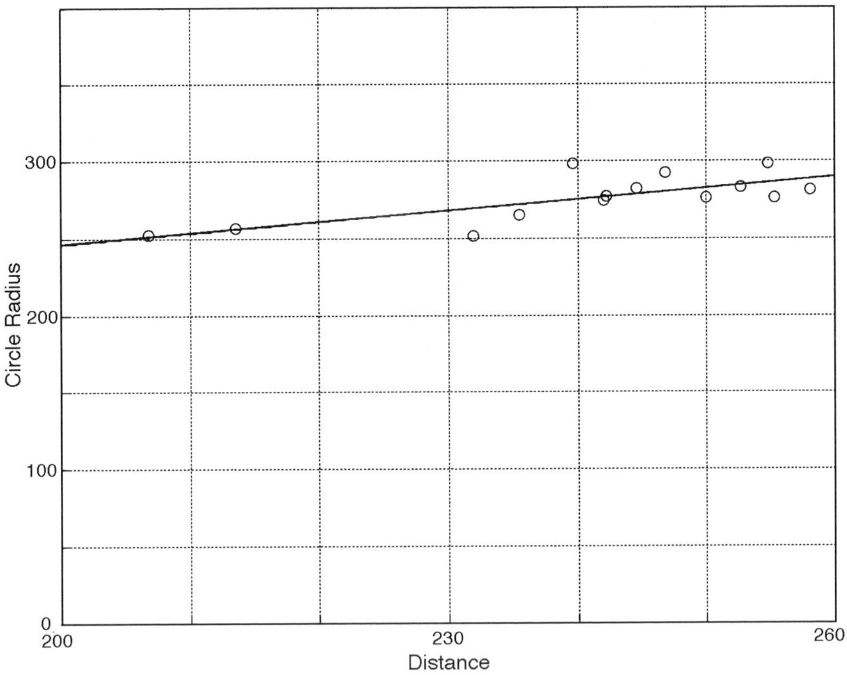

Fig. 17. Radius v. distance of the circles shown in Figure 16 (first circle is left out).

trend, a northward average flow direction is obtained (towards the left in the figure). The alternate solution, towards South, is dismissed for regional geological reasons. The extracted point bars also give a very good idea of their average size as well as their volumes. These parameters are important inputs for stochastic models of such reservoirs and are summarized in Table 1.

Discussion and conclusions

Modern analogues and an adaptation of stratal slicing in the form a the GeoTime cube were used to develop a methodology for describing and parameterizing channels and point bars extracted from a HRS data set. The calculated shape and curvature parameters of the channels allow prediction and

Fig. 18. Circles fitted to successive accretion surfaces of the high-sinuosity river from Figure 7. Circle centres are shown by white dots, with the straight lines fitted to them indicating the axes of accretion of each point bar.

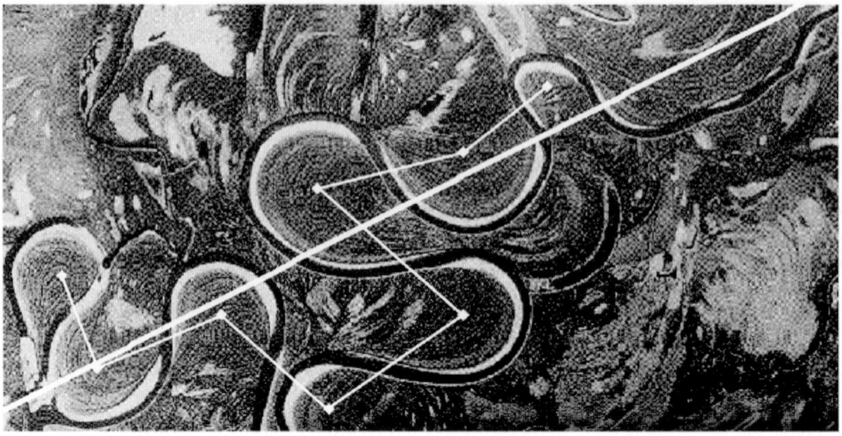

Fig. 19. Connecting the centres of the most recent accretion surfaces of the meander loops (thin white lines) and averaging them gives the regional flow direction of the river system (thick white line).

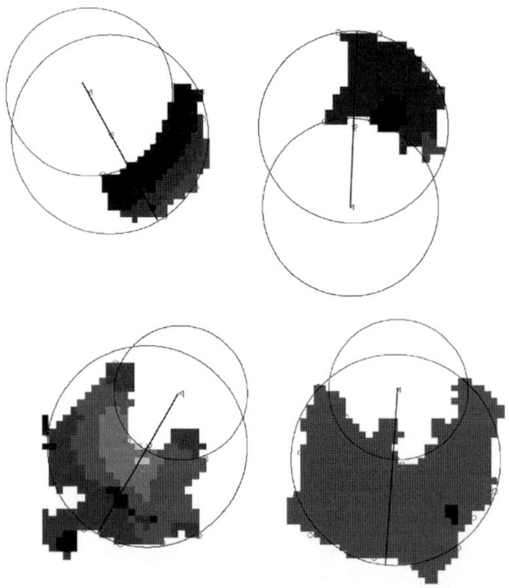

Fig. 20. Parameterized ancient point bars from the HRS data in the Tambaredjo field, with grey shades indicating the thickness. North is left.

extrapolation of them beyond the seismic survey. A simple point bar model is derived from the curvature parameters obtained from ancient channels and recent analogues. It is applied to point bars extracted from the seismic data set and forms the basis for stochastic modelling of similar reservoirs. From the accretion lines of the point bars the average drainage trend of the fluvial system can be determined. Along this trend point bars can be placed outside the control area to build a reservoir model. The volumes of point bars obtained form an important basis for the calculation of hydrocarbon reserves.

A next step in this work is to determine the lithofacies of the objects and analyse the relationship between the intrinsic petrophysical properties and the geomorphology of these geological bodies. The large number of wells available can be used as 'ground truth'. The approach can also be extended to include crevasse splays and other potential reservoir elements. This work can also be extended to include other depositional environments in which the individual building blocks of the sedimentary succession can be identified from HRS.

Table 1. *Main parameters for the four point bars in Figure 20 (numbered from upper left to lower right)*

	Point bar 1	Point bar 2	Point bar 3	Point bar 4
Volume (m^3)	20.5×10^3	23.9×10^3	39.1×10^3	69.2×10^3
Surface area (m^2)	6.9×10^3	8.1×10^3	13.2×10^3	23.4×10^3
Radius of the inner circle (m)	68	74	55	58
Radius of the outer circle (m)	82	81	80	85
Alignment (deg)	337–157	2–182	28–208	3–183

We are grateful to the Netherlands Institute of Applied Geosciences (TNO-NITG) for financial support and to Staatsolie Maatschappij Suriname N.V. for supplying the data. Schlumberger provided the Inside Reality software used in our 3D visualization centre for interpreting the seismic data and its derivatives. Rick Donselaar, Aernout Schram de Jong and Hillbrand Haverkamp are thanked for their contributions. Joe Cartwright and an unknown reviewer provided valuable suggestions for improving the manuscript.

References

ALEXANDER, J. 1992. A discussion on the use of analogues for reservoir geology. *In*: ASHTON, A. (ed.) *Advances in Reservoir Geology*. Geological Society, Special Publications, **69**, 175–194.

AMINZADEH, F., BARHEN, J. & GLOVER, C. 1999. Estimation of reservoir parameters using a hybrid neural network. *Journal of Petroleum Science & Engineering*, **24**(1), 49–56.

BAKKER, P. 2002. *Image structure analysis for seismic interpretation*. Ph.D. thesis Delft University of Technology, Delft University Press.

FARIN, G. E. 1997. *Curves and Surfaces for Computer-aided Geometric Design. A Practical Guide*. Academic Press, San Diego, CA.

GONZALEZ, R. C., WOODS, R. E. & READING, M. A. 1992. *Digital Imaging Processing*. Addison-Wesley.

HARDAGE, B. A., CARR, D. L. *ET AL.* 1996. 3-D seismic imaging and seismic attribute analysis of genetic sequences deposited in low-accommodation conditions. *Geophysics*, **61**(5), 1351–1362.

HASSIBI, M. & ERSHAGI, I. 1999. *Reservoir heterogeneity mapping using an automated pattern recognition approach*. Society of Petroleum Engineers, paper 56818.

KHARAB, A. & GUENTHER, R. B. 2002. *An Introduction to Numerical Methods: a Matlab Approach*. Chapman & Hall, London.

KOENDERINK, J. J. 1984. The structure of images. *Biological Cybernetics*, **50**, 363–370.

LEE, E. T. Y. 1989. Choosing nodes in parametric curve interpolation. *Computer-Aided Design*, **21**, 63–370.

LEEDER, M. R. 1993. Tectonic controls upon drainage basin development, river channel migration and alluvial architecture: implications for hydrocarbon reservoir development and characterization. *In*: NORTH, C. & PROSSER, J. (eds) *Characterisation of Fluvial and Aeolian Reservoirs*. Geological Society, Special Publications, **73**, 7–22.

MIALL, A. M. 1996. *The Geology of Fluvial Deposits, Sedimentary Facies, Basin Analysis, and Petroleum Geology*. Springer, Berlin.

MYERS, L. M. & BRINKLEY, J. F. 1995. Visualization of Brain Surface Features Using Registered Partially Segmented MRI Scans. *Proceedings SPIE Medical Imaging: Image Display*, 43–52.

POSAMENTIER, H. W., DORN, G. A., COLE, M. J., BEIERLE, C. W. & ROSS, S. P. 1996. Imaging elements of depositional systems with 3-D seismic data: a case study. *17th Annual Conference, SEPM Foundation, Gulf Coast Section*, 213–228.

POULTON, M. M. 2001. *Computational Neural Networks for Geophysical Data Processing*. Elsevier, Amsterdam.

READING, H. G. 1986. *Sedimentary Environments and Facies*. Blackwell, Oxford.

RIVERA RABELO, I., LUTHI, S. M. & VLIET, L. J. VAN. 2005. *Parameterization of Point Bars in High Resolution Seismic Data*. Proceedings 67th EAGE Conference, Madrid.

SCHUMM, S. A. 1977. *The Fluvial System*. Wiley-Interscience, New York.

SOILLÉ, P. 1999. *Morphological Image Analysis: Principles and Applications*. Springer, Berlin.

STARK, T. J. 2004. Relative geologic time (age) volumes – relating every seismic sample to a geologically reasonable horizon. *The Leading Edge*, **23**, 928–932.

VLIET, L. J. VAN 1993. *Grey-scale measurements in multidimensional digitised images*. Ph.D. thesis Delft University of Technology.

WEICKERT, J. 1999. Coherence-enhancing diffusion filtering. *International Journal of Computer Vision*, **31**, 111–127.

WEISSTEIN, E. W. 2002. *Concise Encyclopedia of Mathematics*. CRC Press, Boca Raton, FL.

WITKIN, A. 1983. Scale-space filtering. *Proceedings of the International Joint Conference on Artificial Intelligence*, **83**, 1019–1023.

WONG, T. E. 1976. *Tertiary stratigraphy and micropaleontology of the Guiana Basin*. Ph.D. Thesis University of Utrecht. Also in: *Meded. Geol. Mijnb. Dienst Suriname*, **25**, 13–107.

WONG, T. E. 1998. *The History of Earth Sciences in Suriname*. KNAW, Amsterdam.

YILMAZ, O. 1987. *Seismic Data Processing*. Investigations in Geophysiscs. Society of Exploration Geophysics.

ZENG, H., BACKUS, M. M., BARROW, K. T., TYLER, N. 1998*a*. Stratal slicing, part I: realistic 3-D seismic model. *Geophysics*, **63**, 502–513.

ZENG, H., HENRY, B. C., RIOLA, J. P. 1998*b*. Stratal slicing, part II: real 3-D seismic data. *Geophysics*, **63**, 514–522.

Curvature for visualization of seismic geomorphology

B. S. HART & J. A. SAGAN*

Earth and Planetary Sciences Department, McGill University, Montreal, Canada

**Present address: Devon Canada Ltd, Calgary, Canada.*

Abstract: We illustrate the use of curvature attributes for defining stratigraphic features of interest on horizons mapped in three-dimensional seismic data. Curvature is a two-dimensional property of a curve that quantifies how much the curve deviates from a straight line. Many different types of curvature may be defined for a surface, and these can be more useful than dip, azimuth or even 'conventional' (i.e. second derivative) curvature analyses for defining subtle structural or stratigraphic features of interest. In our workflow, we drape curvature over rendered three-dimensional surfaces and adjust lighting to highlight stratigraphic and structural features of interest. The examples we present are derived from clastic and carbonate settings of various ages, and include applications of curvature analyses to multibeam bathymetry and digital elevation model data.

The advent of three-dimensional (3D) seismic technology has provided geoscientists with unsurpassed opportunities for identifying stratigraphic and structural features of interest in the subsurface. Many different visualization techniques have been developed for viewing 3D data volumes, including vertical transects, timeslices, horizon slices, stratal slices, voxel rendering and others (e.g. Brown 1999; Hart 2000). Seismic images derived this way have proven to be useful for identifying stratigraphic bodies such as channels or channel elements in clastic settings (e.g. Posamentier & Allen 1999; Kolla *et al.* 2001; Miall 2002). Applications in carbonate settings are less common, although examples are presented in Eberli *et al.* (2004).

Maps of horizons picked in 3D seismic data may be presented and analysed in various ways, including structure maps (in time or depth converted), surface rendering, trend surfaces and various horizon attributes such as dip, azimuth and curvature. Originally, these horizon attributes were employed to detect subtle structures such as faults or fractures that might have an impact on reservoir performance (e.g. Hesthammer & Fossen 1997; Townsend *et al.* 1998; Dalley *et al.* 1989; Hart *et al.* 2000, 2002; Marroquin & Hart 2004). Current interest in identifying and understanding stratigraphic features, for lithology prediction, reconstructions of geologic history, geohazard prediction or other purposes, has led interpreters to examine the use of horizon attributes for examining stratigraphic features.

In this paper we describe and illustrate the use of curvature attributes, a type of horizon attribute, to study horizons picked in 3D seismic datasets. Our objective is to demonstrate the utility of this technique for seismic geomorphology studies. The examples presented include both clastic and carbonate settings that range in age from Precambrian to Late Tertiary. We also illustrate curvature images derived from multibeam bathymetry and digital elevation data. We present visualization techniques that have helped us to define subtle stratigraphic and structural features in these data.

What is curvature?

Curvature is a two-dimensional property of a curve that quantifies how much the curve deviates from a straight line (Roberts 1998, 2001). Mathematically, it is also the reciprocal of the radius of curvature, i.e. the curvature is high for a small radius of curvature. In two dimensions, positive curvature describes an 'upward' inflection of a line or a convex shape, while negative curvature is a 'downward' inflection or concave shape.

In three dimensions, there is an infinite number of ways in which the curvature can be measured at a given point on a surface. However, the most meaningful measures are those that are defined by planes orthogonal to the surface, called 'normal curvatures' (Fig. 1). Even amongst the family of normal curvatures, there are some that are of more interest than others. For example, as geologists we may be able to identify the strike and dip orientations at a particular point on a surface. We can then derive the curvature in those two mutually perpendicular orientations. The maximum and minimum curvatures (sometimes referred to as 'principal curvatures') on a surface will also be mutually perpendicular, but

From: Davies, R. J., Posamentier, H. W., Wood, L. J. & Cartwright, J. A. (eds) *Seismic Geomorphology: Applications to Hydrocarbon Exploration and Production*. Geological Society, London, Special Publications, **277**, 139–149.
0305-8719/07/$15.00

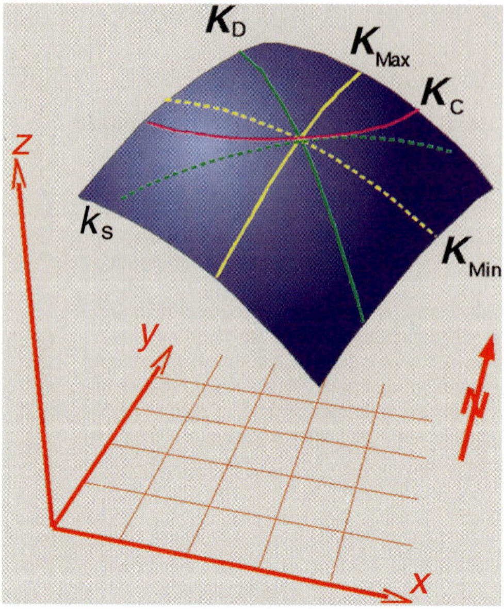

Fig. 1. Curvature is a two-dimensional property of a surface. Among the many different types of curvature that may be derived are curvature in the strike (K_s) and dip (K_d) directions, maximum (K_{max}) and minimum (K_{min}) curvature and curvature along structural contours (contour curvature; K_c). Adapted from Roberts (1998, 2001).

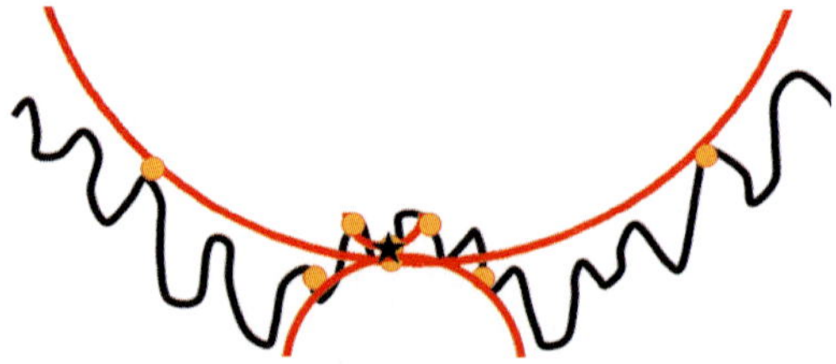

Fig. 2. The way curvature is defined from a surface influences the scale of feature that will be observed. By changing the aperture of the curvature analysis, i.e. how far apart the points used to define curvature are, features of different scales or wavelengths may be imaged. This image shows the differences that might be observed when deriving curvature for a surface with superimposed scales of curvature using a small aperture (concave up), medium aperture (concave down) and wide aperture (concave up).

not necessarily aligned with the strike and dip directions. The product of the maximum and minimum curvatures is known as the total curvature (or Gaussian curvature; Lisle 1994; Lisle & Robinson 1995).

Roberts (1998, 2001) compiled formulae for these and other types of curvature, discussed their meaning and usefulness for defining structural features, and showed how these curvatures could be derived from regularly gridded surfaces such as horizons mapped in 3D seismic data. We use the equations compiled by Roberts but have added the ability to derive curvatures of different wavelengths by changing the aperture of the curvature calculation (Fig. 2; Stewart & Podolski 1998; Stewart & Wynn 2000; Roberts 2001). In structural analyses these different wavelengths of curvature are sometimes sought in order to define fracture orientations, the rationale being that different tectonic fabrics, and therefore fracture orientations, sometimes express themselves as curvatures of different wavelengths. Our experience with stratigraphic analyses is that wider apertures are sometimes effective smoothers for 'noisy' surfaces. Caution must be exercised in the application of this or any other smoothing technique because, although they can reduce noise, they can also smear out small-scale stratigraphic and structural features of interest. Stewart & Podolski (1998) discuss various other aspects of deriving curvature from gridded surfaces.

Curvature is better than the conventional dip and azimuth measures for defining subtle structural and stratigraphic features. Curvature analysis of surfaces removes the effects of regional dip, thus emphasizing small-scale features that might be associated with primary depositional features or small-scale faults. A tilted planar surface has dip but no curvature. When a surface is steeply dipping, for example on the limbs of a fold, high dips can obscure subtle features on the surface. This problem is commonly referred to as 'dip saturation'. Curvature analysis identifies deviations from a planar surface, regardless of whether the surface is horizontal or tilted. The sign (positive or negative) of the curvature also conveys shape information (convex-up v. concave-up, respectively). As discussed by Roberts (2001), the different types of curvature convey different information.

Figure 3 shows the benefits of using curvature rather than dip for defining structural or stratigraphic features. When dip is derived from a simple, regularly folded surface (Fig. 3a), the steeply dipping limbs are highlighted whereas the crest and troughs of the folds are indistinguishable from the dip alone. Curvature measures, such as dip curvature (Fig. 3b), allow the crests and troughs to be distinguished based on the sign of the curvature. If the folded surface is superimposed on a larger fold with a hingeline at right angles to the original fold hingelines, surface dip becomes a poor indicator of the small-scale features (Fig. 3c). Dip curvature (Fig. 3d) is better, but is affected by the curvature in the y-axis direction. Mean curvature

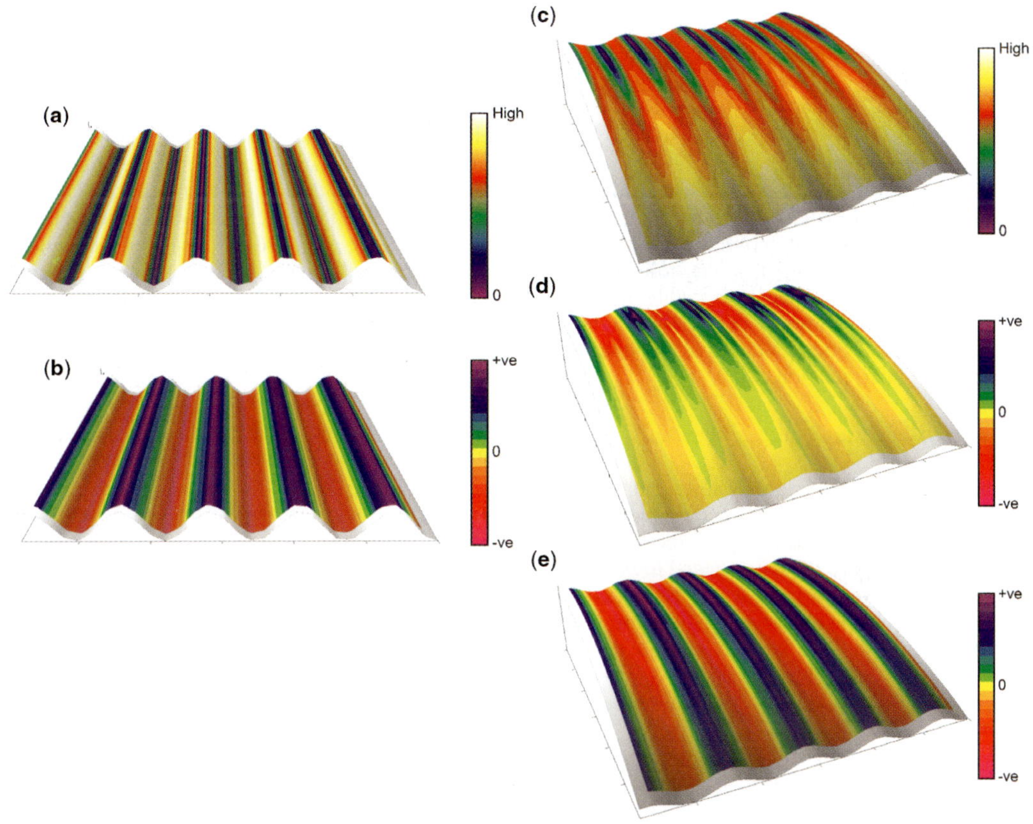

Fig. 3. Differences between surface dip and curvature derived from different surfaces. (**a**) Surface dip derived from a flat-lying, regularly folded surface (surface has been tilted for viewing purposes). The relatively steeply dipping flanks are highlighted (yellow and white). The crests and troughs on the surface are indistinguishable based on surface dip alone (blue). (**b**) The same surface in (a) but with a dip-curvature overlay. The blue and purple colours indicate the concave-down crests (positive curvature) and the red and magenta areas indicate the concave-up troughs (negative curvature). (**c**) When a larger fold is superimposed on the surface (fold axis perpendicular to the strike of the small-scale folds), dip becomes a less useful indicator of surface morphology. The small-scale crests and troughs are still indistinguishable on the crest of the larger fold, and the small-scale features are difficult to distinguish on the steeply dipping limb of the large-scale feature (a problem known as 'dip saturation'). (**d**) Dip curvature better defines the small-scale structures but is affected by the larger fold. (**e**) Mean curvature is almost completely unaffected by the presence of the larger structure and clearly distinguishes the smaller-scale folds.

(Fig. 3e) is almost completely unaffected by the larger structure. It is clear that curvature measures better detect the small-scale structures and that not all curvature measures yield the same result.

Both Roberts (2001) and Stewart & Podolski (1998) critiqued various previously existing (commercial or otherwise) implementations of curvature calculations. For example, although some software applications offer second-derivative extraction, this measure is only equal to the curvature if the dip on the surface is zero (Stewart & Podolski 1998).

Examples

In this section we illustrate a variety of cases where we have used curvature to define the morphology of stratigraphic features. In these examples, derived from 3D seismic data, multibeam bathymetry and digital elevation model (DEM) data and comprising clastic and carbonate deposits of varying ages, the curvature analyses were generally undertaken not as an end in themselves, but to help us address other problems. Although Roberts (2001) suggested different uses for each of the curvature types he discussed, our experience suggests that it is commonly difficult to predict in advance which type of curvature will be most useful in any given situation (although some curvatures, like the shape index or Gaussian curvature seem to be seldom, if ever, useful for defining small-scale stratigraphic or structural features). Owing to the rapidity with which the different types of curvature may be derived, with different apertures, we typically test many different combinations of curvature and aperture. If an interesting result is produced, we drape curvature over rendered 3D surfaces, then adjust color bars and illumination angle to detect and highlight features of interest. Dip curvature, because it highlights relief on a surface, often produces usable images for stratigraphic interpretation.

Deep-water channel system

We mapped a Pliocene deep-water channel system using a 3D seismic data volume covering approximately 2900 km^2. Because of the rectangular bin size (22.5 × 12.5 m), the data needed to be gridded in order for curvature analyses to be undertaken. Two shingled channel-levee complexes are present in this area (Fig. 4). The channels extend for approximately 21 km within our survey, but their true length is not known because they extend beyond the limits of our data. Figure 5(a) shows a time-structure map of the top of the stratigraphically highest channel–levee complex. Note the regional dip of the seafloor, and the channel, with cuspate margins in places. Outside of the channel, the contours suggest the presence of various curvilinear features but their origin cannot be ascertained from the image.

Figure 5(b, c) shows dip curvature draped over time structure, with illumination to further help emphasize subtle features on the surface. In these views, portions of the surface with no curvature in the dip direction are shown in yellow and areas of concave-up (orange and red) and concave-down (green and blue) curvature are highlighted. The general depositional dip to the left is de-emphasized with the result that edges associated with smaller-scale features become more visible. Note how the curvature visualization displays better define the faults, as well as details of the morphology within the channel itself. It is clear that the channel margins have been enlarged by meander loop migration. An erosional inner thalweg was subsequently cut, at least locally. Faults are clearly visible outside of the channel. Comparison of Figure 5(b, c) shows how changing the viewing angle, zoom and surface illumination angle may be used to emphasize different scales and types of feature. Ideally the zoom, viewing angle and illumination can be adjusted interactively during the interpretation process.

Devonian carbonate buildups, Western Canada

We present here two curvature examples that show very different effects. The first is from a small (56 km^2) 3D survey in the Williston Basin of SE Saskatchewan. Most production in this area is from Mississippian carbonates, although the deeper Ordovician Red River Formation is also considered to be prospective in this area. Unfortunately, Devonian pinnacle reefs of the Winnipegosis Formation are, at least in places, responsible for pull-up at the Red River level and so some operators have found it difficult to distinguish between Ordovician structures that are real (and therefore potential drilling targets) and those that are related to velocity artifacts (e.g. Fig. 6a). To study this problem, Pearson *et al.* (2003) mapped pinnacles and proposed a seismic inversion-based methodology for identifying and correcting for velocity pull-up.

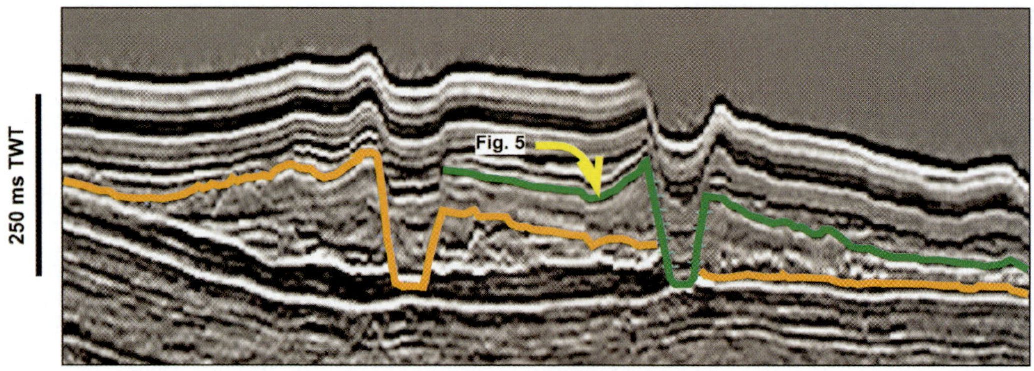

Fig. 4. Seismic transect from a deep-water area showing two shallowly buried leveed–channel complexes. The top of the upper complex (green) is used in the images shown in Figure 5.

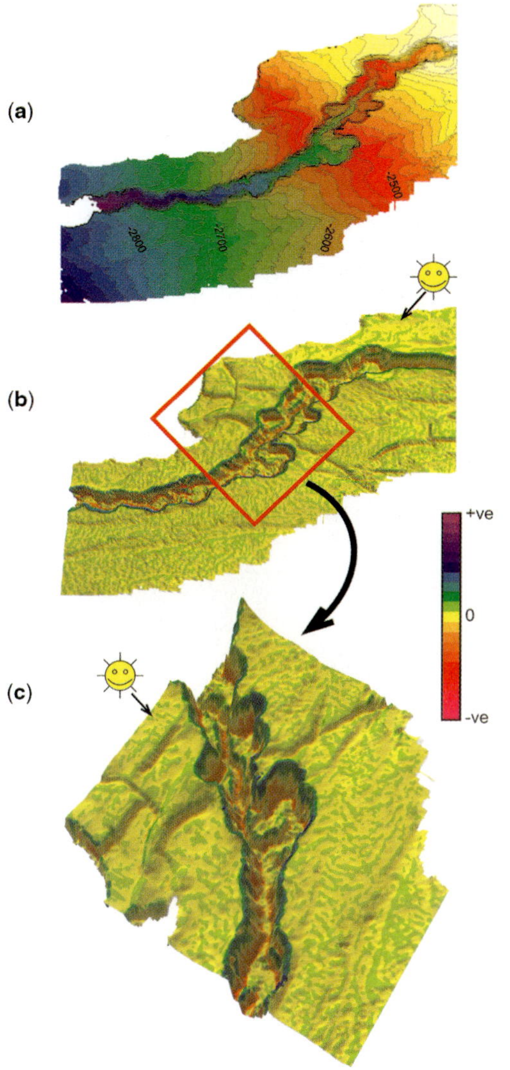

Fig. 5. (**a**) Time–structure map of the top of the stratigraphically highest leveed-channel complex shown in green in Figure 4. (**b**) Subtle structural and stratigraphic features are emphasized when dip curvature is overlain over the surface and directional lighting is applied. (**c**) Expanded and rotated view of the dip curvature overlay showing some fine-scale morphological features that are emphasized by curvature visualization. Note the changes in sharpness of the channel margin (green and blue) along its length, and the presence of an incised inner channel (shown by red colours indicating negative curvature) in the upper portion of this image.

Figure 6(b) shows a time-structure map of the Winnipegosis in the 3D area. The image shows a regional dip to the SW as well as the presence of many pinnacle reefs. Figure 6(**c**) shows the same surface, as a 3D visualization with dip curvature draped over time structure and illumination. The regional dip is removed using the curvature attribute, thus emphasizing depositional features along the surface. The broad aperture (five bins, bin size 30 × 30 m) we used in this example helps to emphasize the flanks of the pinnacles, and so to define the areas below which pull-up might be expected. Note the excellent distinction between concave-up and concave-down (negative and positive curvature respectively) segments of the pinnacles, a result that is not defined using dip or azimuth alone. This particular viewing angle emphasizes an apparent alignment of the buildups and might suggest that the larger features are composite reefs that have amalgamated during growth.

The second Devonian example comes from the Deep Basin area of north-central Alberta. The focus of the study (Nodwell & Hart 2006) was to define the controls on the location of productive shoreface conglomerates in the Lower Cretaceous Falher Member of the Spirit River Formation. The conglomerates form linear trends and have excellent porosity and permeability (locally in the Darcy range) whereas the surrounding sandstones of the Falher are tight and considered to be unproductive. A map of one of the conglomerate trends (based on core and well control) is shown in Figure 7(a).

Several authors have previously suggested, using logs and core and in the absence of supporting 3D seismic data, that the location of productive linear conglomerate trends at the Falher level has been influenced by the presence of underlying structures. Nodwell & Hart (2006) sought to test this hypothesis using a 3D seismic dataset to look at horizons below the Falher interval. Figure 7(b) shows a seismic transect that illustrates the relative positions of the Falher and the Wabumun Formation, an underlying Devonian carbonate approximately 1500 m (based on well control) below the Falher. Note the subtle upward inflection of the Wabumun near the middle of the image. Figure 7(c) shows a rendered time–structure map of the Wabumun with a dip-curvature overlay. Note the utility of the combination of curvature and illumination for defining the morphology of this rimmed platform. Areas of positive and negative curvature (shape information) are clearly highlighted by the colours. Nodwell & Hart (2006) show that the productive conglomerates of the Falher F trend (Fig. 7a) directly overlie this platform rim and suggest how subtle depositional relief associated with compactional drape over the underlying structure may have led to their concentration in a shoreface setting.

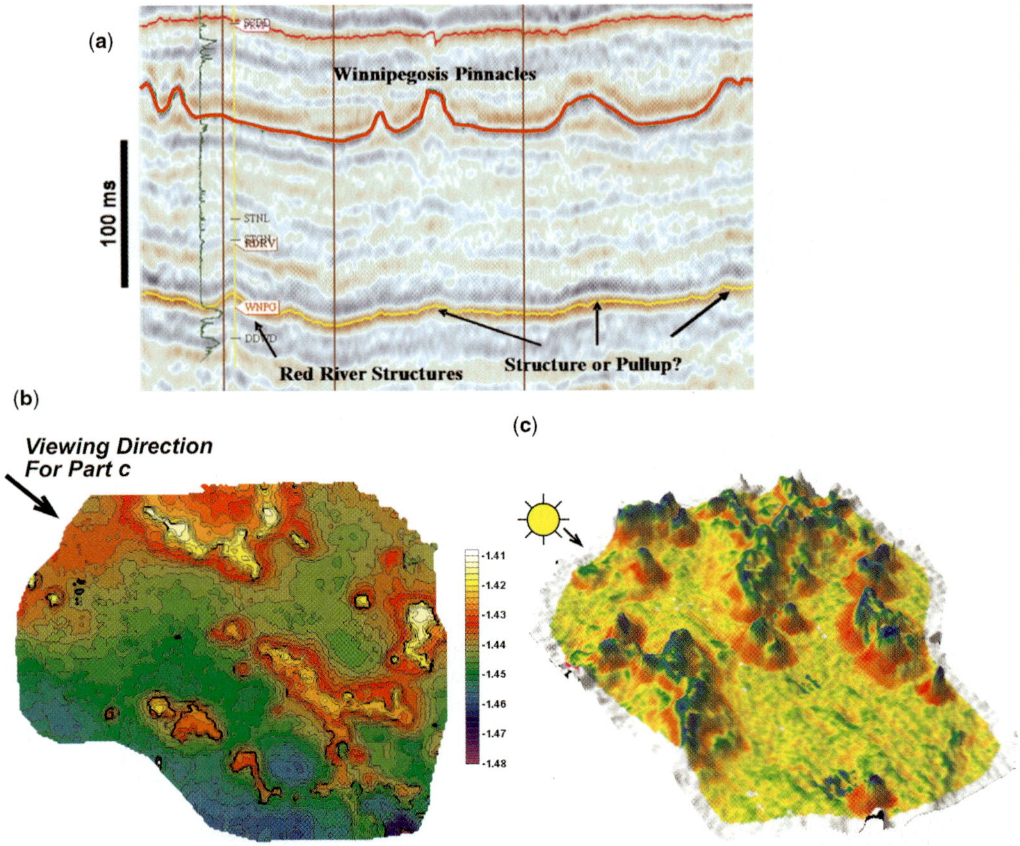

Fig. 6. (**a**) Seismic transect from the Williston Basin (SE Saskatchewan, Canada) showing (Devonian) pinnacle reefs at the Winnipegosis level and subtle apparent structures at an underlying (Ordovician) level that could be either real or related to velocity pull-up beneath the pinnacle reefs (drilling in the area has encountered both possibilities). (**b**) Time-structure map showing Devonian pinnacle reefs identified in (a). Note the subtle regional dip to the south (bottom of the page). (**c**) The same surface shown in 3D with a dip curvature drape and illumination. This image removes the effects of regional dip and highlights the morphology of the individual pinnacles.

Basement morphology, San Juan basin

The 3D seismic data illustrated in this example were used in two studies to understand the controls on reservoir productivity in Cretaceous 'tight-gas' and coalbed methane reservoirs. Hart & Chen (2004) studied the Cenomanian Dakota Formation (a succession of shelf, littoral and coastal plain clastics), whereas Marroquin & Hart (2004) studied the Fruitland Coal. Both reservoirs have stratigraphic (thickness of sandstone or coal) and structural (fracture/cleat density, orientation, and connectivity) components. Throughout the basin, NE–SW striking fractures and cleats (in coal) are known or suspected to be present in Cretaceous reservoirs in the San Juan Basin based on outcrop mapping, seismic analyses and engineering analyses (e.g. Hart *et al.* 2000, 2002; Lorenz & Cooper 2003; Marroquin & Hart 2004). These NE–SW trends, and other trends, are visible in curvature maps of Cretaceous horizons in the study area (Marroquin & Hart 2004). In the San Juan Basin and elsewhere, it has been suspected that reactivation of basement structures might have helped to generate fractures in the overlying strata, and so techniques that can identify basement morphologic features are of considerable interest. Although the Cretaceous section is extensively drilled, no wells penetrate to basement within the study area. The general Palaeozoic stratigraphy is known from outcrops and other boreholes in the basin.

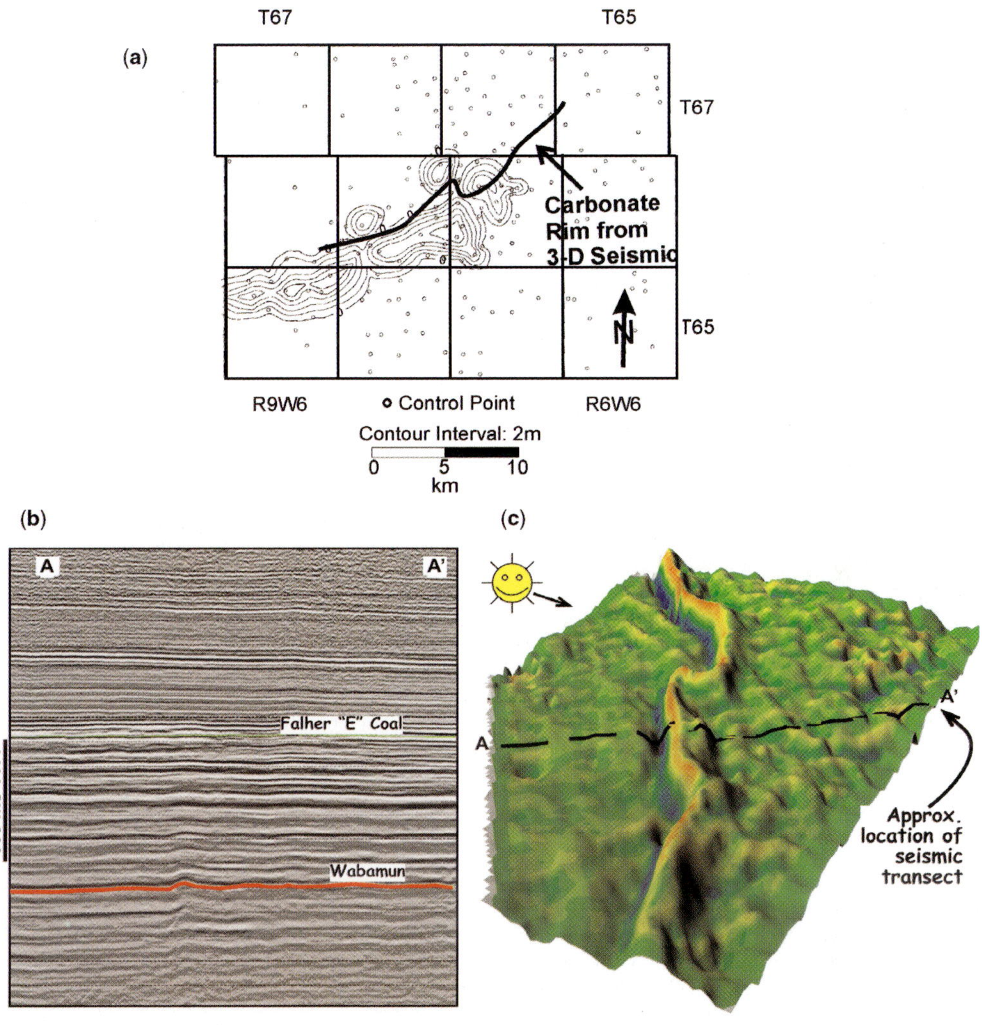

Fig. 7. (**a**) Net porosity map (meters) of the Falher F trend (lower Cretaceous, Deep Basin of western Canada) showing a SW–NE linear trend associated with the presence of productive and porous shoreface and beachface conglomerates (cf. Nodwell & Hart, 2006). (**b**) Seismic transect showing the relative positions of the Lower Cretaceous Falher Member (here defined using a reflection from a prominent coal in the Falher) and the Devonian Wabamun Formation in the Deep Basin, central Alberta (Canada). Approximate location of seismic transect shown in (**c**). (**c**) Dip curvature display, draped over illuminated rendered surface of the Wabamun Formation. The seismic data image the northwestern flank of the Gold Creek Reef trend. Note the presence of a well-defined rim. The conglomerate trend shown in (a) directly overlies the rim of the platform.

Figure 8(a) shows a NE–SW seismic transect through the data volume, showing the relative positions of the Dakota and Fruitland intervals with respect to Precambrian basement. A strong trough that is present between the Mississippian Leadville Limestone and Precambrian basement is a useful proxy for mapping structure on the pre-Palaeozoic unconformity. Figure 8(b) shows minimum curvature draped over a rendered time-structure image of the unconformity. High values of minimum curvature are typically associated with faulted surfaces (Roberts 2001). Notice the good definition of a north–south trending, slightly sinuous channel incised into the unconformity. Two curvilinear NW–SE striking down-to-the-north normal faults are also imaged by minimum curvature, but not by some other types of curvature (e.g. strike curvature, not shown). The orientation and location of these two faults does not

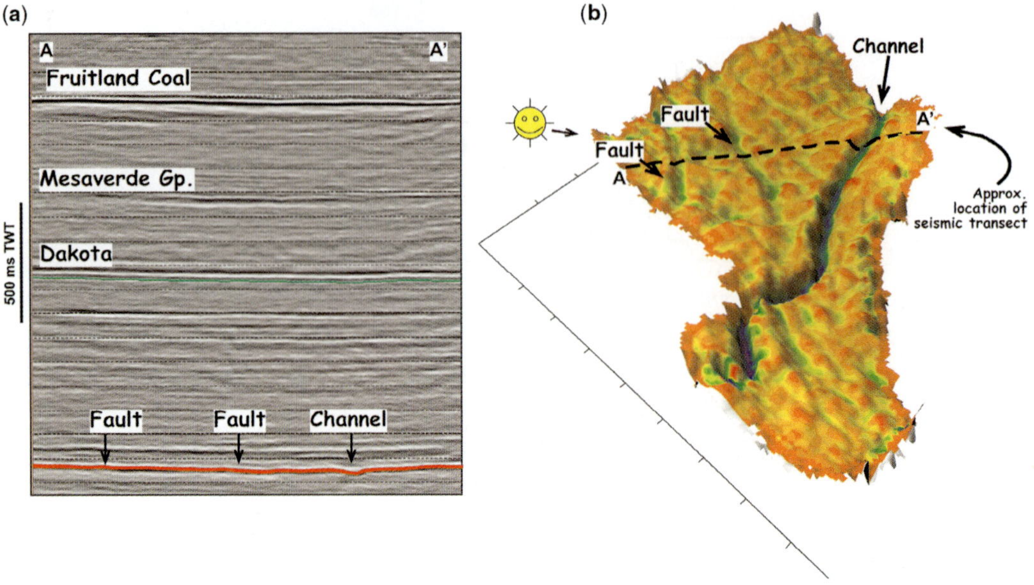

Fig. 8. (**a**) Seismic transect from the San Juan Basin showing the relative positions of the Cretaceous Dakota Formation, Mesaverde Group and Fruitland Coal, and the Basement. The Fruitland Coal is the world's most prolific coalbed methane reservoir, and the Mesaverde and Dakota intervals produce gas from low-permeability ('tight') sandstones. Together these three intervals produce in excess of 800 BCF/year in the basin. (**b**) Minimum curvature overlay on the Basement seismic horizon. The curvilinear faults are part of a NW–SE striking set of faults that has been mapped regionally in the basin. Note the presence of the low-sinuosity incised meandering channel, several 100 m wide.

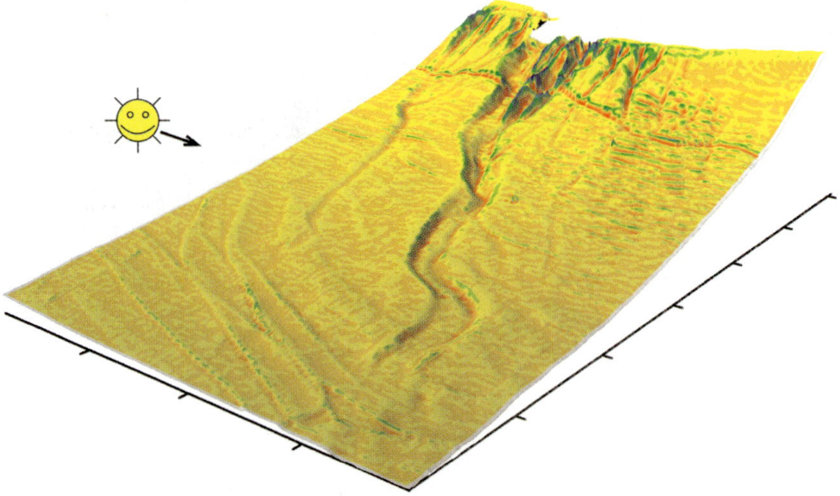

Fig. 9. Dip curvature overlay on top of multibeam bathymetry data from the Fraser Delta slope, seaward of the main fluvial distributary. Image shows approximately 200 m of seafloor relief. Ticks on the x- and y-axes are in km. Note the presence of a variety of submarine channels, the largest of which extends to the base of slope. The bathymetry data have had only partial geometric corrections, and data-collection artefacts are obvious in several places.

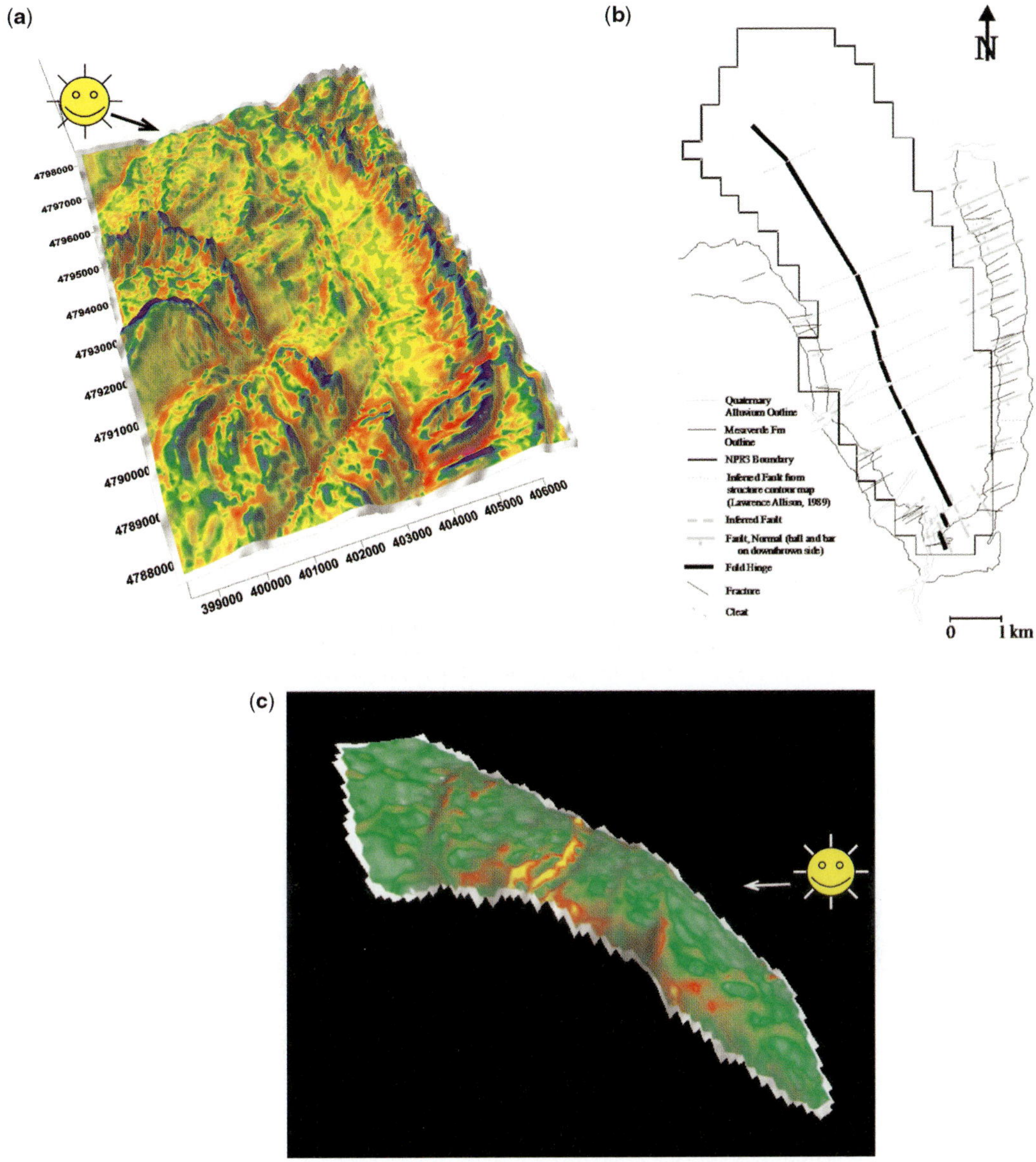

Fig. 10. Teapot Dome, Wyoming. (**a**) Dip curvature overlay on digital elevation data. (**b**) Map of surface geology and hinge-parallel fractures (modified from Cooper *et al.* 2003). (**c**) 3D seismic-based time-structure map, with dip-curvature overlay, of the Cretaceous Dakota Formation at Teapot Dome.

correspond to the faulting seen at the Cretaceous level, suggesting that reactivation of the basement faults is not responsible for the production-enhancing faults (and associated fractures/cleats) higher up-section. This means that the distribution of the productive fracture swarms will be difficult to predict from basement images.

Multibeam bathymetry and DEM examples

A submarine channel/failure complex that has developed on the present Fraser Delta slope seaward of the main fluvial distributary was studied using high-resolution seismic ('boomer') data, side-scan sonar data and coring by Hart *et al.* (1992), largely for submarine hazard detection. Hart *et al.* (2003) revisited this area as a possible modern analogue for shelf-margin deltas and presented newly acquired, but not fully processed, multibeam bathymetry data showing some of the morphological features on the delta slope. Figure 9 shows a rendered, illuminated view of the bathymetry data with a dip-curvature overlay. Clearly the geometric corrections have not been fully applied in this dataset, however it is still possible to clearly define the main, slightly meandering submarine channel that extends from the delta front to a depth of approximately 200 m. Details of morphological features within the channel are also visible, as are: (a) a series of smaller channels on either side of the main channel; and (b) an area of shallow rotational slides on the upper and middle part of the delta slope to the south of the main submarine channel. Time-lapse multibeam surveying of the seafloor here, incorporating curvature analyses of the successive surveys, would be useful for monitoring the evolution of this system; differences in channel morphology are apparent between the multibeam data and the side-scan/high-resolution seismic dataset collected over 10 years earlier.

Figure 10(a) shows curvature analysis of DEM data from Teapot Dome, Wyoming. Production at this basement-cored uplift is from Cretaceous and underlying levels, and faults and fractures in the subsurface have an impact on production. A comparison of subsurface and surface fault and fracture trends was presented by Cooper *et al.* (2003). Sandstones of the Cretaceous Mesaverde Group form a ridge at the surface that encircles the southern part of this doubly plunging uplift. The curvature display identifies this ridge, and breaches in it that are associated with surface drainage networks and faults. Figure 10(b) shows the location and orientation of faults and fractures mapped at the surface. Some of the drainage network cutting across the Mesaverde in the south and SW is clearly following faults, whereas NE–SW striking structures crossing the eastern rim of the dome are clearly imaged by the curvature visualization. This trend is parallel to some faults that may be defined by curvature in a 3D seismic dataset from this area (Fig. 10c).

Discussion and conclusions

The initial interest in curvature analyses of surfaces mapped in 3D seismic data was to identify subtle structures (e.g. faults, folds, fractures) that could influence reservoir behaviour. For this purpose, curvature of surfaces related to stratigraphic bodies was regarded as unavoidable 'contamination' (Stewart & Podolski, 1998) that might obscure the definition of structural features. In this paper we illustrate that one person's poison is another person's meat, that curvature analyses of horizons mapped in 3D seismic data can be highly useful to the seismic geomorphologist.

Curvature is not a single attribute of a surface but a class of attributes that are generally more helpful than conventional dip and azimuth displays for detecting surface features. Some types of curvature are more useful than others in this regard. Although it is possible to define relationships between surface shape (saddles, domes, ridges, etc.) and curvature from first principles, our experience has been that it is not always been possible to predict which type of curvature will best define geomorphologic or structural features. As such, and given the ease and rapidity with which the different types of curvature may be generated, we encourage seismic geomorphologists to experiment with the different types of curvature. Draping curvature over the corresponding rendered 3D surface is a useful technique for identifying and highlighting features of interest. Other forms of visualization might also be developed.

We thank past and present members of the McGill Seismic Research Group who have contributed to our understanding of curvature, including R. Pearson, B. Nodwell, and Ivan Marroquin. Development of the curvature freeware we use was financed by US Department of Energy Contract Number DE-FC26-98FT40486. The work described herein was funded by a NSERC Discovery Grant to Hart. Data used in this paper were made available by Nexen, Millenium Seismic, Anadarko, Williams Energy, the Geological Survey of Canada and the Rocky Mountain Oilfield Testing Center. Seismic interpretation software used in this paper was provided by Landmark Graphics Corporation and Paradigm Geophysical. We thank all of these people and groups for their efforts.

References

BROWN, A. R. 1999. *Interpretation of Three-dimensional Seismic Data*, 4th edn. American Association of Petroleum Geologists Memoirs, **42**.

COOPER, S. P., HART, B. S., LORENZ, J. C., GOODWIN, L. B. & MILLIKEN, M. 2003. Outcrop and seismic analysis of natural fractures, faults and structure at Teapot Dome, Wyoming. *In*: HORN, M. S. (ed.) *Wyoming Basins/Reversing the Decline*. Wyoming Geological Society Field Guidebook 2002/2003, 63–74.

DALLEY, R. M., GEVERS, E. C. A., STAMPFLI, G. M., DAVIES, D. J., GASTALDI, C. N., RUIJTENBERG, P. A., & VERMEER, G. J. O. 1989. Dip and azimuth displays for 3D seismic interpretation. *First Break*, **7**, 86–95.

EBERLI, G. P., MASAFERRO, J. L. & SARG, J. F. (eds). 2004. *Seismic Imaging of Carbonate Reservoirs and Systems*. American Association of Petroleum Geologists Memoirs, **81**.

HART, B. S., 2000, *3-D Seismic Interpretation: a Primer for Geologists*. SEPM Short Course Notes, **48**.

HART, B. S. & CHEN, M. A. 2004. Understanding seismic attributes through forward modeling. *The Leading Edge*, **23**, 834–841.

HART, B. S., PRIOR, D. B., BARRIE, J. V., CURRIE, R. A. & LUTERNAUER, J. L. 1992. A river mouth submarine landslide and channel complex, Fraser Delta, Canada. *Sedimentary Geology*, **81**, 73–87.

HART, B. S., PEARSON, R., HERRIN, J. M., ENGLER, T. & ROBINSON, R. L. 2000. Horizon attributes and fracture-swarm sweet spots in low-permeability gas reservoirs. **SPE 63207**.

HART, B. S., PEARSON, R. A. & RAWLING, G. C. 2002. 3-D seismic horizon-based approaches to fracture-swarm sweet spot definition in tight-gas reservoirs. *The Leading Edge*, **21**, 28–35.

HART, B. S., BARRIE, J. V. & HILL, P. R. 2003. Submarine slope bypassing of sands through submarine channels on a modern lowstand delta, Fraser Delta, Canada. *In*: ROBERTS, H. H., FILLON, R. H. & ANDERSON, J. B. (eds) *Shelf Margin Deltas and Linked Down Slope Petroleum Systems*, 23rd Annual GCSSEPM Foundation Bob F. Perkins Research Conference, 281–307.

HESTHAMMER, J. & FOSSEN, H. 1997. Seismic attribute analysis in structural interpretation of the Gullfaks Field, northern North Sea. *Petroleum Geoscience*, **3**, 13–26.

KOLLA, V., BOURGES, P., URRUTY, J. M. & SAFA, P. 2001. Evolution of deep-water Tertiary sinuous channels offshore Angola (West Africa) and implications for reservoir architecture. *AAPG Bulletin*, **85**, 8, 1373–1405.

LISLE, R. J. 1994. Detection of zones of abnormal strains in structures using Gaussian curvature analysis. *AAPG Bulletin*, **78**, 1811–1819.

LISLE, R. J. & ROBINSON, J. M. 1995. The Mohr circle for curvature and its application to fold description. *Journal of Structural Geology*, **17**, 739–750.

LORENZ, J. C. & COOPER, S. P. 2003. Tectonic setting and characteristics of natural fractures in Mesaverde and Dakota reservoirs of the San Juan Basin. *New Mexico Geology*, **25**, 3–14.

MARROQUIN, I. D. & HART, B. S. 2004. Seismic attribute-based characterization of coalbed methane reservoirs: an example from the Fruitland Formation, San Juan Basin. *American Association of Petroleum Geologists Bulletin*, **88**, 1603–1621.

MIALL, A. D. 2002. Architecture and sequence stratigraphy of Pleistocene fluvial systems in the Malay Basin, based on seismic time-slice analysis. *American Association of Petroleum Geologists Bulletin*, **86**, 1201–1216.

NODWELL, B. & HART, B. 2006. Deeply-rooted topographic control on the deposition of the Falher F conglomerate trend, Wapiti Field, Deep Basin, Alberta. *Bulletin of Canadian Petroleum Geology*, **54**, 1–21.

PEARSON, R., HART, B. & TEBO, J. 2003. Eliminating pull-up below Devonian pinnacle reefs. *Joint CSPG/CSEG Annual Convention*, Calgary, CD-ROM abstract volume.

POSAMENTIER, H. W. & ALLEN, G. P. 1999. Siliciclastic Sequence Stratigraphy – Concepts and Applications. *SEPM Concepts in Sedimentology and Paleontology*, **7**.

ROBERTS, A. 1998. Curvature analysis: 'new' attributes for the delineation of faults, map lineaments and surface shape. *1998 AAPG Annual Convention Program with Abstracts* (CD-ROM), Salt Lake City, 17–28 May.

ROBERTS, A. 2001. Curvature attributes and their application to 3D interpreted horizons. *First Break*, **19**, 85–99.

STEWART, S. A. & PODOLSKI, R. 1998. Curvature analysis of gridded geological surfaces. *In*: COWARD, M. P., DALTABAN, T. S. & JOHNSON, H. (eds) *Structural Geology in Reservoir Characterization*. Geological Society of London Special Publications, **127**, 133–147.

STEWART, S. A. & WYNN, T. J. 2000. Mapping spatial variation in rock properties in relationship to scale-dependent structure using spectral curvature. *Geology*, **28**, 691–694.

TOWNSEND, C., FIRTH, I. R., WESTERMAN, R., KIRKEVOLLEN, L., HÅRDE, M. & ANDERSEN, T. 1998. Small seismic-scale fault identification and mapping. *In*: JONES, G., FISHER, Q. J. & KNIPE, R. J. (eds) Faulting, Fault Sealing and Fluid Flow in Hydrocarbon Reservoirs. Geological Society Special Publications, **147**, 1–25.

Imprints of former ice streams, imaged and interpreted using industry three-dimensional seismic data from the south-western Barents Sea

K. ANDREASSEN, C. M. ØDEGAARD & B. RAFAELSEN

Department of Geology, University of Tromsø, N-9037 Tromsø, Norway (e-mail: karin.andreassen@ig.uit.no)

Abstract: Former ice-stream activity is shown from industry three-dimensional (3D) seismic data from the south-western Barents Sea. Although designed for deeper targets, the data allow, due to high spatial sample rate and three-dimensional migration techniques, construction of detailed plan view images. The integration of sea-floor geomorphology with stratigraphy documents the importance of glacial processes in the seascape evolution of this area. Fast-flowing ice streams occupying the cross-shelf troughs during the Late Weichselian glaciation caused large-scale erosion, and also left their imprints in the form of mega-scale glacial lineations on the sea floor as indicators of ice-flow direction. Various types of 3D seismic attributes, combined with detailed geomorphology and seismic stratigraphy, are used to investigate the 2–3 km of stratigraphic record that corresponds to over a million years of ice-stream activity. The appearance of mega-scale glacial lineations on various 3D seismic attribute maps indicates, together with other characteristics of ice streams, that they are formed by erosion beneath fast-flowing grounded ice. Bedform records of former ice streams may, however, be related only to the final stages of ice-streaming, immediately prior to shut down. Because we here have preserved up to several hundred metres of sediments between the buried, glacially eroded surfaces, we have the opportunity to study ice-stream imprints and associated processes covering longer time spans than just the last stages. Seismic volumetric attribute maps reveal that megablocks and rafts, often aligned in chains, commonly occur within the till units, implying that glaciotectonic erosion by fast-flowing ice streams was an important process in the transfer of sediments from the continental shelf to the Bjørnøya Trough Mouth Fan and the deep sea during the Plio-Pleistocene glaciations.

The south-western Barents Sea (Fig. 1a) has been explored for hydrocarbons since 1979. Most of the wells drilled have failed to prove significant hydrocarbons, and all reservoirs seem to have experienced partial or completely drainage of hydrocarbons. This is commonly attributed to the erosion of up to 500–2000 m of rocks and sediments from the Barents Shelf during the Cenozoic and subsequent uplift (e.g. Riis & Fjeldskar 1992; Doré & Jensen 1996). The result is expansion of gas, tilting of reservoirs, and hydrocarbon leakage. Timing and amount of erosion are important for basin modelling and understanding of geological evolution of the area. Since the early 1990s it has been discussed how much of this erosion was glacial and Plio-Pleistocene in age, and how much was pre-glacial and older (Vorren *et al.* 1991; Eidvin *et al.* 1993; Sættem *et al.* 1994; Faleide *et al.* 1996; Vorren & Laberg 1996). Results from the three-dimensional (3D) seismic data presented here provide evidence that a large part of this erosion was caused by grounded glaciers, most probably by ice streams. These are large (>20 km wide and >150 km long) rapidly flowing corridors within a glacier, where the ice is moving faster, typically around 100–800 m per year, than the surrounding part of the ice sheet, which is more typically moving around 5 m per year.

Previous studies of the Cenozoic sediments and environments of the western Barents Sea margin (e.g. Laberg & Vorren 1995; Fiedler & Faleide 1996; Kuvaas & Kristoffersen 1996; Vorren & Laberg 1996) were based mainly on analysis of two-dimensional (2D) seismic profiles. We here investigate the imprints of former ice streams in the south-western Barents Sea using three semi-regional industry 3D seismic data sets (Fig. 2a). With dense spatial sampling, three-dimensional migration techniques and advanced computer-techniques for interpretation and visualization, the 3D seismic data clearly add another dimension to the geological interpretation, opening the opportunity of investigating detailed geomorphology of the beds and the internal structure of units

From: Davies, R. J., Posamentier, H. W., Wood, L. J. & Cartwright, J. A. (eds) *Seismic Geomorphology: Applications to Hydrocarbon Exploration and Production*. Geological Society, London, Special Publications, **277**, 151–169.
0305-8719/07/$15.00

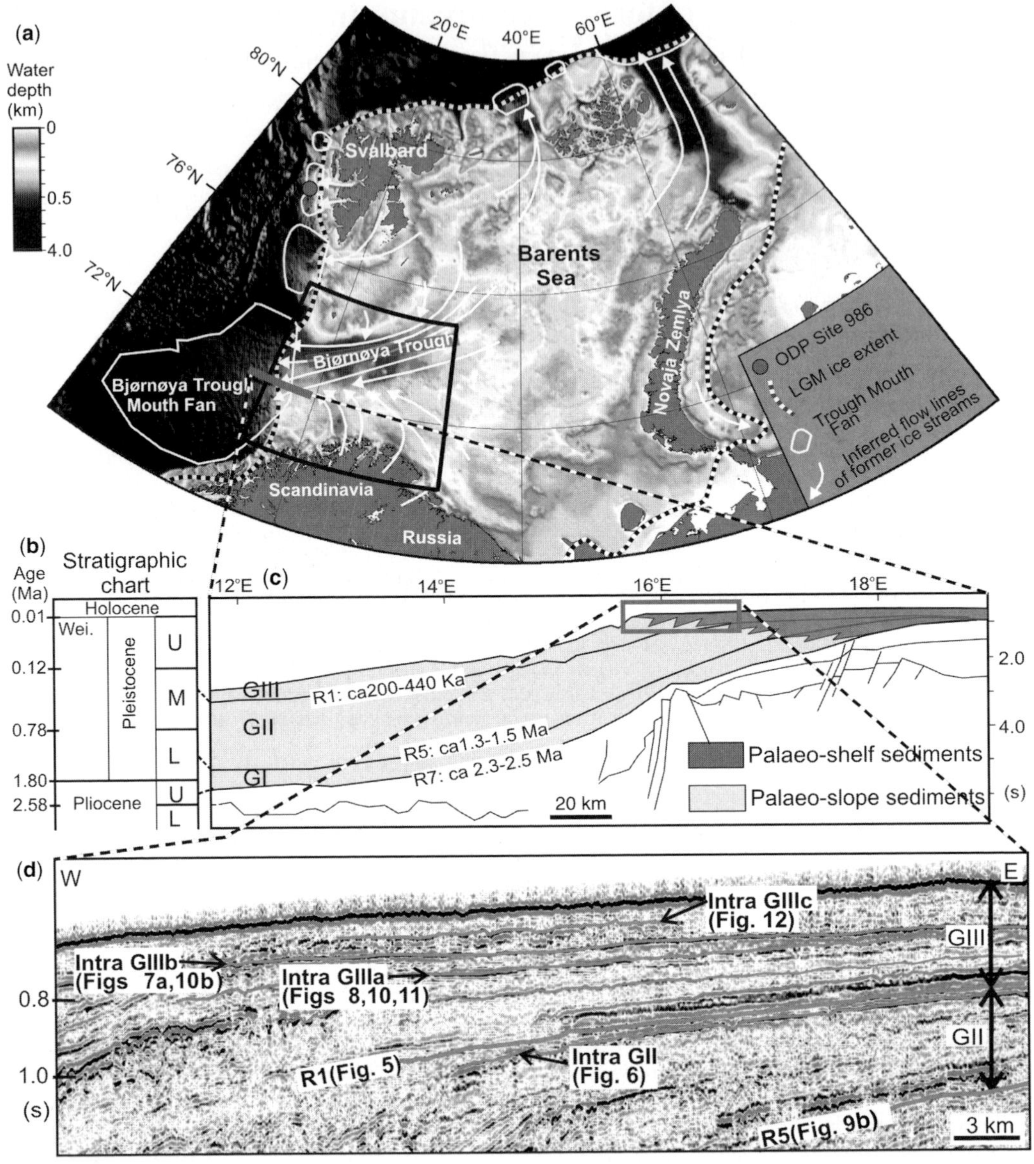

Fig. 1. (**a**) Location map with relevant information and study area indicated by the black frame. LGM: last glacial maximum (around 28–20 ^{14}C ka). The black frame indicates location of Figure 2. (**b**) Stratigraphic chart. Wei.: Weichselian. (**c**) Geoseismic profile, location indicated by the red line of (a). The rectangle indicates location of the 3D study area at the western Barents Sea margin. (**d**) Seismic profile, location indicated by the red rectangle of (c). (s) indicates two-way travel time.

left behind by former ice streams, revealing detailed information about the processes involved.

Identifying beds of former ice streams

Ice streams may discharge the majority of ice and sediment within an ice sheet (Bennett 2003), and they are being regarded as potential forcing mechanisms for high-frequency climate changes (Broecker 1994). It was recognized during the mid-1980s that fast motion of the West-Antarctic ice streams is taking place within the subglacial sediments (Alley *et al.* 1986; Boulton & Hindmarsh 1987), but the nature of this sediment deformation

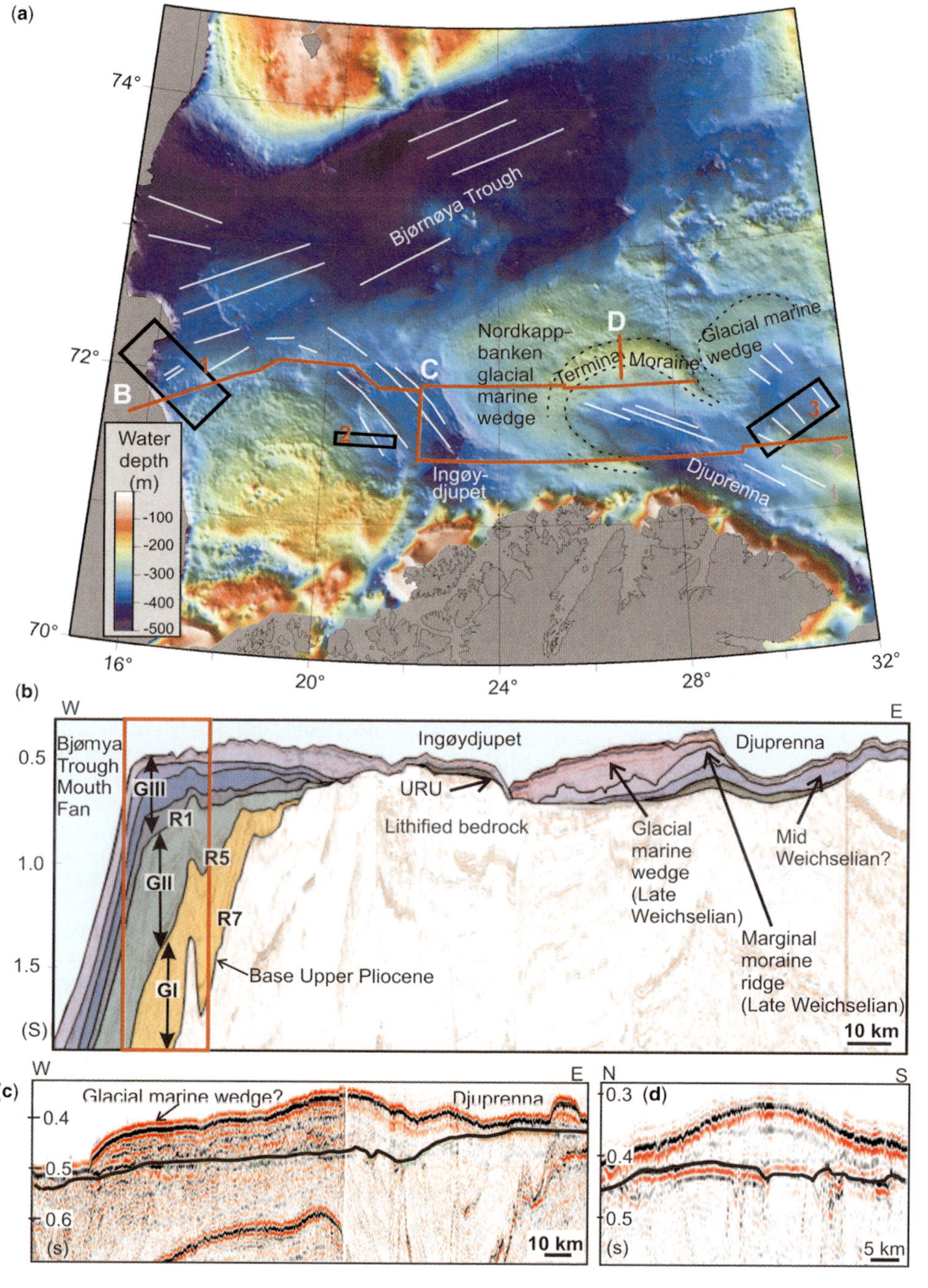

Fig. 2. (**a**) Illuminated shaded relief image of the south-western Barents Sea seafloor. The bathymetry, obtained from Statoil, is compiled by Mauring at the Norwegian Geological Survey in 2003. White lines indicate orientation of mega-scale glacial lineations. The three rectangles numbered 1, 2 and 3 indicate location of the 3D-seismic areas used in this study. (**b**) Geoseismic profile with location as indicated in (a). The red frame indicates location of 3D area 1. URU: upper regional unconformity. (**c**) Seismic profile crossing inferred glacial marine wedge. (**d**) Seismic profile crossing inferred terminal moraine. Location of the seismic profiles of (c) and (d) is indicated in (a). (s) indicates two-way travel time.

is not well understood (Kamb 1991; Tulaczyk *et al.* 2000; Binschadler *et al.* 2003; Bennett 2003). The processes operating at the base of ice streams often produce distinctive subglacial bedforms, appearing as parallel large-scale elongated streamlined ridge-groove features, called mega-scale glacial lineations. Sets of such lineations, formed by former ice streams were first described by Clark (1993) from satellite images of palaeo-ice sheet beds, and regarded as indicators of fast ice-stream flow and a common signature of ice-stream beds in both terrestrial (Stokes & Clark 1999, 2002; Clark *et al.* 2003) and marine (Canals *et al.* 2000; Ottesen *et al.* 2002) settings. Individual lineations have reliefs of typically up to 15 m, they are hundreds of metres wide, several kilometres long, and have high elongation (length–width) ratios (exceeding 10:1; Stokes & Clark 2002). Among other geomorphological criteria for identifying former ice streams are indications of convergent onset zones, abrupt lateral margins (<2 km), deformable bed conditions and focused sediment delivery, like the occurrence of a trough mouth fan (Stokes & Clark 1999). It is, however, highly unlikely that all these criteria will be found in one location, but together they represent a landscape assemblage that an ice stream may be expected to produce (Stokes & Clark 1999).

It seems to be a consensus that mega-scale glacial lineations are formed by some sort of sub-glacial sediment deformation, but it is an ongoing discussion whether they are predominantly depositional landforms (Stalker 1960; Clark 1993), whether they are mainly erosional features formed by ploughing of ice keels at the base of ice streams (Tulaczyk *et al.* 2001; Clark *et al.* 2003), or whether they may be a combination of the two (Ó Cofaigh *et al.* 2005).

Mega-scale glacial lineations differ, both in morphological appearance and formation from iceberg plough marks. Ploughing or scouring of the sea floor by icebergs is a common feature in front of marine glaciers (Barnes & Lien 1988). Icebergs, detached from a glacier terminus may be dragged by winds and currents for long distances before finally melting, and on their way, keels at their base may erode the sea floor. The iceberg plough marks are typically more curved than the mega-scale glacial lineations formed by ice streams. Some of them may be straight, but because each iceberg makes just one, or at most a few furrows, these furrows do not appear as sets of parallel lineations being tens of kilometres long and several kilometres wide, such as the grooves eroded at the base of fast-flowing ice streams. Relict iceberg plough marks, which are common sea-floor features in the south-western Barents Sea (Fig. 3), were probably formed during the Late Weichselian deglaciation (around 20–15 ^{14}C ka), when the Barents Sea and Scandinavian ice sheets (Fig. 1a) retreated rapidly through intense iceberg calving.

Study area and geologic setting

During glacial periods, former ice sheets and their associated ice streams expanded across high-latitude continental shelves, where an imprint of their activity is preserved as wide bathymetric cross-shelf troughs with streamlined subglacial landforms (Sejrup *et al.* 2000; Canals *et al.* 2000; Ottesen *et al.* 2002, 2005). Where ice streams reached the shelf breaks, large volumes of sediments were often deposited at the continental slope, building up fan-shaped sediment accumulations, trough mouth fans (Vorren & Laberg 1997; Barker *et al.* 1999; Sejrup *et al.* 2000). The Bjørnøya (Bear Island) Trough Mouth Fan (Fig. 1a), the largest of the Polar North Atlantic trough mouth fans, contains up to 3–4 km of Plio-Pleistocene sediments (Fig. 1c). It extends from the shelf break, where the average water depth is 500 m, to the abyssal plain, and is in areal extent comparable with the Amazon and Mississippi Fans, although the drainage areas of the two latter fans are 5–10 times greater than that of the Bjørnøya Fan (Solheim *et al.* 1996).

The existence of a major ice stream in the Bjørnøya Trough, that during glacial maxima delivered sediments to the fan at it's mouth (Fig. 1a) has been inferred from bathymetry and ice-sheet geometry (Denton & Hughes 1981; Stokes & Clark 2001), glacial flutes on the sea floor (Solheim *et al.* 1990), investigations of the Bjørnøya Trough Mouth Fan (Vorren & Laberg 1997) and from mega-scale lineations on several buried surfaces at the outer Bjørnøya Trough (Andreassen *et al.* 2004). This ice stream developed in an area of strongly convergent ice flow from the Svalbard-Barents Sea and the Scandinavian ice sheets (Fig. 1a). It is modelled to have flowed at about 800 m a^{-1}, to have produced about 30 km^3 a^{-1} of icebergs during full glacial periods, and was the largest single outlet from the Eurasian Arctic Ice Sheet (Siegert & Dowdeswell 2002).

3D seismic data

Results from three industry 3D seismic data sets from the south-western Barents Sea (rectangles numbered 1–3 in Fig. 2a) are presented here. Beds of Late Weichselian ice streams are investigated using sea-floor images constructed from the two easternmost 3D areas, whereas the data set at the Barents Sea margin, where a 2–3 km record of glacigenic sediments is preserved due to subsi-

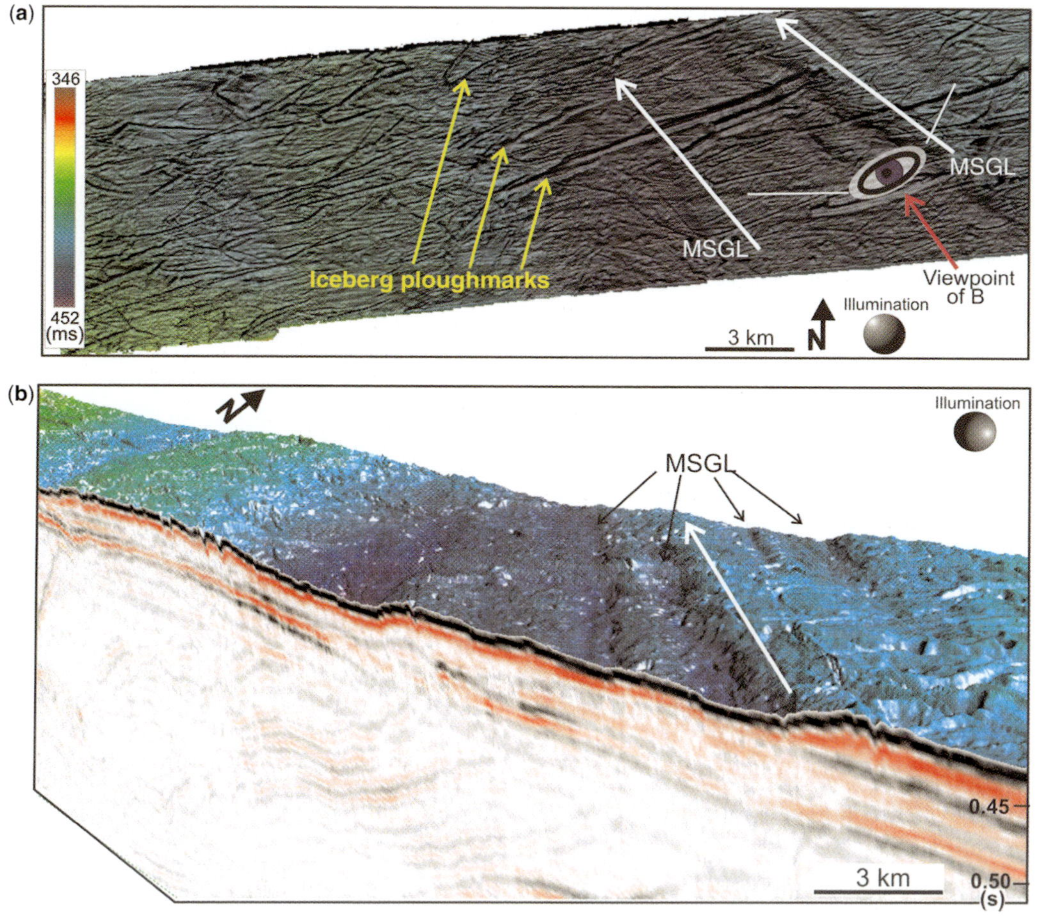

Fig. 3. (**a**) Illuminated shaded relief time structure map of the sea floor constructed from the 3D at the western flank of Ingøydjupet (rectangle no. 2 on Fig. 2a). The vertical scale is 8 × exaggerated. (**b**) Three-dimensional perspective view from the southeastern end of (a) with vertical scale 15 × exaggerated. The north-northwest trending grooves (orientation indicated by white arrows) are interpreted to be mega-scale glacial lineations, and the more irregularly oriented, smaller grooves are iceberg plough marks. (s) indicates two-way travel time.

dence, is used to investigate ice-stream imprints from older pre-Weichselian glaciations. The vertical resolution of the 3D seismic data sets is, with a dominant frequency around 40 Hz for the Barents Sea Margin 3D in theory around 12 m (assuming a velocity of 2000 m s^{-1}), and slightly better for the other two 3Ds. This is not very good compared with the high-resolution seismic data that is typically used for studies of glacigenic sediments (e.g. Vorren & Laberg 1997). The horizontal resolution is however dramatically improved compared to that of 2D seismic due to high spatial sample rate (12.5 m) and 3D migration techniques, and is in theory around 12 m for the studied data sets. The 3D data therefore allow construction of detailed plan view maps, and have proven to be very well suited for detecting images of ice-stream imprints in form of mega-scale glacial lineations on buried surfaces and chains of mega-scale sediment blocks between glacially eroded horizons (Andreassen *et al.* 2004).

Seismic stratigraphic framework and chronology

Three main sediment packages (GI, GII and GIII), separated by regionally correlatable reflectors R_7, R_5 and R_1 (Fig. 1c) have been identified within the Plio-Pleistocene sedimentary succession along the western Barents Sea–Svalbard margin (Faleide *et al.* 1996; Butt *et al.* 2000). With reference to Cenozoic formations commonly identified on the Norwegian Shelf by the oil industry, the sediment packages GI–GIII correspond to the *Naust Formation* of the *Nordland Group*, when used as a succession of Upper Pliocene to Recent, including glacial and interglacial sequences. The Ocean Drilling Program (ODP) Site 986 west of Svalbard (Fig. 1a) is a key borehole for age constraints of the seismic stratigraphy of the western Barents Sea–Svalbard margin. Linear interpolation between the maximum age of 2.6 Ma at the base of this hole and the base of the Olduvai palaeomagnetic event gives a tentative age of 2.3–2.5 Ma for R_7, supported by biostratigraphic and Sr-isotope data (Butt *et al.* 2000). Although uncertain, this is in accordance with age estimates from seismic correlation to commercial wells in the south-western Barents Sea (Faleide *et al.* 1996; Ryseth *et al.* 2003). R_5 is assigned the interpolated age 1.3–1.5 Ma at Site 986, supported by biostratigraphic and Sr-data (Butt *et al.* 2000). R_1 is, based on amino acid analysis (Sættem *et al.* 1992) indicated to be younger than 440 ka. Extrapolation of calculated sedimentation rates in piston cores on the Svalbard margin has given an approximate age of 200 ka of R_1 in this area (Elverhøi *et al.* 1995). A likely age between 440 and 200 ka is thus suggested for R_1.

Mega-scale lineations from sea-floor morphology

Description

The present day topography of the Barents Sea is influenced to a major degree by Late Cenozoic glacial erosion and deposition, although the underlying bedrock and structure is also important. A large-scale shaded relief time-structure map of the sea floor in the south-western Barents Sea (Fig. 2a) provides a regional overview of the sea floor geomorphology, providing evidence of different glacial regimes. Our focus here is on submarine landforms associated with former ice streams.

Large-scale elongated ridge-groove structures characterize the morphology of the major cross-shelf troughs of the south-western Barents Sea. These linear features are especially well developed in the Bjørnøya Trough, Ingøydjupet and Djuprenna where they appear as lineations on the regional image (white lines of Fig. 2a). The lineations in the Bjørnøya Trough are up to 180 km long, 2–5 km wide, 5–7 m deep, and have elongation ratios of up to 40:1. Sea-floor images with higher resolution, constructed from the 3D seismic data (Figs 3 and 4) show more morphological details of the linear features in Ingøydjupet and Djuprenna. The lineations at the western flank of Ingøydjupet appear on the 3D-seismic sea-floor image as well-developed depressions (orientation indicated by white arrows on Fig. 3), which are up to 120 km long, 0.5–3.5 km wide, 4–5 m deep and have elongation ratios of up to 33:1. These features are barely recognisable on vertical seismic sections, but can be well imaged on shaded relief images by strong exaggeration of the vertical scale (Fig. 3). The more small-scale sea-floor morphology in 3D area 2 is dominated by irregularly oriented curved furrows (Fig. 3a; yellow arrows), which are up to 300 m wide and up to 10 m deep. These are interpreted to be iceberg plough marks.

Mega-scale lineations are well-developed also in the Djuprenna trough, and are well visualised on the sea-floor image from a 3D covering the north-eastern flank of this trough (Fig. 4). These lineations are in general narrower (less than 850 m) than those in Ingøydjupet, and they have an elongation ratio of 85:1. They also have a higher relief (up to 9 m), but they are so closely spaced that it difficult to decide whether they are mainly positive or negative features (Fig. 4).

Interpretation

The sea-floor mega-scale lineations described above have a high parallel conformity and are also parallel to the long axes of the cross-shelf troughs in which they occur. Their length vary from several km to up to 180 km, and their elongation ratios are well beyond the minimum 10:1 ratio characteristic of mega-scale glacial lineations formed by fast-flowing ice streams. The onset of the lineations and the troughs occur in zones where inferred flow lines of former glaciers converge (Fig. 1a). The Bjørnøya Trough is inferred to have been the main drainage area of glaciers from the former ice sheets of Svalbard and Barents Sea in the north and east, and from the Scandinavian mainland and north-western Russia in the south and south-east (Vorren & Laberg 1997; Denton & Hughes 1981; Stokes & Clark 2001; Ottesen *et al.* 2005). The troughs of Ingøydjupet and Djuprenna (Fig. 2a) have their onsets in convergence zones of inferred ice streams from the Scandinavian mainland (Punkari 1997; Stokes & Clark 2001). The sea-floor lineations described above were discussed by Ottesen *et al.* (2005), who concluded, based on their appearance and setting, that they were the result of soft-sediment deformation at the base of fast-flowing ice streams draining large ice sheets. Our results support this conclusion. The mega-scale lineations in the Bjørnøya Trough are interpreted to be formed by an ice stream that drained to the shelf break during the last, Late Weichselian glacial maximum (inferred maximum ice extent is indicated in

Fig. 4. (**a**) Shaded relief time structure map of the sea floor constructed from the 3D north-east of Djuprenna (rectangle no. 3 on Fig. 2a) with vertical scale 8× exaggerated. The white arrows indicate orientation of mega-scale lineations. The black dotted line in the southwest indicates orientation of the footprint artifact caused by acquisition geometry (Marfurt *et al.* 1998), which is especially bad in this survey. (**b**) Three-dimensional perspective view from the south-western part of (a) with vertical scale 15× exaggerated. (**c**) Three-dimensional perspective view from the north-eastern part of (a) with vertical scale 12× exaggerated.

(Fig. 1a); marked LGM). The Ingøydjupet lineations bend westwards where they meet the Bjørnøya Trough (Fig. 2a), suggesting that the Ingøydjupet ice stream flowed into that of the Bjørnøya Trough.

The integration of sea-floor morphology with seismic stratigraphy, as illustrated in Figure 2, documents the importance of glacial processes in the seascape evolution of the south-western Barents Sea. Ice streams draining from the Scandinavian ice sheet out Ingøydjupet and Djuprenna have caused large erosion and left a relatively thin succession of sediments above the upper regional unconformity in the trough areas (Fig. 2b; URU). Earlier work has shown that this unconformity separates unlithified, glacigenic sediments from underlying lithified bedrock (Solheim & Kristoffersen 1984). The former ice streams have also left their imprints in the form of mega-scale glacial lineations on the sea floor as indicators of their orientation during their last phase of ice-streaming (white lines on Fig. 2a). Local sediment accumulations up to 120 m thick (using a velocity of 1700 m s^{-1}) were deposited at the north-western end of the over-deeped Djuprenna, causing the build-up of Nordkappbanken (Fig. 2). Seismic stratigraphic correlation suggests that Nordkappbanken developed mainly during the last glaciation (Late Weichselian). An arcuate-shaped ridge at the southern part of this bank area (Fig. 2a, c, d) is interpreted to be a terminal moraine depost-ed by a glacier draining out Ingøydjupet at a period of still-stand or re-advance of the ice-margin during the deglaciation. The well-developed mega-scale lineations in Djuprenna (Figs 2a and 4) suggest that this was a fast-flowing ice stream. The sediment accumulation proximal to the terminal moraine (Fig. 2a, c) is interpreted to be mainly ice-proximal glacimarine sediment associated with this glacial event.

Mega-scale glacial lineations on buried surfaces

The thick Plio-Pleistocene deposits of the Bjørnøya Trough Mouth Fan depocentre at the western Barents Sea margin are indicated on the geoseismic profile of Figure 2b. The westernmost of our studied 3Ds (no. 1 in Fig. 2a) is strategically located where ice streams reached the shelf break during the last glacial maximum, as well as during earlier periods of maxium glaciation. Palaeo-shelf units in the eastern part of this 3D data set continue westwards into mass-movement and glacial-marine palaeo-slope sediments (Fig. 1b). The focus of this study is on palaeo-shelf sediments, and because the lowermost sediment package GI within the 3D area consists mainly of palaeo-slope sediments, our focus here is on the palaeo-shelf deposits of the upper two sediment packages GII and GIII (rectangles in Figs 1b and 2b). The palaeo-shelf units of these two sediment packages are characterised by a chaotic seismic reflection configuration and common occurrence of high-amplitude reflection segments (Fig. 1c).

Mega-scale lineations occur on most of the buried horizons of the palaeo-shelf part of GII and GIII and are well imaged by illuminated shaded relief maps of the actual seismic horizons (Figs 5a and 6a). These lineations have reliefs of up to10 m, widths from 50 to 360 m, and lengths that in some areas exceed 38 km. Based on their high elongation rates (maximum >100:1) and large size, their parallel conformity and their location at the mouth of the Bjørnøya Trough that leads to the Bjørnøya Trough Mouth Fan, we argue that these buried lineations must be formed by fast-flowing ice streams, similar to the mega-scale lineations on the sea floor.

The image of Figure 6a is especially interesting as we here see four different sets of well-developed large-scale lineations, numbered 1–4. The four sets show a high internal parallel conformity of lineations, and each set has a different orientation from the others. A vertical seismic section (Fig. 6d) crossing two well-developed lineations of orientation 1 (Fig. 6a) suggests that these lineations

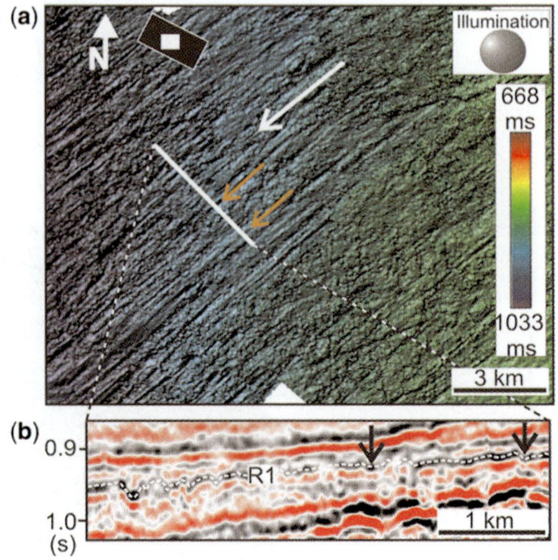

Fig. 5. (**a**) Shaded relief image of the seismic reflection R1, which separates sediment package GII from GIII (stratigraphic location is shown in Fig. 1d). White arrow indicates orientation of mega-scale glacial lineations. The small white box in the upper right corner shows the location of the image within the Sørvestsnaget 3D (rectangle no. 1 on Fig. 2a). (**b**) Seismic section with location as indicated in (a); (s) indicates two-way travel time. Orange arrows in (a) and black arrows in (b) indicate locations of pronounced mega-scale troughs.

represent grooves rather than positive relief ridges. An attribute plot showing seismic amplitude of the reflection from this surface (Fig. 6b) reveals clearly the same lineations as Figure 6a. It is striking that the lineations with orientation 1, which are well-developed grooves on the shaded relief image of this surface (Fig. 6a) are barely visible on the amplitude plot (Fig. 6b), whereas the lineations of orientation 3 and 4 are especially well defined on the amplitude-plot (Fig. 6b). A vertical seismic transect crossing two of the well-developed lineations with orientation 3 (Fig. 6c) indicates that they represent zones where the seismic amplitudes are close to zero. It is our preliminary interpretation that the seismic reflection of this Intra GII surface (Fig. 6) is caused by interference from the top and base of a layer (is not the reflection from a single surface), and that the grooves of orientation 3 and 4 are associated with complete removal of this layer, whereas the grooves of orientation 1 and 2 have not eroded deep enough for this to happen. Because of their highly internal parallel conformity and persistence for tens of kilometres, we argue that all four sets of mega-scale lineations on this intra GII surface (Fig. 6) are grooves eroded by fast-flowing ice streams, and thereby indicate flow directions of

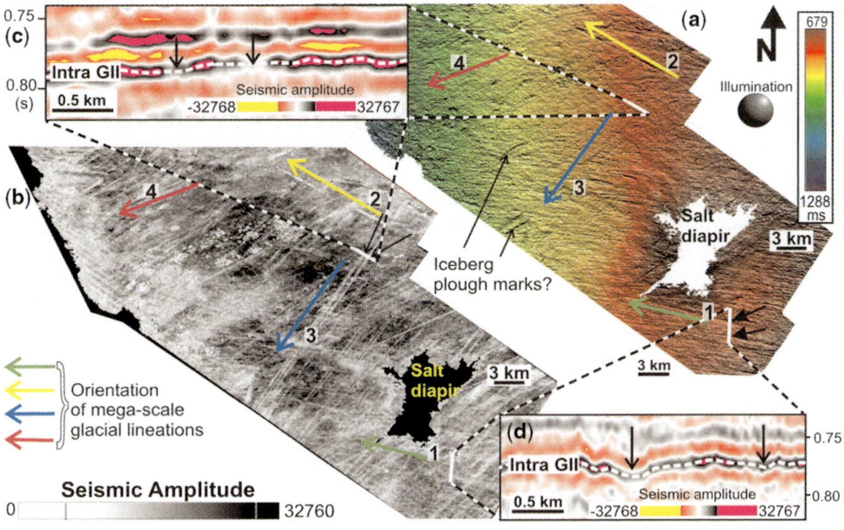

Fig. 6. (**a**) Shaded relief image of Intra GII at the upper part of sediment package GII (stratigraphic location is indicated Fig. 1d, marked Intra GII). (**b**) Reflection amplitude of seismic horizon Intra GII measured along the peak amplitude. (**c**) and (**d**) Seismic sections with locations as indicated. (s) indicates two-way travel time. The colored arrows in (a) and (b) numbered 1–4 indicate the orientations of four sets of mega-scale glacial lineations, all with different orientation.

former ice streams. The different orientations of the four sets of mega-scale glacial lineations are most probably related to reorganisation of ice-stream flow. A few shorter depressions (<3 km long and elongation rates around 3:1) on this surface (Fig. 6a) are most likely iceberg plough marks.

Linkage between mega-scale glacial lineations and underlying chains of sediment blocks

The Sørvestsnaget 3D (no. 1 in Fig. 2), which is located at the south-western Barents Sea margin reveals, because of high subsidence of the margin, a 2–3 km thick stratigraphic record (Fig. 2b) that corresponds to over one million years of glacial activity. This 3D is therefore well suited also for investigating imprints of ice streams older than the last, Late Weichselian glaciation. The Barents Sea shelf has, in contrast been exposed to extensive glacial erosion, nothing of sediment packages GI and GII seems to be left, and only parts GIII are preserved here (Fig. 2b).

Mega-scale glacial lineations are well preserved on most of the buried surfaces within and between sediment packages GII and GIII, suggesting erosion of former ice streams. It should, however, be acknowledged that the bedform records of former ice streams may be related only to the final stages of ice-stream operation (i.e. conditions immediately prior to shut down; Stokes & Clark 2001). Because we here have preserved up to several hundred meters of sediments between the buried, glacially eroded surfaces, we have the opportunity to study ice-stream imprints and associated processes covering longer time spans than just the last stage. Recently, this same 3D seismic survey was used to document the existence of long chains of megablocks and rafts, inferred to be eroded, transported and deposited by former ice streams (Andreassen *et al.* 2004). Here we illustrate how different 3D seismic attributes can provide additional information about such chains of sediment blocks, and reveal knowledge about their relation to the more well-known ice-stream related mega-scale glacial lineations.

Figure 7 shows how root-mean-square (RMS) amplitude calculated for a stratigraphically defined window (shaded band of the seismic profile of Fig. 7a) may reveal the internal architecture of these sediments, in this case the distribution of sediment blocks (Fig. 7b). RMS amplitude can be thought of as an average amplitude of a window. First a window is defined (e.g. shaded band of Fig. 7a). The software then squares all the amplitudes within this window (making them all positive), calculates the average of the squared amplitude values, and finally,

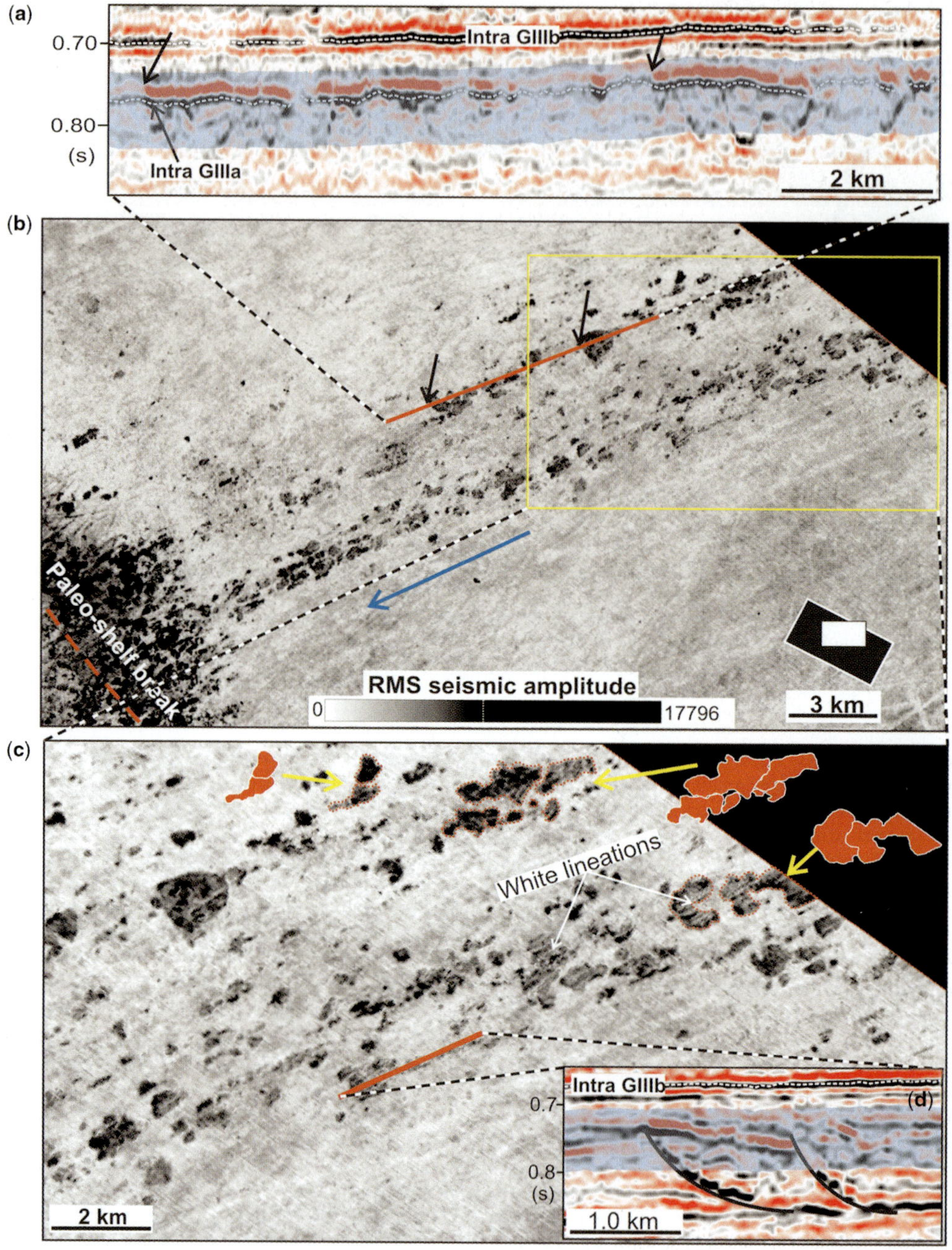

Fig. 7. (**a**) Seismic profile showing characteristic high-amplitude seismic reflections within sediment package GIII, which represent mega-scale sediment blocks or rafts (dark grey areas in b) of different sediment type (much higher product of velocity and density) than the surrounding till material. (**b**) Root-mean-square (RMS) amplitude of the volume indicated by the shaded zone of (a). This volume is a 90 ms thick interval 40–130 ms beneath reflection Intra GIIIb. RMS amplitude can be thought of as an average amplitude within a window, but because the amplitude values are squared before averaged and taken square root of, high seismic amplitudes are enhanced. The dark areas represent sediment blocks, and the blue arrow indicates orientation of the chains of sediment blocks. (**c**) Detail of (b). The red pieces illustrate that many of the sediment blocks fit together like parts of a jig-saw puzzle. (**d**) Seismic section parallel to the orientation of the sediment blocks; (s) indicates two-way travel time.

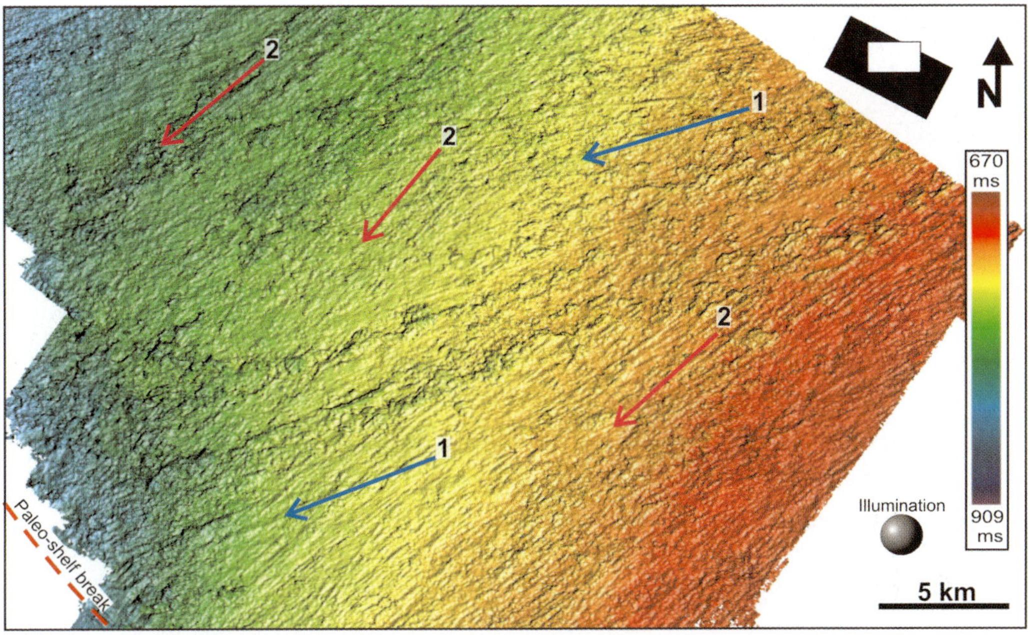

Fig. 8. Illuminated shaded relief image of reflection Intra GIIIa (stratigraphic location is indicated in Figs 1d and 7a). The arrows indicate orientation of two sets of mega-scale glacial lineations with different orientation.

the square root of that number. This procedure is done for every trace within in the 3D area. Although it gives an 'average value', the RMS procedure highlights high amplitudes more than just the average absolute amplitude. The RMS amplitude map of Figure 7(b) provides a similar distribution of sediment blocks to that obtained from the 3D volumetric attribute of integrated magnitude of seismic amplitudes by Andreassen *et al.* (2004). Details of the sediment blocks are shown in Figure 7(c, d). The sediment blocks, which stratigraphically occur in the middle of sediment package GIII, are aligned in two 1–2 km-wide northeast-trending chains that are over 50 km long. The larger sediment blocks, with areal extent exceeding 1 km^2 are, following Aber *et al.* (1989) classified as medium-sized megablocks, whereas the smaller sediment bodies (less than 1 km^2 each) are classified as rafts. The blocks and rafts are well imaged on the RMS amplitude map (Fig. 7) because they consist of a material with much higher product of velocity and density than the surrounding till material, and give thereby rise to high-amplitude seismic reflections. The two sediment chains have the same south-west orientation as a set of mega-scale glacial lineations on the surface at the top of or slightly above the sediment blocks (Fig. 8; orientation 1; in the further discussion this surface is called the 'top block surface') and as lineations on the surface at the base of the till unit containing the sediment blocks (Andreassen *et al.* 2004). The two sediment chains (Fig. 7b) are, from their orientation, location between surfaces eroded by ice streams, their morphology and internal structure interpreted to have been picked up, transported and deposited by an ice stream that drained trough the Bjørnøya Trough to the shelf break during a pre-Weichselian glaciation (younger that 440 ka years). Seismic profiles parallel to the orientation of the Intra GIII sediment chains suggest that at least some of these sediment blocks are back-tilted, imbricated, sub-horizontal sediment slabs that have been displaced from the north-east along a series of shear planes (Fig. 7d), suggesting sediment compression in the form of low-angle thrust faults. A similar pattern has been observed for sediment blocks at other stratigraphic levels of GII and GIII, where the blocks have been large enough to reveal the relationship between them on vertical seismic transects. A good example of this is shown in Figure 9, revealing details from sediment blocks from two different stratigraphic levels close to seismic reflection R5 at the base of sediment package GII (stratigraphic location is shown in Fig. 1c).

The detail image of the Intra GIII sediment chains clearly reveals that many of the sediment blocks fit together like parts of a jig-saw puzzle (red pieces in Fig. 7c). This suggests that these sediment blocks

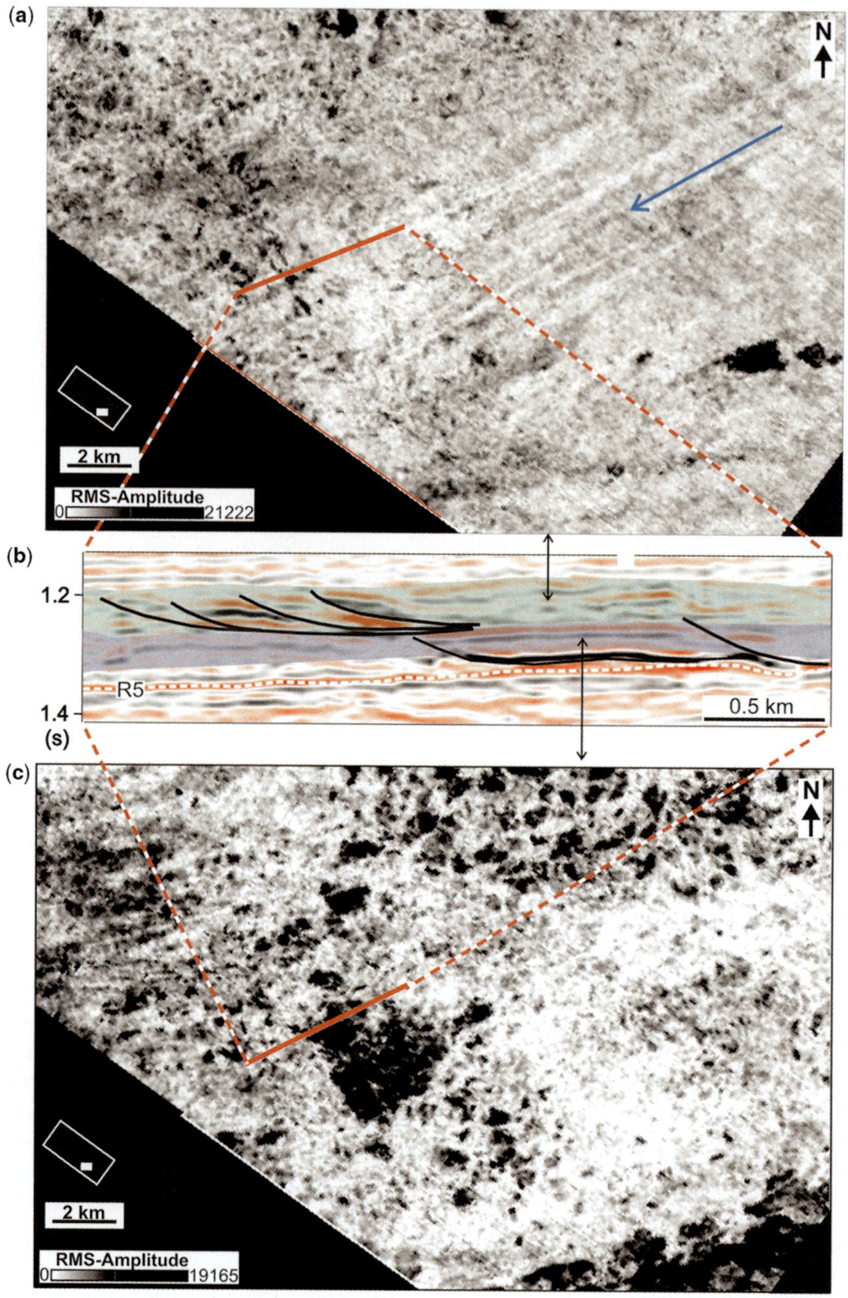

Fig. 9. (**a**) Root-mean-square (RMS) amplitude of the 60 ms 'thick' upper shaded interval of (b). The blue arrow indicates orientation of mega-scale glacial lineations at a surface close to the top of the shaded interval. (**b**) Seismic profile showing characteristic high-amplitude segments at two stratigraphic levels close to the seismic reflection R5 at the base of sediment package GII. These high-amplitude reflection segments represent sediment blocks and rafts (dark grey areas in a and c) of different sediment type (much higher product of velocity and density) than the surrounding till material. (s) indicates two-way travel time. (**c**) Root-mean-square (RMS) amplitude of the 70 ms 'thick' lower shaded interval of (b). RMS amplitude can be thought of as an average amplitude within a window, but because the amplitude values are squared before averaged and taken square root of, high seismic amplitudes are enhanced. The two shaded intervals of (b) which are used to calculate the maps of (a) and (c) are stratigraphically parallel to R5.

have been transported as larger blocks that later have been ripped apart. Their location now, up to several hundred meters apart suggests that they have been exposed to extensional forces. It was suggested by Andreassen *et al.* (2004) that light (low-amplitude) lines appearing on the amplitude image of the sediment blocks (corresponding to the white lineations indicated on Fig. 7c) represent northeast-oriented fault planes, and it was noticed that these were parallel to a set of mega-scale glacial lineations on the 'top block surface' (Fig. 8; orientation 2). To investigate this further, we constructed an image where the RMS amplitude map showing the sediment blocks (Fig. 7b) is draped over the 'top-block surface' (Fig. 8). High RMS amplitudes are displayed white and orange, and low amplitudes are dark grey on this new combination plot (Fig. 10) in order to obtain a better visualization of the morphological features involved. This new image suggests that the ice stream with flow orientation 1, which transported and deposited the two chains of sediment blocks (orange in Fig. 10a) has also eroded mega-scale furrows of this orientation in the 'top block surface'. It also reveals that what appear as light (low-amplitude) lines on many of the sediment blocks of Figure 7(b), and interpreted by Andreassen *et al.* (2004) to be vertical shear zones parallel to the ice flow, are in fact mega-scale glacial furrows. This is better illustrated when integrating the combined amplitude-shaded relief map of Figure 10(a) with a vertical seismic transect oriented transverse to the lineations (Fig. 10b; yellow arrows indicate furrows). A second set of mega-scale furrows, with orientation 2 has also affected both the 'top block surface' and the mega-scale sediment blocks (Fig. 10a, c). Ice-stream event 1 seems to have been responsible for picking up and transporting the sediment blocks, depositing them in two chains within a thick till unit and modifying them by furrow erosion, while ice-stream event 2 seems mainly to have eroded mega-scale glacial furrows into the sediment blocks and 'top block surface'. We therefore infer that the mega-scale glacial lineations with orientation 2 are associated with a younger ice-streaming event than those with orientation 1. This change in ice stream direction is most probably related to a reorganisation of ice-stream flow.

The two orientations of ice-streaming associated with the 'top-block surface' are also well imaged in other 3D seismic attribute plots of the reflection from 'the top block surface' Intra GIIIa (Fig. 11), because the top of most of the sediment blocks coincide with this surface. The surface-based amplitude (Fig. 11a) is straight forward a measure of seismic amplitude-value along the defined horizon, and the dip magnitude attribute (Fig. 11b) measures dip changes from trace to trace. Both lineation sets 1 and 2 appear on Figure 11(a) either as white lines of low-amplitude reflections, which might reflect that they are associated with complete removal of the layer that gives rise to this reflection, or they appear as dark lines of high-amplitude reflections, which might be because erosion of the grooves causes tuning of reflections from the top and base of the layer that produces this reflection. Mega-scale glacial grooves appear on the horizon attribute map of dip magnitude (Fig. 11b) as white lines with no dip flanked by dark lines representing high dips, probably because they have a flat bottom and steep sides. The two chains of sediment blocks are also well imaged on this map, characterised by two wide zones of high dip magnitudes, probably related to steep sides of sediment blocks. What is striking and new on Figure 11(b) is the appearance of two 2–5 km-wide zones with high dip magnitude at the north-western part of this map, with the same orientation as lineation set 2 mentioned above. These two zones seem to represent troughs in the 'top block surface', they do not seem to be associated with sediment blocks, and a very preliminary interpretation is that they may represent areas where the ice stream has picked up sediment blocks, leaving troughs behind.

The existence of mega-scale sediment blocks associated with mega-scale glacial lineations occur at many different stratigraphic levels of the sediment packages GII and GIII. Images from the stratigraphically shallowest level at which such features have been observed, close to the top of sediment package GIII (Fig. 1d), are shown in Figure 12, whereas images from the stratigraphically deepest level, close to the base of GII were shown in Figure 9. Single mega-scale sediment blocks, well-visualized on map-view images (Figs 9a and 12a) appear on vertical transects as high-amplitude seismic reflection segments (Figs 9b and 12b; indicated by arrows). Smaller, more closely spaced sediment rafts occur typically on vertical transects as a zone of chaotic seismic reflection configuration and high reflection amplitudes (Fig. 12b).

Conclusions and implications

The integration of regional scale sea-floor geomorphology with seismic stratigraphy (Fig. 2) and detailed geomorphology from industry 3D seismic data (Figs 3 and 4) documents the importance of glacial processes in the seascape evolution of the western Barents Sea. Fast-flowing ice streams occupying the cross-shelf troughs during the Late Weichselian glaciation (28–15 ^{14}C ka) have caused large-scale erosion, in some instances in the form of mega-scale glacial lineations on the sea floor as indications of ice-flow directions.

The Sørvestsnaget 3D at the western Barents Sea Margin (Fig. 2a; 3D area 1) is strategically located

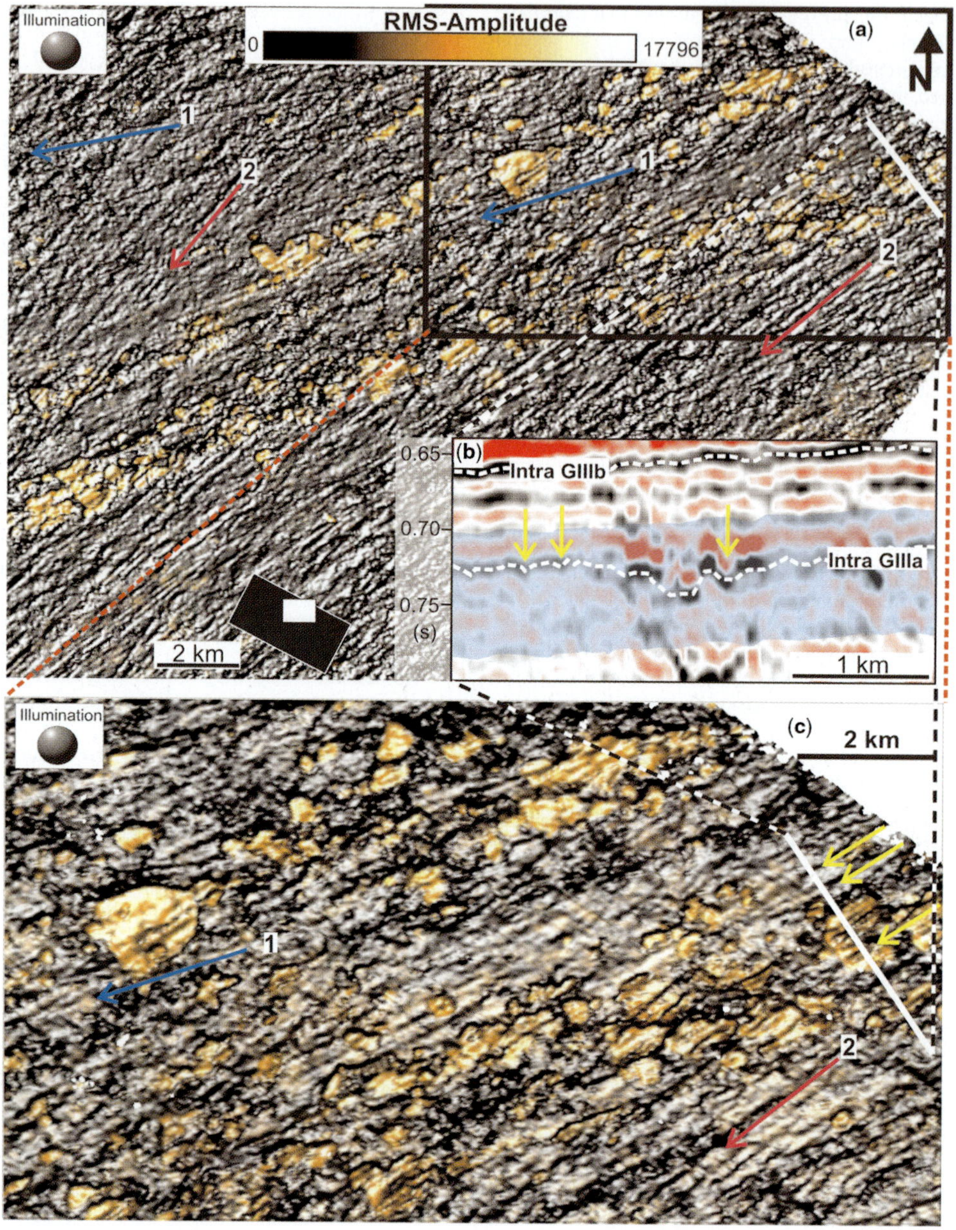

Fig. 10. (**a**) RMS amplitude map of Figure 7(b) draped over the 'top block surface' of Figure 8. Note that the high RMS amplitudes (of the sediment blocks) are here displayed white and orange, and low amplitudes are dark grey in order to obtain a better visualisation of the morphological features involved. The arrows numbered 1 and 2 indicate orientation of two sets of mega-scale glacial lineations with different orientation. (**b**) Seismic profile oriented transverse to the lineations of (a) and (c). (**c**) Detail of (a). The yellow arrows of (b) and (c) indicate pronounced furrows on reflection Intra GIIIa. Stratigraphic location of Intra GIIIa is indicated in Fig. 1D.

where a thick sucession of the Plio-Pleistocene sediment packages GI, GII and GIII occur (Fig. 1c). Except from local sediment accumulations of the Weichselian, upper part of sediment package GIII on the bank areas, very little is left of the Plio-Pleistocene sediments on the southern Barents Sea shelf (Fig. 2b).

The palaeo-shelf part of the 3D at the Barents Sea margin (Fig. 1c, d) reveals the occurrence of

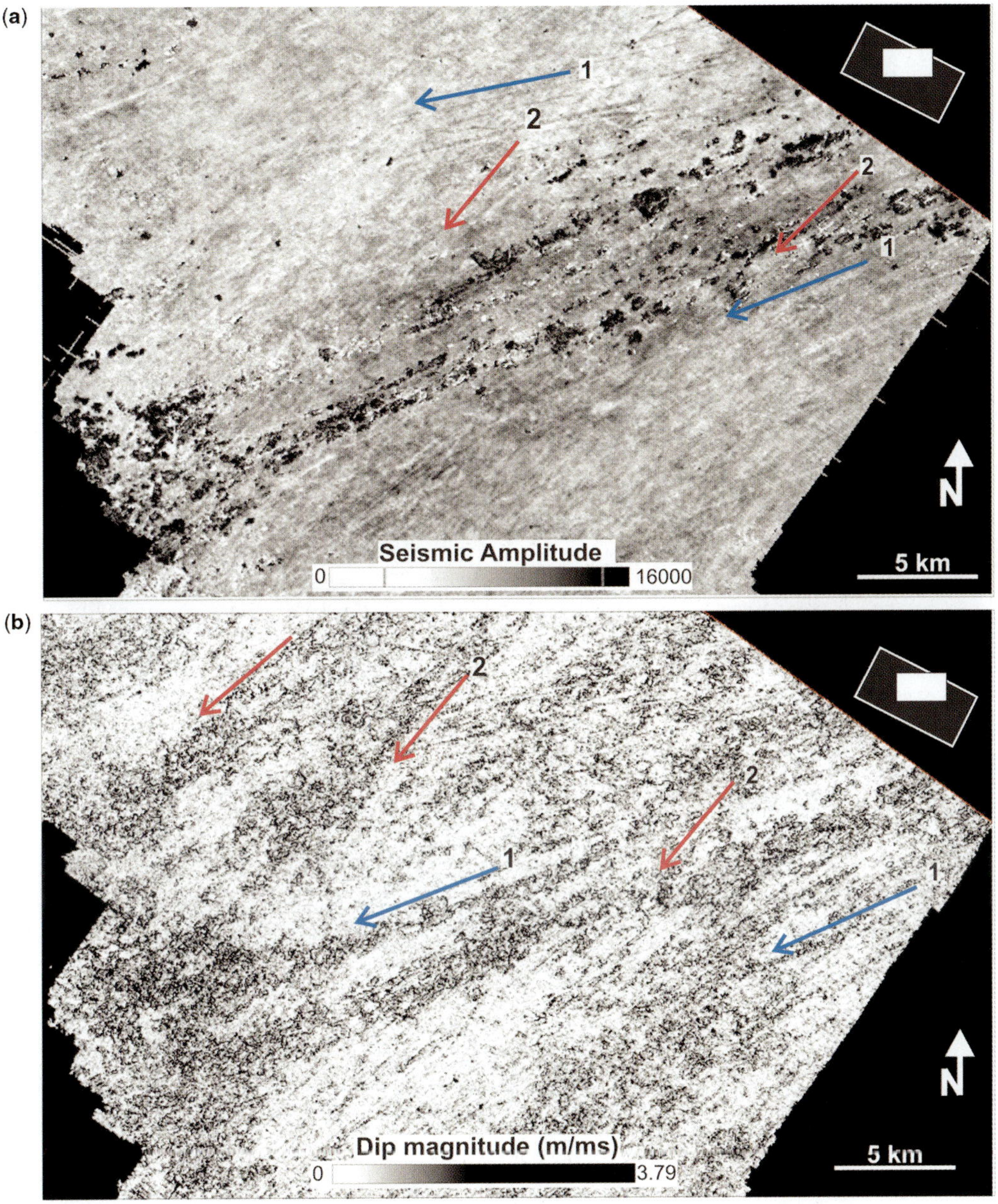

Fig. 11. (**a**) Surface-based amplitude-plot of seismic reflection from Intra GIIIa. (**b**) Dip magnitude plot of horizon Intra GIIIa. Stratigraphic location of Intra GIIIa is indicated in Figure 1d. The arrows numbered 1 and 2 indicate orientation of two sets of mega-scale glacial lineations with different orientation.

mega-scale glacial lineations on most of the buried GII and GIII horizons, and chains of megablocks and rafts in till units between these glacially eroded surfaces. This documents that grounded ice extended to the south-western Barents Sea shelf break several times during last million years, probably as fast-flowing ice streams. The glacigenic palaeo-shelf units of GII and GIII continue westwards into palaeo-slope units (Fig. 1c), which are also inferred to represent a glacigenic environment. An archive of palaeo-shelf sediments is left at the western Barents Sea shelf break. The sediments have here, due to high subsidence of the Bjørnøya Trough Mouth Fan, been partly shielded from later ice-stream erosion (Fig. 2b). Large chains of mega-blocks and rafts within all the glacigenic sequences document

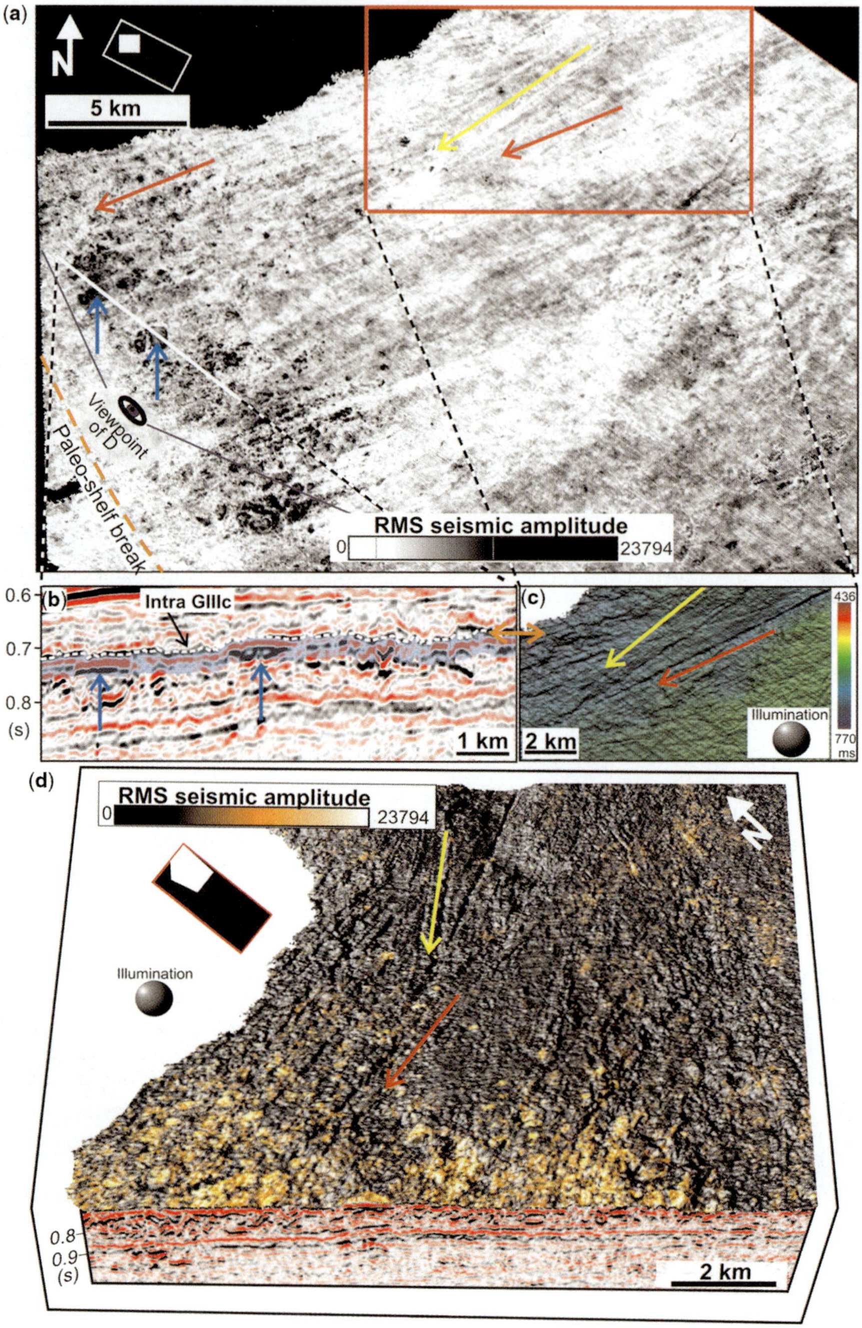

Fig. 12. (**a**) Root-mean-square (RMS) amplitude of the shaded interval of (b), revealing location of sediment blocks just beneath reflection Intra GIIIc. (See Fig. 1d for stratigraphic location.) (**b**) Seismic section showing high-amplitude segments (blue arrows), representing sediment blocks in (a) (dark areas with high RMS amplitudes). The shaded area indicates the zone of 50 ms just beneath Intra GIIIc, for which the RMS amplitudes of (a) were calculated. (**c**) Illuminated shaded relief map of seismic horizon Intra GIIIc at top of the sediment blocks of (a), see (b) for stratigraphic location. (**d**) Three-dimensional perspective view showing the RMS amplitude map of (a) draped over the shaded relief image of (c). The red and yellow arrows indicate orientation of two sets of mega-scale glacial lineations with different orientation.

the large erosive capacity of the former ice streams of this area, not only by sub-glacial scouring, but also their ability to remove square kilometre-large sediment blocks from the shelf and transport them to the shelf break. Based on the results above, we infer that the large amounts of the palaeo-slope sediments of the GII and GIII sediment packages consist of sediments that have been eroded from the Barents Sea shelf, brought to the shelf edge by glaciers, and further mobilised down-slope by mass-movement processes. A major part of this erosion has thus most probably taken place during the last million years, although the chronology is somewhat uncertain.

Various types of 3D seismic attributes, combined with detailed geomorphology and seismic stratigraphy are used to investigate the 2–3 km of stratigraphic record that reflects over a million year of ice-stream activity at the western Barents Sea margin. The appearance of mega-scale glacial lineations on various 3D seismic attribute maps indicates that they are formed by erosion beneath fast-flowing ice streams. The observations are consistent with the groove-ploughing theory, where ice keels at the base of ice streams are carving elongated grooves and deforming subglacial material up into intervening ridges (Tulaczyk *et al.* 2001; Clark *et al.* 2003). Significant amounts of cross-cutting by different ice-stream flow sets are most probably related to reorganisation of ice-stream flow patterns.

Seismic volumetric attribute maps reveal that mega-blocks and rafts, often aligned in chains, commonly occur within the till units between the glacially eroded horizons, implying that glaciotectonic erosion by fast-flowing ice streams was an important process in the transfer of sediments from the continental shelf to the Bjørnøya Trough Mouth Fan and the deep sea during the Plio-Pleistocene glaciations.

We infer that both mega-scale glacial lineations and chains of mega-scale sediment blocks and rafts are products of fast-flowing ice-steams, and that they probably represent different modes of ice-stream erosion. Mega-scale glacial lineations are most probably related to deformation of subglacial unfrozen sediments (Stokes & Clark 2002), whereas glacial transportation of mega-scale sediment blocks is usually interpreted to be related to freezing onto the base of the glacier (Aber *et al.* 1989). It has been proposed that palaeo-ice streams might have undergone periods of basal freezing near their margins, maybe associated with fast downward advection of cold surface ice (Christoffersen & Tulaczyk 2003), and that frozen sediment blocks were entrained by the glacier and transported along thrust planes before being re-deposited at the margin (Andreassen *et al.* 2004). Basal freezing is known to cause over-consolidation of sub-glacial sediments (Sættem *et al.* 1996), and can produce a rheological contrast between the frozen sediments at the ice base and the less consolidated sediments with depth. This process may focus sediment thrust deformation at the transition between the over-consolidated and less consolidated sediments.

We are grateful to K. Hogstad who made us aware of the fantastic potential of industry 3D seismic surveys, also for much shallower targets than they were designed for, and to L. Kuilman for enthusiastically shearing his expertise in 3D seismic interpretation. The University of Tromsø acknowledges GeoQuest for seismic interpretation software, and Norsk Hydro ASA and Statoil for kindly providing data, technical support and cooperation. Particular thanks are due to S. Schwencke and S. Buenz for keeping the GeoQuest software running, to Maarten Vanneste for helping us with the Generic Mapping Tool software, and to L. C. Nilssen for making some of the figures. We thank H. W. Posamentier and B. Hart for careful reviews that improved the manuscript. This is a contribution to the Strategic University Program 'Sedimentary Processes and Palaeoenvironment on the Northern Continental Margins (SPONCOM)' funded by the Research Council of Norway.

References

ABER, J. S., CROOT, D. G. & FENTON, M. M. 1989. *Glaciotectonic Landforms and Structures*. Kluwer Academic, Dordrecht.

ALLEY, R. B., BLANKENSHIP, D. D., BENTLEY, C. R. & ROONEY, S. T. 1986. Deformation of till beneath ice stream B, West Antarctica. *Nature*, **322**, 57–59.

ANDREASSEN, K., NILSSEN, L. C., RAFAELSEN, B. & KUILMAN, L. 2004. Three-dimensional seismic data from the Barents Sea margin reveal evidence of past ice streams and their dynamics. *Geology*, **32**, 729–732.

BARKER, P. F., CAMERLENGHI, A. *ET AL*. 1999. *Proceedings of the Ocean Drilling Program, Initial Reports*, **178**.

BARNES, P. W. & LIEN, R. 1988. Iceberg rework shelf sediments to 500 m off Antarctica. *Geology*, **16**, 1130–1133.

BENNETT, M. R. 2003. Ice streams as the arteries of an ice sheet: their mechanics, stability and significance. *Earth-Science Reviews*, **61**, 309–339.

BINSCHADLER, R. A., KING, M. A., ALLEY, R. B., ANANDAKRISHNAN, S. & PADMAN, L. 2003. Tidally Controlled Stick–Slip Discharge of a West Antarctic Ice. *Science*, **301**, 5636, 1087–1089.

BOULTON, G. S. & HINDMARSH, R. C. A. 1987. Sediment deformation beneath glaciers; rheology and geological consequences. *Journal of Geophysical Research*, **92**, B9, 9059–9082.

BROECKER, W. S. 1994. Massive iceberg discharges as triggers for global climate change. *Nature*, **365**, 143–147.

BUTT, F. A., ELVERHØI, A., SOLHEIM, A. & FORSBERG, C. F. 2000. Deciphering late Cenozoic development of the western Svalbard margin from ODP Site 986 results. *Marine Geology*, **169**, 373–390.

CANALS, M., URGELES, R. & CALAFAT, A. M. 2000. Deep sea-floor evidence of past ice streams off the Antarctic Peninsula. *Geology*, **28**, 31–34.

CHRISTOFFERSEN, P. & TULACZYK, S. 2003. Signature of palaeo-ice-stream stagnation: till consolidation induced by basal freeze-on. *Boreas*, **32**, 114–129.

CLARK, C. D. 1993. MSGLs and cross-cutting ice-flow landforms. *Earth Surf. Processes Landforms*, **18**, 1, 1–29.

CLARK, C. D., TULACZYK, S. M., STOKES, C. R. & CANALS, M. 2003. A groove-ploughing theory for the prediction of mega-scale glacial lineations, and implications for ice-stream mechanics. *Journal of Glaciology*, **49**, 165, 240–256.

DENTON, G. H. & HUGHES, T. J. 1981. The Arctic Ice Sheet: An outrageous hypothesis. *In*: DENTON, G. H. & HUGHES, T. J. (eds) *The Last Great Ice Sheets*. Wiley, New York.

DORÉ, A. G. & JENSEN, L. N. 1996. The impact of late Cenozoic uplift and erosion on hydrocarbon exploration: offshore Norway and some other uplifted basins. *Global and Planetary Change*, **12**, 415–436.

EIDVIN, T., JANSEN, E. & RIIS, F. 1993. Chronology of Tertiary fan deposits off the western Barents Sea: implications for the uplift and erosion history of the Barents Shelf. *Marine Geology*, **112**, 109–131.

ELVERHØI, A., SVENDSEN, J.-I., SOLHEIM, A., MILLIMAN, J. D., MANGERUD, J. & HOOKE, R. L. 1995. Late Quarternary sediment yield from the high Arctic Svalbard area. *Journal of Geology*, **103**, 1, 1–17.

FALEIDE, J. I., SOLHEIM, A., FIEDLER, A., HJELSTUEN, B. O., ANDERSEN, E. S. & VANNESTE, K. 1996. Late Cenozoic evolution of the western Barents Sea-Svalbard continental margin. *Global and Planetary Change*, **12**, 53–74.

FIEDLER, A. & FALEIDE, J. I. 1996. Cenozoic sedimentation along the southwestern Barents Sea margin in relation to uplift and erosion of the shelf. *Global and Planetary Change*, **12**, 75–93.

KAMB, B. 1991. Rheological nonlinearity and flow instability in the deforming-bed – mechanisms of ice stream motion. *Journal of Geophysical Research*, **96B**, 16585–16595.

KUVAAS, B. & KRISTOFFERSEN, Y. 1996. Mass movements in glaciomarine sediments on the Barents Sea continental slope. *Global and Planetary Change*, **12**, 287–307.

LABERG, J. S. & VORREN, T. O. 1995. The Middle and Late Pleistocene evolution of the Bear Island Trough Mouth Fan. *Global and Planetary Change*, **316**, 1–22.

MARFURT, K. J., SCHEET, R. M., SHARP, J. A. & HARPER, M. G. 1998. Suppression of the acquisition footprint for seismic sequence attribute mapping. *Geophysics*, **62**, 6, 1774–1778.

Ó COFAIGH, C., DOWDESWELL, J. A., ALLEN, C. S., HIEMSTRA, J. F., PUDSEY, C. J., EVANS, J. & EVANS, D. J. A. 2005. Flow dynamics and till genesis associated with a marine-based Antarctic palaeo-ice stream. *Quaternary Science Reviews*, **24**, 709–740.

OTTESEN, D., DOWDESWELL, J. A., RISE, L., ROKOENGEN, K. & HENRIKSEN, S. 2002. Large-scale morphological evidence for past ice-stream flow on the mid-Norwegian continental margin. *In*: DOWDESWELL, J. A. & Ó COFAIGH, C. (eds) *Glacier-influenced Sedimentation on High-latitude Continental Margins*. Geological Society, London, Special Publications, 245–258.

PUNKARI, M. 1997. Glacial and glacifluvial deposits in the interlobate areas of the Scandinavian Ice sheet. *Quaternary Science Reviews*, **16**, 741–753.

OTTESEN, D., DOWDESWELL, J. A. & RISE, L. 2005. Submarine landforms and the reconstruction of fast-flowing ice streams within a large Quaternary ice sheet: the 2,500 km-long Norwegian-Svalbard margin (57°–80°N). *GSA Bulletin*, **117**, 1033–1050.

RIIS, F. & FJELDSKAR, W. 1992. On the magnitude of the Late Tertiary and Quaternary erosion and its significance for the uplift of Fennoscandia and the Barents Sea. *In*: LARSEN, R. M., BREKKE, LARSEN, B. T. & TALLERAAS, E. (eds) *Structural and Tectonic Modelling and its Application to Petroleum Geology*. Norwegian Petroleum Society Special Publications, **1**. Elsevier, Amsterdam, 163–185.

RYSETH, A., AUGUSTSON, J. H., *ET AL*. 2003. Cenozoic stratigraphy and evolution of the Sørvestsnaget Basin, southwestern Barents Sea. *Norwegian Journal of Geology*, **83**, 107–130.

SÆTTEM, J., POOLE, D. A. R., ELLINGSEN, L. & SEJRUP, H. P. 1992. Glacial geology of outer Bjørnøyrenna, southwestern Barents Sea. *Marine Geology*, **103**, 15–51.

SÆTTEM, J., BUGGE, T., *ET AL*. Cenozoic margin development and erosion of the Barents Sea: core evidence from southwest of Bjørnøya. *Marine Geology*, **118**, 257–281.

SÆTTEM, J., RISE, L., ROKOENGEN, K. & BY, T. 1996. Soil investigations, offshore mid Norway: A case study of glacial influence on geotechnical properties. *Global and Planetary Change*, **12**, 271–285.

SEJRUP, H. P., LANDVIK, J., KING, K., HAFLIDASON, H. & NESJE, A. 2000. Quaternary glaciations in southern Fennoscandia: evidence from southwestern Norway and the northern North Sea region. *Quaternary Science Reviews*, **19**, 667–285.

SIEGERT, M. J. & DOWDESWELL, J. A. 2002. Late Weichselian iceberg, meltwater and sediment production from the Eurasian High Arctic ice sheet: results from numerical ice-sheet modelling. *Marine Geology*, **188**, 109–127.

SOLHEIM, A. & KRISTOFFERSEN, Y. 1984. Distribution of sediments above bedrock and glacialhistory in the western Barents Sea. *Norw. Polarinst. Skr.*, Norsk Polarinstitutt Skinfter (Norwegian Polar Research Institute) **179B**.

SOLHEIM, A., RUSSWARM, L., ELVERHØI, A. & NYLAND BERG, M. 1990. Glacial geomorphic features: direct evidence for grounded ice in the northern Barents Sea and implications for the pattern of deglaciation and late glacial sedimentation. *In*: DOWDESWELL, J. A. & SCOUSE, J. D. (eds) *Glacimarine Environments: Processes and Sediments*. Geological Society, London, Special Publication, **53**, 253–268.

SOLHEIM, A., ANDERSEN, E. S., ELVERHØI, A. & FIEDLER, A. 1996. Late Cenozoic depositional history of the western Svalbard continental shelf, controlled by subsidence and climate. *Global and Planetary Change*, **12**, 135–148.

STALKER, A. M. 1960. *Ice-pressed drift forms and associated deposits in Alberta*. Bulletin, Geological Survey of Canada.

STOKES, C. R. & CLARK, C. D. 1999. Geomorphological criteria for identifying Pleistocene ice streams. *Annals of Glaciology*, **28**, 67–74.

STOKES, C. R. & CLARK, C. D. 2001. Palaeo-ice streams. *Quaternary Science Reviews*, **20**, 1437–1457.

STOKES, C. R. & CLARK, C. D. 2002. Are long subglacial bedforms indicative of fast ice flow? *Boreas*, **31**, 239–249.

TULACZYK, S. M., KAMB, B. & ENGELHARDT, H. 2000. Basal mechanics of Ice Stream B: I. Till mechanics. *Journal of Geophysical Research*, **105**, 463–481.

TULACZYK, S. M., SCHERER, R. P. & CLARK, C. D. 2001. A ploughing model for the origin of weak tills beneath ice streams: a qualitative treatment. *Quaternary International*, **86**, 59–70.

VORREN, T. O. & LABERG, J. S. 1996. Late glacial air temperature, oceanographic and ice sheet interactions in the southern Barents Sea region. *In*: ANDREWS, J. T., AUSTIN, W. E. N., BERGSTEN, H. & JENNINGS, A. E. (eds) *Late Quarternary Palaeoceanography of the North Atlantic Margins*. Geological Society Special Publication, 303–321.

VORREN, T. O. & LABERG, J. S. 1997. Trough mouth fans – Palaeoclimate and ice-sheet Monitors. *Quaternary Science Reviews*, **16**, 865–882.

VORREN, T. O., RICHARDSEN, G., KNUTSEN, S.-M. & HENRIKSEN, E. 1991. Cenozoic erosion and sedimentation in the western Barents Sea. *Marine and Petroleum Geology*, **8**, 317–340.

Geometry and seismic geomorphology of carbonate shoreface clinoforms, Jurassic Smackover Formation, north Louisiana

C. R. HANDFORD[1] & L. R. BARIA[2]

[1]*Strata-Search, LLC, 10209 Brimfield Drive, Austin, TX 78726, USA*
(e-mail: Handford@strata-search.com)

[2]*Jura Search Inc., PO Box 997, Jackson, MS 39208, USA*
(e-mail: bearlear@bellsouth.net)

Abstract: Across much of the Gulf Coast basin of the USA, the Smackover Formation consists of a nearly 100 m-thick shoaling upward cycle capped by oolitic/oncolitic packstones and grainstones. It has long been interpreted as a homoclinal ramp succession, which was analogous to the modern example in the southeastern Arabian Gulf. In a three-dimensional seismic survey in north Louisiana, the shoaling-upward cycle is imaged as basinward progradational clinoforms (4–7° inclination) with well-defined toplap and downlap surfaces. In map view, amplitude slices show that the clinoform bodies are strike-oriented and continuous. The inclination and width of the clinoform bodies indicate that water depths of up to 90 m were present within 1 km of the shoreline. Such characteristics indicate that the Smackover Formation cannot be classified as a homoclinal ramp in north Louisiana and that the Arabian Gulf is not analogous to the Smackover.

Jurassic carbonates of the Smackover Formation (Fig. 1) have produced hydrocarbons in the US Gulf Coast since the 1920s. With its long exploration history, Smackover exploration has undergone a transition into a very mature play. However, recent advancements in three-dimensional (3D) seismic survey technology have resulted in renewed exploration interest for mature reservoirs, such as the Smackover, because of improved visualization of stratigraphic details. Improved visualization is leading to new exploration opportunities and previously unseen stratigraphic details, which are suggesting that a new depositional and stratigraphic model is needed.

Recent development of sophisticated visualization techniques allows geoscientists to produce seismic amplitude displays, which can be used to interpret palaeogeomorphological features. Many of the recently published examples of palaeogeomorphological interpretations were derived from near-surface and shallow subsurface siliciclastic strata in offshore surveys. In general, image quality deteriorates and palaeogeomorphological details diminish with deeper burial and increasing compaction, especially in onshore areas and where carbonates are present. In these cases, interpretation of seismic geomorphology is challenging.

This paper presents an interpretation of oolitic carbonate shoreface deposits in the Smackover Formation as imaged in a 3D seismic survey in north Louisiana (Fig. 2) where burial depths are more than 2.4 km (1.5 miles). The survey revealed the presence of numerous progradational clinoform bodies, which were subsequently used to develop a revised depositional and stratigraphic model of the Smackover Formation.

Overview of Smackover Formation

The long and storied history of Smackover exploration includes a large number of investigations, which have addressed Smackover stratigraphy, lithofacies and depositional environments (Ahr 1973; Akin & Graves 1969; Baria *et al.* 1982; Bishop 1968, 1971*a*, 1973; Budd & Loucks 1981; Chimene 1991; Harris & Dodman 1982; Moore 1984; Troell & Robinson 1987, and many others). These studies have documented the regional distribution of lithofacies (Fig. 2) and the well-known shallowing-upward facies succession (Fig. 3), both of which were key to the development of a general depositional model of the Smackover Formation (Fig. 4). The model, which has been modified only slightly during the past 20–30 years, suggests that the Smackover Formation accumulated on a ramp with a low-gradient slope (<1°) from shoreline to basin (Ahr 1973; Baria *et al.* 1982; Budd & Loucks 1981; Read 1985). A series of oolitic shoals, islands and tidal bars developed in the inner ramp across northern Louisiana and south Arkansas and prograded basinward as shingled clinoforms (Baria *et al.* 1982). The high-energy oolitic facies passes updip into sabkha to salina evaporites and redbeds

From: Davies, R. J., Posamentier, H. W., Wood, L. J. & Cartwright, J. A. (eds) *Seismic Geomorphology: Applications to Hydrocarbon Exploration and Production*. Geological Society, London, Special Publications, **277**, 171–185.
0305-8719/07/$15.00

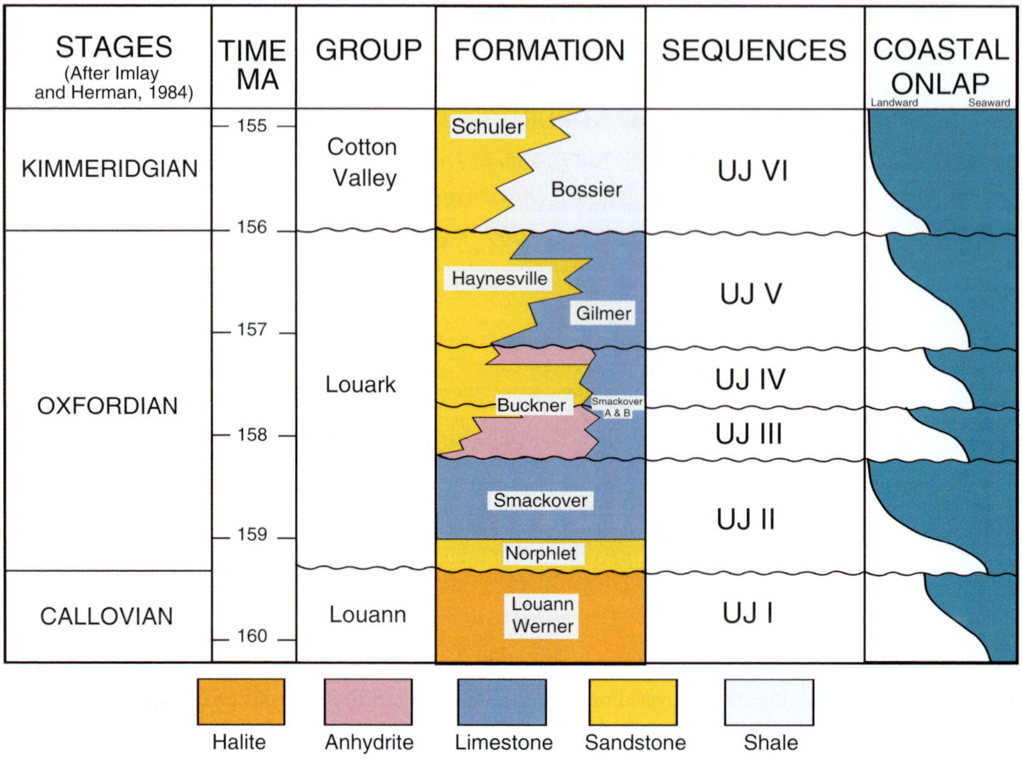

Fig. 1. Jurassic stratigraphic chart as interpreted by Moore (2001). Reprinted by permission of the AAPG, whose permission is required for further use.

(Bishop 1971*b*) and downdip into middle-outer ramp wackestones and mudstones (Budd & Loucks 1981).

Regional studies by Ahr (1973) and Bishop (1968) have shown that the Smackover Formation is made up of a series of strike-oriented, relatively narrow lithofacies belts across Texas, Arkansas, Louisiana, and Mississippi (Fig. 2). These belts include depositionally strike-oriented evaporites (salina-sabkha) and redbeds in the north, which change basinward into a belt of oolitic facies (inner-ramp beaches and shoals), followed by a peloidal facies belt (mid-outer ramp) and laminated mudstones (basin). In some areas of northeastern Louisiana to eastern Mississippi, the upper Smackover contains terrigenous sandstones (Ahr 1973; Bishop 1968) deposited in shoreface and eolian settings (Tedesco *et al.* 2002).

Studies of Walker Creek, Chalybeat Springs, and Shongaloo fields (Troell & Robinson 1987; Chimene 1991; Demiss & Millikin 1993) have shown with well-log correlations that the upper Smackover reservoirs consist of a series of offlapping, or progradational, porous grainstone units, which pinch out updip into evaporites and/or muddy facies. Unfortunately, the frequency and signal quality of most of the published seismic data (e.g. Demiss & Millikin 1993) for the Smackover Formation are too low to show the offlapping relationships. However, Burchette & Wright (1992) published a line tracing of a seismic line showing Smackover clinoforms, which had prograded approximately 60 km basinward. The dimensions and geometries of the offlapping units were not discussed.

Although the Smackover ramp has been a popular hypothesis, Wilson (1975) first suggested that the Smackover limestones may not represent a ramp because he recognized that the facies belts are relatively narrow, thus indicating a steeper slope than many other ramp carbonates. This important observation, however, seems to have been ignored because the ramp model was consistently invoked for more than 25 years following Wilson's (1975) observation. This trend began to change, however, when Moore (2001) proposed that a shelf margin system developed above a basement structure in north Louisiana and south Arkansas.

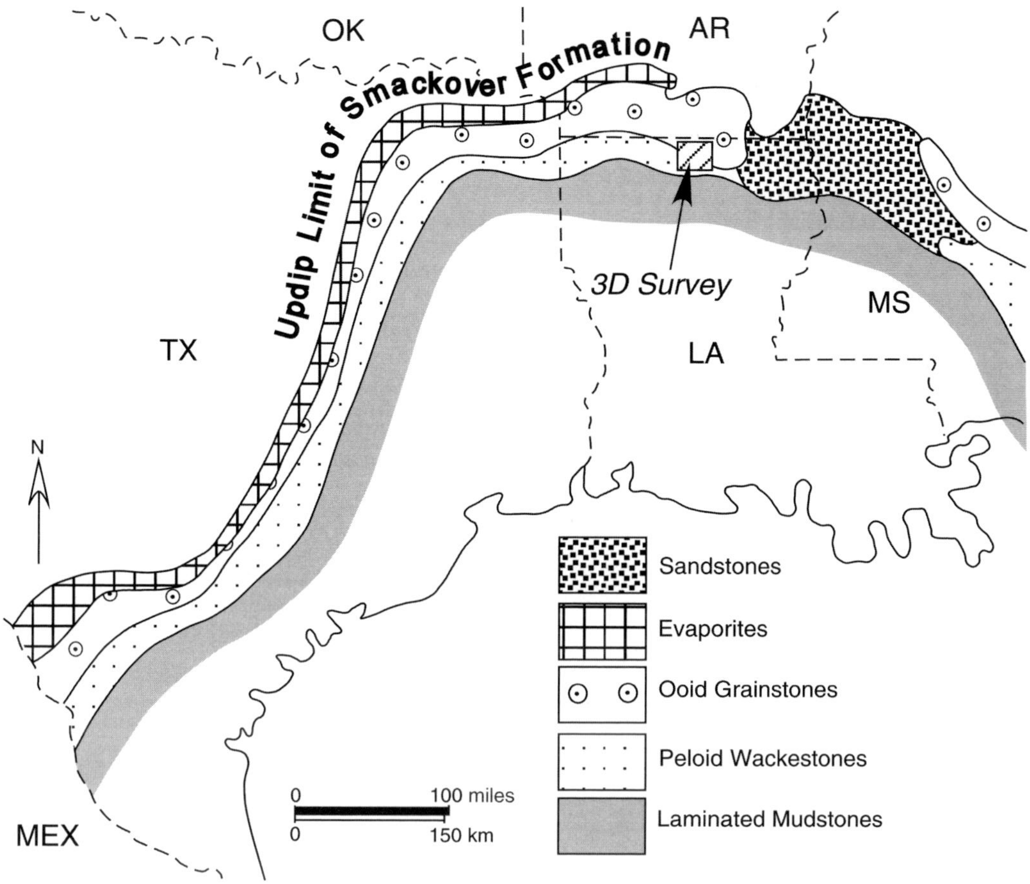

Fig. 2. Regional map of Smackover lithofacies belts in the U.S. Gulf Coast basin and location of 3D seismic survey. Map modified from Ahr (1973) and Bishop (1968). Reprinted by permission of the AAPG, whose permission is required for further use.

Clinoform geometry

A high-frequency 3D seismic survey (approximately 77 km^2), which was shot in northern Louisiana (Fig. 2), has revealed stratigraphic details within the Smackover formation that were previously unavailable. Images have been generated from the seismic data, which are fostering a more accurate reconstruction of the depositional systems and coastal palaeogeomorphology. Depositional dip-oriented (northeast–southwest) arbitrary lines (Fig. 5) reveal the presence of well-developed progradational clinoform units in the upper Smackover. These clinoform bodies have prograded basinward at least 10.5 km and the entire extent of the 3D survey.

Each clinoform unit shows well-defined toplap and downlap relationships (Fig. 5). The downlap surface corresponds to the top of the laminated dense lower Smackover carbonates and the toplap surface corresponds to the top of the Smackover and base of the Buckner evaporites. These two surfaces are respectively interpreted as a maximum flooding surface and upper sequence boundary (Fig. 5) to either one of Moore's (2001) UJ II or UJ III (Smackover A and B) sequences (Fig. 1). Regional isopach maps of Moore (2001) suggest that the Smackover of this study belongs to UJ III. Thus, the lower Smackover carbonates in the seismic data set probably correspond to the lowstand and transgressive systems tracts, and the overlying progradational units of the middle-upper Smackover represent the highstand systems tract. Middle Smackover beds are typically bioturbated peloidal wackestones and packstones. These are transitionally overlain by upper Smackover ooid grainstones, which are well sorted and commonly cross stratified to planar-laminated. These characteristics indicate deposition in a foreshore to berm, or beach, and possibly

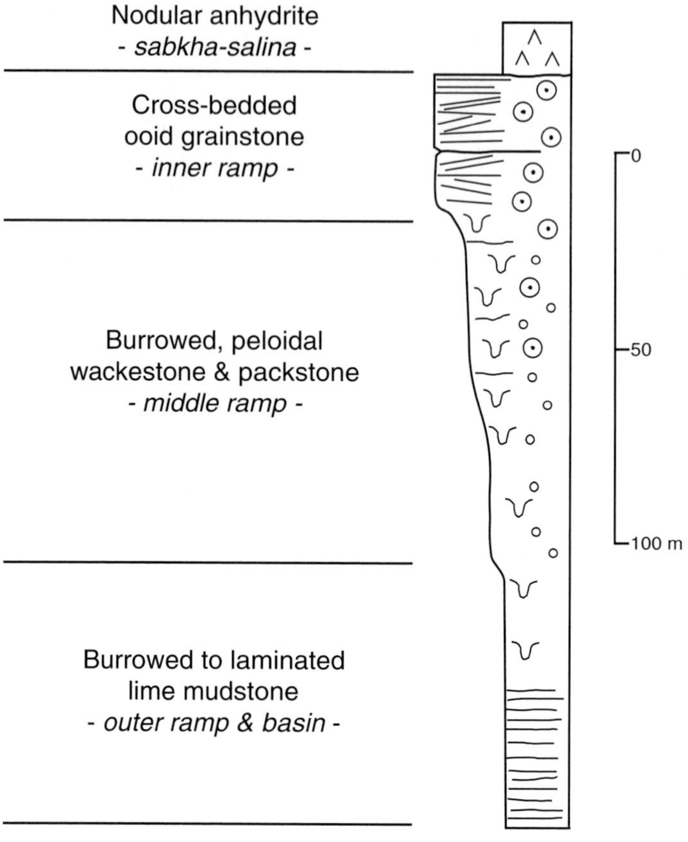

Fig. 3. Schematic diagram showing typical vertical succession of Smackover lithofacies in the northern Gulf Coast area of Louisiana and Arkansas.

coastal eolian dune environments. We also expect this sequence boundary to exhibit evidence of subaerial exposure (palaeosol) similar to that documented by Moore (2001) at the top of the UJ II sequence.

Although clinoform dimensions vary, the general ranges of dimensions of Smackover clinoform bodies are as follows (Fig. 6):

1. Clinoform length (lateral distance from toplap to downlap pinchout) is 0.5–1.1 km.

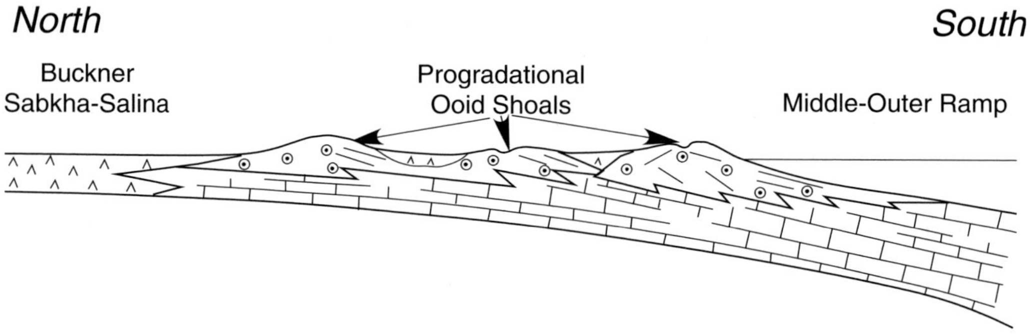

Fig. 4. Schematic depositional model of the Smackover-Buckner formations is based upon Ahr (1973), Baria *et al.* (1982), Harris & Dodman (1982) and others.

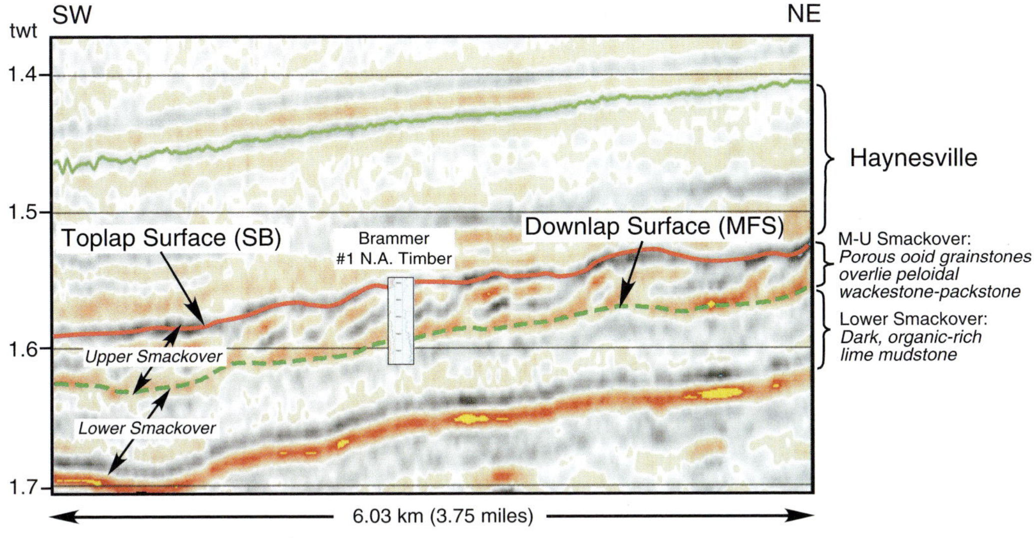

Fig. 5. Arbitrary NE–SW line through seismic survey shows Smackover-Haynesville formations. Clinoforms lie within the middle to upper Smackover Formation. Downlap and toplap surfaces are identified and a general 'tie' with a Smackover well log that lies along the line. No synthethic seismogram was available for the Brammer well (inset). All seismic data are in two-way travel time (twt) in seconds.

2. Clinoform height (vertical distance from toplap surface to downlap surface) is 50–70 m. This figure was calculated from interval velocity data.
3. Based upon the above dimensions, the maximum slope angle, or inclination, of the clinoform units ranges from 4–7°.

The height of each clinoform unit must closely represent the water depth during highstand progradation, with the toplap surface forming at or near highstand sea level and the downlap surface representing the seafloor during clinoform progradation. Thus, the clinoform dimensions presented above indicate that water depths

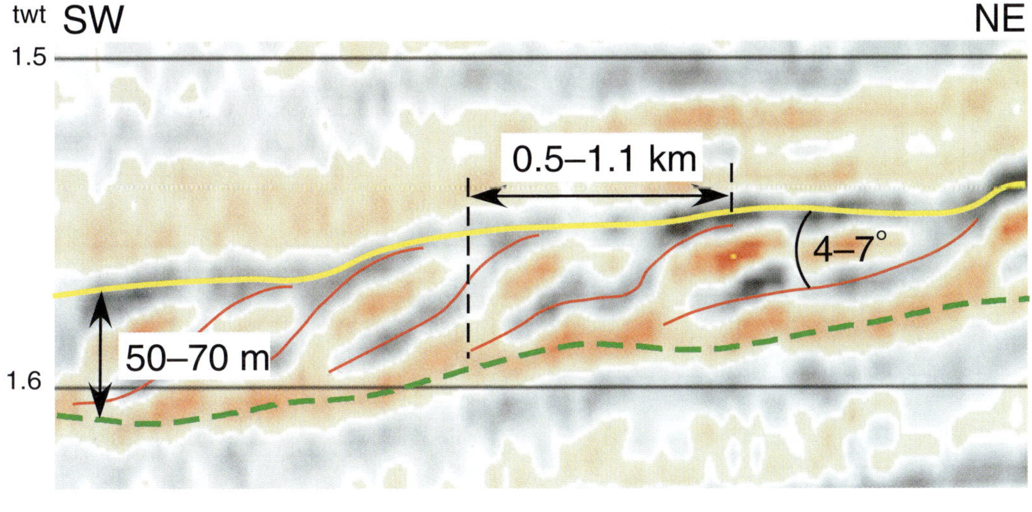

Fig. 6. Closeup of arbitrary seismic line shown in Figure 5 shows the calculated dimensions (from interval velocity data) of the Smackover clinoform bodies.

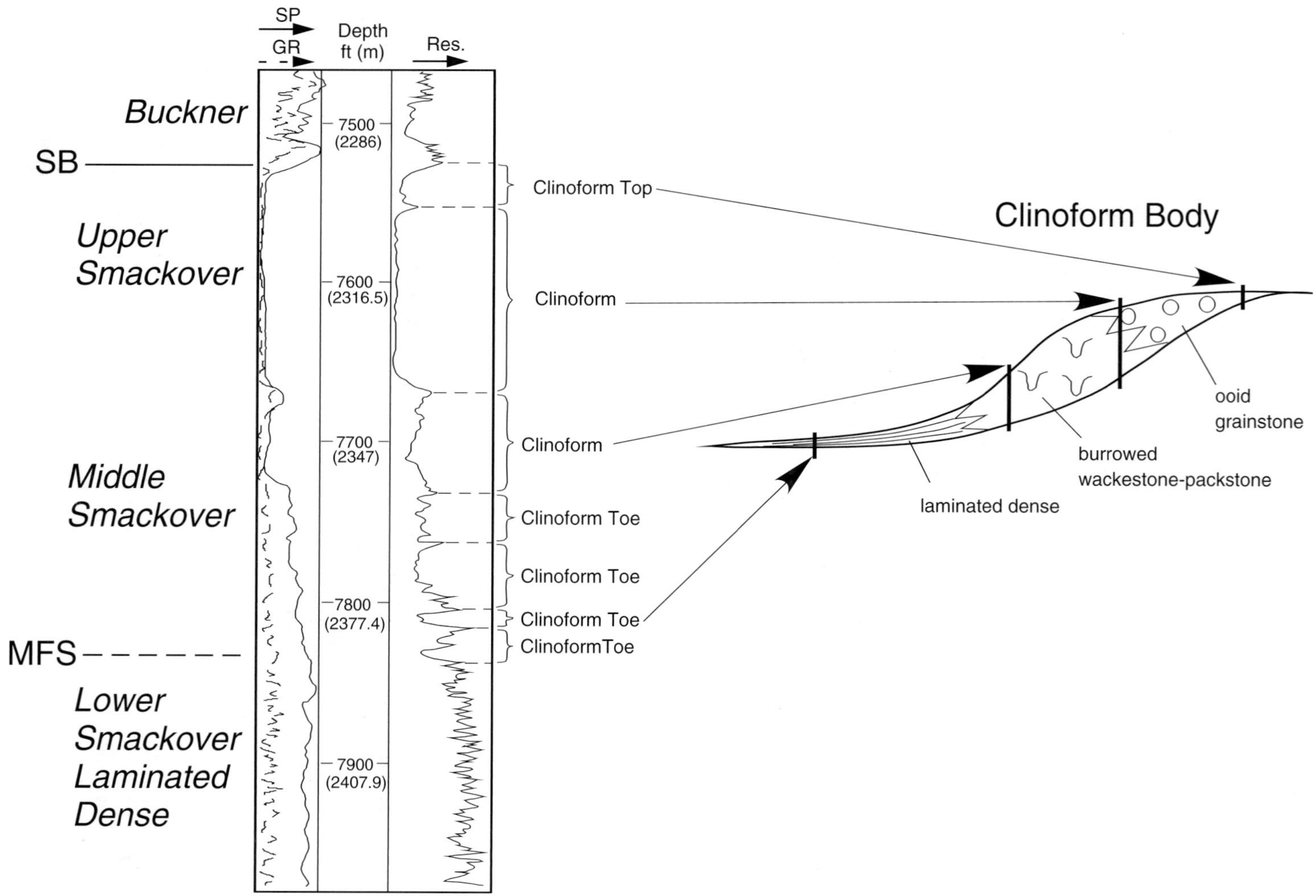

Fig. 7. Well log-derived stacking pattern interpretation of the Brammer no. 1 North American Timber well and a schematic diagram showing the relationship between clinoform position and the well log.

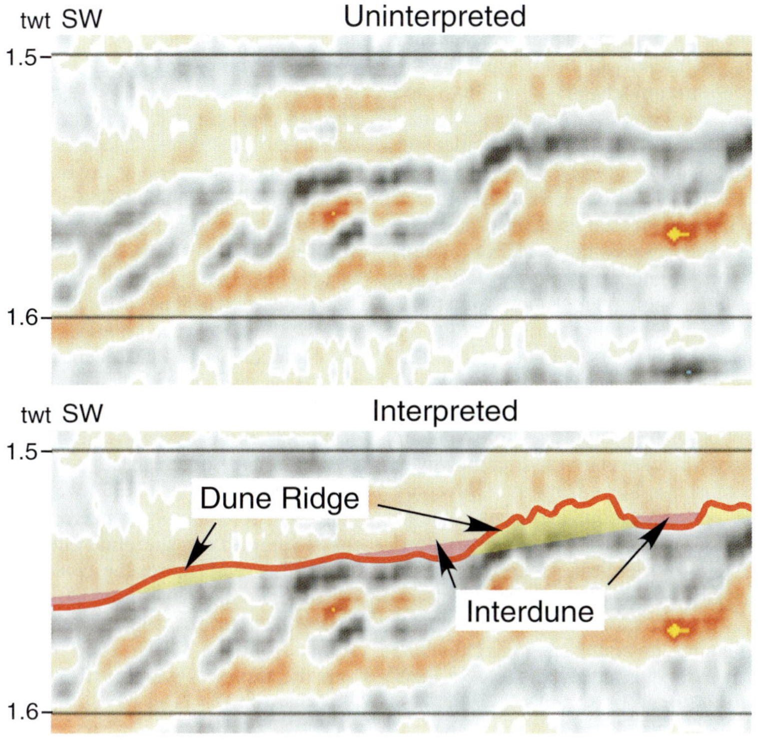

Fig. 8. Closeup of arbitrary seismic line shows irregular toplap surface, which may indicate coastal dune–interdune topography.

of 50–70 m were present less than 1 km from offshore.

Independent evidence of water depth comes from spontaneous potential (SP), gamma-ray (GR) and resistivity (Res) logs in the Brammer 1 North American Timber well, which is located in the 3D seismic survey (Figs 5 and 7). Serrated, relatively high gamma-ray and serrated high-resistivity log responses are present in the lower Smackover dense facies. Above this facies, however, the gamma ray and SP logs display a funnel-shaped, or 'coarsening-upward', pattern through the middle and upper Smackover. The boundary between the serrated lower Smackover and the overlying middle-upper Smackover with a funnel-shaped log pattern is considered to correspond to the maximum flooding surface, or downlap surface, as expressed in the seismic data. Thus, the middle-upper Smackover clinoform strata belong to the highstand systems tract and its thickness should indicate water depth in the vicinity of the well. The measured thickness of the highstand systems tract is 91 m, which is greater than the water depth (70 m) calculated from seismic interval velocity data. This difference probably indicates that the interval velocity data are inexact. Thus, the well-log interpretation is considered more accurate. In either case, remarkably deep water (70 or 90 m) was present at a distance of 1 km or less from the shoreline.

Resistivity log patterns suggest that the Brammer well penetrated several highstand clinoform bodies. Highstand strata are marked by an upward decrease in resistivity but several high-resistivity spikes interrupt the trend (Fig. 7). These spikes probably represent parasequence, or bedset, boundaries between stacked clinoform bodies. The thickness of these bodies changes systematically upward through the exploratory well from thin to thick and thin again, as follows (Fig. 7).

1. Thin (3–12 m) units in the lower highstand strata are interpreted to represent clinoform toeset strata.
2. Thicker (18–34 m) units in the middle-upper highstand strata are interpreted to represent the thickest parts of clinoform foreset strata.
3. A thin (9 m) unit in the upper highstand strata just below the interpreted sequence boundary is believed to represent topset strata.

The observed vertical change in thickness is what would be encountered by a well penetrating several stacked, progradational clinoform bodies.

An irregular topographic surface of ridges and swales is present at the toplap surface, or sequence boundary, of the progradational clinoforms (Fig. 8). The ridges are approximately 10–20 m high and 200–700 m wide. They are separated by low areas, or swales, which are 200–600 m wide. Irregular surfaces such as this can be produced by (1) fluvial incision, (2) submarine erosion or (3) depositional processes. As will be discussed later, the amplitude displays of time slices show no evidence of dip-oriented erosional features, which would be associated with fluvial incision and submarine tidal-channel erosion. The scale and location of the ridges and swales along the carbonate strandline suggest that they represent coastal eolian dunes separated by interdune sabkhas and/or salinas. We are not aware of any published references to carbonate eolian dunes in the Smackover Formation; however, an unpublished Master's thesis from the University of New Orleans reported vadose meniscus cements in a strike-oriented, well-sorted ooid grainstone (eolian) at the top of the Smackover Formation in southern Arkansas (W. C. Ward, pers. comm. 2005).

Carbonate eolianites are common along Holocene oolitic shorelines and we propose that some of them are analogous to the interpreted Smackover dunes. For example, coastal carbonate dunes, 5–16 m high, are relatively common in the Bahamas (Harris 1979; Lloyd *et al.* 1987), Cancun (Ward & Brady 1973) and Abu Dhabi (Loreau & Purser 1973). Of these, perhaps the closest analogue to the Smackover example is present on Caicos Bank, which is located at the southeastern end of the Bahamas banks. Lloyd *et al.* (1987) documented a 0.5 km wide belt of seaward-stepping Holocene oolitic dunes up to 16 m high adjacent to the oolitic beach of West Caicos Island (Fig. 9). A gypsum-bearing salina, which may be a modern analogue to Buckner evaporites overlying the Smackover carbonates, is present between the Holocene and Pleistocene dunes. Lloyd *et al.* (1987) showed that the ooids are generated at the beach by intertidal wave swash associated with wind-driven onshore waves. The West Caicos beach is constantly subjected to persistent southeasterly trade winds, and shoaling wind waves. Thus, an intertidal ooid factory rapidly produces the grains that accrete to the shoreline allowing it to prograde seaward. Furthermore, the beach is also the source of ooids for the eolian dunes. The strongest tradewinds, storms and hurricanes sweep the ooid grains from the swash zone across the berm and into the dunes. Owing to the sediment producing efficiency of the ooid factory, the growth of eolian

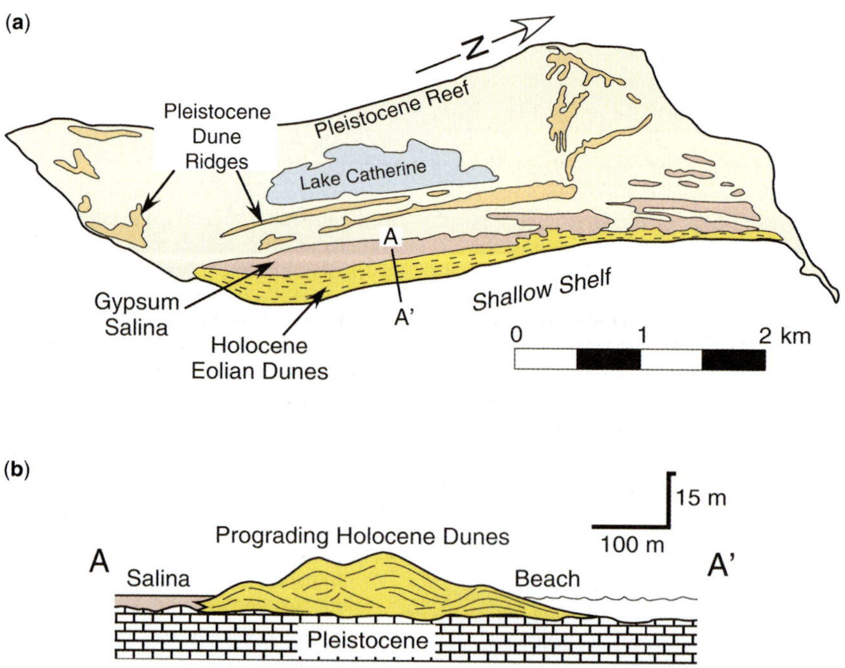

Fig. 9. (**a**) Map of West Caicos Island showing location of gypsum salinas, Pleistocene eolian dune ridges, Holocene dunes and cross section A–A'. (**b**) Schematic cross section (A–A') shows prograding Holocene beach–dune complex along the eastern shore of West Caicos island. Modified from Lloyd *et al.* (1987). Reprinted by permission of the SEPM, whose permission is required for further use.

dunes is driven by deposition rather than wind deflation of pre-existing deposits.

Onshore wind and wave processes probably were active along the Smackover strandline. Smackover ooids were generated along the high-energy shoreline by wave activity and the onshore winds transported the fine-grained ooids into the backshore where they accumulated as eolian dunes parallel to the coastline. Alternating dune and interdune areas may have formed as the strandline prograded seaward in 'jumps' caused by the interaction between varying longshore currents, storm frequency, etc. Steady progradation would have formed amalgamated dunes but no interdunes, a situation similar to the Holocene dune and beach system on West Caicos island.

Wave- and wind-dominated shoreline

A seismic amplitude map of the clinoform interval was constructed by flattening on the top of the Haynesville Formation, approximately 100 ms above the top of the Smackover (Fig. 10a). The amplitude map (Fig. 10b) clearly shows that the clinoform bodies are oriented NW–SE along depositional strike throughout the entire area of the 3D seismic survey. Note, too, that this orientation conforms to the regional lithofacies belts (Fig. 2). Furthermore, individual clinoform bodies are remarkably continuous along strike with some bodies extending at least 9 km in a northwest to southeast direction across the 3D survey (Fig. 10b). We interpret the amplitude map as a record of the palaeo-geographic distribution of the clinoforms and shoreline migration. The clinoforms were capped by beach ridges and strike-oriented eolian dunes along a carbonate strandplain, which was prograding to the southwest. Dip-oriented erosional and depositional features, which could indicate tidal channels and spillover lobes, such as those documented in modern settings by Ball (1967) and Hine (1977) are not visible in the amplitude map. As a result, we believe that tidal processes were minimal. Furthermore, the absence of landward-directed spillover lobes indicates that a lagoon was not present behind the clinoform belt within the confines of the 3D seismic survey. Their absence coupled with the lateral continuity and regularity of the clinoform bodies suggests that the Smackover shoreline was wave-dominated and probably had a strong component of longshore currents.

A remarkably similarly modern analogue in terms of appearance, scale and setting (strandplain) is present along the coast of Nayarit, Mexico (Fig. 11). Although this is siliciclastic rather than a carbonate shoreline, it is dominated by wave and longshore current processes. The shoreline has prograded approximately 8 km since sea level stabilized at the end of the Holocene transgression.

Reevaluation of the Smackover ramp

The seismic and well-log observations and interpretations discussed above necessitate a reevaluation of the long held ramp-model for the upper Smackover in north Louisiana. The clinoform interpretation, which suggests water depths of at least 50–70 m and probably as much as 90 m within 1 km of the presumed shoreline, is astonishing given the longstanding view that the Smackover was deposited in a gently sloping ramp without a marked break in slope. The geometry and morphology of the Smackover clinoforms show that the nearshore depositional profile was relatively steep and narrow (1 km). The nearshore zone was a fringing ooid bank-margin, which passed from the shoreline into the deeper subwave-base shelf floor (Fig. 12).

The well-documented Holocene carbonate ramp of the Persian (Arabian) Gulf has been frequently called a modern analogue to many ancient ooid grainstone and evaporite-dominated successions, including the Smackover (Budd & Loucks 1981; Read 1985). The Smackover Formation and the homoclinal ramp of the Persian (Arabian) Gulf (Fig. 13) have been compared because: (1) both are characterized by an ooid grain-dominated shoreline backed by sabkha evaporites; and (2) shallow wave-agitated facies in the nearshore zone pass offshore into low-energy, muddy carbonate facies below wave base (Loreau & Purser 1973).

Read (1985) considered that the Persian (Arabian) Gulf ramp has two basic types of shoal complexes (Fig. 14): (1) fringing ooid-shoal complex along the coastline and (2) barrier ooid/pellet shoal complex. Sand shoals complexes, consisting of rippled and megarippled ooid sands, in fringing ooid-shoal are 2–3 m deep and 0.5–5 km wide and they pass offshore into skeletal packstones and wackestones at 10 m depth on the shallow, lower energy ramp floor (Read 1985). Barrier ooid-pellet shoal complexes contain shore-parallel beach ridge/dune barriers and subtidal shoals, cut by tidal channels (10 m deep) with ooid tidal delta com-plexes on the seaward sides.

Although the Persian (Arabian) Gulf and the Smackover Formation share similar facies, the gradients and water depths in the nearshore Gulf are entirely different from what has been interpreted for the Smackover Formation. The overall gradient between the Persian (Arabian) Gulf shoreline near Abu Dhabi and the deepest (80 m) part of the Gulf, which lies approximately 180 km from the Abu Dhabi shoreline, is only 0.03° (Fig. 13). The Smackover depositional profile reached 80 m of water depth in less than 1 km before flattening out. Thus, the homoclinal ramp profile, as exhibited by southeastern Persian (Arabian) Gulf, is not a good analogue for the much steeper shoreline profile Smackover of north Louisiana.

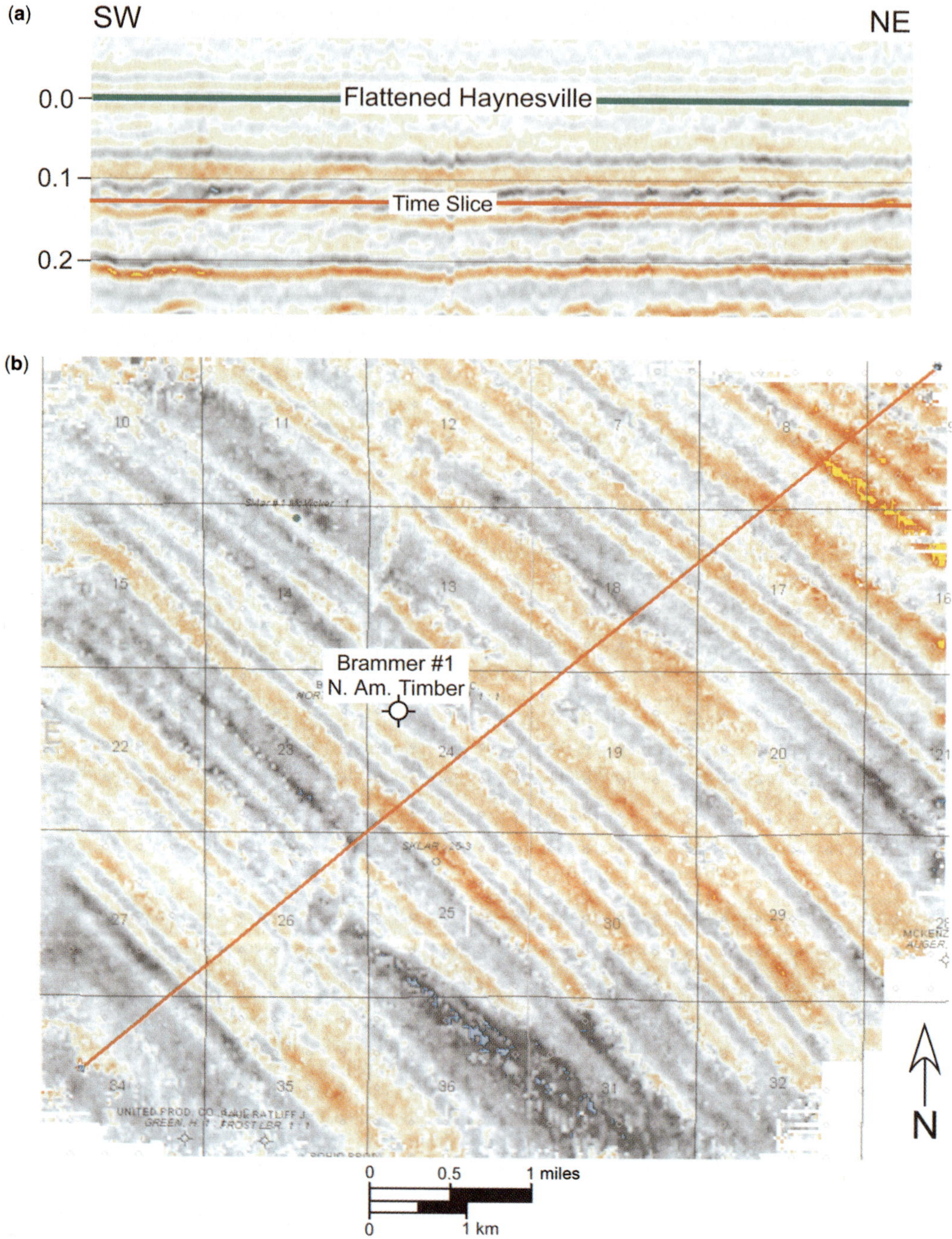

Fig. 10. (**a**) Arbitrary line shows flattened section and time-slice for amplitude map. (**b**) Seismic amplitude map of 3D survey demonstrates the lateral, strike-wise continuity of clinoform bodies and the progradation to the southwest. Note the absence of dip-oriented (northeast–southwest) features such as channels and spillover lobes.

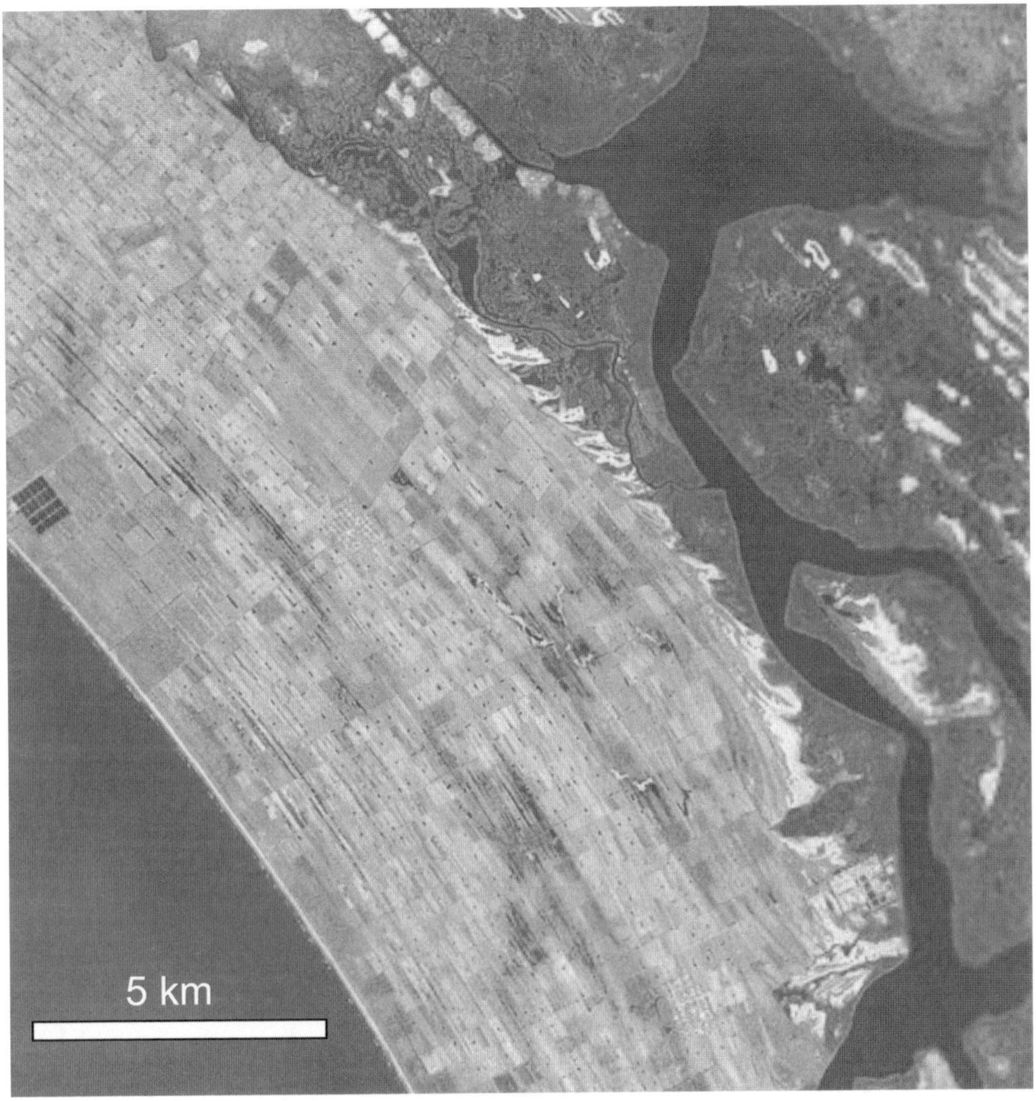

Fig. 11. Satellite image (Google Earth) of progradational beach ridges and Laguna Agua Brava along the Pacific coast of Nayarit, Mexico. Note the similarity of the beach ridge trends and progradation distance to that of the Smackover shoreline.

Although we have not found any modern carbonate grain-shoal setting with a nearshore profile exactly the same (4–7° inclination and 90 m of water depth) as that interpreted for the Smackover clinoforms, some modern settings share similar features with the Smackover profile. For example, skeletal carbonate banks, which parallel the shoreline in Shark Bay, Western Australia, have seaward margin slopes of 20–30° but the maximum water depths at the toe of slope are only 10 m (Davies 1970; Hagan & Logan 1974; Read 1974, 1982). Hine and Neumann (1977), who studied sedimentation along the Little Bahama Bank margin, documented Holocene clinoform-shaped sand bodies, which downlap buried Holocene reefs (Fig. 15). These sand bodies are about 800–1000 m wide and up to 21 m thick. Though similar in scale and shape to the Smackover clinoforms, the Little Bahama Bank examples accumulated by offbank transport rather than onshore transport, and they are not composed of ooids. Furthermore, these sand bodies are perched along the high-relief platform margin with

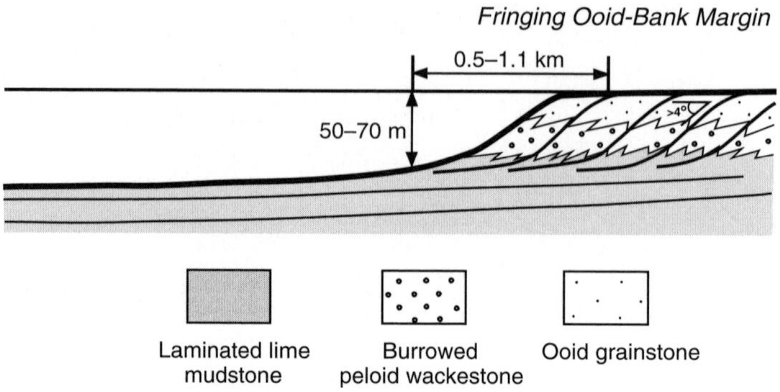

Fig. 12. Schematic Smackover depositional profile reconstructed from analysis of 3D seismic data and well log. The geometry and dimensions of the Smackover clinoforms indicate a fringing bank margin rather than a homoclinal ramp.

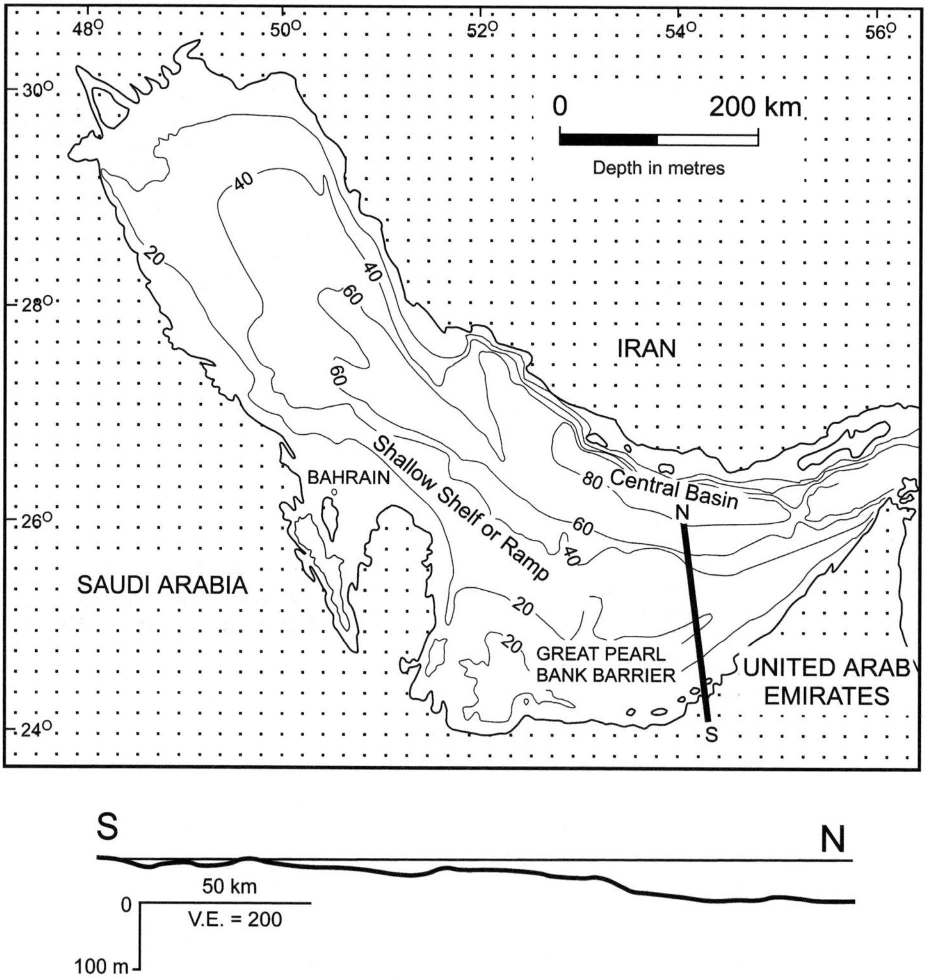

Fig. 13. Map of the Persian (Arabian) Gulf shows the shallow bathymetry and line of section (N–S). Lower part of diagram is a profile (N–S) from Abu Dhabi to the deepest part of the Gulf. Profile modified from Read (1985). Reprinted by permission of the AAPG, whose permission is required for further use.

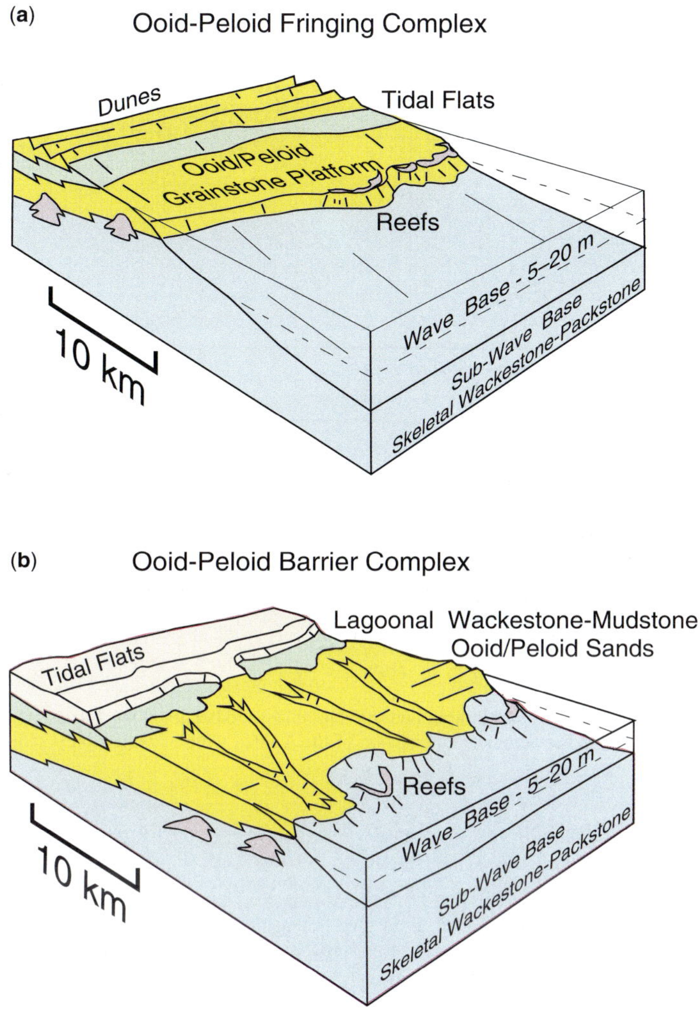

Fig. 14. Depositional models of fringing ooid shoals and barrier ooid/pellet shoals were proposed for the Persian (Arabian) Gulf by Read (1985). Reprinted by permission of the AAPG, whose permission is required for further use.

steep slopes, which was not the case with the Smackover Formation.

Conclusions

The Smackover Formation has been commonly cited as the product of ramp deposition. Furthermore, the Smackover ramp and lithofacies are usually compared with those of the southeastern Persian (Arabian) Gulf near Abu Dhabi. However, recent examination of 3D seismic reflection data in north Louisiana shows that the Persian (Arabian) Gulf is an unsatisfactory analogue, at least in terms of the depositional profile. The 3D data show that numerous shingled-sigmoidal clinoform strata are present in the middle to upper Smackover Formation. The clinoform bodies are about 90 m thick but they extend just 1 km from toplap to downlap pinchout points. Thus, they are remarkably steep (4–7°) and narrow. These dimensions indicate that water depths up to 90 m deep were present only 1 km or less from the shoreline. By comparison, similar water depths in the Arabian Gulf lie 180 km offshore. Although no modern carbonate settings are presently known with the same nearshore profile as the Smackover example, several, such as the Little Bahama Bank and Shark Bay, may share more in common with the Smackover than the Arabian Gulf.

The results from this study indicate that the long-standing idea of a Smackover-Arabian Gulf homoclinal-ramp analogy must now be questioned throughout the U.S. Gulf Coast basin. Question-ing

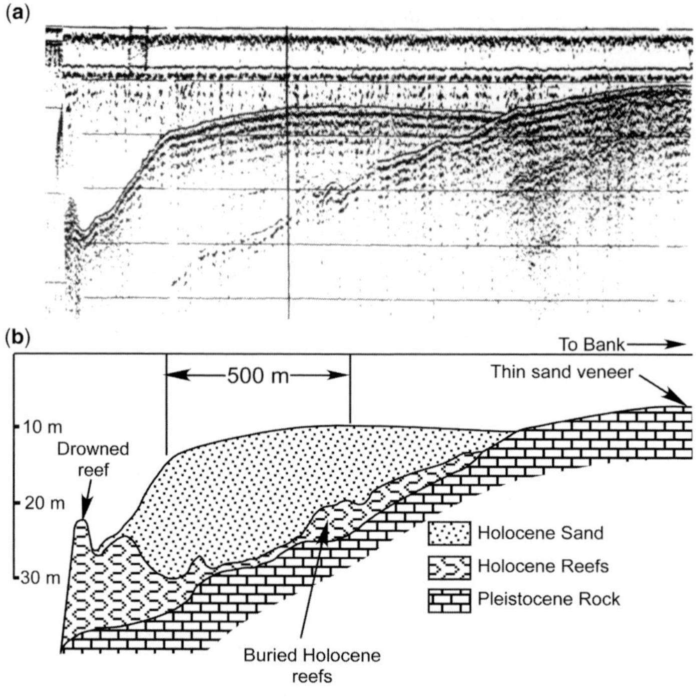

Fig. 15. (**a**) Subbottom profile and (**b**) stratigraphic interpretation of sand bodies along Little Bahama bank margin (Hine & Neumann 1977). Reprinted by permission of the AAPG, whose permission is required for further use.

the paradigm is essential to developing new ideas for hydrocarbon exploration, and those new ideas will likely come by combining the following methodologies for analysing subsurface data sets:

1. seismic stratigraphic analysis;
2. high-frequency well-log sequence stratigraphic analysis;
3. core-based depositional facies and stacking pattern analysis; and
4. seismic geomorphological analysis.

The authors express thank Drs U. P. Baaske and V. P. Wright for their insightful reviews and suggestions to improve the manuscript.

References

AHR, W. M. 1973. The carbonate ramp – an alternative to the shelf model. *Gulf Coast Association of Geological Societies Transactions*, **23**, 221–225.

AKIN, R. H. & GRAVES, R. W. 1969. Reynolds oolite of southern Arkansas. *American Association of Petroleum Geologists Bulletin*, **53**, 1909–1922.

BALL, M. M. 1967. Carbonate sand bodies of Florida and the Bahamas. *Journal of Sedimentary Petrology*, **37**, 556–591.

BARIA, L. R., STOUDT, D. L., HARRIS, P. M. & CREVELLO, P. D. 1982. Upper Jurassic reefs of Smackover Formation, United States Gulf Coast. *American Association of Petroleum Geologists Bulletin*, **66**, 1449–1482.

BISHOP, W. F. 1968. Petrology of upper Smackover limestone in North Haynesville field, Claiborne Parish, Louisiana. *American Association of Petroleum Geologists Bulletin*, **52**, 92–128.

BISHOP, W. F. 1971*a*. Geology of a Smackover stratigraphic trap. *American Association of Petroleum Geologists Bulletin*, **55**, 51–63.

BISHOP, W. F. 1971*b*. Geology of upper member of Buckner Formation, Haynesville Field area, Claiborne Parish, Louisiana. *American Association of Petroleum Geologists Bulletin*, **55**, 566–580.

BISHOP, W. F. 1973. Late Jurassic contemporaneous faults in north Louisiana and south Arkansas. *American Association of Petroleum Geologists Bulletin*, **57**, 566–580.

BUDD, D. A. & LOUCKS, R. G. 1981. *Smackover and lower Buckner formations, south Texas. Depositional systems on a carbonate ramp*. Bureau of Economic Geology, The University of Texas at Austin, Report of Investigations, **112**.

BURCHETTE, T. P. & WRIGHT, V. P. 1992. Carbonate ramp depositional systems. *Sedimentary Geology*, **79**, 3–57.

CHIMENE, C. A. 1991. Walker Creek field. *In*: FOSTER, N. H. & BEAUMONT, E. A. (eds) *Stratigraphic Traps*

II. Treatise of Petroleum Geology, Atlas of Oil and Gas Fields. American Association of Petroleum Geologists, 55–116.

DAVIES, G. R. 1970. Carbonate bank sedimentation, eastern Shark Bay, Western Australia. *In*: *Carbonate Sedimentation and Environments, Shark Bay, Western Australia*. American Association of Petroleum Geologists Memoirs, **13**, 85–168.

DEMISS, W. D. & MILLIKIN, J. V. 1993. Shongaloo field: a recent Smackover (Jurassic) discovery in the Arkansas-Louisiana State Line graben. *Gulf Coast Association of Geological Societies Transactions*, **43**, 109–119.

HAGAN, G. M. & LOGAN, B. W. 1974. Development of carbonate banks and hypersaline basins, Shark Bay, Western Australia. *In*: *Evolution and Diagenesis of Quaternary Carbonate Sequences, Shark Bay, Western Australia*. American Association of Petroleum Geologists Memoirs, **22**, 61–139.

HARRIS, P. M. 1979. *Facies anatomy and diagenesis of a Bahamian ooid shoal*. Sedimenta 7, Comparative Sedimentology Laboratory, University of Miami.

HARRIS, P. M. & DODMAN, C. A. 1982. Jurassic evaporites of the U.S. Gulf Coast: the Smackover-Buckner contact. *In:* HANDFORD, C. R., LOUCKS, R. G. & DAVIES, G. R. (eds) *Depositional and Diagenetic Spectra of Evaporites*. Society of Economic Paleontologists and Mineralogists Core Workshops, **3**, 189–212.

HINE, A. C. 1977. Lily Bank, Bahamas: history of an active oolite sand shoal. *Journal of Sedimentary Petrology*, **47**, 1554–1581.

HINE, A. C. & NEUMANN, A. C. 1977. Shallow carbonate-bank-margin growth and structure, Little Bahama Bank, Bahamas. *American Association of Petroleum Geologists Bulletin*, **61**, 376–406.

LLOYD, R. M., PERKINS, R. D. & KERR, S. D. 1987. Beach and shoreface ooid deposition on shallow interior banks, Turks and Caicos Islands, British West Indies. *Journal of Sedimentary Petrology*, **57**, 976–982.

LOREAU, J. P. & PURSER, B. H. 1973. Distribution and ultrastructure of Holocene ooids in the Persian Gulf. *In*: PURSER, B. H. (ed.) *The Persian Gulf*: Springer, New York, 279–328.

MOORE, C. H. 1984. The upper Smackover of the Gulf rim: depositional systems, diagenesis, porosity evolution and hydrocarbon production. *In*: VENTRESS, W. P. S., BEBOUT, D. G., PERKINS, B. F. & MOORE, C. H. (eds) *The Jurassic of the Gulf Rim: Society of Economic Paleontologists and Mineralogists, Gulf Coast Section, Third Annual Research Conference Proceedings*, 283–307.

MOORE, C. H. 2001. *Carbonate Reservoirs – Porosity Evolution and Diagenesis in a Sequence Stratigraphic Framework*. Developments in Sedimentology, **55**. Elsevier, Amsterdam.

READ, J. F. 1974. Carbonate bank and wave-built platform sedimentation, Edel Province, Shark Bay, Western Australia. *In*: *Evolution and Diagenesis of Quaternary Carbonate Sequences, Shark Bay, Western Australia*. American Association of Petroleum Geologists Memoirs, **22**, 1–60.

READ, J. F. 1985. Carbonate platform facies models. *American Association of Petroleum Geologists Bulletin*, **69**, 1–21.

TEDESCO, W. A., MAJOR, R. P. & BARIA, L. R. 2002. An eolian facies within the upper Jurassic Smackover Formation, Tchula field, Mississippi. *American Association of Petroleum Geologists Bulletin*, **86**, 485–503.

TROELL, A. R. & ROBINSON, J. D. 1987. Regional stratigraphy of the Smackover limestone (Jurassic) in south Arkansas and north Louisiana, and the geology of Chalybeat Springs oil field. *Gulf Coast Association of Geological Societies Transactions*, **37**, 225–262.

WARD, W. C. & BRADY, M. J. 1973. High-energy carbonates on inner shelf, northeastern Yucatan Peninsula. *Gulf Coast Association of Geological Societies Transactions*, **23**, 226–238.

WILSON, J. L. 1975. *Carbonate Facies in Geologic History*. Springer, Heidelberg.

Seismic geomorphology of Palaeozoic collapse features in the Fort Worth Basin (USA)

E. C. SULLIVAN[1], K. J. MARFURT[2], C. BLUMENTRITT[2], & M. AMMERMAN[3]

[1]*Pacific Northwest National Laboratory, Richland, WA 99352, USA*
(e-mail: charlotte.sullivan@pnl.gov)
[2]*AGL, University of Houston, TX, USA*
[3]*Devon Energy, 20 North Broadway, Oklahoma City, 73102, USA*

Abstract: Modern multi-trace geometric attributes produce three-dimensional volumes that can facilitate the recognition of karst geomorphology by avoiding the need to pre-interpret irregular horizons and by enhancing subseismic lateral variations in reflectivity. These geometric attributes include the well-established coherence technology, coupled with recent developments in spectrally limited estimates of volumetric curvature. Coherence measures lateral changes in waveform, and as such, is often sensitive to joints, small faults, sinkholes and collapse features. The many components of reflector curvature, including the most negative, most positive, Gaussian curvature and related shape indices (e.g. valleys, saddles, domes), are complimentary to coherence measures. Short wavelength estimates of curvature will illuminate small-scale lineaments while longer wavelength estimates of curvature illuminate more subtle flexures and compaction features. We show the results of applying a variety of multi-trace geometric attributes to a three-dimensional seismic volume from the Fort Worth Basin, where a collapse system extends vertically some 800m from the Ordovician Ellenburger carbonates through the dominantly siliciclastic Mississippian– Pennsylvanian interval. The collapse features in our data set appear as rounded, sinkhole-like appearances on time and horizon slices in the Pennsylvanian Marble Falls Limestones and the Ellenburger horizon displays features that can be interpreted as cockpit karst, dolines and frying pan valleys. Although a variety of palaeocave breccia facies in core and image logs indicate that the Ellenburger surface has been karsted, these breccias are not confined to the mega collapse features visible in seismic. The large (up to 700 m diameter) collapse chimneys can be shown in multi-spectral curvature attributes to have elongate rhombohedral shapes associated with intersections of Pennsylvanian age, field-scale to basin-scale, basement lineaments and faults. Isochores indicate greatest tectonic growth on faults from Mississippian until early Pennsylvanian, coincident with thickest fill of collapse features. Thus we interpret the origin of the chimneys to be primarily tectonic. The multi-trace geometric attributes permit better imaging of the three-dimensional shapes of the collapse features, provide better constraints on timing of their formation, allow us to begin to separate karst processes from tectonic processes and provide a means of predicting most likely locations of fluid movement along faults.

Collapse chimneys, visible in the Palaeozoic section of 3D seismic from the northern Fort Worth Basin (Fig. 1), extend vertically some 800 m from the Ordovician Ellenburger Formation to the middle Pennsylvanian Caddo Limestone (Figs 2 and 3). Vertical collapse features in carbonates commonly compartmentalize reservoirs (Kerans 1990; Bagdan & Pemberton 2004), and in parts of the Fort Worth Basin, they have persisted as topographic features (Fig. 4), influencing the distribution and reservoir behaviour of Pennsylvanian sandstone reservoirs (Hardage *et al.* 1996*a*). Because collapse features can result from combinations of processes related to subaerial karst, subsurface cavern collapse, tectonic movement and hydrothermal brecciation and dissolution (Berger & Davies 1999; McClay & Boora 2001; Loucks *et al.* 2004; Sagan & Hart 2004), it is important to determine the relative contribution and sequence of each process. Multi-trace geometric attributes facilitate the recognition of collapse geomorphology and the processes of collapse formation by eliminating the need to accurately pick irregular horizons and by enhancing subseismic lateral variation in reflectivity. The objectives of this paper are to demonstrate the use of multi-trace seismic attributes to image collapse chimneys and geomorphologic surfaces, and to unravel the processes of formation of the vertically extensive collapse features in the Palaeozoic section of the northern Fort Worth Basin. We begin with a brief overview of the geometric attributes used in this paper, and the geologic setting of the study area. We then examine the geomorphology of the karst and collapse chimneys in the three-dimensional (3D) survey area by integrating our new attributes with conventional seismic displays. Finally, we use these new images to support our hypothesis that the formation of

From: DAVIES, R. J., POSAMENTIER, H. W., WOOD, L. J. & CARTWRIGHT, J. A. (eds) *Seismic Geomorphology: Applications to Hydrocarbon Exploration and Production*. Geological Society, London, Special Publications, **277**, 187–203.
0305-8719/07/$15.00

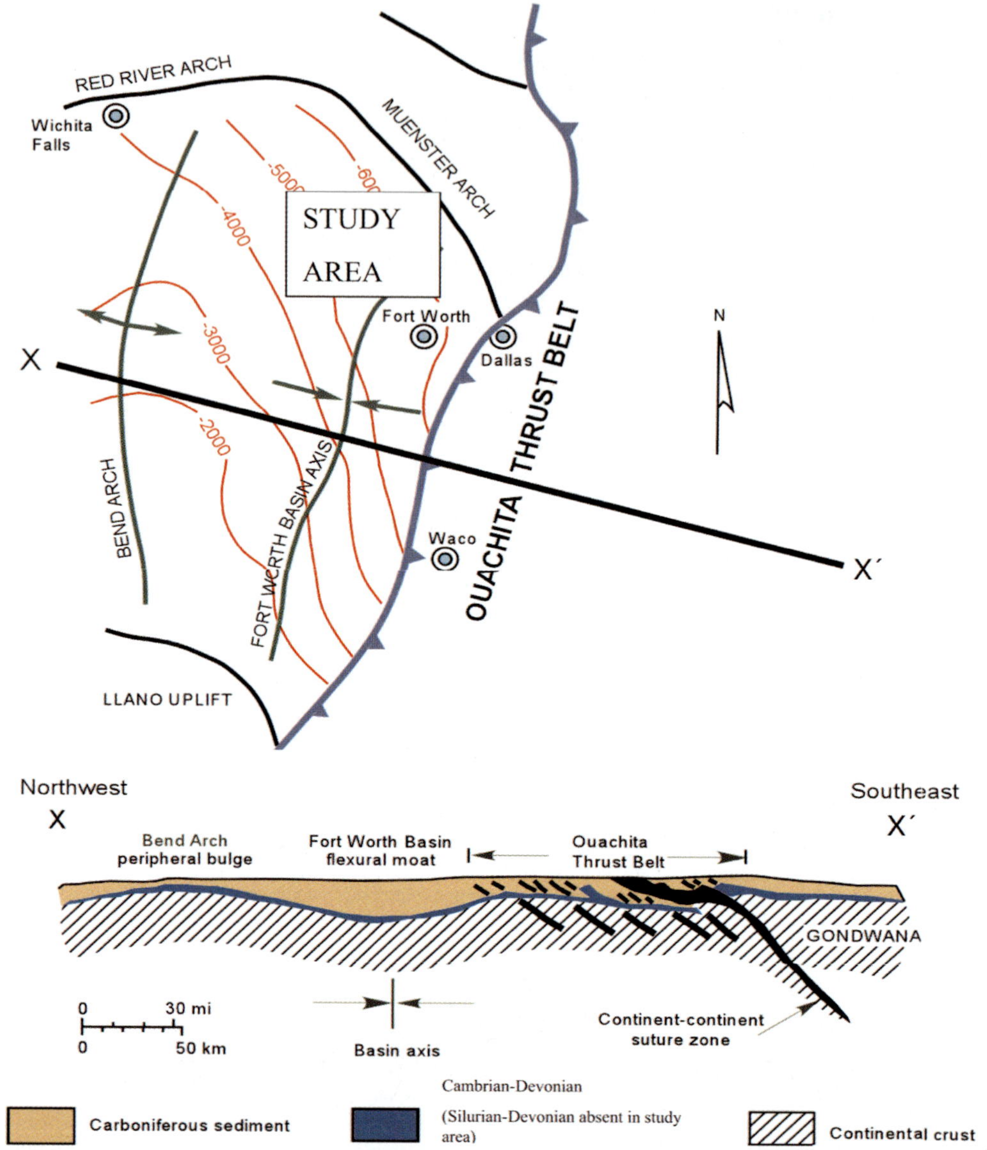

Fig. 1. The Fort Worth Basin formed as a foredeep to the Ouachita fold and thrust belt. Collapse chimneys in the Palaeozoic section are most pronounced in areas where erosion exposed the Ellenburger along peripheral bulges, prior to the deposition of upper Mississippian deepwater shales. The basin underwent compression and wrench faulting during the Pennsylvanian, followed by at least two periods of extension. Contours are structure on the Ellenburger (after Hardage *et al.* 1996*a*).

the collapse features is controlled more by tectonics than by subaerial karst or hydrothermal processes.

Multi-trace geometric attributes

Seismic attributes belong to three families: waveform, reflector shape, and amplitude (Marfurt 2006). In this paper, we focus on waveform and reflector shape. Multi-trace geometric attributes exploit the mathematical relations between reflector events in time and spatial domains. These attributes, which include reflector curvature, coherence and reflector rotation, provide improved technology for imaging small-scale and subtle geologic features and for

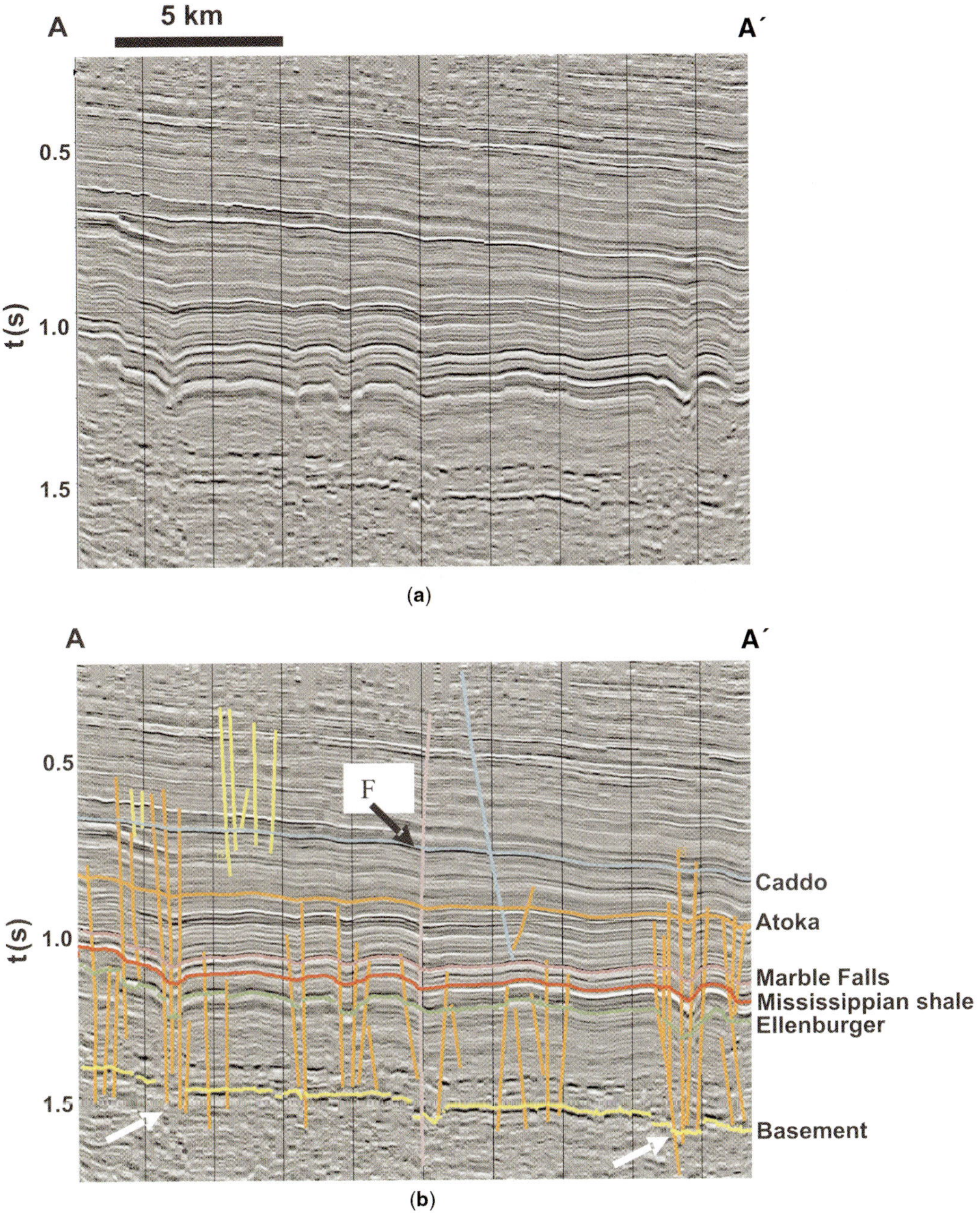

Fig. 2. North–south line AA′ through the horizon extraction shown in Figure 3; (**a**) without and (**b**) with interpretation. The cyan is the Pennsylvanian Caddo limestone; the green horizon is the Lower Ordovician Ellenburger limestone. Orange faults penetrate the basement (white arrows); yellow faults are confined to strata above the basement. Vertically extensive collapse features indicated by white arrows can be tracked from the basement to post Caddo horizons, over some 800 m. Fault 'F' is well imaged in time slice and horizon extractions, and displays dip–slip and possible strike–slip motion.

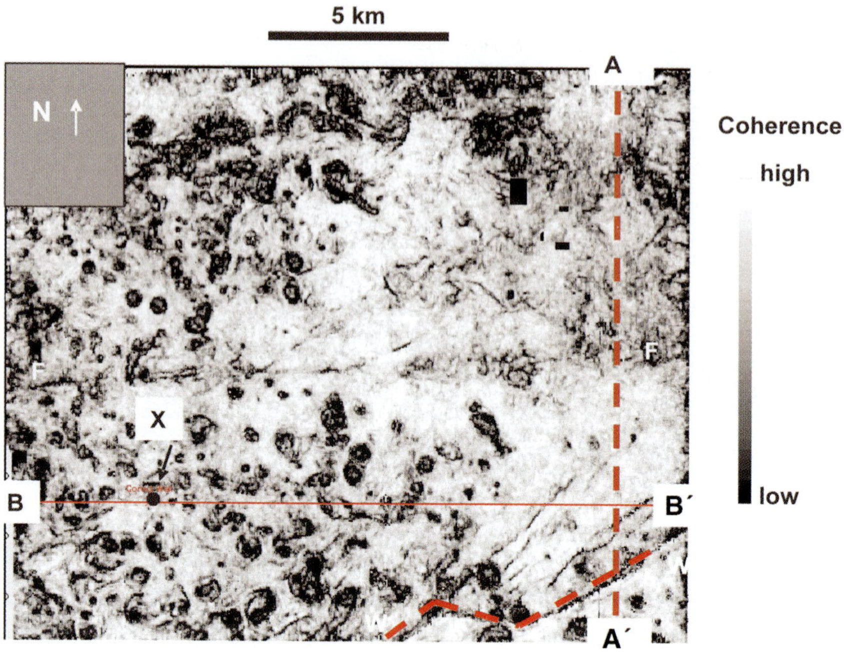

Fig. 3. Coherence extraction along the Ellenburger horizon. Darker shades indicate less coherent or broken reflectors. The dark elongate and circular features are collapse chimneys, many of which extend vertically from the Precambrian crystalline basement to the Pennsylvanian interval. These collapse features tend to be aligned in conjugate NW–SE and SW–NE trends. A–A′ is the seismic cross-section shown in Figure 2; line B–B′ is shown in Figure 14. A regional wrench fault (W) is visible in the southeast corner, and the east–west lineament in the centre is the fault 'F' in Figure 2. Ellenburger core from Well X contains a variety of karst fabrics.

tracking changes in these features through time. Reflector curvature (Fig. 5) calculated from discrete interpreted horizons is well correlated to fracture intensity (Lisle 1994; Roberts 2001). We have expanded this curvature technology to volumetric applications and to prediction of azimuth of open fractures (Blumentritt *et al.* 2006), and to multi-spectral estimates of volumetric reflector curvature, which allow interpreters to view long (800 m) and short (30 m) wavelength geologic features (al-Dossary & Marfurt 2006). Coherence is a well-established technology that measures lateral changes in waveform and is sensitive to breaks in reflectors that include joints systems, small/faults and sinkholes. Components of reflector curvature, including the most negative, most positive, Gaussian curvature and related shape indices, are complimentary to coherence measures. Reflector rotation and combinations of coherence, curvature, dip and azimuth, can be used to show subtle components of wrenching along faults, which may localize fluid flow.

Geologic setting

The Fort Worth Basin developed during the Mississippian as a foreland basin westward of the advancing Ouachita fold and thrust belt, which is associated with oblique convergence of the Laurentian and South American plates (Walper 1982). The shallow water carbonates of the lower Ordovician Ellenburger Group were regionally exposed and karsted prior to the deposition of upper Ordovician Viola limestones (Kerans 1990; Franseen *et al.* 2003), and were subjected to a second generation of erosion and deep karstification during the Mississippian (Grayson & Merrill 1991; Montgomery *et al.* 2005) along the peripheral bulge of the developing Fort Worth Basin. Subsequent subsidence placed upper Mississippian organic-rich, deep-water shales in direct contact with the karsted Ellenburger (Fig. 6) over large areas of the Basin (Bowker 2003). The shallow water siliciclastics and carbonates of the Marble Falls and Atoka record accommodation-limited basin filling during the early to middle Pennsylvanian (Walper 1982; Grayson & Merrill 1991; Hardage *et al.* 1996*b*).

Continental collision and continued westward advance of the Ouachita fold-trust belt resulted in late Pennsylvania–early Permian structural inversion and erosion of part of the Pennsylvanian basin fill (Ball & Perry 1996; Pollastro *et al.* 2003). Structures

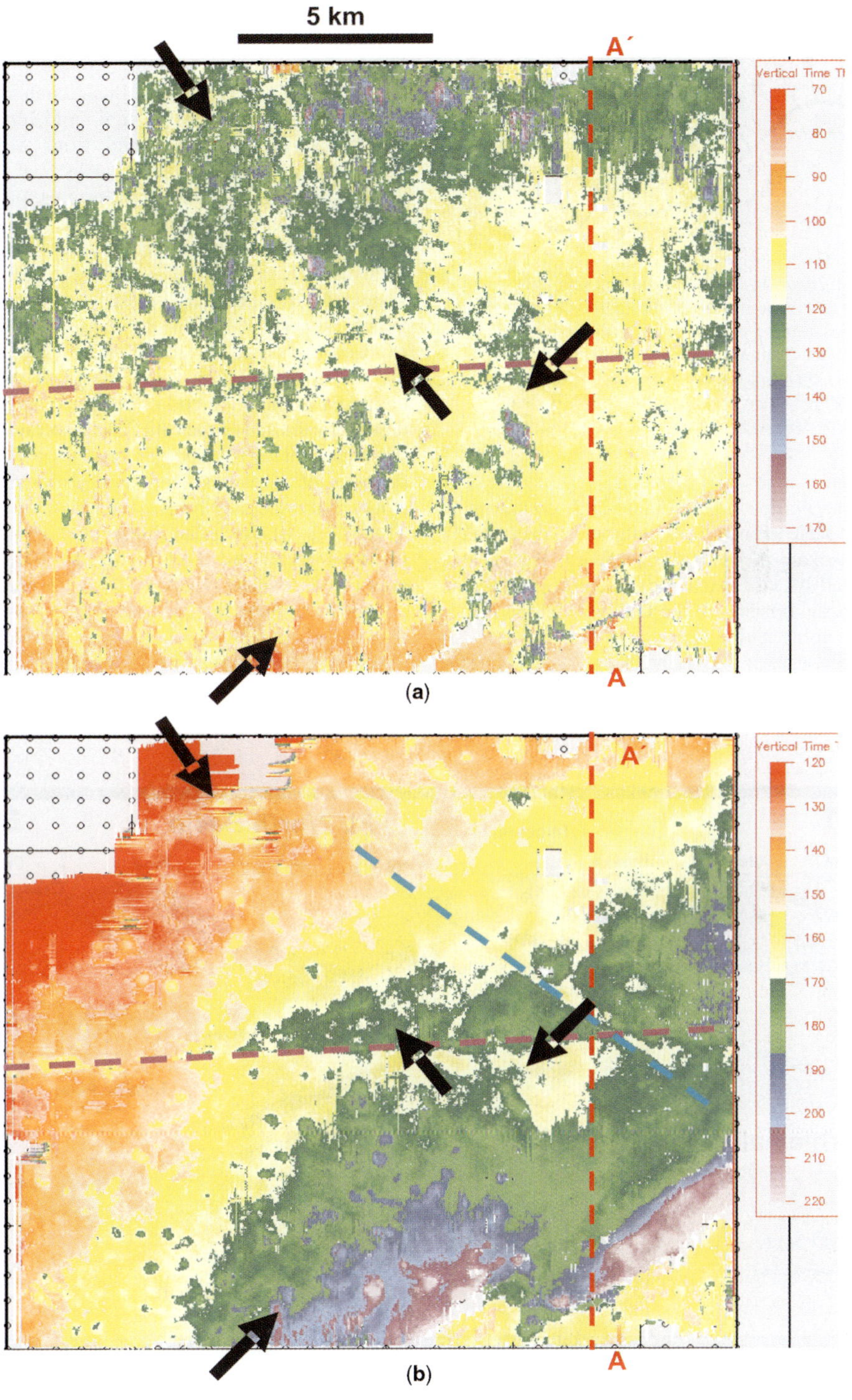

Fig. 4. Isochores between (**a**) the Ellenburger and Marble Falls horizons and (**b**) Marble Falls and Atoka horizons. Hot colours indicate thins, cool colours indicate thicker time intervals. Arrows show collapse features. Note differential thickness reflecting fault activation, and continued chimney collapse after Marble Falls deposition. Line AA′ corresponds to the vertical section in Figure 2 (after Sullivan *et al.* 2006).

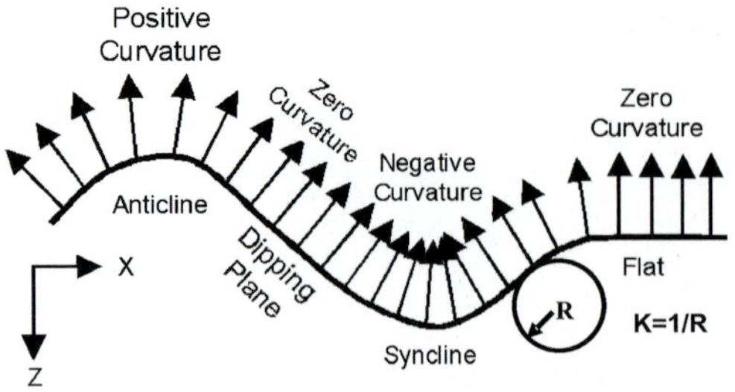

Fig. 5. Curvature of a two dimensional surface. Curvature (K) is the inverse of the radius (R) of a circle that is tangent to the surface at any point. By convention, positive curvature is convex and negative curvature is concave; flat surfaces and uniformly dipping surfaces have zero curvature (after Roberts 2001; Blumentritt *et al.* 2006).

formed during this time indicate both compression and basin scale wrenching (Montgomery *et al.* 2005). The Palaeozoic rocks of the Ft Worth Basin were overprinted by late extensional tectonics related to the Mesozoic opening of the Gulf of Mexico and to Miocene uplift and formation of the down-to-the coast Balcones and Mexia/Talco fault systems (Hoskins 1982). Evidence for hydrothermal and basinal scale fluid flow associated with these tectonic events is recorded in cements in cores and outcrops (Kupecz & Land 1991; Montgomery *et al.* 2005). Ellenburger carbonate breccias exposed at the southern margin of the Fort Worth Basin contain baroque dolomite precipitated from high-temperature fluids during the Pennsylvanian and from warm-water fluids during the Cretaceous (Loucks *et al.* 2004).

The geomorphology of karst terrains may contain distinct features such as sinkholes, cockpit landforms (Fig. 7) and round-ended 'frying pan' valleys (Cansler & Carr 2001). Cave and potential collapse systems most commonly develop at or above water tables, generally within the upper 100 m of an exposed carbonate surface (Kerans 1990). Most of the buried Ellenburger cave systems of West Texas collapsed prior to the end of the Ordovician, as evidenced by the age of their fill. The associated cave fill deposits contain Ellenburger breccias and stratified deposits of transgressive upper Ordovician sandstones that regionally overlie the Ellenburger in West Texas (Fig. 8). Irregular Ellenburger topography is often completely filled by the first 20–50 m of transgressive deposition (Kerans 1990). In contrast, the collapse chimneys in the Fort Worth Basin persist through about 800 m of section (Hardage *et al.* 1996*a*, and this study).

The collapse chimneys in the Fort Worth Basin may record a complex history, similar to the history

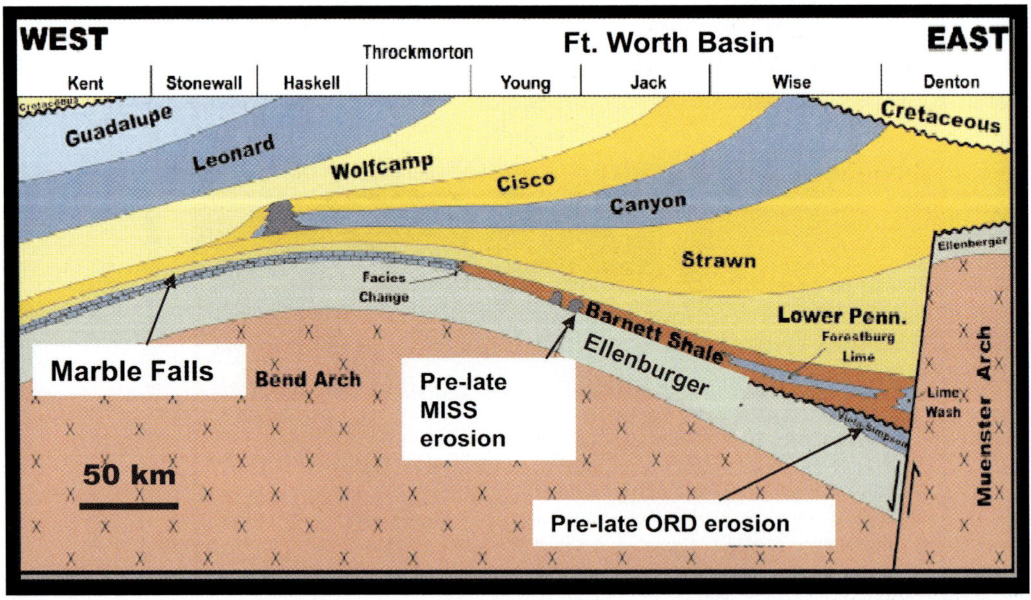

Fig. 6. Schematic west–east cross section across the northern Fort Worth Basin, with relation of a migrating peripheral bulge to erosion of the Ordovician carbonates. Mississippian erosion along the bulge stripped the upper Ordovician Viola and exposed the Ellenburger to a second time of subaerial karst processes. Subsequent subsidence and deposition placed upper Mississippian deepwater shales in direct contact with the karsted Ellenburger over large parts of the basin (after Pollastro *et al.* 2003).

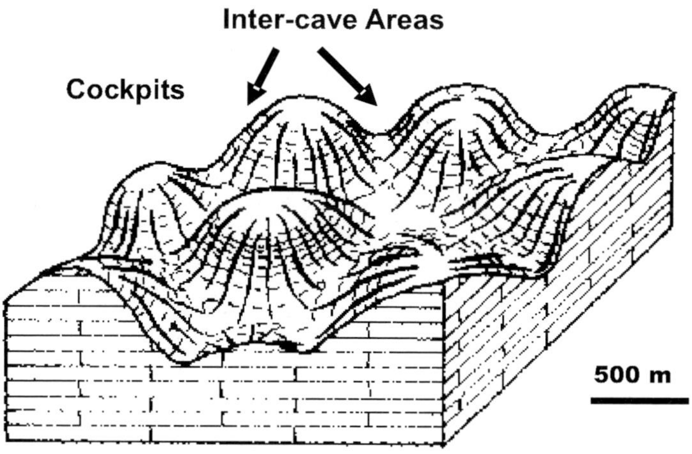

Fig. 7. Mature 'cockpit' karst geomorphology results from a combination of dissolution and preburial collapse of cave systems. In this process, well cemented, low porosity inter-cave areas persist as rounded knolls and hills (after Cansler & Carr 2001).

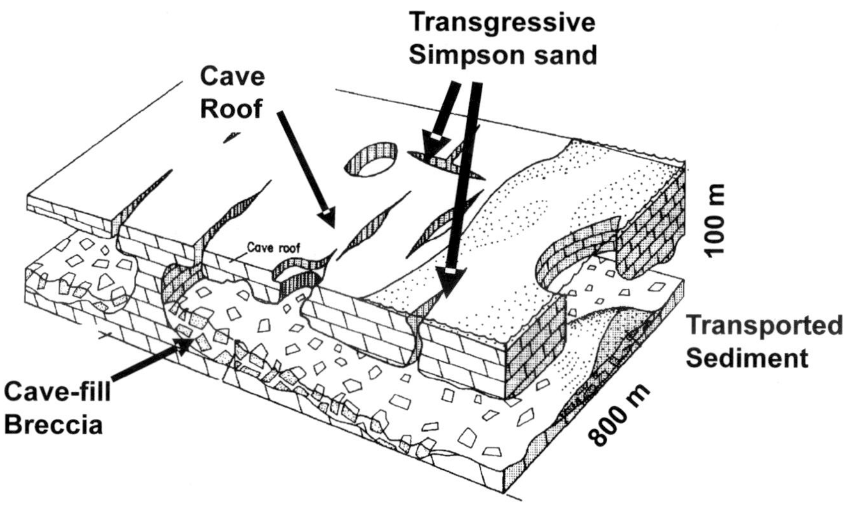

Fig. 8. Karst features associated with subaerial weathering and erosion of West Texas Ellenburger carbonates. These palaeocave systems most commonly fill with collapse breccias and sediments associated with subsequent marine transgressions (after Kerans 1990).

interpreted by Lucia (1996) for outcrops at the McKelligon Sag outcrop, near El Paso, Texas (Fig. 9). These outcrops record a vertical collapse system of 760 m that includes collapse of individual Ordovician, Silurian and Devonian cave systems over a period of 100 my. Here again, once collapse ended, the surface topography quickly healed with about 16 m of additional sediment deposition. Outcrops of Ellenburger limestones immediately south of the Fort Worth Basin in the Llano uplift also record coalesced palaeocave systems (Loucks *et al.* 2004). These collapse systems contain Devonian and Mississippian conodonts but do not contain evidence of extensive vertical chimneys. Unlike the McKelligon outcrops, carbonates overlying the Ellenburger in the Llano uplift (the Marble Falls) do not display pronounced palaeokarst surfaces or vertically extensive collapse chimneys (Kier 1980; Loucks *et al.* 2004).

Methodology

The conventional P-wave seismic surveys used in this study were acquired and processed by a petroleum company through a standard commercial workflow.

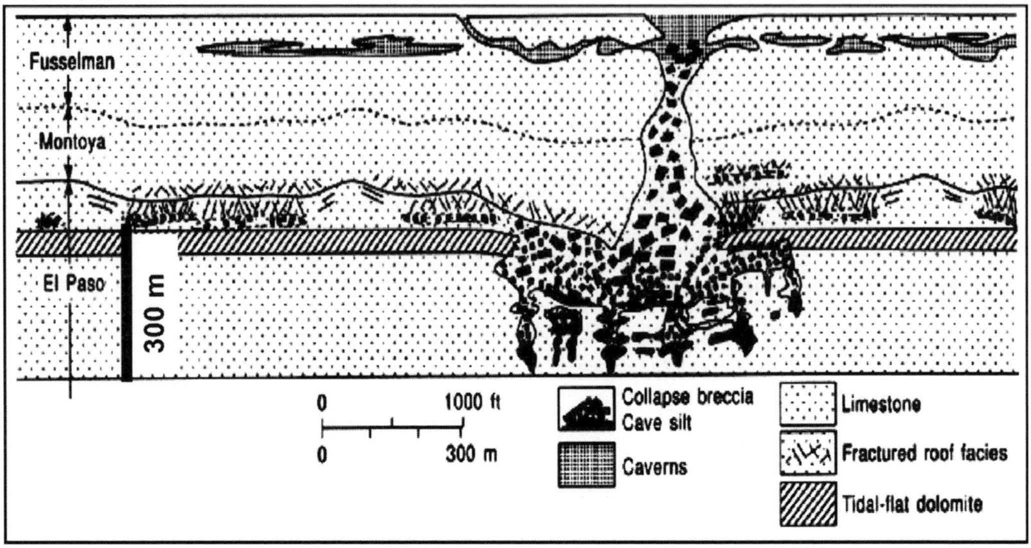

Fig. 9. Diagrammatic view of the complex lower Palaeozoic collapse structure at McKelligon Sag, near El Paso in West Texas. Fossils and field evidence indicate this feature formed from the collapse of individual Ordovician, Silurian and Devonian cave systems over some 100 million years (after Hardage *et al.* 1996*b*, based on Lucia 1996). Kupecz and Land (1991) state that isotopes in the associated baroque dolomite cements indicate a minor Pennsylvanian hydrothermal overprint.

We applied edge-preserving principal component filtering (al Dossary *et al.* 2002; al Dossary & Marfurt 2003) to the post stack data volume (Fig. 2). This filtering, which suppresses random noise and sharpens event terminations, has minimal impact on waveform and amplitude. Next we generated a coherence volume (Marfurt *et al.* 1999) and a suite of geometric seismic attribute volumes, along with more conventional single trace attributes. For these multi-trace coherence and dip/azimuth attribute volumes, we used nine overlapping 9-traces, and a vertical analysis window of ± 10 ms. To calculate the multispectral curvature and rotation volumes, we used circular analysis windows between 13 traces (for short wavelength calculations) and 78 traces (for long wavelength calculations). We found the long wavelength most positive and most negative curvatures, and the principal component estimate of coherence to be especially useful in our interpretation of faults fractures and karst features.

Data analysis

In Figure 2 we observe that collapse features extend from within the Precambrian metamorphic basement (yellow) through the Ordovician Ellenburger (green), early Pennsylvanian Marble Falls (magenta) and middle Pennsylvanian Caddo limestone (cyan). Productive sandstones within the Atoka and lower Caddo are localized and compartmentalized by these features. We indicate faults that penetrate the deeper part of the section in orange and those that are limited to the shallower section in yellow, and note that most of the faulting does not continue higher than the top of the Caddo. The magenta fault (F) is a persistent feature that displays minor dip slip and possibly strike–slip motion.

The geomorphology of the Ellenburger surface, displayed in Figure 3 as a horizon extraction through the coherence volume, is dominated by elongate and circular depressions, but it is not obvious which features are due to subaerial karst processes and which ones are caused by tectonic collapse. Curvilinear features in the northeast quadrant may be remnants of palaeodrainage. The collapse features tend to be aligned in conjugate NW–SE and SW–NE trends. A regional wrench fault crosses the southeast corner of the map, and the east–west magenta lineament is the fault 'F' in Figure 2.

The time isochore map between the Ellenburger and Marble Falls (Fig. 4a) indicates a gentle increase in thickness to the north and pronounced local thickening in some of the collapse features. In contrast, the time isochore for the Marble Falls–Atoka interval (Fig. 4b) shows dramatic thickening to the north of the magenta fault and along the fault in the southeast corner of the map, and thickening along NW and SE lineaments, indicating increased tectonic activity and growth of faults.

The mean curvature attribute, extracted along the top of the Caddo Limestone (Fig. 10) presents a detailed view of a surface affected by minor erosion and by compaction. The most striking feature is the hummocky topography related to compaction over the collapse features; only a few sinkhole shapes appear to actually break the Caddo surface (white arrow). Post Caddo movement along the east–west fault is obvious, overprinted by even later conjugate NW–SE lineaments. A very linear channel incises the Caddo surface in the northwest quadrant, and highly sinuous channels are present in the southeast. The wrinkled surface appearance in the southeast is due to compaction over slump features in the underlying Atoka siliciclastics.

A time slice through the same Mean Curvature attribute volume near the top of the Ellenburger shows the striking vertical extent and shape retention of the three collapse chimneys (white arrow). Features in the extreme western and southern part of the survey resemble cockpit topography of mature karst landscape. Although the composite feature at C is similar to a 'frying pan' karst valley, these features align, in fact, with a lineament that cuts the Caddo surface.

While coherence horizon extractions are particularly valuable for mapping smooth stratigraphic features, we find time slices often provide a less biased view of irregular or rugose surfaces. A time slice through the coherence volume near the top of the Marble Falls (Fig. 11) shows a complex system of lineaments and collapse features. There is no record of extensive subaerial karst of the Marble Falls, either in outcrop or subsurface, and many of the collapse chimneys align with those in the Ellenburger in Figure 10(b). A time slice at the same level through a multi-attribute volume that combines coherence, dip and azimuth provides insight into rotation along lineaments and faults that might be conduits for fluids. Here we see evidence of north dip along the central fault, possible rotation along the active wrench fault in the southeast, and two directions (NW–SE and NE–SW) of long wavelength, subtle folds.

We use the criteria of Loucks *et al.* (2004) to identify palaeocave facies (Fig. 12) in core and image log from an Ellenburger well, located on the seismic line in Figure 13. Note that the well is not located in one of the large collapse features. The image log (Fig. 15) reveals over 50 m of fabric and textures indicative of palaeocave facies related to subaerial karst processes, and reveals no indications of pervasive hydrothermal overprint.

To better understand the collapse patterns and their expression in the curvature attributes, we display a folded multi-attribute display (Fig. 16) through the filtered vertical seismic cube corresponding to line AA′ in Figure 12, coupled with a time slice at $t = 1.2$ s, through the most negative (Fig. 16a) and most positive curvature (Fig. 16b) attribute volumes. It is important to remember that while the value of the most negative curvature is always less than the value of the (orthogonal) most positive curvature at any analysis point, both attributes can have negative values (describing a bowl shape) and both attributes can have positive values (describing a dome). Careful examination of Figure 16 confirms this relation. Features that are domes will have a positive value of negative curvature and will show up as red in Figure 16(a). Collapse features have a negative value of positive curvature and show up as green in Figure 16(b). We note in Figure 16 that our collapse features are often linked together, suggestive of cockpit karst and karst collapse features described in Figures 7 and 8. While we expect there is a subaerial karst component to the polygonal features observed in the seismic data, some of the collapse features extend into the Precambrian metamorphic basement, where karst processes cannot operate.

Finally, in Figure 17, we compare the most negative curvature time slices for the Caddo, Marble Falls, Ellenburger and Basement. We clearly image the central fault and a complex system of NW–SE and NE–SW conjugate faults and joints. Many of the collapse features have elongate rhombohedral shapes, some of which are associated with intersections of lineaments that cut the Pennsylvanian interval. Lineaments in curvature volumes are amenable to quantitative analysis, and we note a marked change in the distribution of these lineaments with age, which we show as rose diagrams. While the collapse features are pronounced from basement to just below the Caddo horizon, the tectonic stresses have changed direction over geologic time.

Discussion

The horizontal expression of the vertical collapse features in the Fort Worth data is readily seen on conventional seismic and on isochore maps (Fig. 4), but the relation to faulting is more clearly presented in time slices and horizon extractions through coherence and curvature attribute volumes. We note that the collapse features are intense in the Marble Falls and Ellenburger intervals and cut into the Precambrian crystalline basement in some areas, but are greatly subdued at the Caddo Limestone level. The Caddo horizon (Fig. 10a) shows few active collapse features and only minor erosive features, implying that the hummocky Caddo topography is due to compaction over the collapse features, or due to dissolution from below, rather than to top-down, subaerial karst of the Caddo limestone. Isochrons between the Ellenburger and

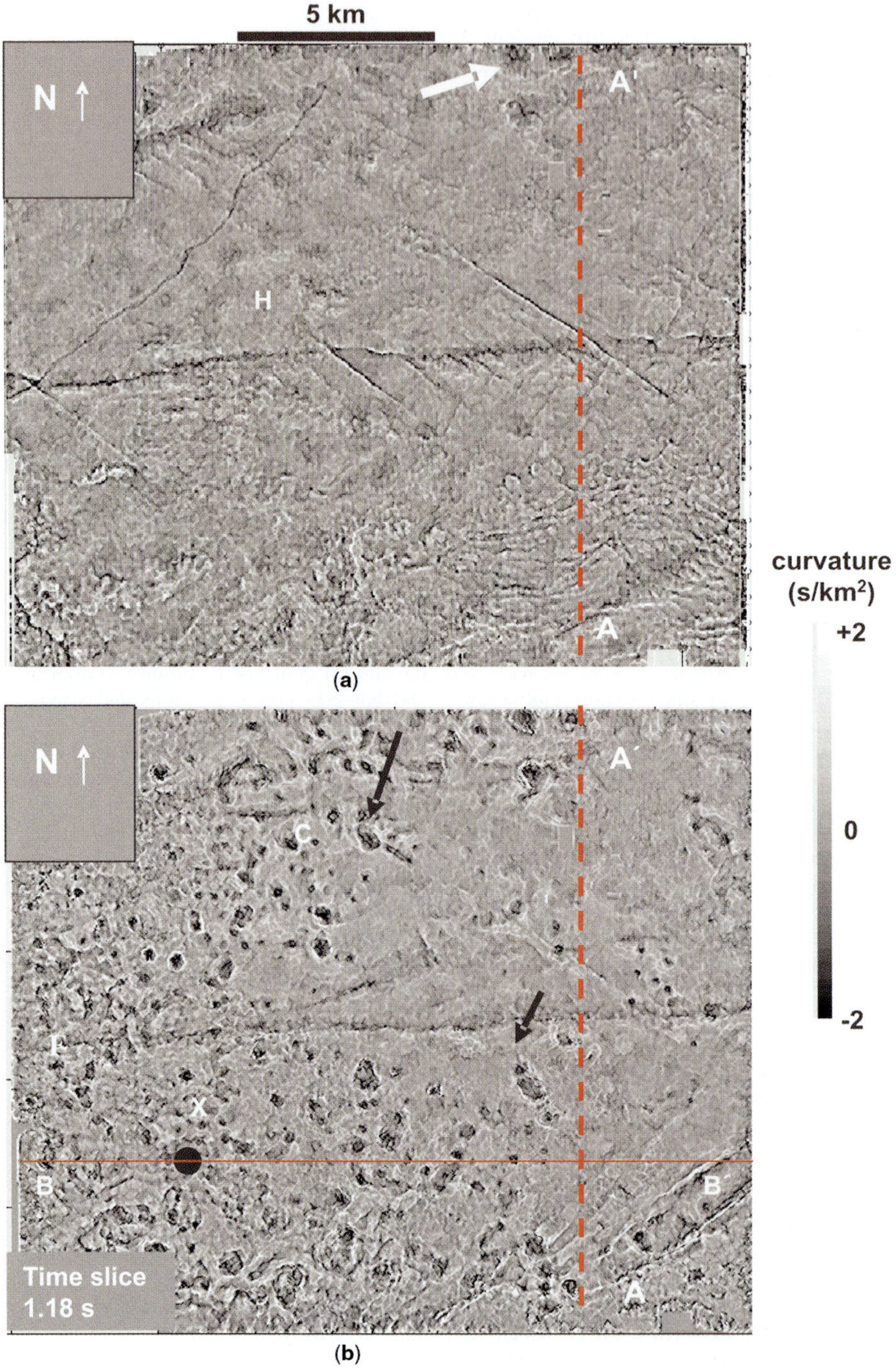

Fig. 10. Mean curvature extracted along the Caddo Horizon (**a**), and on a time slice (**b**) at 1.180 s, near the top of the Ellenburger. Note persistence of collapse chimneys (black arrows) from the Ellenburger to the middle Pennsylvanian Caddo. The hummocky geomorphology (H) of the Caddo is due to compaction over these collapse features. Features in the western part of the Ellenburger time slice resemble cockpit topography of mature karst landscape. The arrow at C indicates a composite feature similar to a 'frying pan' karst valley. Cross Section A–A′ is shown in Figure 2; cross-section B–B′ is shown in Figure 14. A section of the resistivity-based image log from the Ellenburger at Well X (black dot) is shown in Figure 15.

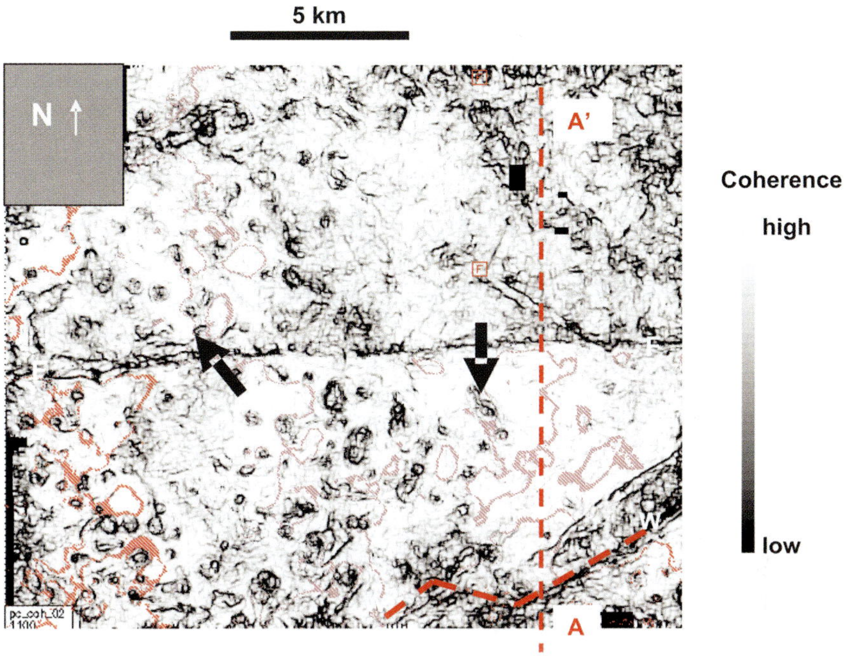

Fig. 11. Time slice through the coherence volume at 1.1 s, near the top of the Marble Falls, using a ±10 ms, nine-trace analysis window. Arrows indicate collapse features, organized along NW–SE and SW–NE trends. The Magenta colour is the top of Marble Falls horizon, red is the top of the underlying Mississippian shale. Note the east–west lineament in the centre of figure, marking a down-to-the-north fault (F) and lineaments in the SE that mark the regional wrench fault (W). Line A–A′ corresponds to the vertical section in Figure 2. Dark areas indicate lack of coherence.

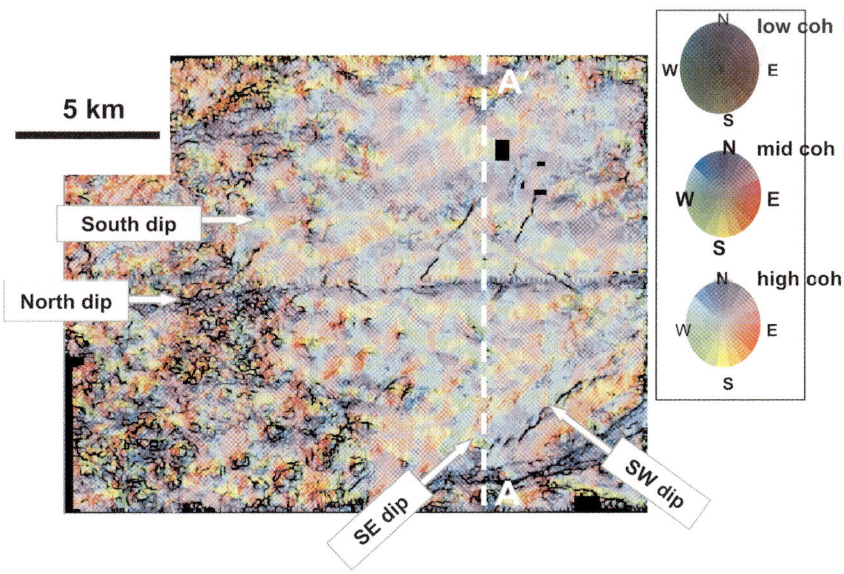

Fig. 12. Multi-attribute time slice combining coherence, dip and azimuth at 1.0 s, near the top of the Marble Falls. This combination of attributes clearly delineates the low coherence of the collapse features in the SW, the North dip along the fault crossing the centre of view, and broad wavelength folds and compaction features that have southeast and southwest dip. Cross section A–A′ is shown in Figure 2.

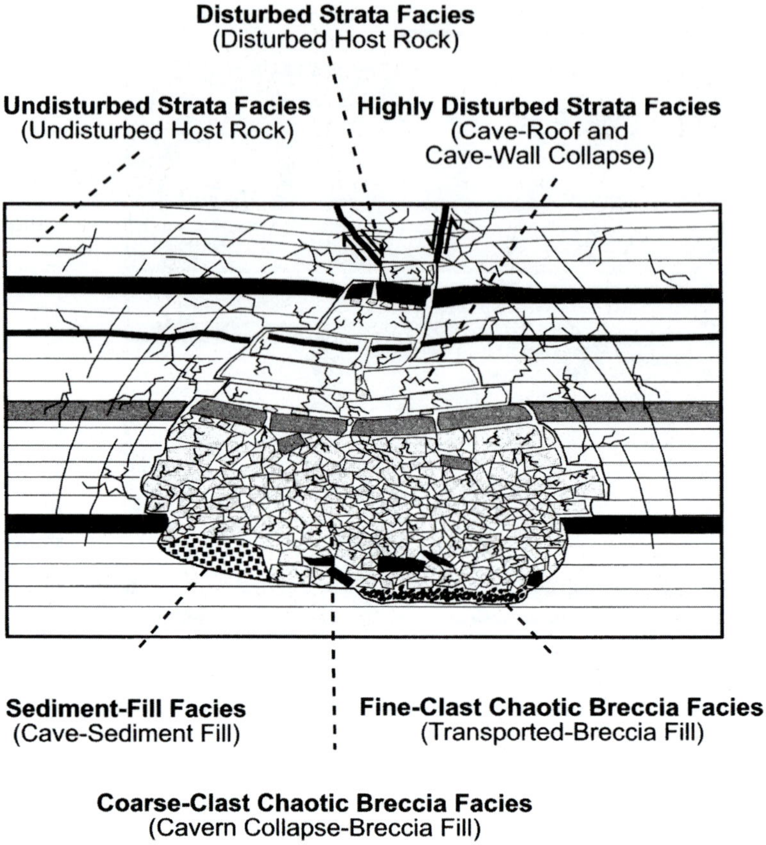

Fig. 13. Textures, sediments and cave facies associated with collapse and fill of palaeocave systems (after Loucks *et al.* 2004). Examples of these cave facies are present in core and image log (Fig. 15) from well 'X' shown in Figure 14.

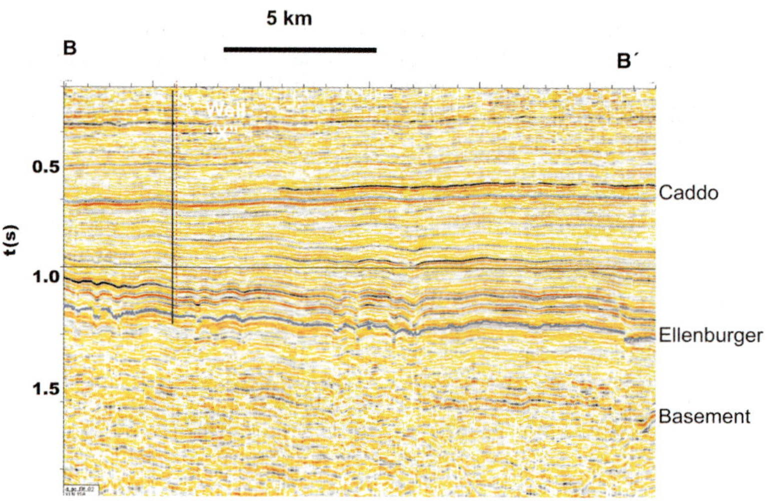

Fig. 14. East–west line B–B′ through the time slice shown in Figure 10(**b**). Core and image logs (Fig. 15) from the Ellenburger interval of well 'X' display over 50 m of breccias that we interpret as palaeocave facies.

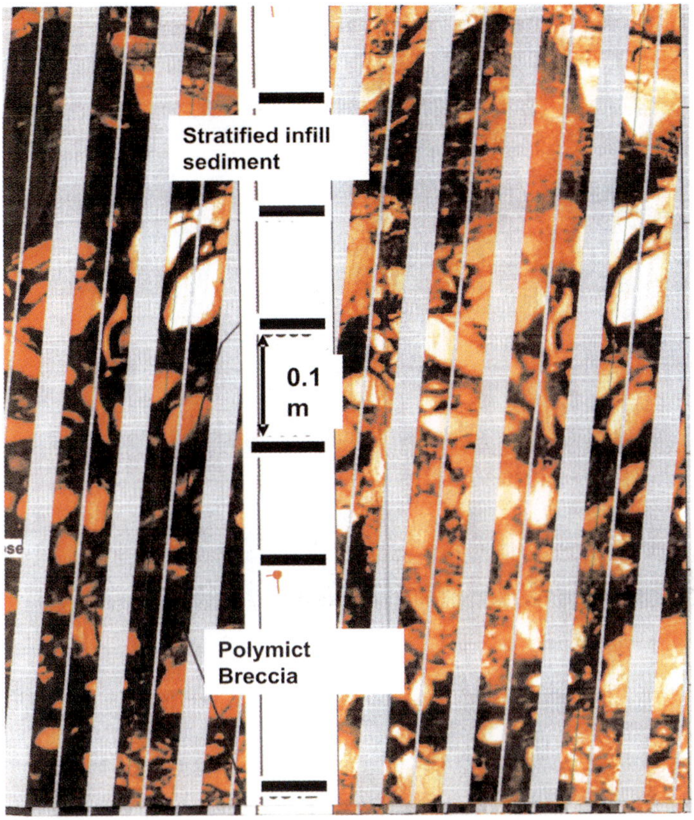

Fig. 15. Image log from the upper part of the Ellenburger interval in Well X, with rotated and rounded clasts (**a**) in a polymictic carbonate breccia, and (**b**) stratified sediment infill. We interpret these breccias, which extend over about 50 m, to be palaeocave deposits.

Marble Falls (Fig. 4a) and between the Marble Falls and Atoka (Fig. 4b) show greatest development of collapse features and fault growth between deposition of the Marble Falls and Atoka, although the wrench fault continued to be active until early Caddo deposition.

Inspection of other 3D surveys from the Fort Worth Basin indicates that regionally, collapse chimneys are larger and more common where the Ellenburger subcrops below the Mississippian than where it subcrops below the Upper Ordovician Viola Limestone. The Mississippian erosion of the Viola and Ellenburger is coincident with the formation of a foreland peripheral bulge, prior to the deposition of Mississippian marine shales. Textures and fabrics diagnostic of palaeocave facies are present in Ellenburger cores within the 3D survey, but the age of the karst is unknown at this time. The large (up to 700 m horizontal and 800 m vertical) collapse structures we observe do not appear to be compound coalescing cave systems that result from multiple subaerial karst episodes. Outcrop data indicate no prolonged subaerial exposure on the Marble Falls, and regional seismic data show no large-scale karst features on the Caddo limestone.

The role of hydrothermal brecciation and dissolution in the formation of the collapse chimneys is unquantified. Ellenburger breccias exposed at the southern margin of the basin contain baroque dolomites precipitated by high-temperature fluids during the Pennsylvanian and by warm-water fluids during the Cretaceous (Kupecz & Land 1991; Loucks *et al.* 2004). Native copper, dolomite and calcite are present in hairline fractures in cores of subsurface Mississippian shales but have not been dated (Montgomery *et al.* 2005). No large-scale hydrothermal fabrics have yet been reported from cores, and hydrothermal minerals are known only from fracture filling cements.

The alignment of Ellenburger collapse features with post-Caddo lineaments and joints, as revealed by our multi-trace attributes, suggests a relatively

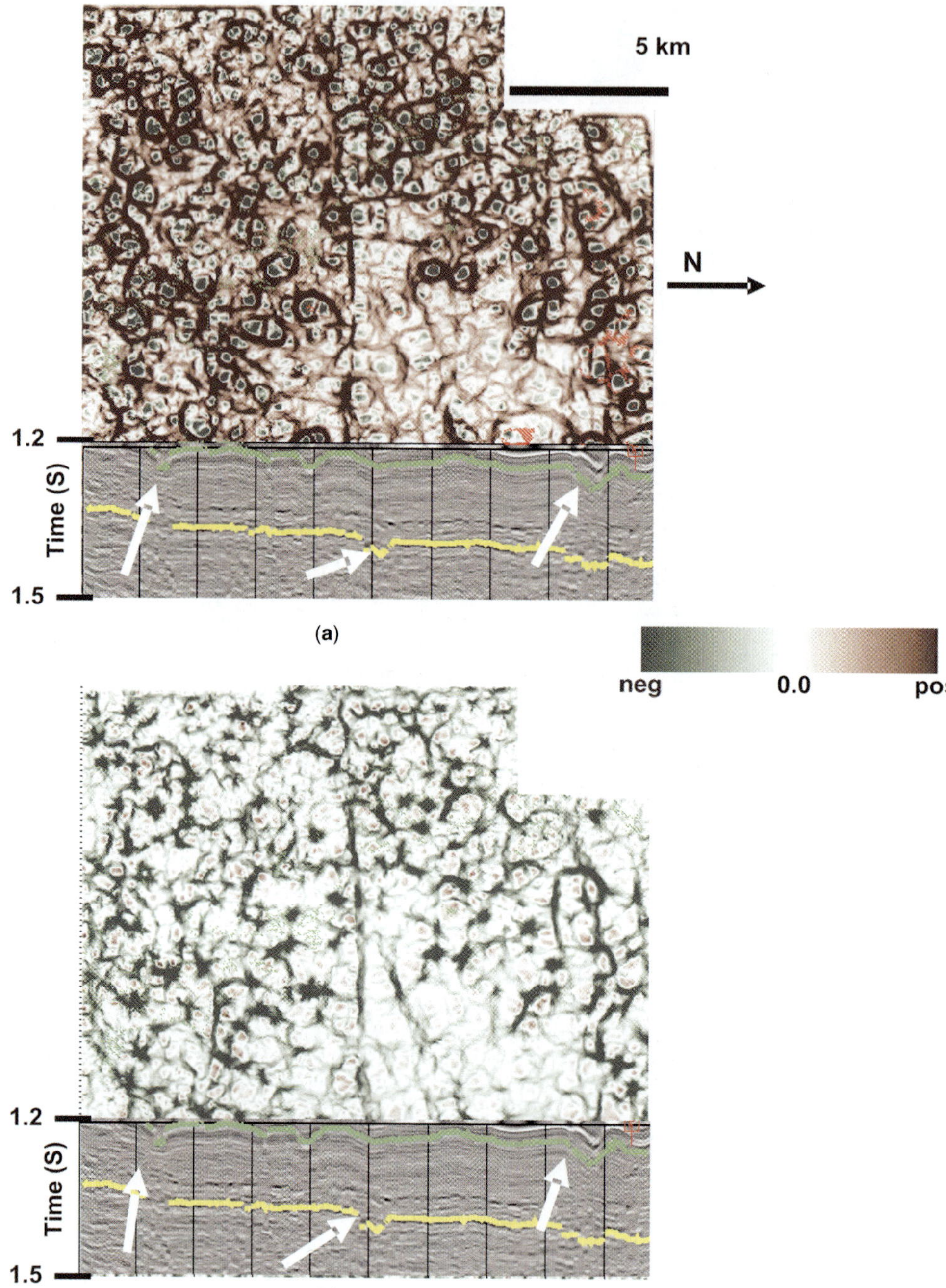

Fig. 16. Fold away multi-attribute images along line A–A′ shown in Figure 12. Seismic data is displayed on the vertical face. Long wavelength (**a**) most negative curvature and (**b**) most positive curvature are shown on the time slices at 1.2 s. Positive values (red) on the most negative curvature slice correspond to domes, while negative values (green) on the most positive curvature slice correspond to bowl shapes. The red horizon present in map view is the Mississippian shale. The white arrows mark places where the collapse extends through the non-carbonate, Precambrian crystalline basement (yellow). We interpret these collapse features to be tectonically controlled, linked by a complex system of faults and joints (after Sullivan *et al.* 2006).

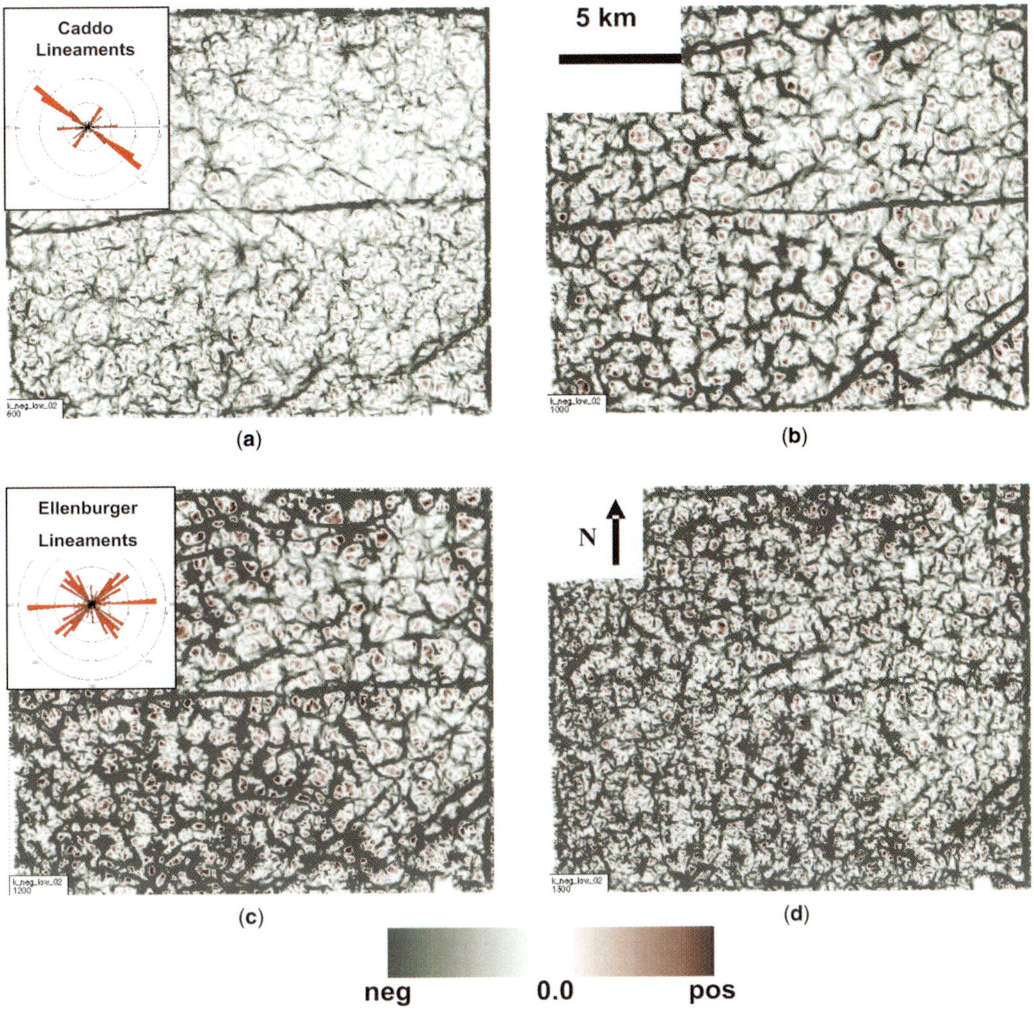

Fig. 17. Slices through the Most Negative Curvature volume at (**a**) near the top of the Caddo limestone, (**b**) within the upper Marble Falls limestone, (**c**) within the upper Ellenburger limestone, and (**d**) within the upper part of the Precambrian metamorphic basement. We used a long wavelength calculation with an analysis window of ±10 ms and 75 traces. The colour bar is identical to that used in Figure 16. Domes are red. Bowls and valleys are green. Time slices from the Marble Falls and Ellenburger show the formation of the collapse chimneys at the junction of intersecting lineaments. Lineament azimuths change between the Ellenburger and Caddo levels.

late reactivation of these features. Current maximum horizontal stress in the subsurface Mississippian strata near the study area is N40E (Siebrits *et al.* 2000). Within the study area, open joint systems in surface outcrops of Upper Pennsylvanian rocks are dominated by NNE and NNW sets, which Hoskins (1982) interprets as related to extension along the trans Texas Miocene-age, down-to-the-coast, Balcones fault system (although we speculate there is Mesozoic extension as well). Fractures formed or re-opened during these extensional events may have allowed hot or warm burial fluids to migrate along the collapse chimneys.

Conclusions

Multitrace geometric attributes provide improved imaging of the seismic geomorphology of collapse chimneys in Palaeozoic strata of the Ft Worth Basin, through better horizontal detection of subtle features and by eliminating the need to prepick irregular and rugose surfaces that are prone to operator error. Coherence based attributes detect lateral discontinuities down to one/tenth of a wavelength, and spectral curvature attributes permit separate analysis of short (30 m) to long wavelength (300 m) geomorphologic features. The combination of volumetric rotation attributes and coherence, curvature, and dip azimuth through the use of hue, light, and saturation highlights changes in dip along subtle faults and lineaments. This detection of stratal rotation along faults, combined with estimations of temporal change in stress/strain regime through attribute-based lineament analysis, hold the potential to predict probability of fluid migration pathways. The presence of subaerial karst in the Ellenburger is suggested by seismic geomorphology and supported by features in core and resistivity-based image logs of the upper Ellenburger over at least 50 m. However, we conclude that the largest collapse features are tectonically controlled for the following reasons: (1) many of the chimneys coincide with deep basement faults and with Pennsylvanian and younger lineaments; (2) no regional unconformity is associated with the observed collapse features in the lower Pennsylvanian Marble Falls limestone; and (3) horizon slices on the Middle Pennsylvanian Caddo Limestone lack exposure features of significant magnitude to produce top-down karst through 800 m of mostly siliciclastic section. Finally, the geometries of the collapse features suggest that they may be small tectonic pull-apart features at intersections of regional fault and fracture systems, perhaps similar to restraining stepovers described by McClay & Borora (2001).

The application of our new seismic attributes to 3D seismic volumes allows detection of normally subseismic features and demonstrates great potential for determining timing and nature of features that may be related to karst, tectonic or hydrothermal collapse and for improved mapping of reservoir bodies, permeability fairways and various surfaces associated with weathered and fractured carbonates.

Funding for this research was provided through the State of Texas Advanced Technology Project 003652-0321-2003 and the sponsors of the Allied Geophysical Laboratories, University of Houston. We thank Devon Energy for use of their data.

References

AL-DOSSARY, S. & MARFURT, K. J. 2003. Fracture-preserving smoothing. *73rd Annual International Meeting Society of Exploration Geophysicists, Expanded Abstracts*, 378–381.

AL-DOSSARY, S. & MARFURT, K. J. 2006. 3-D Volumetric multispectral estimates of reflector curvature and rotation. *Geophysics*, **71**(5), P41–P51.

AL-DOSSARY, S. MARFURT, K. J. & LUO, Y. 2002. 3-D edge preserving smoothing for seismic edge detection. *72nd Annual International Meeting Society of Exploration Geophysicists Expanded Abstracts*, 524–527.

BAGDAN, C. A. & PEMBERTON, S. G. 2004. Karst Breccia and Bank Collapse Breccia; implications for reservoir characterization of the McMurray Formation, Alberta, Canada. [Abstract.] *AAPG Bulletin*, **88**, 13 (supplement).

BALL, M. M. & PERRY, W. J. 1996. *Bend Arc – Fort Worth Basin.* USGS Report for Province, **45**.

BERGER, Z. & DAVIES, G. 1999. The development of linear hydrothermal dolomite (HDT) reservoir facies along wrench or strike slip fault systems in the Western Canadian Sedimentary Basin. *Reservoir*, **26**(1), 34–38.

BLUMENTRITT, C. H., SULLIVAN, E. C. & MARFURT, K. J. 2006. Volume based curvature computations illuminate fracture orientations, lower Palaeozoic, Central Basin platform, West Texas. *Geophysics*, **71**(5), B159–B166.

BOWKER, K. A. 2003. Recent developments of the Barnett Shale Play, Fort Worth Basin, West Texas. *Geological Society Bulletin*, **42**, 6, 4–11.

CANSLER, J. R. & CARR, T. R. 2001. P*aleogeomorphology of the sub-Pennsylvanian unconformity of the Arbuckle Group (Cambrian-Lower Ordovician)*. Kansas Geological Survey Open file Report 2001-55.

FRANSEEN, E. K., BYRNES, A. P., CANSLER, J., STEINHAUFF, D. M., CARR, T. R. & DUBOIS, M. K. 2003. Geologic characteristics of Arbuckle Reservoirs. *In*: *Kansas: Proceedings of the Fifteenth Oil Conference, Tertiary Oil Recovery Project*, The University of Kansas.

GRAYSON, R. C. JR, & MERRILL, G. K. (eds). 1991. Carboniferous geology and tectonic history of the Southern Fort Worth (foreland) Basin and Concho Platform, Texas. *AAPG, 1991 Annual Convention, Dallas Geological Society Guidebook*, Field Trip No. **13**, 3–67.

HARDAGE, B. A., CARR, D. L., LANCASTER, D. E., SIMMONS JR, J. L., ELPHICK, R.Y., PENDELTON, V. M. & JOHNS, R. A. 1996*a*. 3-D seismic evidence for the effects of carbonate karst breccia collapse on overlying clastic stratigraphy and reservoir compartmentalization. *Geophysics*, **61**, 1336–1350.

HARDAGE, B. A., CARR, D. L., *ET AL.* 1996*b*. 3-D seismic image and seismic attribute analysis of genetic sequences deposited in low-accommodation conditions. *Geophysics*, **61**, 1351–1362.

HOSKINS, B. W. 1982. Fracture analysis of Pennsylvanian carbonate bank systems in the Graford Formation North Central Texas. *In*: MARTIN, C. A. (ed.) *Petroleum Geology of the Fort Worth Basin and Bend Arch Area.* Dallas Geological Society, 179–192.

KERANS, C. 1990. *Depositional systems and karst geology of the Ellenburger Group (Lower Ordovician), subsurface West Texas*. The University of Texas at Austin, Bureau of Economic Geology Report of Investigations, **193**.

KIER, R. S. 1980. Depositional history of the marble falls formation of the llano region. Central Texas. *In*: *Geology of the Llano Region, Central Texas, West Texas Geological Society Guidebook*, **80–73**, 59–75.

KUPECZ, J. A. & LAND, L. S. 1991. Late stage dolomitization of the Lower Ordovician Ellenburger Group, west Texas. *Journal of Sedimentary Petrology*, **61**, 551–574.

LISLE, R. J. 1994. Detection of zones of abnormal strains in structures using Gaussian curvature analysis. *AAPG Bulletin*, **78**, 1811–1819.

LOUCKS, R. G., MESCHER, P. K. & MCMECHAN, G. A. 2004. Three-dimensional architecture of a coalesced, collapsed-palaeocave system in the Lower Ordovician Ellenburger Group, central Texas. *AAPG Bulletin*, **88**, 545–564.

LUCIA, F. J. 1996. Lower Palaeozoic cavern development, collapse, and dolomitization, Franklin Mountains, El Paso, Texas. *In*: STOUDT, E. (ed.) *Precambrian–Devonian geology of the Franklin Mountains, West Texas – Analogs for Exploration and Production in Ordovician and Silurian Reservoirs in the Permian Basin. West Texas Geological Society 1996 Guidebook*, 175–196.

MCCLAY, K. & BORORA, M. 2001. Analog models of restraining stopovers in strike–slip fault systems. *AAPG Bulletin*, **85**, 2, 233–260.

MARFURT, K. J. 2006. Robust estimates of 3-D reflector dip and azimuth. *Geophysics*, **71**(4), P29–P40.

MARFURT, K. J., SUDHAKAR, V., GERSZTENKORN, A., CRAWFORD, K. & NISSEN, S. 1999. Coherency calculations in the presence of structural dip. *Geophysics*, **64**, 104–111.

MONTGOMERY, S. L., JARVIE, D. M., BOWKER, K. A. & POLLASTRO, R. M. 2005. Mississippian Barnett Shale, Fort Worth Basin, north-central Texas: gas-shale play with multi-trillion cubic foot potential. *AAPG Bulletin*, **89**, 155–176.

POLLASTRO, R. M., HILL, R. J., JARVIE, D. M. & HENRY, M. E. 2003. Assessing undiscovered resources of the Barnett–Palaeozoic total petroleum system, Bend Arch–Fort Worth Basin Province, Texas. *AAPG Online Search and Discovery* article.

ROBERTS, A. 2001. Curvature attributes and application to 3D interpreted horizons. *First Break*, **19**, 85–99.

SAGAN, J. & HART, B. 2004. Seismic and structural analysis of a Trenton-Black River Hydrothermal Dolomite Reservoir. *AAPG Search and Discovery* article 40129. unpaginated.

SIEBRITS, E., ELBEL, J. L., *ET AL*. 2000. Refracture reorientation enhances gas production in the Barnett Shale tight gas sand. *SPE 2000 Annual Technical Conference and Exhibition, Dallas*, SPE paper 63030.

SULLIVAN, E. C., MARFURT, K., LACAZETTE, A. & AMMERMAN, M. 2006. Application of new seismic attributes to collapse chimneys in the Ft. Worth Basin. *Geophysics*, **71**(4), B111–B119.

WALPER, J. 1982. Plate Tectonic evolution of the Fort Worth Basin. *In*: MARTIN, C. A. (ed.) *Petroleum Geology of the Fort Worth Basin and Bend Arch Area*. Dallas Geological Society, 237–251.

Fluvial seismic geomorphology: a view from the surface

F. G. ETHRIDGE & S. A. SCHUMM

Department of Geosciences, Colorado State University, Fort Collins, CO 80523-1482, USA (e-mail: fredpet@cnr.colostate.edu)

Abstract: Three-dimensional seismic data enable geoscientists to image the stratigraphic record along selected stratal or time slices. These slices provide detailed images of the planform geometry of ancient depositional systems and environments. In this presentation we attempt to provide a partial answer to the question: what kind of information do nonmarine channel and valley patterns and parameters convey regarding sediment load, channel stability, structural or climatic history of the area, and palaeochannel and reservoir architecture? Geomorphologists have long recognized four basic channel patterns: straight, meandering, braided and anastomosing. They have also developed a classification based on aerial pattern and sediment load. This relationship seems straightforward; however the patterns form a continuum with a great deal of complexity related to degree and character of sinuosity, braiding and anabranching. For example an equal-width sinuous pattern suggests moderate to high lateral stability, and low bed load to suspended load deposits. A wide bend sinuous pattern suggests low to moderate lateral stability, and higher bed load to suspended load deposits. Because of convergence and divergence, interpretation of controls on palaeochannel patterns is difficult. Nevertheless, a better understanding of the various controls on modern river morphology and dynamics can produce useful information regarding palaeo-systems. For example, river type (straight, braided, meandering and anastomosing) depends upon upstream controls such as geologic history (glaciated v. non-glaciated systems), tectonics (relief), lithology (sediment type) and climate (hydrology and vegetation), and downstream controls such as base-level (up and down) and length (avulsion). In addition, factors such as bedrock, active tectonics, floods and vegetation cause considerable pattern variability. Only by taking into account the effect of these controls and by using all available data can there be more predictive and interpretive explanations of seismic data.

The purpose of this paper is to review what is known about the plan view geometry of modern alluvial river channel and channel networks and how this information can be used to assess the sediment load, channel stability, structural or climatic history, and reservoir architecture of ancient fluvial systems. The motivation for this review is the explosion in the use of three-dimensional (3D) seismic data by the petroleum industry and the opportunity that this procedure offers in imaging elements of a depositional system, especially the plan view geometry of discrete palaeo landforms (Posamentier 2000). The characteristics and limitations of subsurface data sets in interpreting and predicting the character of ancient fluvial systems are reviewed by North (1996) and Bridge & Tye (2000). In his 1996 paper North concludes that only in very favorable circumstances can 3D seismic be used to fill in details of fluvial stratigraphy, and he reviews the traditional techniques of wireline logs and cores. In contrast Bridge and Tye state that amplitude analysis of 3D seismic horizon slices is the only method of directly yielding width of channel belts and images of the channel patterns in the subsurface. Their research, however, was based on the use of wireline logs and cores in assessing the dimensions of ancient fluvial deposits. With sufficiently close well spacing relative to the size of inferred meandering fluvial channels, a few studies have demonstrated the ability to construct a plan view map of these deposits and to estimate fluvial variables such as channel width, radius of curvature, meander wavelength and meander belt width (Fig. 1; Cornish 1984; Ebanks & Weber 1987). Swanson (1976) accomplished much the same thing for modern meandering rivers using aerial photographs. Padgett & Ehrlich (1976) demonstrated how stream characteristics could be determined from exhumed outcrops of Carboniferous meander belt deposits in Morocco. These types of exposures are, however, extremely rare and do not solve the problem of subsurface control. With 3D seismic time, horizon or proportional slices it is possible to distinguish between meandering, braided and anastomosing channels and their deposits, to distinguish channels from valleys, to recognize underfit streams, and to estimate the size of river systems from direct measurements (Fig. 2; Fachmi & Wood 2003). With this knowledge comes the important ability to understand better the evolution of stratigraphic sequences and to document the filling of sedimentary basins.

From: Davies, R. J., Posamentier, H. W., Wood, L. J. & Cartwright, J. A. (eds) *Seismic Geomorphology: Applications to Hydrocarbon Exploration and Production*. Geological Society, London, Special Publications, **277**, 205–222.
0305-8719/07/$15.00

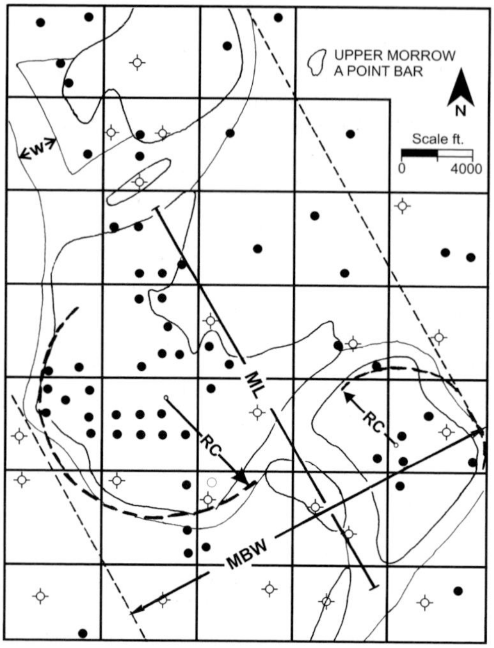

Fig. 1. Meanderbelt characteristics of stream width (W), meander wavelength (ML), radius of curvature (RC) and meanderbelt width (MBW) estimated from closely spaced wireline logs and cores; Pennsylvanian Upper Morrow A Sandstone, Beaver County, Oklahoma (modified from Cornish 1984).

In this paper major river types will be identified and attempts to classify rivers will be reviewed. In addition, controls on river variability, both longitudinally and through time will be examined and related to 3D seismic images. The discussion will be limited to adjustable alluvial rivers with sediment loads composed primarily of sand, silt and clay and to valleys. These reviews will be followed by a discussion of possible interpretations of channel patterns and properties, the limitations imposed by the dual problem of convergence and divergence, and some examples of how these interpretations have been applied to 3D seismic slices to date. Convergence refers to the situation when different controls produce similar results, for example, braiding can be the result of valley slope, sediment type or flow variability. Divergence is the situation where a single control produces different results, for example, active tectonic steepening can result in increased sinuosity, a change from meandering to braided, or incision depending on the magnitude and original conditions.

Alluvial river types and classification

Based on plan view morphology there are four basic channel types: straight, meandering, braided and anastomosing or anabranching (Fig. 3). There is an ongoing debate regarding the use of the terms anastomosing and anabranching. We think of these terms as synonymous, but for the sake of simplicity

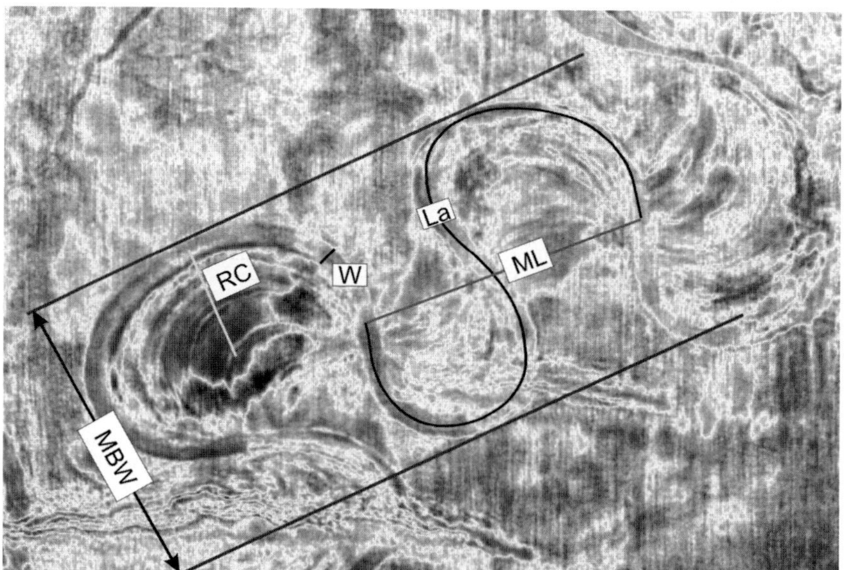

Fig. 2. Channel planform characteristics that can be determined from sinuous rivers visualized by a seismic time or horizon slice, Belanak, west Natuna Basin, Indonesia. MBW = meander belt width; RC = radius of curvature; W = channel width; ML = meander wavelength; La = length of channel for one wavelength; sinuosity = La/ML (interpreted slice provided by Wood; from Fachmi and Wood 2003; data provided by ConocoPhillips).

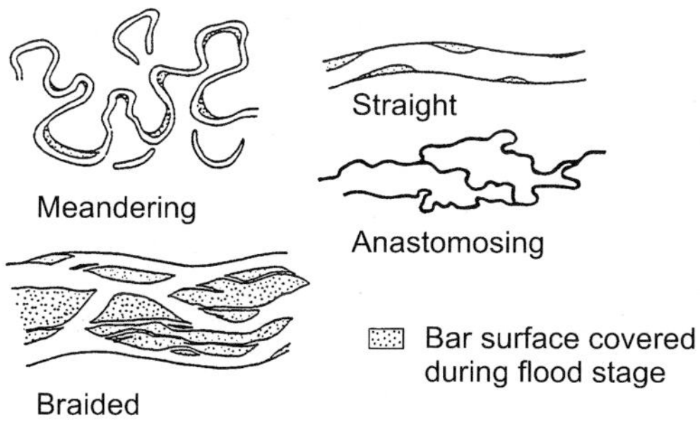

Fig. 3. Spectrum of channel types based on plan view morphology (Miall 1977).

we will exclusively use the term anastomosing to refer to rivers with two or more contemporaneous channels separated by wide semi-permanent, vegetated islands or floodplain segments. Multi-channel braided rivers differ from anastomosing rivers in that they consist of multiple channels separating ephemeral bars only at low water stage. At high water they may appear as a single channel.

The type of river is determined by four upstream controls and two downstream controls. Upstream controls include geologic history, tectonics (relief), lithology (sediment type) and climate (hydrology and vegetation; Fig. 4). Geologic history includes changing relief and climate through time but is considered separately from relief and climate by Schumm (2005). Lithology determines ease of erosion and therefore quantity and type of sediment load. Climate variability from arid to humid controls vegetation and discharge characteristics and therefore sediment load characteristics. Downstream controls include base-level (up and down) and length (avulsion). Local controls can change sediment load and type, hydrology and valley gradient, thus causing considerable pattern variability.

An attempt to determine if the morphologic pattern of alluvial rivers is related to sediment type resulted in the classification scheme of Schumm (1977), which is based on channel pattern and sediment load (Fig. 5). This classification is based on a database limited to sand-bed steams from the North American Great Plains and the Murrumbidgee River, Australia and does not include anastomosing channels. These basic patterns represent end members in a broad spectrum of channel types. Using aerial photographs from across the USA, Brice (1975, 1981, 1982, 1983) expands on this scheme by recognizing that six variables stand out as important in classifying rivers. They are degree and type of sinuosity, degree and type of braiding, and degree and type of anastomosing (anabranching). He uses these six variables to define a variety of channel patterns which are illustrated in Figure 6. Degree of sinuosity is determined as the ratio of the stream length measured along the centerline of the stream to the meander belt centerline (Fig. 2). For character of sinuosity he uses the word phase from Hollard (1973). A single-phase distribution suggests a unimodal distribution, whereas a two-phase distribution infers either a sinuous low-water, underfit channel in a wider sinuous bankfull channel or a bimodal distribution of loop sizes (radius of curvature; Fig. 2) for a bankfull channel. The character of sinuosity also concerns channel width in bends relative to channel width at straight reaches and/or crossings. Sinuous channels are equiwidth if the average width at a bend is not greater than 1.5 $\times$ the average width in a crossing or straight

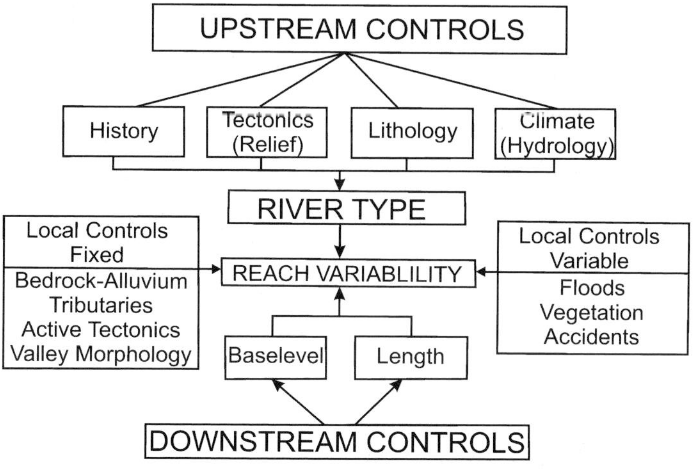

Fig. 4. Flow chart showing various controls of river morphology and behaviour (after Schumm 2005).

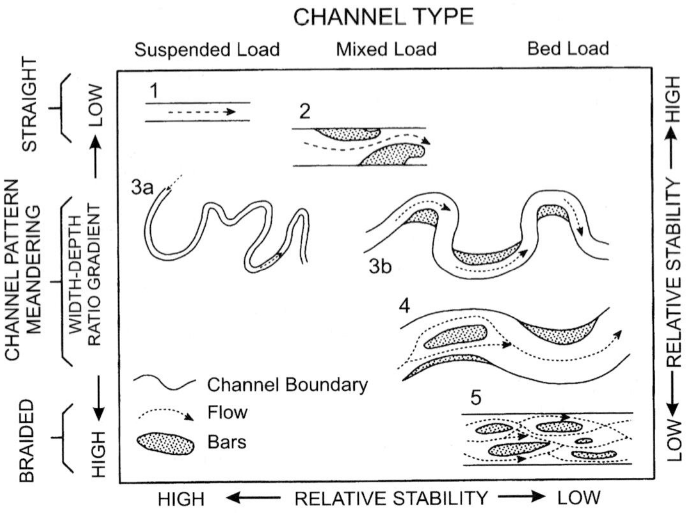

Fig. 5. Classification of channel types based on sediment load (Schumm 1977).

reach (Fig. 6a). The degree of braiding refers to the percent of a channel reach that is divided by one or more bars or islands (Fig. 6b). The character of braiding is a function of the relative abundance of bars and islands. Since islands are vegetated and the distinction between non-vegetated bars and vegetated islands is impossible on the basis of 3D seismic data character of braiding is not considered important for our purposes. Brice uses the term anabranching (not anastomosing) to refer to the division of a river into individual islands whose width is greater than 3 × water depth at normal discharge. His degree of anastomosing refers to the total percent of a given reach of the river that is characterized by multiple channels (Fig. 6c). The character of anastomosing refers to the sinuosity and orientation of anabranch channels relative to each other.

Brice (1982) suggested that there are four major types of alluvial streams and that they can be differentiated on the basis of variability of width, nature of point bars, and degree of braiding (Figs 7 and 8). He also emphasized that each stream type may be anastomosing locally or regionally. For this classification scheme to work he emphasizes that the reach classified must be longer than 50 channel widths. Two important aspects of this scheme are the distinction between stable equiwidth sinuous channels and unstable wide-bend, point-bar meandering channels and between braided point-bar channels and braided channels without point bars (Figs 7 and 8). While understanding that lateral stability was the primary emphasis for these distinctions, Brice also gives a good indication of how channel width, valley slope, channel sinuosity and sediment load vary systematically with pattern type. Additional emphasis on the spectrum of channel types is given by Schumm (1977), who recognized five basic bed-load and mixed-load channel patterns and three types of suspended-load patterns that reflect an increase followed by a decrease in sinuosity as valley slope and bedload increases (Fig. 9). Bed-load channels have the highest width–depth ratio. Suspended load channels are the narrowest and deepest and may have the highest sinuosity, and are very stable. Mixed load patterns show a wide range of sinuosity and bar types, ranging from truly meandering channels with point bars (pattern 8, Fig. 9) to wandering sinuous-braided channels (pattern 9, Fig. 9), which is similar to Brice's braided point-bar channel pattern (Fig. 7). The database for these patterns is analogue experimental studies in flumes at Colorado State University (Schumm 1977).

Brice shows that each of his single channel patterns has a corresponding multiple channel type, Fig. 7). Anastomosing rivers, which form by avulsions, are composed of two or more channels separated by vegetated, semi-permanent alluvial islands or enclosing floodbasin deposits (Nance & Knighton 1996; Makaske 2001). Some of the requirements for establishing an anastomosing system include a flood-dominated regime, resistant river banks and aggradation of the active channel belt. Nance & Knighton (1996) recognized six different types of anastomosing river patterns on the basis of stream energy, sediment size, and morphologic characteristics. These include:

type 1 – cohesive sediment rivers with low w/d ratio stable channels;

type 2 – sand-dominated, island-forming rivers;

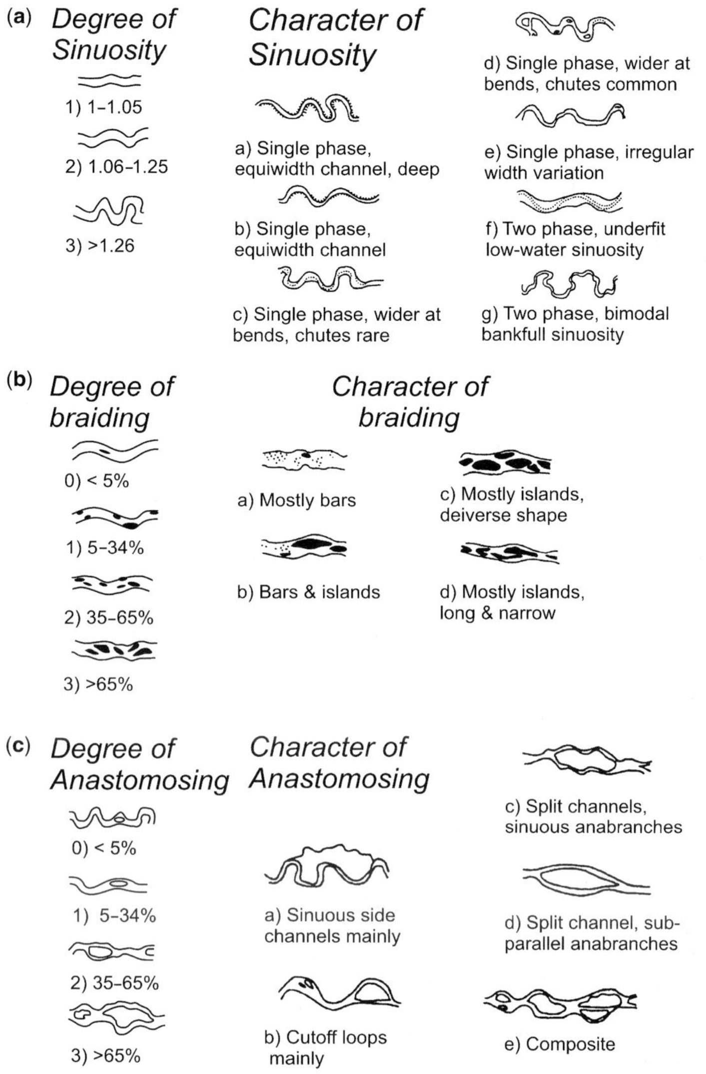

Fig. 6. Channel pattern classification based on degree and character of sinuosity (**a**), braiding (**b**) and anabranching (**c**) (modified from Brice 1975).

- type 3 – mixed-load meandering rivers,
- type 4 – sand-dominated, ridge-forming rivers,
- type 5 – gravel-dominated, meandering to braided rivers,
- type 6 – gravel-dominated, island-form rivers.

Based on plan view morphology alone only five types are distinguished (Fig. 10a). Types 1, 2 and 6 are similar enough in plan view that they cannot be distinguished on this criterion alone. The five anastomosing river patters can be further subdivided into those with stable channels (laterally inactive) and those that are unstable (laterally active; Fig. 10a). Examples of modern laterally inactive sinuous, laterally active meandering and laterally active braided anastomosing channels are shown in Figure 10b–d).

In summary, it is clear that a spectrum of channel types exist in nature and that any attempt to pigeonhole rivers into a classification scheme based on plan view morphology will vary depending on the interest and perspective of the investigator. One common theme of all of the classifications examined here is the distinction between stable (laterally inactive) and unstable (laterally active)

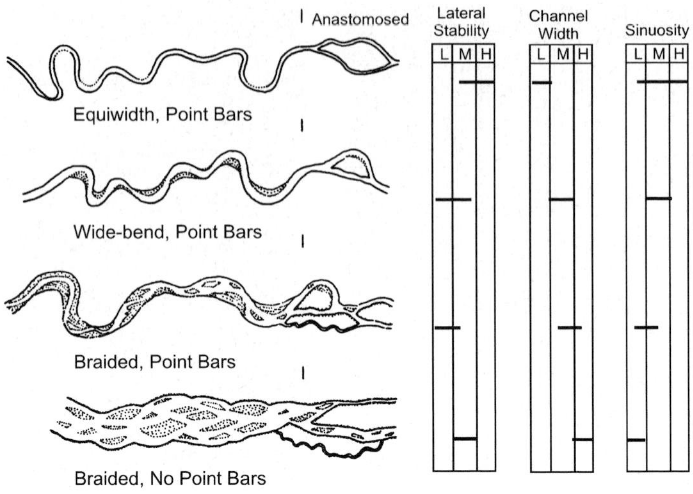

Fig. 7. Summary of alluvial river types. L, M, and H stand for low, medium and high in terms of lateral stability, channel width, and sinuosity (modified from Brice 1982).

channels and multiple channels. This theme is obviously important to geomorphologists and engineers working on modern rivers and may be equally important to sedimentologists and stratigraphers interpreting the preserved rock record.

Causes of variability and change in fluvial systems

For Holocene rivers we can examine the major controlling factors that determine river type and reach variability (Fig. 4). Schumm (2005) reviews the three types of controlling factors which are: (1) upstream controls of history, tectonics, lithology and climate; (2) local controls that dominate at the reach scale and which are divided into fixed local controls (bedrock-alluvium, tributaries, active tectonics, and valley morphology) and variable controls (floods, vegetation and accidents); and (3) downstream controls of base level and length. These controls determine the type of rivers that occur and the longitudinal variability of a river. In spite of this variability some rivers are in regime, that is, relatively stable through time. However, when these variables change, a river is no longer in regime, and it changes through time. Non-regime channels are those that are responding to altered conditions by eroding, depositing, avulsing and changing pattern. With the advent of 3D seismic slices we can now examine the changing conditions of fluvial systems through time, and speculate on the possible conditions that might have brought about these changes.

Although the geologic history, regional tectonics, bedrock lithology and climate are all considered important upstream variables that determine river type and size (Fig. 4), at least one of these controls, climate, is rarely used in explanations of the evolution of pre-Quaternary fluvial systems (Blum & Törnqvist 2000). This is unfortunate given that climate is a significant part of the glacio-eustatic model, and it has a clear correlation with base-level change, a major downstream control. Ethridge *et al.* (1998) suggests that correlation between climate change and fluvial activity is possible, but prediction of the geomorphic effect is difficult at best. Blum & Törnqvist (2000) add that 'clear recognition and interpretation of fluvial response remains difficult'. Most recently White *et al.* (2005) differentiated base-level and climate signals in alluvial palaeosols from the margins of the Western Cretaceous Interior Seaway, North America. However, since their technique involves the oxygen isotope values from sphaerosiderites, it will not be reviewed here.

If the upstream controls determine the river type, it might seem logical to assume that a uniform morphology exists downstream. It usually does not. Braided rivers become anastomosing and meandering rivers become straight, etc. Variability along a river course is the result of local and downstream controls (Fig. 4). For example, a tributary can introduce a different type of sediment load and active tectonics can modify the valley floor. These controls are fixed in location in contrast to local controls that can shift position, such as local floods, accidents (landslides, dams, etc.) and variable vegetation types. Schumm & Galay (1994) and Schumm *et al.* (2000) discuss local river channel changes along the Nile River and relate these changes to active

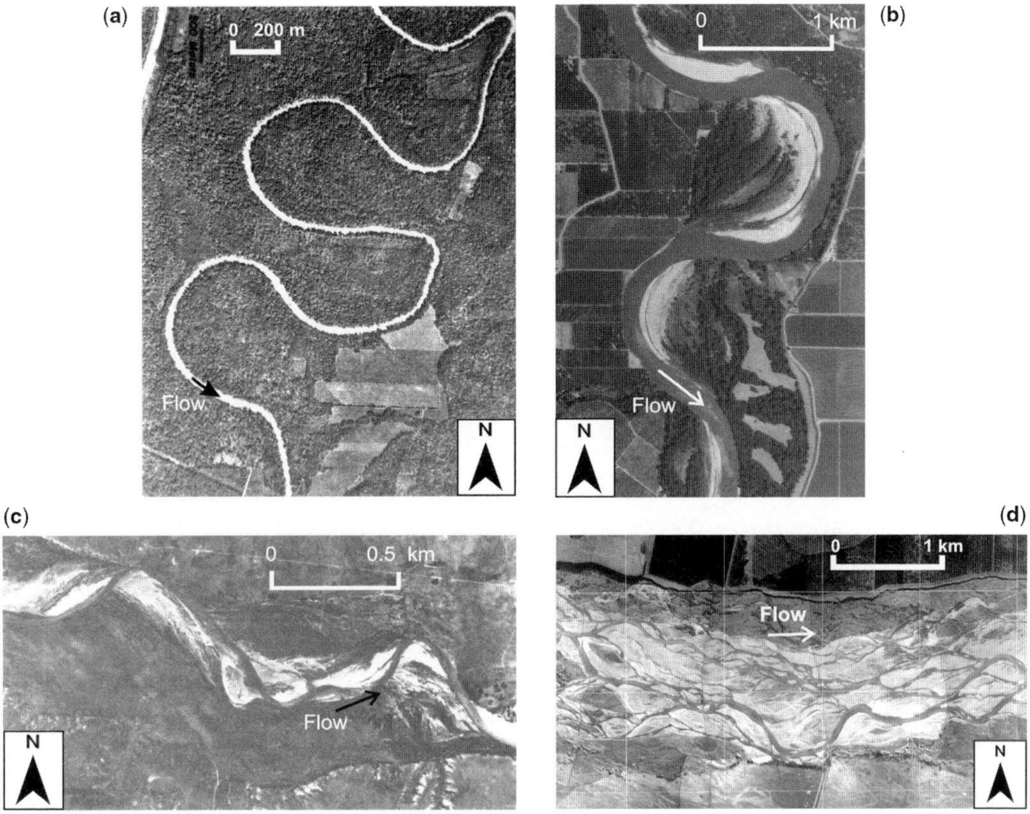

Fig. 8. Aerial photographs of modern alluvial river types based on classification of Brice (1982). (**a**) Red River of the North near Perley, MN; equiwidth meandering stream incised into cohesive lake bed deposits and glacial till (modified from Brice 1982; 1966 USDA photo). (**b**) Sacramento River near Butte City, CA, USA; single-phase meandering river wider at bends with point bars (modified from Lagasse *et al.* 2004; photo from Owens Ayres & Associates Inc., Fort Collins, CO, USA). (**c**) North Canadian River near Guymon, OK; a braided point-bar stream on the semiarid Great Plains (modified from Brice 1982; 1966 USDA photo). (**d**) Waimakariri River, South Island, New Zealand; braided river with a high degree of braiding and mostly bars.

deformation. For example, at El Manshah and Assiut steeepening of the valley floor by faulting has resulted in a significant increase in sinuosity of the Nile channel (Fig. 11). The increase in both localities is followed by a decrease in sinuosity and meander amplitude in a downstream direction. A plot of sinuosity v. valley slope suggests that the sinuosity at El Manashah is too high for the valley slope and that the Nile should straighten in this reach with time. Stream capture, base-level lowering, meander cutoff, avulsion, sediment storage and lateral shift of a channel all cause local steeepening of a channel gradient and can result in incision and or pattern change.

Pattern changes include meander growth and shift, island and bar formation, river or channel cutoff and avulsion, and river metamorphosis (i.e. straight to braided, braided to meandering, etc.). Avulsion is the relatively abrupt change of the course of a river (i.e. the abandonment of one channel and the subsequent formation of a new one). Avulsions include local changes such as neck and chute cutoffs on meandering rivers as well as major river changes that affect hundreds of kilometers (Fisk 1944). The most common underlying causes of avulsion are related to the ratio of the slope of the potential avulsion (Sa) channel to the slope of the existing channel (Se) (Sa/Se; Ethridge *et al.* 1999; Jones & Schumm 1999). A decrease in the slope of an existing channel will reduce its ability to carry water and sediment, increasing the ratio and increasing the chance for an avulsion. Similarly the slope of the potential avulsion surface can increase as the result of active tectonics (Peakall *et al.* 2000). If a river is near an avulsion threshold because of a decrease in the slope of an existing channel, a small event such as a localized ice jam may trigger an avulsion (Ethridge *et al.* 1999). Avulsions can also result from causes unrelated to

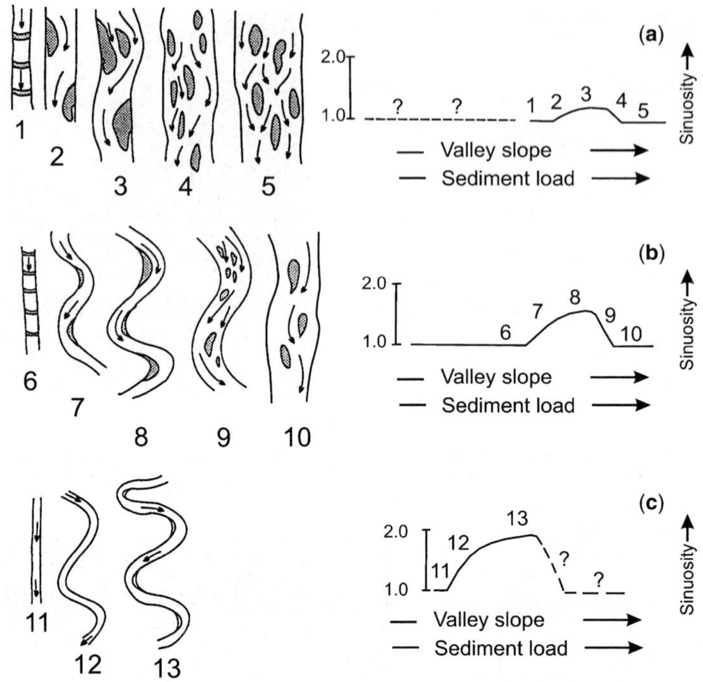

Fig. 9. A range of alluvial channel patterns for (**a**) bed-load channels, (**b**) mixed-load channels and (**c**) suspended-load channels. Patterns within each group are related to changes of valley slope, stream power and sediment load (from Schumm 1977).

this ratio such as vegetation encroachment and animal trails (McCarthy *et al.* 1992), change in peak discharge, increase in sediment load, and log jams. Aslan & Blum (1999) recognized three styles of avulsion associated with different stages of valley filling during base-level rise in rivers on the Texas coastal plain. These included: (a) reoccupation of abandoned channel courses during slow aggradation during an early stage of valley filling; (b) avulsion initiated by crevassing followed by diversion onto floodplains during later stages of rapid valley filling; and (c) avulsion by channel reoccupation during highstand. Each style can be recognized by distinctive plan view morphologies which are recognizable on aerial photographs and perhaps on 3D seismic time and/or horizon slices.

Base-level change is considered a major cause of incision and aggradation in the downstream portions of river systems. Just how far upstream the effects of base-level change extend is a subject of much debate (Koss *et al.* 1994; Shanley & McCabe 1994; Leeder & Stewart 1996; Blum & Törnqvist 2000; Ethridge *et al.* 2005). Fisk (1944) concluded that Pleistocene sea-level lowering caused erosion of alluvium and bedrock up to 1000 km upstream from the present-day river mouth of the Mississippi River. In contrast Saucier (1994) concluded that the extent was only about a quarter of that suggested by Fisk. Blum & Törnqvist (2000) suggest that the distance upstream (coastal onlap) that is affected by post-Pleistocene sea-level rise is directly related to hinterland sediment supply and inversely related to the gradient of the floodplain surface. It appears that the impact of base-level change on fluvial systems depends on a number of variables summarized by Schumm (2005), such as: direction, magnitude, rate and duration of base-level change; lithology, structure, and character of the alluvium; gradients of exposed surfaces (Summerfield 1985); and river morphology and ability to adjust. The amount of incision or aggradation can be buffered by a change in sinuosity or plan view pattern (Schumm 1993), and base-level fall has a more significant impact than does base-level rise. It has also been noted that the lower portion of a fluvial system may be out of phase with the upper portions during and after a base-level change. Germanoski (1990) observed this phenomenon in a flume study of the effects of base-level change on experimental braided streams. During base-level lowering the downstream portion of the experimental flume was undergoing incision while the upstream portion was aggrading as a result of sediment input (Fig. 12).

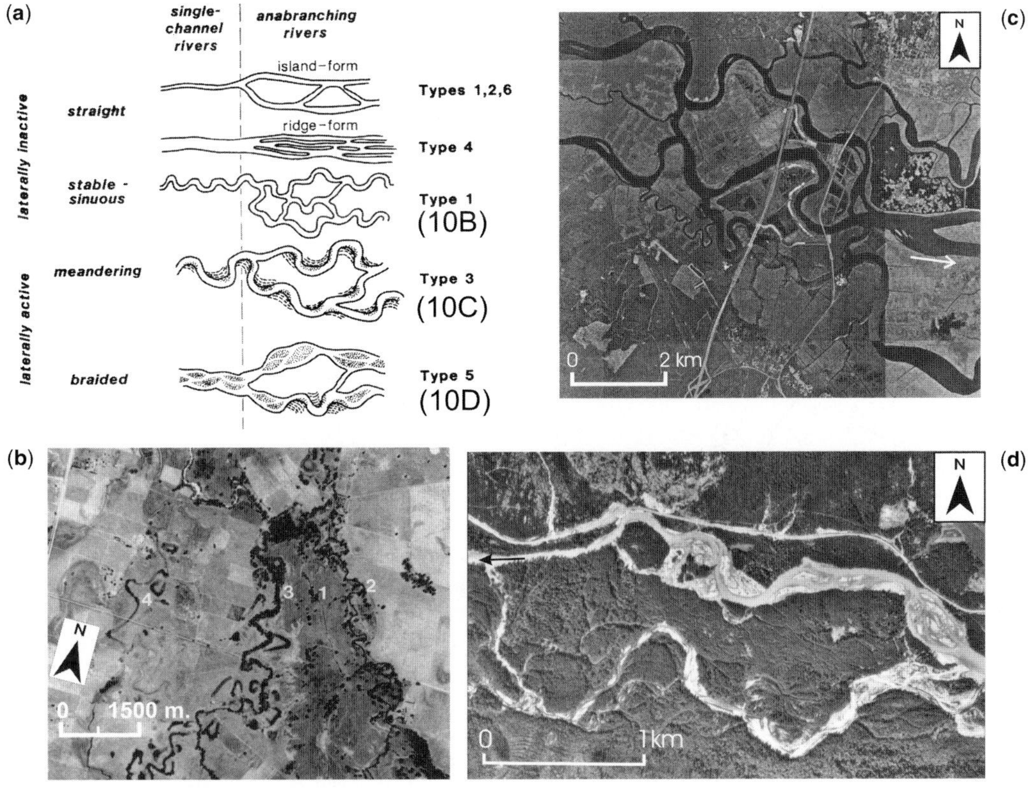

Fig. 10. (**a**) Classification of river patterns including both single-channel and anastomosing types. Note that laterally inactive channels consist of straight and sinuous forms whereas laterally active channels consist of meandering and braided forms (from Nance Knighton 1996). (**b**) Stable, sinuous channels of the anastomosing King River near Docker, Victoria, Australia; numbers equal relative ages of channels with 4 being the oldest inactive channel. Note that all channels are not active at the same time (from Schumm et al. 1996). (**c**) Laterally inactive to active meandering channels of the anastomosing Altamaha River near Darien, GA, USA (Lagasse et al. 2004; photo from Microsoft TerraServer Website). (**d**) Laterally active braided anastomosing Bella Coola River in British Columbia, Canada (photo complements of Mike Church, similar to photo shown in Nance & Knighton 1996).

Anastomosing rivers are favored by a low floodplain gradient which might occur as a result of a rapid rise in base level. They are more common in the distal reaches of river systems. In this setting preservation potential of anabranching river deposits should be high (Makaske 2001). The fact that anastomosing river systems have not been commonly recognized in the rock record may be the result of the difficulty in determining if the multiple channels observed in the rock record are contemporaneous or not. The advent of high resolution 3D seismic data should provide a basis for recognition of anastomosing rivers deposits in the rock record. Anastomosing channel sandstone bodies generally have a ribbon shape and a flat top, and are commonly associated with crevasse splay and natural levee deposits (Makaske 2001).

Examples of 3D seismic studies of ancient fluvial systems: lessons to date

A number of questions arise regarding the use of 3D seismic time, horizon or proportional slices in evaluating ancient fluvial deposits in relation to what is known regarding modern fluvial systems. These questions include: can the basic river types be differentiated? Can laterally stable and unstable meandering channels be differentiated? Can multi-channel braided and anastomosing river systems be differentiated? Can underfit streams be recognized? Can time and/or horizon slices permit better evaluation of the causes of variability in ancient fluvial deposits? To attempt to answer these and other questions we will review the limited published literature in this section.

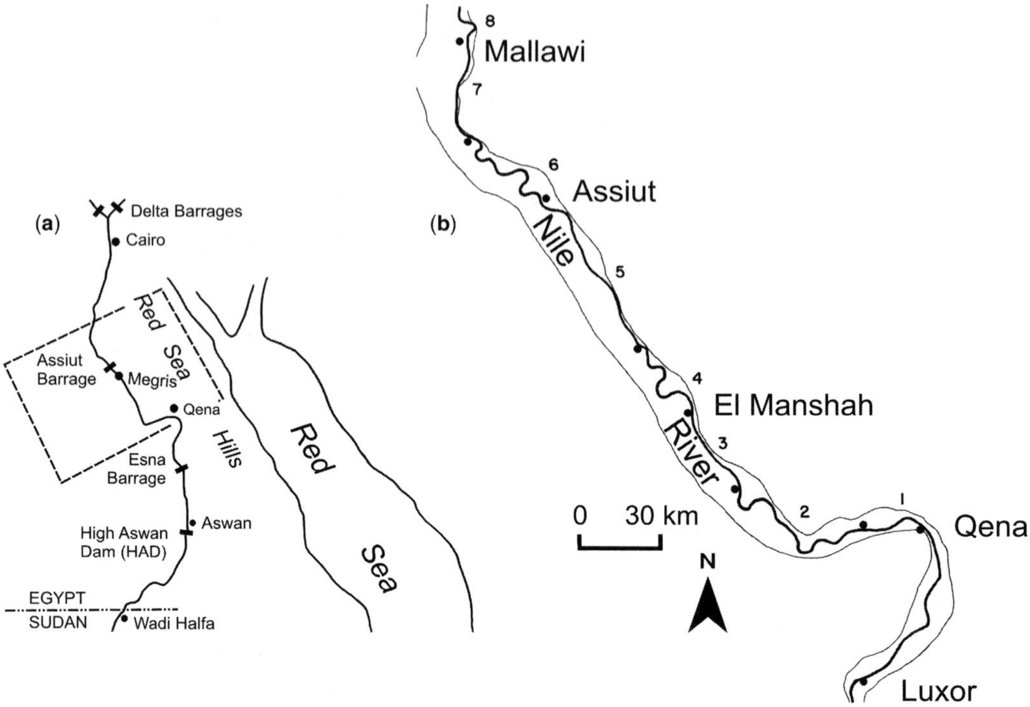

Fig. 11. (**a**) Course of Nile River through Egypt. Dashed rectangle shows portion of river shown in (b). (**b**) Course of Nile River downstream from Luxor showing zones of high and low sinuosity (modified from Schumm & Galay 1994).

Although the value of imaging time slices in 3D seismic data has been know for some time (Brown 1985, 1991; Isa *et al.* 1992), the number of refereed publications using this technique to document and interpret ancient depositional systems has been limited prior to 2000. One important exception is the edited volume on applications of 3D seismic to exploration and production (Weimer & Davis 1996). Research papers documenting deep-water turbidite and submarine fan deposits are fairly numerous (Posamentier *et al.* 2000; Kolla, *et al.* 2001; Posamentier & Kolla 2003; Deptuck *et al.* 2003*a*, *b*; Steffens *et al.* 2004). Three-dimensional seismic investigations of deep-water deposits will not be reviewed here. Recently several review papers (Posamentier, 2002, 2004, 2005) have documented the practical applications of using 3D seismic visualization techniques in identifying depositional elements (fluvial and submarine channels, valleys, shelf ridges, reefs, etc.). The emphasis in these reviews is on the acquisition of images and their use in developing sequence stratigraphic models for providing a better understanding of the evolution of sedimentary sequences and for predicting the distribution of lithologies and hence reservoirs.

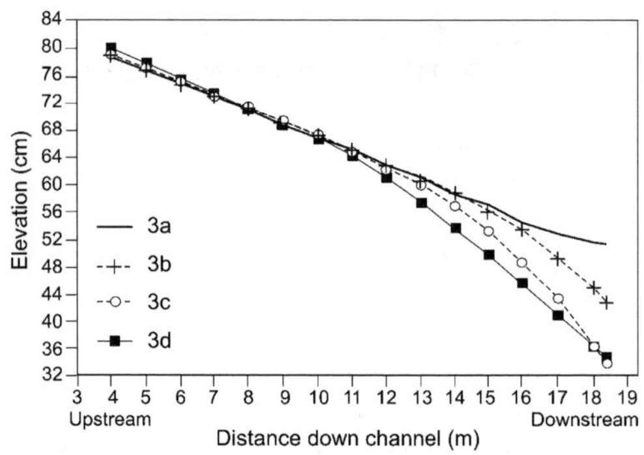

Fig. 12. Longitudinal profiles of an experimental braided stream during base-level lowering. Profiles indicate stages of profile adjustment during the experiment from **a** to **d** (Germanoski 1990). Note incision in lower portions and aggradation in the upper portions of the profile.

Papers that involve an evaluation of seismic slices of subsurface fluvial systems include those by Isa *et al.* (1992), Hardage *et al.* (1994), Burnett, (1996), Saudi Arabian Oil Company (1996), Posamentier (2001), Miall (2002) and Martinez *et al.* (2004). Isa *et al.* (1992) utilized 3D seismic and site survey data to identify a Pleistocene meandering channel 80 m deep and 500 m wide in the

Malay basin. Borings into the channel identified the fill of the channel as clay with significant organic material above a thin basal unit of fine sand with pebbles. Interestingly enough their primary interest was to determine if the channel deposits constituted a hazard in locating the drilling platform. An important observation was that channels of this type are commonly observed on 3D site surveys along the axis of the Malay basin. Hardage *et al.* (1994) documented narrow meandering channel reservoirs as thin as 3 m (10 ft) in the Frio Formation, at Stratton field, South Texas. Imaging of reservoirs this thin at 1800 m (6000 ft) below the surface was possible because the 3D seismic data was calibrated using vertical seismic profiling (VSP). Various channel, crevasse splay and floodplain deposits were documented in four reservoirs. With the incorporation of additional geologic and reservoir engineering pressure data they were able to define compartment boundaries within these heterogeneous reservoirs. Burnett (1996) used 3D seismic to image fluvial systems of the Upper Pennsylvanian Cisco system of West Central Texas at depths of 914–1829 m (3000–6000 ft). Sinuous channels are clearly imaged in the time slices shown in Burnett's paper; however, he also speculates on the presence of mid-channel sandstone bodies as well as levee and overbank deposits from seismic cross sections. Unfortunately he presents no evidence for these secondary interpretations. A well was drilled to test a point bar identified in the 3D seismic analysis. The well was successful in finding a point bar deposit, but not in finding oil. A paper by Saudi Arabian Oil Company (1996; compiled by Simms) reviews the impact of 3D seismic data on development drilling in the Ghinah and Umm Jurf fields, Saudi Arabia. Two wireline log cross sections and two time slices are shown that purport to show braided fluvial deposits of the productive Permian-Carboniferous deposits in the field area. Even if this conclusion is correct, neither type of data presented offers convincing evidence to support it, and there is no discussion of why the patterns shown should be interpreted as braided stream deposits.

Posamentier (2001) investigated alluvial systems ranging in age from Miocene to late Pleistocene on the Java Sea shelf NW of Java. His paper represents one of the first serious attempts to interpret the results of multiple imaged fluvial depositional elements in time slices from 3D seismic surveys. Both incised valleys and un-incised channel systems were recognized in the images evaluated. Incised valleys were clearly recognized based on the presence of small incised tributary valleys to the main trunk valleys (Fig. 13). Possible mechanisms for the formations of these valleys are discussed in general and specifically with reference to the study area. It was concluded that the incised valleys were the exception rather than the rule in the Pleistocene section and formed only when sea level was lowered to the extent that the shelf was fully exposed. When sea-level fall did not expose the entire shelf a low-stand alluvial bypass channel system and shelf delta developed. Two related observations of importance were made in this study: (1) the fact that not all lowstands result in valley incision has serious implications for correlating periods of lowstand with deep-water sedimentation; and (2) the preservation of features such as point bars, meandering channels, and fluvial terraces and bars within incised valleys suggests rapid transgression and/or low coastal energy during transgression. Similar observations concerning the transport of coarse-grained sediments to the deep basin and the preservation potential of the coastal wedge of sediments during slow and rapid transgressions were made in flume experiments by Wood *et al.* (1993), Koss *et al.* (1994), and Ethridge *et al.* (2005). However one should not rule out the possibility that the differences between the incised valleys and the unincised sinuous channels might simply reflect the fact that the low valleys were cut by major bedload rivers carrying sediment from a distance source area and that the unincised, high sinuosity channels represent minor mixed to suspended-load systems sourced locally on the shelf (see below for further discussion).

Miall (2002) investigated the architecture and sequence stratigraphy of Pleistocene fluvial systems in the Malay Basin. On eight time slices ranging from 220 to 136 ms he recognized a variety of river patterns and types including classic meander belt, braided, incised valley, low sinuosity and small sinuous underfit stream systems. The dimensions of these fluvial systems are documented, modern analogues for most are identified, their evolution through time is discussed and they are placed in a sequence stratigraphic framework. Given the distal coastal plain to exposed shelf setting of the study area, Miall interprets the changes in fluvial style throughout the succession in terms of variations in accommodation caused by changes in subsidence rate and sea level. One important aspect of this paper is the first-time recognition of large braided rivers in seismic time slices. This inferred river, up to 4.2 km wide, is shown in the upper-left-hand corner the time slice (Fig. 14a; not shown in the original paper) and the interpretation diagram (Fig. 14b). According to Miall the size of this inferred braided system is comparable to the Red River (Schwartz 1978) along the Oklahoma–Texas border. We suggest further that the inferred braided river is a low-sinuosity (1 on Fig. 14b) bedload river carrying sediment from a distant source area (probably outside the coastal plain), that the high sinuosity river (2 on Fig. 14b) is a minor, mixed-load stream carrying sediment from a local source

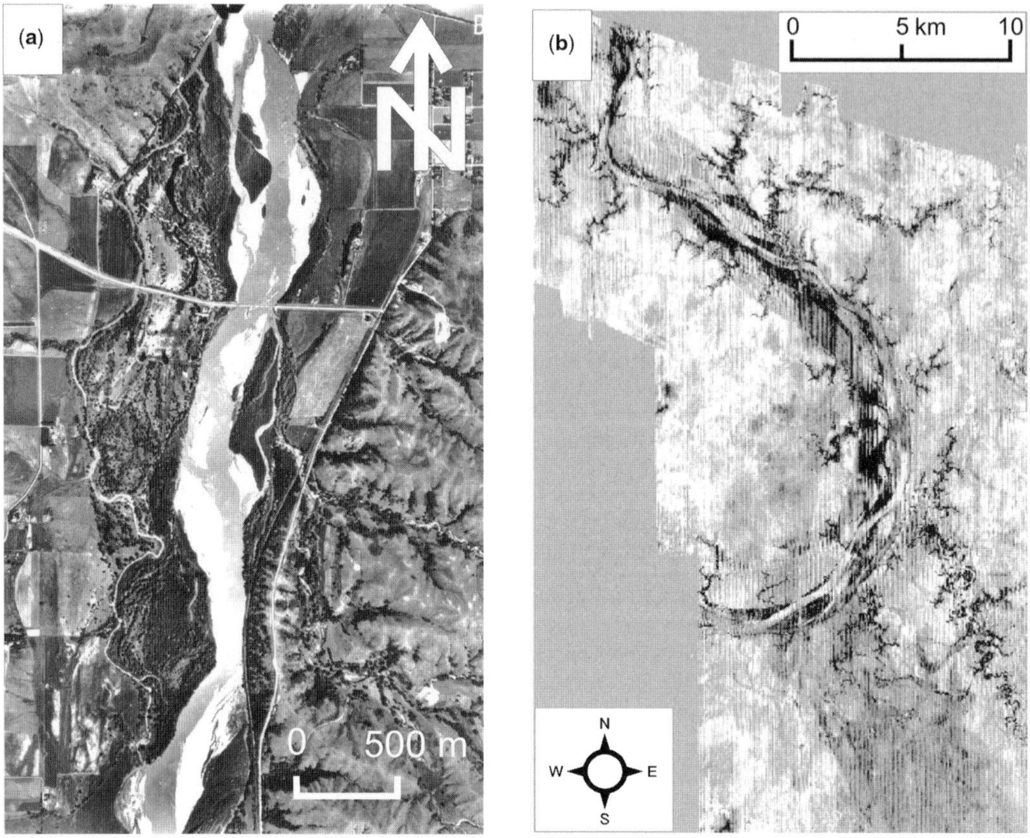

Fig. 13. (**a**) Modern Niobrara River, northeastern Nebraska showing small tributary drainages on interfluve. (**b**) An 84 ms time slice through 3D seismic volume, Java Sea Shelf, offshore NW Java (modified from Posamentier 2001). Note the same type of small tributary drainages on interfluves adjacent to the trunk incised valley.

(the coastal plain) and that the small sinuous patterns (3 on Fig. 14b) were minor, locally sourced suspended-load streams. In addition to Miall's interpretations of the 196 ms slice (his Fig. 2), we add the following based on interpretations of Brice and Schumm's work on modern rivers. The meander belt system (Fig. 15a) is clearly similar in pattern, if not in scale, to the equiwidth Henry's Fork River described by Brice (1982). The Henry's Fork had displayed no evidence of historic migration in spite of the presence of conspicuous meander scrolls. Perhaps the inferred ancient meandering river evolved to a state that was 'in regime' during the period represented by this slice. The incised valley, identified by Miall on the SE side of the seismic slice, is a classic example of an equiwidth incised stream and is similar in geometry if not in scale to the Red River of northern Michigan (Brice 1982; Fig. 15b). The large low to high-sinuosity river identified by Miall (2002) on the left side of the seismic slice (Fig. 15c) has features that suggest that local tectonic controls such as those inferred by Schumm & Galay (1994) may have been the cause of the major changes in sinuosity observed along the course of this inferred river. Other explanations are possible such as the introduction of a significantly different sediment load from a captured tributary, a possibility which comes to mind given the split in the system toward the NW; however, the two channels appear to be of different ages based on the single seismic slice available. Miall also recognized small meanders of underfit streams in several of the slices. These could be similar to modern underfit streams in the Murrumbidgee River of New South Wales, Australia (Fig. 16a, b) or the modern underfit Trinity River on the Texas coastal plain (Blum *et al.* 1995) or simply a preserved low water channel in an alluvial valley. Given the complexity of fluvial styles observed in this study area, Miall's warning, against the use of simplistic models that show somewhat uniform fluvial styles in models of nonmarine oil and gas fields, seems well founded.

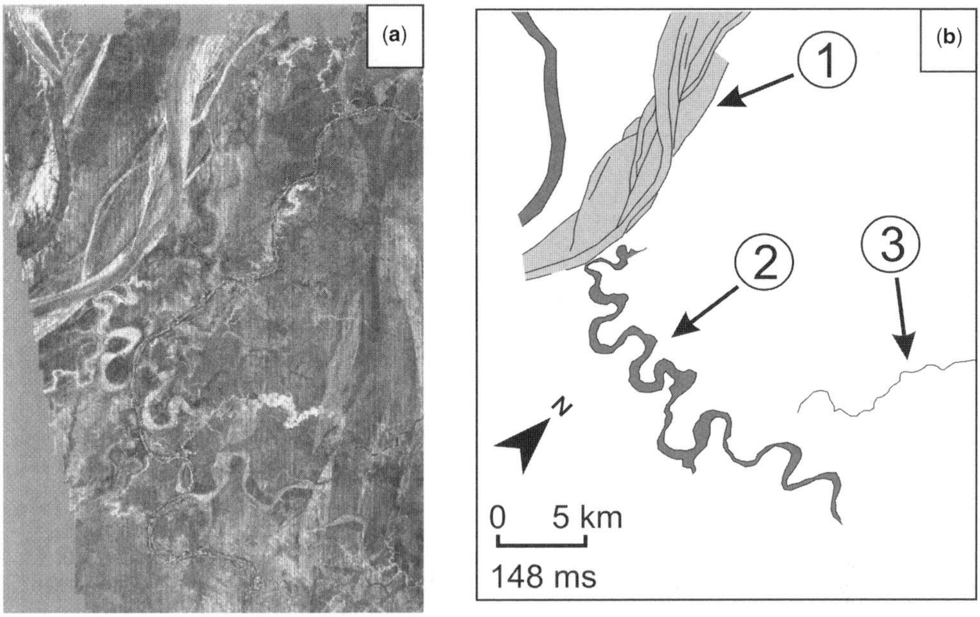

Fig. 14. (**a**) Seismic image showing a portion of the 148 ms slice from Pleistocene deposits in the Malay basin, Gulf of Thailand (Miall 2002). (**b**) Interpreted channels in the portion of 148 ms seismic image shown in (a). The base of the large braided river in the NW corner is interpreted to mark a sequence boundary within the succession (modified from Miall 2002; his fig 8). Numbers 1–3 refer to interpretations discussed in text (seismic slice provided by A. Miall).

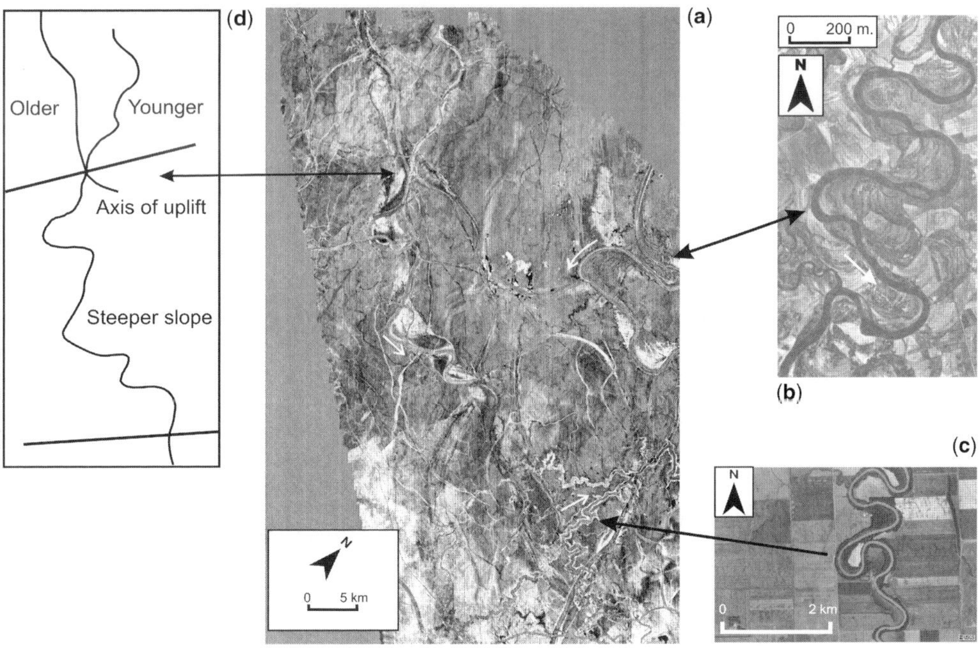

Fig. 15. (**a**) Seismic slice (Miall 2002, his fig. 3). (**b**) Comparison with similar, but smaller modern equiwidth stable, sinuous stream displaying scroll bar topography, Henry's Fork River, Idaho (modified from Brice 1982). (**c**) Comparison with similar, but smaller modern equiwidth, incised stream, Red River of northern Michigan (modified from Brice 1982). (**d**) Tracing of the large imaged channel that shows change from lower to higher sinuosity downstream and interpretation of possible palaeo-tectonic tilting.

Carter (2003) examines the evolution and reservoir characteristics of four, Oligocene–Miocene meandering, fluvial, point bar reservoirs at Widuri field in the Java Sea. His model was deduced from 3D seismic slices, well logs and limited core data. The fluvial system evolved from one in which active channels migrated across the entire preserved meander belt to one which had a high avulsion frequency relative to lateral migration. The resulting reservoirs from the oldest to youngest varied from wide, superimposed meander belt sandstones to narrow, isolated sandstones. Carter tied the changes in reservoir stacking pattern and preserved reservoir plan forms to significant variations in hydrocarbon production drive mechanisms, with the lower reservoirs having strong water drives and the upper reservoirs having poor aquifer support. Details of these deposits suggest that they are part of a lowstand bypass channel system and are not part of an incised valley fill. Details observed in the 3D seismic data set allow for the deduction of flow direction in the original fluvial system where truncated point bar deposits are seen on the upstream side of preserved meander forms; an observation also made by Posamentier (2001) and Miall (2002).

Martinez *et al.* (2004) document the complex three-dimensional geometry of the Late Miocene Messinian unconformity east of the present-day Ebro delta on the Castellon Shelf in the Mediterranean. A detailed map of the unconformity surface reveals a complex dendritic drainage pattern which resembles badland topography from semi-arid climate zones in the NW portion of the study area and a smooth dipping surface to the SW. The

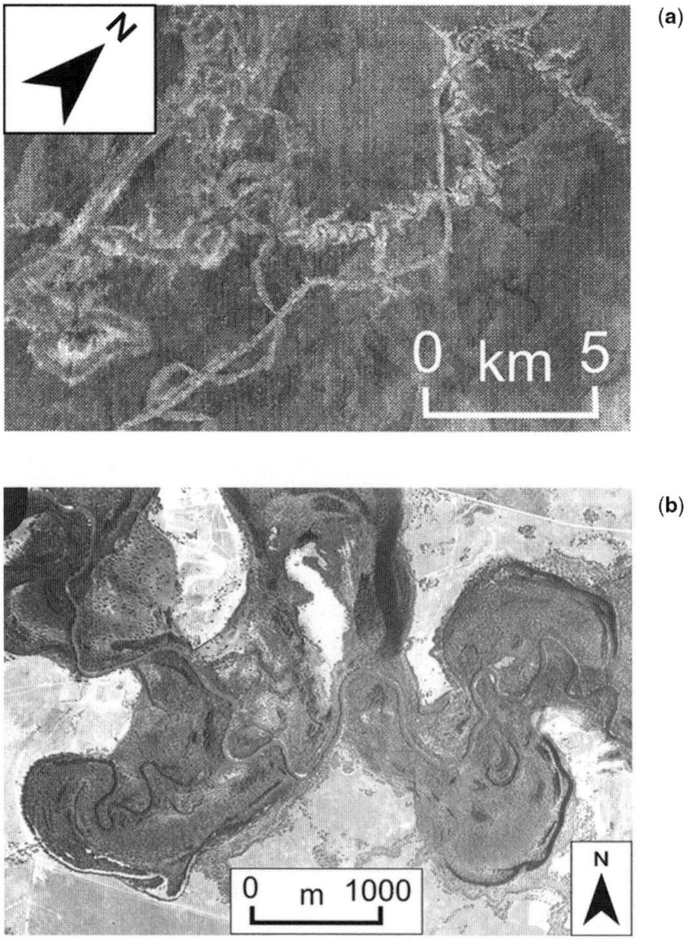

Fig. 16. (**a**) Portion of a seismic time slice interpreted as an underfit stream by Miall (2002; his fig. 6), Pleistocene deposits in the Malay basin, Gulf of Thailand. (**b**) Modern underfit Murrumbidgee channel (70 m wide) and ancient channel (250 m wide), New South Wales, Australia (seismic slice provided by A. Miall).

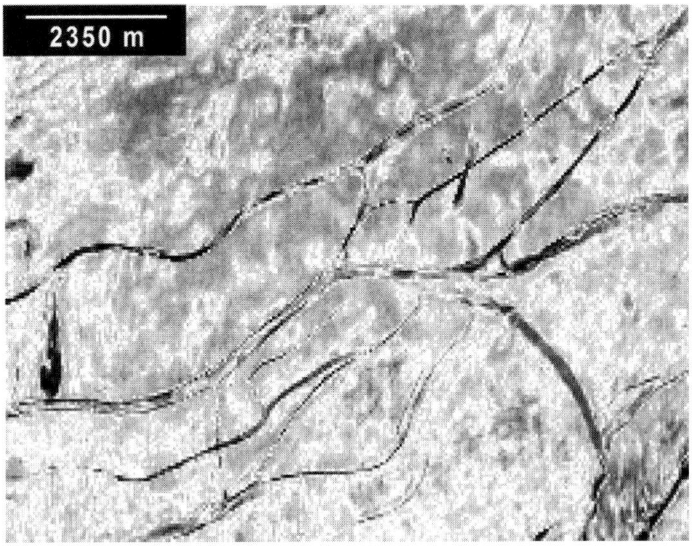

Fig. 17. A proportional slice from the West Naturna basin, Indonesia showing inferred anastomosing channels that are tens of metres wide each. This proportional slice was taken at approximately 1436 ms and represents 4 ms of time thickness. The average velocity for a nearby well for this interval is about 1700 m s^{-1}, which equates to approximately 6–7 m of resolution. Frequency for the volume of data ranges from 10–70 Hz and averages 53Hz. As is the case with the modern King river (see Fig. 10b above) all channels shown were probably not active at the same time. Seismic slice provided by Wood (2005; data provided by Conoco-Phillips).

difference in topography reflects differences in the lithologies of the subcropping units. Sinuous plan view channel forms are recognized on the surface in the southeast portion of the study area. Incised valleys in the NW portion are 1 km wide, 6–7 km long, and show a palaeo-relief of 400 m, providing a minimum value for the sea-level drop associated with this unconformity.

In summary, most of the basic river types, meandering, braided and straight, have been recognized and described using 3D seismic slices. Both equiwidth and wide-bend sinuous channels are clearly visible in published 3D seismic slices, but the distinction made by Brice between these channel types has not been recognized or discussed. Anastomosing channels have been described from seismic slices in the West Natuna basin, Indonesia (Fig. 17; Wood, pers. comm.). Underfit streams have been recognized, but their significance in the evolution of specific ancient fluvial systems has not been discussed. Examples of the possible local effects of active deformation on stream sinuosity have also been documented. Posamentier (2001) and Miall (2002) used multiple 3D slices to elucidate the evolutionary history of Pleistocene sequences in Java shelf and Malay basin respectively. Carter (2003) documents the evolution of four, Oligocene–Miocene meanderbelt sandstone packages at Widuri Field in the Java Sea and correlates this evolution to reservoir geometry, characteristics and hydrocarbon production drive. Martinez *et al.* (2004) provide unusually high detail regarding the character of the late Miocene Messinian unconformity in the Mediterranean and suggest reasons for the erosional and depositional processes associated with that base-level change. Clearly we are in the beginning stages of a revolution in our ability to visualize and interpret the evolution of ancient fluvial systems.

Conclusions

Understanding how and why geomorphologists have classified modern fluvial systems is important in developing a methodology for describing and interpreting ancient fluvial systems that are visualized in 3D seismic slices. It has long been recognized that there are at least four types of modern rivers: braided, meandering, straight and anastomosing. Schumm (1977) expanded this classification and recorded five types of bedload and mixed-load streams and three types of suspended load streams based on valley slope and sediment load. Brice (1982) examined the degree and type of sinuosity, braiding and anastomosing in a data set of aerial photographs and developed a classification that reflects plan view geometry and lateral stability. Nanson and Knighton expanded Brice's classification by recognizing five types of anastomosing streams based on plan view morphology. Schumm (2005) suggests that overall river pattern is set by upstream controls including history, relief and tectonics, rock type and hydrology and that a variety of local and downstream controls cause longitudinal variability. Changes of any of these variables cause channel adjustment through time.

A review of published 3D seismic literature reveals that most modern river types have been recognized in time/horizon slices and these data provide information for quantifying fluvial sandstone reservoirs and for assessing the evolution of ancient fluvial systems laterally and through time. Because several controls can produce the same effect (convergence) and one control may produce different effects (divergence), unambiguous interpretations are not possible (Schumm 1991). Reasonable hypotheses, however, can be generated based upon transference from modern river characteristics and controls to 3D seismic images. It is clear that we are in the beginning stages of a major enhancement of our ability to characterize and interpret ancient fluvial deposits in the subsurface.

Thanks to Owen Ayres & Associates Inc., Fort Collins, CO and especially B. Spitz for supplying digital images of some of aerial photos from NCHRP Report 533. Thanks to L. Wood for encouraging us to write this review paper, and for providing several digital images used in this paper. Thanks to H. Posamentier for his encouragement and for providing digital versions of his publications and papers in press. Thanks to A. Miall for supplying 3D seismic slice used here and color images of some of his illustrations from the 2002 paper that were used in the oral presentation of this paper. Thanks to M. Church for supplying the digital image of the photo of the anastomosing Bella Colla River in British Columbia. The paper has been significantly improved by helpful critiques by reviewers M. Blum and J. Peakall and co-editor L. Wood.

References

ASLAN, A. & BLUM, M. D. 1999. Contrasting styles of Holocene avulsion, Texas coastal plain. *In*: SMITH, N. D. & ROGERS, J. J. (eds) *Current Research in Fluvial Sedimentology. Proceedings of the 6th International Conference on Fluvial Sedimentology.* International Association of Sedimentologists, Special Publications, **28**, 193–209.

BLUM, M. D. & TÖRNQVIST, T. E. 2000. Fluvial response to climate and sea-level change: a review and look forward. *Sedimentology*, **47** (supplement 1), 2–48.

BLUM, M. D., MORTON, R. A. & DURBIN, J. M. 1995. 'Deweyville' terraces and deposits of the Texas Gulf Coastal Plain. *Gulf Coast Association of Geological Societies Transactions*, **XLV**, 53–60.

BRICE, J. C. 1975. *Air photo interpretation of the form and behavior of alluvial rivers*. Final Report to the U.S. Army Research Office, Durham.

BRICE, J. C. 1981. *Stability of relocated stream channels*. Federal Highway Administration Report no. FHWA/RD-80/021, Washington, DC.

BRICE, J. C. 1982. *Stream channel stability assessment*. Federal Highway Administration Report no. FHWA/RD-82/021, Washington, DC.

BRICE, J. C. 1983. Planform properties of meandering rivers. *In*: ELLIOTT, C. M. (ed.) *River Meandering*. American Society of Civil Engineers, New York, 1–15.

BRIDGE, J. S. & TYE, R. S. 2000. Interpreting the dimensions of ancient fluvial channel bars, channels, and channel belts from wire–line logs and cores. *American Association Petroleum Geologists Bulletin*, **84**, 1205–1228.

BROWN, A. R. 1985. The role of horizontal seismic sections in stratigraphic interpretation. *In*: BERG, O. R. & WOOLVERTON, D. G. (eds), *Seismic Stratigraphy* II. American Association of Petroleum Geologists Memoirs, **42**, 63–74.

BROWN, A. R. 1991. *Interpretation of Three-dimensional Seismic* Data, 3 ed. American Association of Petroleum Geologists Memoirs, **42**.

BURNETT, M. 1996. 3D seismic expression of a shallow fluvial system in west central Texas. *In*: WEIMER, P. & DAVIS, T. L. (eds) *AAPG Studies in Geology no. 42 and SEG Geophysical Development Series 5*. American Association of Petroleum Geologists/Society for Exploration Geophysics, Tulsa, OK, 45–56.

CARTER, D. C. 2003. 3D seismic geomorphology: insights into fluvial reservoir deposition and performance, Widuri field, Java Sea. *American Association of Petroleum Geologists Bulletin*, **87**, 909–934.

CORNISH, F. G. 1984. Fluvial environments and paleohydrology of Upper Morrow 'A' (Pennsylvanian) meanderbelt sandstone, Beaver County, Oklahoma. *Shale Shaker*, **34**, 70–80.

DEPTUCK, M. E., STEFFENS, G. S., BARTON, M. & PIRMEZ, C. 2003*a*. Architecture and evolution of upper fan channel-belts in the Niger Delta slope and in the Arabian Sea. *Marine and Petroleum Geology*, **20**, 649–676.

DEPUCK, M. E., MACRAE, R. A., SHIMELD, J. W., WILLIAMS, G. L. & FENSOME, R. A. 2003*b*. Revised Upper Cretaceous and lower Paleogene lithostratigraphy and depositional history of the Jeanne d'Arc Basin, offshore Newfoundland, Canada. *American Association of Petroleum Geologists Bulletin*, **87**, 1459–1483.

EBANKS, W. J. JR. & WEBER, J. F. 1987. Shallow heavy-oil deposit in a Pennsylvanian fluvial sandstone reservoir. Eastburn field, Missouri. *In*: MEYER, R. F. (ed.) *Exploration for Heavy Crude Oil and Natural Bitumen*. American Association of Petroleum Geologists Studies in Geology, **25**, 327–339.

ETHRIDGE, F. G., WOOD, L. J. & SCHUMM, S. A. 1998. Cyclic variables controlling fluvial sequence development: Problems and perspective. *In*: SHANLEY, K. W. & MCCABE, P. J. (eds) *Relative Role of Eustasy, Climate, and Tectonism in Continental Rocks*. SEPM (Society for Sedimentary Geology) Special Publications, **59**, 17–29.

ETHRIDGE, F. G., SKELLY, R. L. & BRISTOW, C. S. 1999. Avulsion and crevassing in the sandy, braided

Niobrara River: complex response to base-level rise and aggradation. *In*: SMITH, N. D. & ROGERS, J. J. (eds) *Fluvial Sedimentology VI*. International Association of Sedimentologists Special Publications, **28**, 179–191.

ETHRIDGE, F. G., GERMANOSKI, D., SCHUMM, S. A. & WOOD, L. J. 2005. The morphologic and stratigraphic effects of base-level change: a review of experimental studies. *In*: BLUM, M. D., MARIOTT, S. B. & LECLAIR, F. (eds) *Fluvial Sedimentology VII*. International Association of Sedimentologists Special Publications, **35**, 213–241.

FACHMI, M. & WOOD, L. J. 2003. Quantitative seismic geomorphology of belanak and Gabus Fields, West Natuna Basin, Indonesia. *2003 Annual Meeting: American Association of Petroleum Geologists Program and Abstract Volume*, A51–A52.

FISK, H. N. 1944. *Geologic investigations of the alluvial valley of the lower Mississippi River*. US Army Corps of Engineers, Mississippi River Commission, Vicksburg, Mississippi.

GERMANOSKI, D. 1990. Laboratory study of braided channel response to variable rates of base-level change. *Transactions of the American Geophysical Union*, **71**, 513.

HARDAGE, B. A., LEVEY, R. A., PENDLETON, V., SIMMONS J. & EDSON, R. 1994. A 3D seismic case history evaluating fluvially deposited thin-bed reservoirs in a gas-producing property. *Geophysics*, **59**, 1650–1665.

HOLLARD, J. D. 1973. Airphoto interpretation of fluvial features. *Ninth Canadian Hydrology Symposium*, University of Alberta, Edmonton, 1–35.

ISA, Z. M., RICHARDS, F. W. & YUNUS, H. 1992. Integration of 3D and site survey seismic data in analysis of near-surface hazards to platform location at Dulang field, Malay Basin. *Geological Society of Malaysia Bulletin*, **32**, 165–184.

JONES, L. S. & SCHUMM, S. A. 1999. Causes of avulsion: an overview. *In*: SMITH, N. D. & ROGERS, J. J. (eds) *Fluvial Sedimentology VI*. International Association of Sedimentologists Special Publications, **28**, 171–178.

KOLLA, V., BOURGES, Ph., URRUTY, J.-M. & SAFA, P. 2001. Evolution of deep-water Tertiary sinuous channels offshore Angola (West Africa) and implication for reservoir architecture. *American Association of Petroleum Geologists*, **85**, 1373–1405.

KOSS, J. E., ETHRIDGE, F. G. & SCHUMM, S. A. 1994. An experimental study of the effects of base-level change on fluvial, coastal plain and shelf systems. *Journal of Sedimentary Research*, **B64**, 90–98.

LAGASSE, P. F., SPITZ, W. J., ZEVENBERGEN, L. W. & ZACHMANN, D. W. 2004. *Handbook for Predicting Stream Meander Migration*. National Cooperative Highway Research Program Reports, **533**.

LEEDER, M. R. & STEWART, M. D. 1996. Fluvial incision and sequence stratigraphy: alluvial response to relative sea-level fall and their detection in the geologic record. *In*: HESSELBO, S. P. & PARKINSON, D. N. (eds) *Sequence Stratigraphy in the British Isles*. Special Publications of the Geological Society, **103**, 25–39.

MCCARTHY, T. S., ELLERY, W. N. & STANSTREET, I. G. 1992. Avulsions mechanisms on the Okavango fan, Botswana: the control of a fluvial system by vegetation. *Sedimentology*, **39**, 779–796.

MAKASKE, B. 2001. Anastomosing rivers: a review of their classification, origin and sedimentary products. *Earth-Science Reviews*, **53**, 149–196.

MARTINEZ. K. F., CARTWRIGHT, J. A., BURGESS, P. M. & BRAVO, J. V. 2004. 3-D seismic interpretation of the Messinian unconformity in the Valencia basin. Spain. *In*: DAVIES, R. J. CARTWRIGHT, J. A. STEWART, S. A, LAPPIN, M. & UNDERHILL, J. R. (eds) *3-D Seismic Technology: Applications to the Exploration of Sedimentary Basins*. The Geological Society Memoirs, **29**, 91–100.

MIALL, A. D. 1977. A review of the braided river depositional environment. *Earth Science Reviews*, **13**, 1–62.

MIALL, A. D. 2002. Architecture and sequence stratigraphy of Pleistocene fluvial systems in the Malay Basin, based on seismic time-slice analysis. *American Association of Petroleum Geologists Bulletin*, **86**, 1201–1216.

NANCE, G. C. & KNIGHTON, A. D. 1996. Anabranching rivers: their cause, character and classification. *Earth Surface Processes and Landforms*, **21**, 217–239.

NORTH, C. P. 1996. The prediction and modeling of subsurface stratigraphy. *In*: CARLING, P. A. & DAWSON, M. R. (eds) *Advances in Fluvial Dynamics and Stratigraphy*. Wiley, Chichester, 395–508.

PADGETT, G. & EHRLICH, R. 1976. Paleohydrologic analysis of a late Carboniferous fluvial system, southern Morocco. *Geological Society of America Bulletin*, **86**, 1101–1104.

PEAKALL, J., LEEDER, M., BEST, J. & ASHWORTH, R. 2000. River response to lateral ground tilting: a synthesis and some implications for the modeling of alluvial architecture in extensional basins. *Basin Research*, **12**, 413–426.

POSAMENTIER, H. W. 2000. Seismic stratigraphy into the next millennium; a focus on 3D seismic data. *American Association of Petroleum Geologists Annual Conference*, A118.

POSAMENTIER, H. W. 2001. Lowstand alluvial bypass systems: incised vs. unincised. *American Association of Petroleum Geologists Bulletin*, **85**, 1771–1793.

POSAMENTIER, H. W. 2002. Sequence stratigraphy, past, present and future and the role of 3-D seismic data. *Twenty second Annual Gulf Coast Section SEPM Foundation Bob F. Perkins Research Conference*, 37–54

POSAMENTIER, H. W. 2004. Seismic geomorphology: imaging elements of depositional systems from shelf to deep basin using 3D seismic data: implications for exploration and development. *In*: DAVIES, R. J., CARTWRIGHT, J. A., STEWART, S. A., LAPPIN, M. & UNDERHILL, J. R. (eds) *3-D Seismic Technology: Application to the Exploration of Sedimentary Basins*. Geological Society Memoirs, **29**, 11–24.

POSAMENTIER, H. W. 2005. Application of 3D seismic visualization techniques for seismic stratigraphy, seismic geomorphology, and depositional systems analysis: examples from fluvial to deep-marine depositional environments. *In*: DORE, A. G. & VINING, B. A. (eds) *Petroleum Geology: North-West Europe and Global Perspectives – Proceedings of the 6th Petroleum Geology Conference*. Geological Society, London, 1565–1576.

POSAMENTIER, H. W. & KOLLA, V. 2003. Seismic geomorphology and stratigraphy of depositional elements in deep-water settings. *American Association of Petroleum Geologists Bulletin*, 71, 367–388.

POSAMENTIER, H. W., MEASURING, WISEMAN, P. S. & PLAWMAN, T. 2000. Deep water depositional systems–ultra-deep Makassar Strait, Indonesia. *In*: WEIMER, P., SLATT, R. M., COLEMAN, J., ROSEN, N. D., NELSON, H., BOUMA, A. H., STYZEN, M. J. & LAWRENCE, D. T. (eds) *Deep-water Reservoirs of the World. Gulf Coast Society of Society of Economic Paleontologists and Mineralogists Foundation, 20th Annual Research Conference*, 806–816.

SAUCIER, R. T. 1994. *Geomorphology and Quaternary geologic history of the Lower Mississippi Valley*. Mississippi River Commission, Vicksburg, MS.

SAUDI ARABIAN OIL COMPANY. 1996. Impact of 3-D seismic data on development drilling, Ghinah and Umm Jurf fields, central Saudi Arabia. *In*: WEIMER, P. & DAVIS, T. L. (eds) *AAPG Studies in Geology 42 and SEG Geophysical Development Series 5*, 45–56 (compiled by SIMMS, S. C.).

SCHUMM, S. A. 1977. *The Fluvial System*. Wiley, New York.

SCHUMM, S. A. 1991. *To Interpret the Earth: Ten Ways to be Wrong*. Cambridge University Press, Cambridge.

SCHUMM, S. A. 1993. River responses to base-level change. *Journal of Geology*, **101**, 279–294.

SCHUMM, S. A. 2005. *River Variability and Complexity*. Cambridge University Press, Cambridge.

SCHUMM, S. A. & GALAY, V. J. 1994. The river Nile in Egypt. *In*: SCHUMM, S. A. & WINKLEY, B. R. (eds) *The Variability of Large Alluvial Rivers*. American Association of Civil Engineers, 75–100.

SCHUMM, S. A., ERSKINE, W. D. & TILLEARD, J. W. 1996. Morphology, hydrology and evolution of the anastomosing Owens and King Rivers, Victoria, Australia. *Geological Society of America Bulletin*, **108**, 1212–1224.

SCHUMM, S. A., DUMONT, J. F. & HOLBROOK, J. M. 2000. *Active Tectonics and Alluvial Rivers*. Cambridge University Press, Cambridge.

SCHWARTZ, D. E. 1978. Hydrology and current orientation analysis of a braided to meandering transition: The Red River in Oklahoma and Texas, U.S.A. *In*: MIALL, A. D. (ed.) *Fluvial Sedimentology*. Canadian Society of Petroleum Geologists Memoirs, **5**, 231–256.

SHANLEY, K. W. & MCCABE, P. J. 1994. Perspectives on the sequence stratigraphy of continental strata. *American Association of Petroleum Geologists Bulletin*, **78**, 544–568.

STEFFENS, G. S., SHIPP, R. C., PRATHER, B. E., NOTT, J. A., GIBSON, J. L. & WINKER, C. D. 2004. The use of near-seafloor seismic data in deepwater exploration and production. *In*: DAVIES, R. J., CARTWRIGHT, J. A., STEWART, S. A., LAPPIN, M. & UNDERHILL, J. R. (eds) *3-D Seismic Technology: Applications to the Exploration of Sedimentary Basins*. The Geological Society Memoirs, **29**, 35–43.

SUMMERFIELD, M. A. 1985. Plate tectonics and landscape development on the African continent. *In*: MORISAWA, M. & HACK, J. (eds) *Tectonic Geomorphology*. Allen & Unwin, Boston, MA, 27–51.

SWANSON, D. C. 1976. *Meandering Stream Deposits*. Cygnet Group Inc.

WEIMER, P. & DAVIS, T. L. 1996. Applications of 3-D seismic data to exploration and production. *American Association of Petroleum Geologists Studies in Geology 42 and Society of Exploration Geophysicists Geophysical Developments Series 5*.

WHITE, T., WITZKE, B., LUDVIGSON, G. & BRENNER, R. 2005. Distinguishing base-level change and climate signals in a Cretaceous alluvial sequence. *Geology*, **33**, 13–16.

WOOD, L. J., ETHRIDGE, F. G. & SCHUMM, S. A. 1993. The effect of rate of base-level fluctuation on coastal plain, shelf, and slope depositional systems: an experimental approach. *In*: POSAMENTIER, H. W. ET AL. (eds) *Sequence Stratigraphy and Facies Associations*. International Associations. of Sedimentologists, Special Publications **18**, 43–53.

Early evolution of submarine channels offshore Angola revealed by three-dimensional seismic data

M. J. R. GEE[1]* & R. L. GAWTHORPE[1]

[1]*Basin and Stratigraphic Studies Group, School of Earth Atmospheric and Environmental Sciences, The University of Manchester, Manchester M13 9PL, UK (e-mail: rob.gawthorpe@manchester.ac.uk)*

**Present address: Lukoil Overseas Ltd, Moscow, 115035, Russia (e-mail: martinjrgee@yahoo.co.uk)*

Abstract: The deep-water subsurface offshore Angola is characterized by many linear, high-gradient submarine channels typically only tens of metres wide and deep. Larger channel systems (*c.* 3–5 km wide, <300 m deep) with highly sinuous channels at their bases are also common, although they appear to have evolved from small, linear, high gradient systems. Generally, such small linear channels become enlarged by sediment gravity flows and therefore are rarely preserved except in examples where avulsion occurs. These small linear systems are often associated with relatively continuous levees 1–3 km wide flanking the channel. Results presented here suggest that small, linear channels evolve from erosional lineations on the slope generated by large, infrequent turbidity currents. Results also indicate that linear, high-gradient channels also exhibit the most significant and distinctive geometry changes where there is complex topography, such as near salt structures. Sedimentary bodies associated with linear, high-gradient channels often deposit within slope depressions as discreet J- or S-shaped structures in plan view. The dominant control on these sedimentary bodies is interpreted to be seafloor gradient and topography. This paper examines a number of these relatively young channels in terms of their geometry, gradient, levee development and seismic facies. The results improve our ability to predict subsurface channel geometries and recognise key evolutionary trends.

The advance of three-dimensional (3D) seismic acquisition and visualization techniques has created new interest in sedimentary processes that move sediment from shallower to deeper water. Submarine channels have been recognized as the principal depositional element of large submarine fan systems and conduits for sediments being transported basinward (Damuth *et al.* 1988). The study of submarine channels has implications for understanding the origin of deep-water sediments and the development of successful strategies for deep-water hydrocarbons exploration.

Submarine channels often exhibit a wide range of geometries which develop in response to sealevel changes, sediment flux variations, tectonic forcing and climate change and therefore provide a potentially valuable record of ocean margin evolution. Some of the best examples have been described using seismic data from the Gulf of Mexico (Satterfield & Behrens 1990; Roberts and Compani 1996; Badalini *et al.* 2000; Beaubouef & Friedmann 2000), from offshore Angola (Kolla *et al.* 1998, 2001; Sikkema & Wojcik 2000; Mayall & Stewart 2000; Deptuck *et al.* 2003; Fonnesu 2003; Abreu *et al.* 2003; Gee *et al.* 2005), the Nile Delta, Egypt (Samuel *et al.* 2003) and offshore eastern Kalimantan, Indonesia (Posamentier 2003). The extent and evolution of small-scale submarine channels measuring tens of metres wide and deep, is not, in general, well known. The exceptional preservation of small, linear channels offshore Angola makes it an ideal location to study the initiation and evolution of submarine channels in a deep-water setting.

In this paper we use 3D seismic data to show the nature of small scale, linear channel systems. We use observations of a variety of small channels to build a simple model that describes how channels become progressively eroded by sediment gravity flows. Finally, we discuss the significance of these channels in the early evolution of submarine channels.

Geological setting

The Angola continental slope is affected by salt tectonics, which created a complex bathymetry

From: DAVIES, R. J., POSAMENTIER, H. W., WOOD, L. J. & CARTWRIGHT, J. A. (eds) *Seismic Geomorphology: Applications to Hydrocarbon Exploration and Production*. Geological Society, London, Special Publications, **277**, 223–235.
0305-8719/07/$15.00

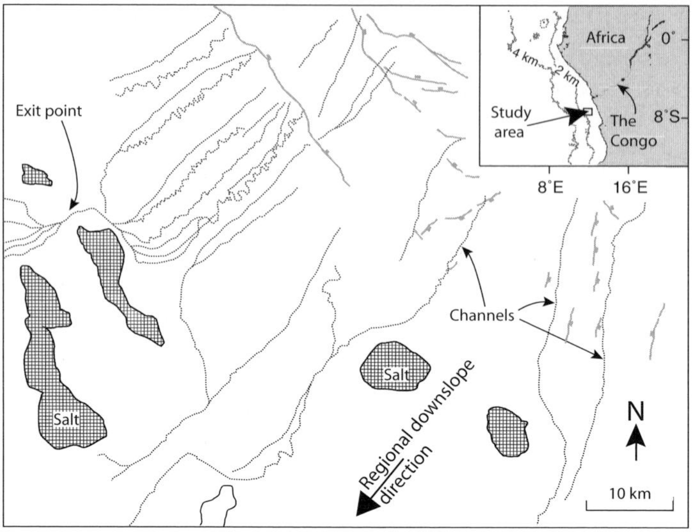

Fig. 1. Basemap of region showing the general network of channel systems within the study area (solid, black lines), faults (grey lines) and salt diapirs (hatchured). Downslope direction is towards the SW.

during the Late Miocene with abrupt gradient changes, slope depressions and barriers to sediment gravity flows travelling downslope. The study area is located 100 km off the Angola coast, in water depths of *c.* 1.5 km (Fig. 1). Margin uplift and a large river-fed sediment flux resulted in progradation of the Angola margin and deposition of the sand-rich Tertiary Congo Fan (Amaral *et al.* 1998; Mayall Stewart 2000; Sikkema & Wojcik 2000; Kolla *et al.* 2001; Lavier *et al.* 2001). During the Late Miocene, a large network of submarine channels transported sediment basinward (Fig. 1). Channel systems are mainly preserved within the upper 500 m of the strata. The channel systems are up to 200 m thick and tens of metres to >5 km wide, and typically sand-rich, confined systems (Abreu *et al.* 2003).

Highly sinuous, incised channels converge into salt-withdrawal depressions and pass through topographic constrictions in upstanding salt structures (Mayall & Stewart 2000). The typical elements of such channel systems may include channel axis, terraces, levees, crevasse splays, sheet-like sands, gravel-rich lag deposits and debris flow deposits. Two early Pliocene linear channel levee systems off West Africa were described by Fonnesu (2003). The channels are <250 m wide, flanked by seismically high amplitude levees of thin-bedded sandstones and associated with high amplitude lobes. Such channels are prone to avulsion and able to deposit sheet sandstones, lobes and form levees. Smaller channels tens of metres wide and tens of metres deep have also been described from the West Africa margin (Mayall & Stewart 2000).

Channels in the Angola margin are dissected by large growth faults trending NW–SE (Fig. 1), which extend to the current seafloor of the continental slope. These faults are related to deformation above a mobile salt layer which reaches thicknesses of *c.* 1000 m beneath the Angolan margin and has resulted in seaward rafting of large sections of stratigraphy (Lavier *et al.* 2001). The salt is Aptian in age and in many places has pierced through the overlying stratigraphy to the current seafloor (e.g. Uchupi 1992). These salt structures (diapiric domes and walls) have resulted in the formation of complex channel geometries, which are highly variable over small distances. Individual salt domes are typically 3–5 km across and are associated with flanking depressions, arcuate normal faults and deformed strata in the surrounding strata. Salt domes tend to affect the strata up to a radius approximately twice the diameter of the structure. On the Angolan margin, salt-related slope depressions typically form circular, downflexed moats which are often associated with a complex pattern of normal faults. Similar but larger minibasins are well known from the Gulf of Mexico and have been extremely efficient sediment traps, into which turbidity currents deposited often highly reservoir prone sands (Beauboeuf & Friedmann 2000 and refs within).

Methodology and database

Channel systems were interpreted from 3D seismic data using Schlumberger Geoframe 4 and Paradigm Geophysical VoxelGeo. The 3D seismic data were acquired for the purposes of oil and gas exploration off Angola and have an inline spacing of 12.5 m. For analysis, *xyz* data generated from the 3D seismic data were gridded at *c.* 15 m using GMT software (Wessel & Smith 1991). Where possible, two-way travel times are converted to metres using a reasonable seismic velocity (unpublished well data) of 2000 m s^{-1}. Channel incision was measured by fitting a least squares, spline curvature across the top of the channel margin and sampling the same points as used for the channel axis below. Incision was calculated as the vertical difference between the top and base of the channel. Long profile channel axis and incision profiles were sampled from the seismic grid at 50 m intervals.

Observations of channel systems

A range of high gradient, low sinuosity channel systems were interpreted in the study area (Fig. 2). These channels vary from simple narrow channel systems <50 m wide and 15 m deep (Fig. 2a), to more incised, sinuous systems >80 m deep with clearly defined levees (Fig. 2b–f).

Figure 2a shows four simple, linear channels which are weakly incised less than 15 m into the slope strata. They are associated with moderately high seismic amplitudes. Two of the channels join to form a tributary pattern. The eastern-most channel truncates the western channel and is therefore younger. Figure 2b shows a channel which has slight sinuosity, is about 30 m deep and 40–60 m across. It has well-developed, relatively continuous and high seismic amplitude levees that diminish in seismic amplitude and relief approximately 1–2 km from the channel axis. The channel axis is slightly sinuous where the channel gradients are reduced. Outside the main channel are downslope trending lineations and a minor channel. Figure 2c depicts two channels <50 m deep which join to form a tributary. The eastern tributary is more sinuous than the western tributary. Both channels appear to have channel axes which are located at the base of linear erosional conduits. Both channels are associated with well-developed and seismically high amplitude levees. The channel system shown in Figure 2d is affected by several small faults that appear to post date development of the channel. The channel width varies from <60 to>300 m and sinuosity from *c.* 1 to 1.5 along its length. Along the channel axis, sinuosity increases markedly in two places. Outside the valley containing the channel, there are some strongly defined and incised out-of-channel lineations. At the lower, southwestern part of the channel axis, the channel system enters a salt withdrawal mini-basin. The two channel systems shown by Figures 2e and f are similar in appearance. They both have valleys several hundred metres wide and channel depths of *c.* 100 m. The valleys appear to have a more irregular morphology compared with the other systems and both channels are associated with well developed and seismically high amplitude levees that exceed 50 m in thickness and extend for >5 km from the channel axis. Lineations running subparallel to the channel are observed within the main valley on both flanks of the slightly sinuous axis shown in Figure 2f. These are interpreted as terraces (see Discussion).

The smallest of the interpreted channel systems are regularly spaced apart (2–3 km) on the slope (Figs 2a and 3). The variance map shows four channels formed by southwest flowing turbidity currents (Fig. 3a). However, the amplitude response indicates the presence of fifth and sixth channel systems which have some minor relief (embryonic channels, Fig. 3d). One of these embryonic channels appears to terminate as an elongate lobe on the slope (Fig. 3d). In addition, the amplitude map (Fig. 3b) shows a finer-scale, downslope-oriented lineated fabric across much of the area. The channels are easily recognized by their relief, and high amplitude response across flanking levees which extend for 1–2 km either side of the channel axis. However, the tributary channel (Fig. 3b) has a low amplitude levee response. The seismic section through the channels shows the seismically high-amplitude levees with slight relief and some incision (<30 m). Some of the channel cross sections are asymmetrical.

More incised linear channel systems are associated with greater levee thicknesses and more complex channel valley morphology. Figure 4 shows one such channel where variations in channel gradient, width, depth and sinuosity appear to correlate with levee development. This channel system, despite having being subject to folding and faulting after formation has a well-preserved planview channel geometry (Fig. 4c). From approximately 9 km down the channel profile, channel depth shows a steady increase from <30 to >50 m. This corresponds with a clear increase in channel sinuosity and levee thickness (Fig. 4). Channel width increases upslope of the sinuous section but then decreases through the sinuous section where the channel is more incised.

Linear, high gradient channels on the Angolan margin exhibit important changes in geometry where they are associated with salt-withdrawal basins. Channel A (Figs 5 and 6), is characterized by a high gradient and a linear narrow channel axis which broadens and thickens as it passes

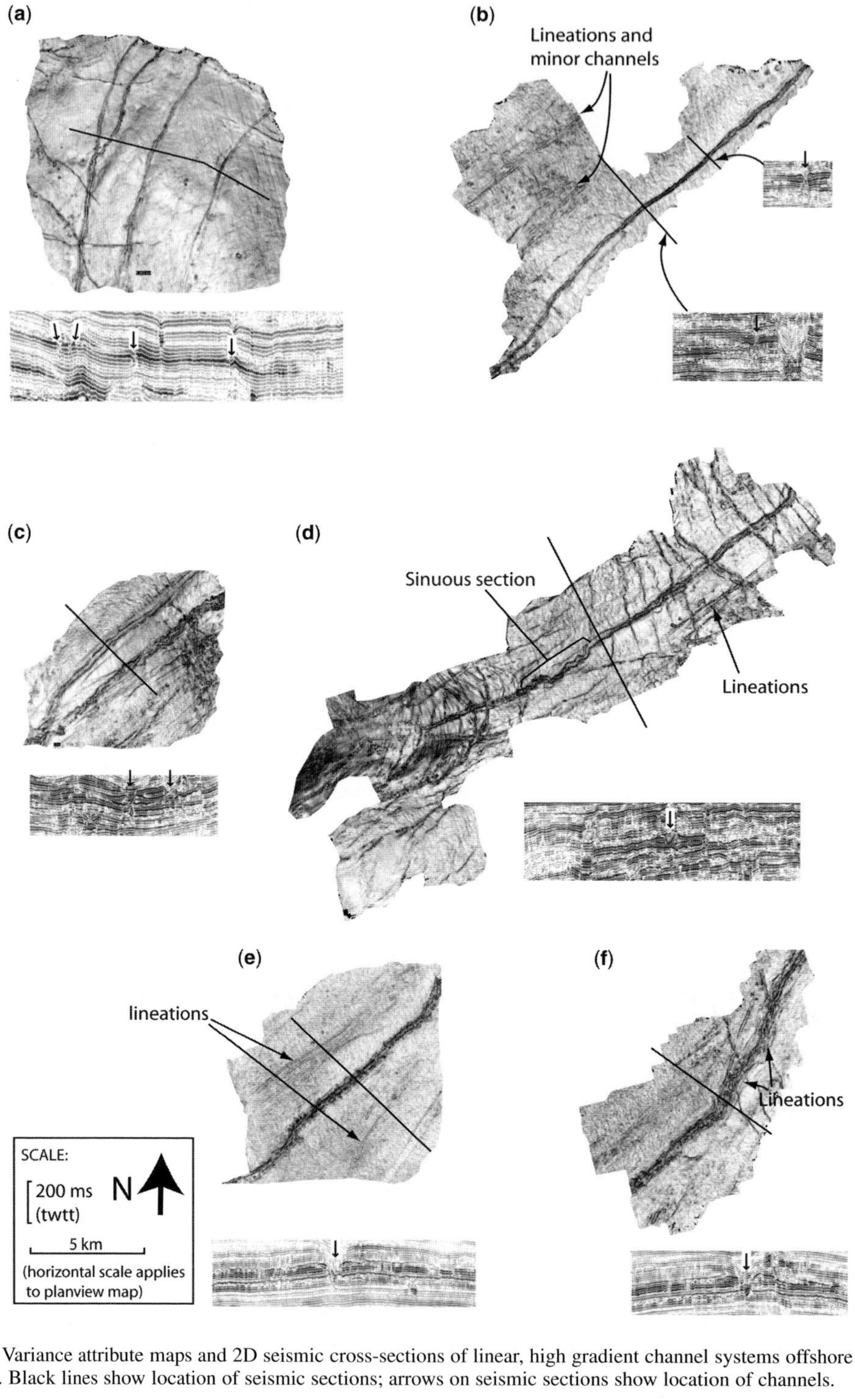

Fig. 2. Variance attribute maps and 2D seismic cross-sections of linear, high gradient channel systems offshore Angola. Black lines show location of seismic sections; arrows on seismic sections show location of channels.

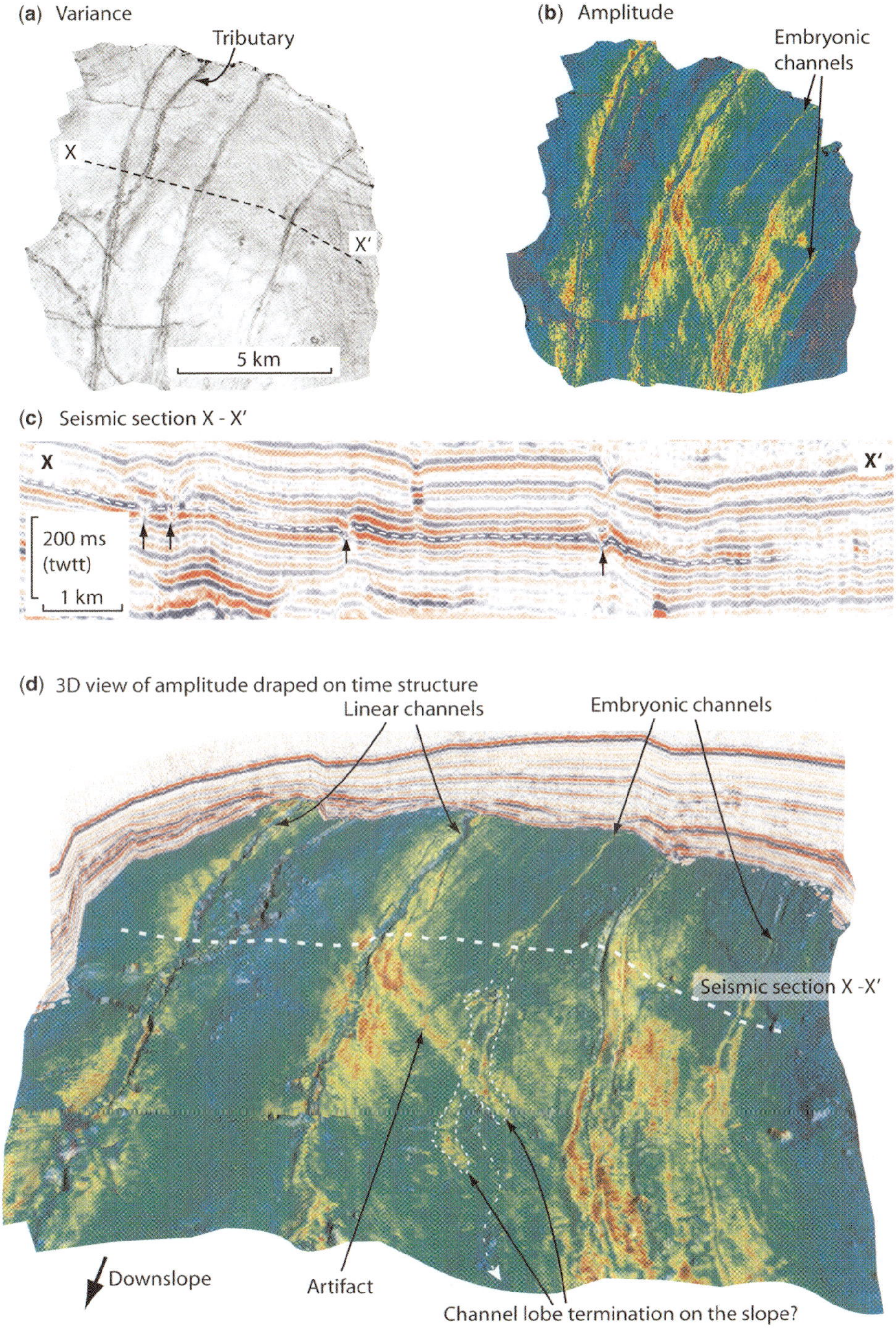

Fig. 3. Simple, linear, high gradient channel systems. (**a**) Variance attribute data showing four simple channels. (**b**) Amplitude attribute data, showing evidence for six simple channels, which include two which are not resolved using variance data. Levee systems associated with channels show high amplitudes red and yellow colours. (**c**) 2D seismic section of channel systems. Note that the two embryonic channel systems visible using amplitude data (b) are not resolved. (**d**) Amplitudes draped on time structure map to produce a 3D perspective view of channels.

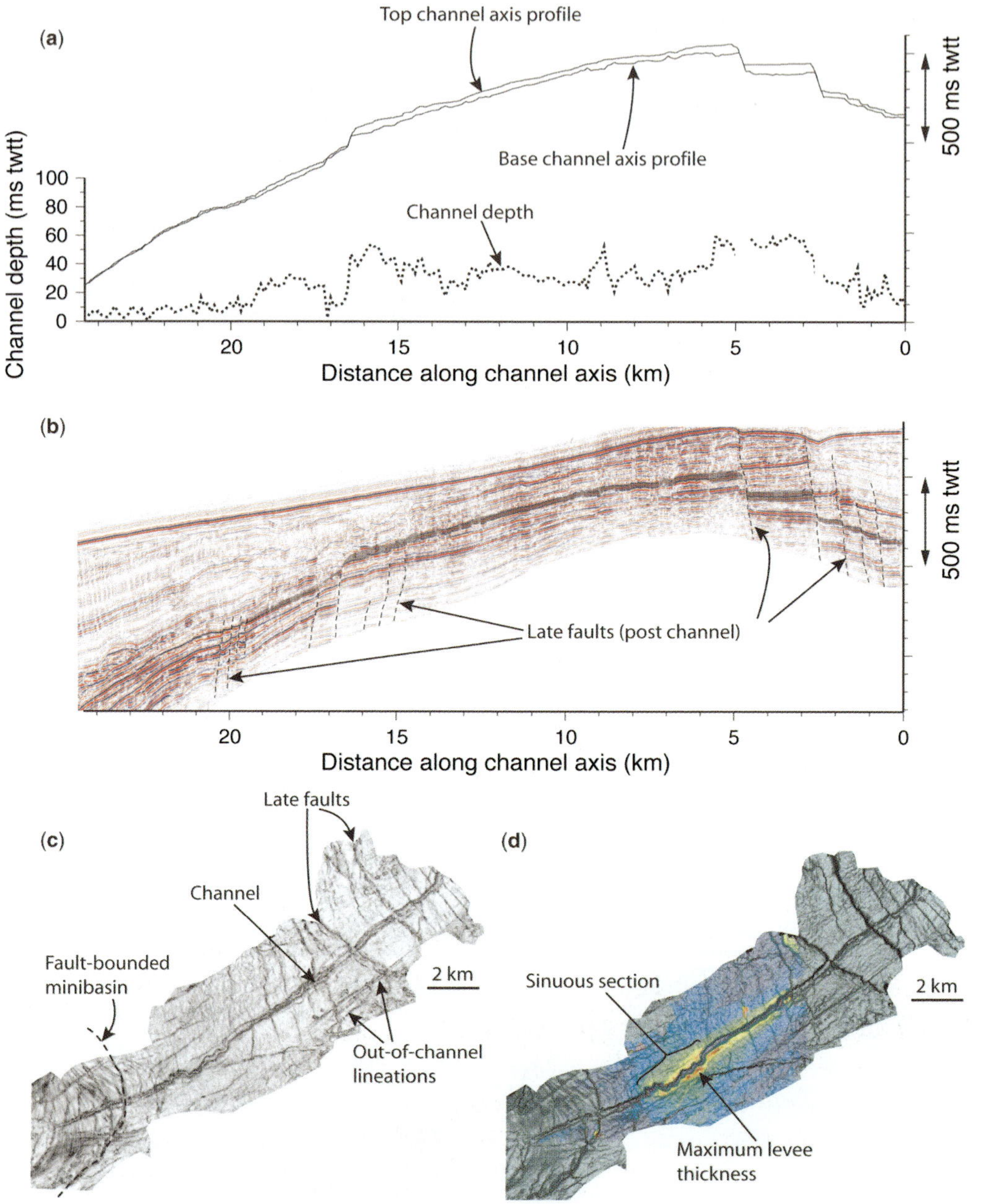

Fig. 4. Linear, high gradient channel system. (**a**) Profiles of channel axis and top of channel sampled every 50 m. Dotted line indicates channel depth along channel. (**b**) Seismic profile along channel axis profile. (**c**) Variance data of channel indicating position of late faults, channel geometry and out-of-channel lineations. (**d**) Time structure map with isochore map of channel interval draped (red and yellow, thicker intervals; blue, thin interval).

through a salt-withdrawal slope depression. Seismic cross sections show that the main axis of channel A is weakly incised (40 m) and has an irregular basal reflection (Fig. 5). There are several smaller incisional notches outside the main channel axis (Fig.5a, b). These smaller notches and the main channel axis correspond in plan view to high amplitude lineations oriented perpendicular to the

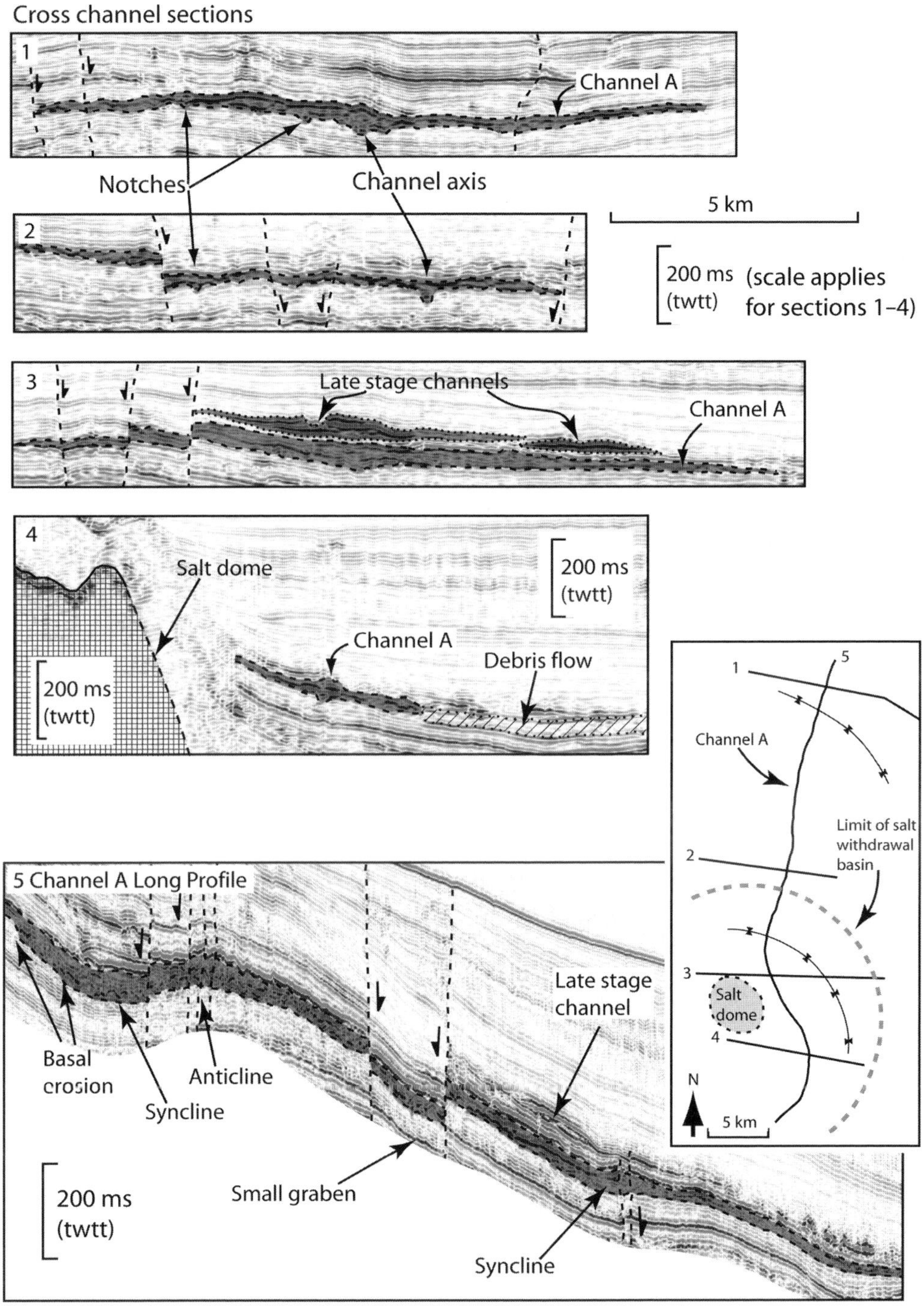

Fig. 5. Seismic cross sections (1–4) of a high gradient channel system. Seismic Section 5 = long profile sampled along the channel axis. See small inset for location. Channels are cut by later normal faults. Note the change in channel geometry from Section 1 to 4 as the channel approach the salt dome. Also see Fig. 6.

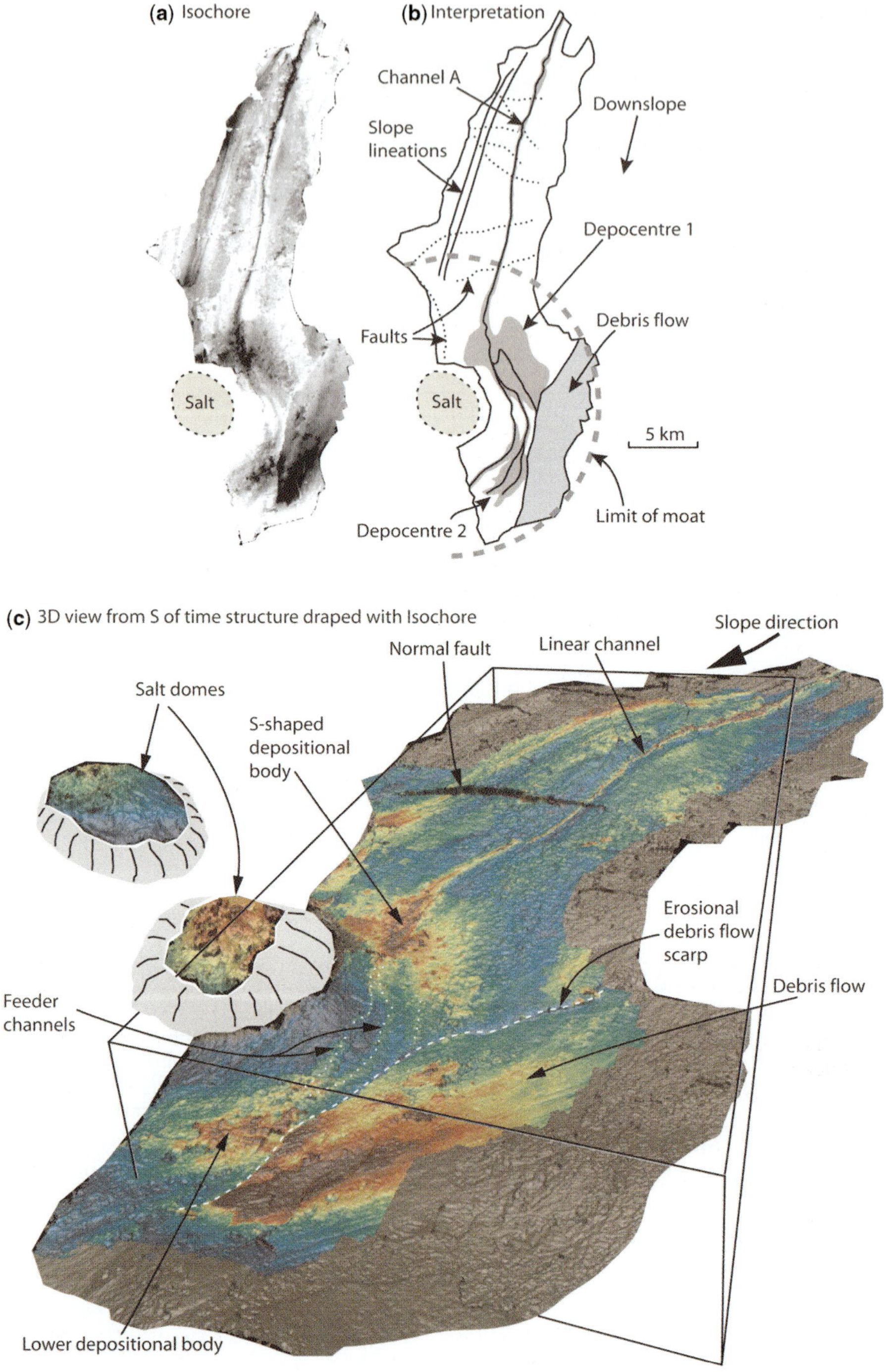

Fig. 6. Map of linear, high gradient channel system (Channel A, Fig. 5) in relation to a salt dome and related slope depression. (**a**) Isochore map of channel sediments (black = thicker sediments). (**b**) Interpretation of channel related features, showing planform geometry changes of channel system in the vicinity of a salt dome. Note the large debris flow which has eroded part of the channel system close to the salt dome. (**c**) 3D perspective view of time structure draped with the isochore of the channel system. Red and yellow colours indicate thicker sediments. Note the diversion of Channel A through the syncline rimming the salt dome and the two discrete depositional bodies and connecting avulsed feeder channels.

slope (Fig. 6a). Channel A, in the upper part of its reach, has a small, well-defined channel axis and its fill reaches a maximum thickness of *c.* 80 m (Fig. 5, Sections 1, 2). Outside the main channel axis, levee thicknesses are <40 m. Levee internal geometry is characterized by high, irregular amplitudes. Downslope, Channel A is more lens-shaped in cross section, gradually decreasing in thickness away from the channel axis midpoint (Fig. 5, Section 3). Farther down the slope, the basal reflector also becomes less irregular and lacks incised notches. A seismic section through the lower part of the salt-withdrawal basin (Fig. 5, Section 4) shows higher amplitudes; to the east a large debris flow has eroded part of the channel levee complex and underlying substrate. The channel is not currently located in the lowest point of the salt-withdrawal basin and has therefore been deformed by later salt movement. Two late-stage, aggradational channel systems with well-formed, high amplitude levees have developed above Channel A, although they appear to have only deposited sediment locally.

Measured from its upslope limit, Channel A has a high gradient over the first 8 km with two steps in its basal profile interpreted as erosional knick points (Fig. 5, Section 5). The channel geometry changes rapidly at *c.* 8 km, where amplitudes increase and sediments begin to thicken into the syncline. Within the syncline there is an asymmetrical deposit *c.* 2–3 km wide and <160 m thick. Exiting the syncline the channel appears to run upslope for a few kilometres, although this appearance is in fact due to post-channel folding (Fig. 5, Section 5). Passing out of the syncline, minor post-channel faults offset the channel sequence. As the channel is traced downslope beyond the anticline, the channel thickness and internal amplitude strength decreases. A small graben has displaced the channel axis post-channel formation. As the channel passes into the syncline depression surrounding the salt dome the channel thickness and internal seismic amplitudes increase dramatically. The late stage channel located above the Channel A sequence is characterized by high amplitude levees and appears to have developed in the low gradients of the syncline sorrounding the salt dome.

Figure 6 shows planview maps of Channel A with channel isochore data, interpretation, and a 3D view of the isochore map draped on the time structure map. It shows the linear, narrow channel system, and significant geometry changes as the channel enters the salt withdrawal depression surrounding the salt dome. Several small normal faults cut the channel, but appear to post-date channel development. Outside the channel there are a number of high amplitude and highly linear features on the slope running parallel to the main channel. Some of these appear to have slightly erosional bases with thin, linear overlying deposits. They are interpreted as embryonic channels and represent the earliest manifestation of channels on the slope imaged in the data. Channel A has a narrow channel axis in its upper reaches with a relatively high gradient. Where the channel encounters the lower gradients around the salt dome, important variations in channel geometry are observed. The channel increases markedly in width from <200 m to >4 km and also in thickness (Figs 5 and 6). This weakly confined section of channel forms a broad S-shaped depositional body in plan view and is limited to the NE sector of the salt-related slope depression. Extending from the lower edge of this S-shaped depositional body are two small channels which lead to a second smaller, possibly older, depositional body in the southern sector of the salt-related slope depression. These small channels are interpreted as avulsed feeder channels (Fig. 6b,c). A debris flow sourced from the NE has removed part of the channel-related sediments in the E and SE sectors of the salt-related slope depression (Fig. 6b,c).

Discussion

General slope model

Salt structures are shown on the lower slope and five different channel systems are illustrated. These channel systems are considered analogues for typical slope reservoir classes evolving on a slope in a high state of disequilibrium. Deposition is focussed in mini-basins and depositional lows. Abrupt changes in channel geometry are often observed as channels enter slope depressions flanking salt intrusions. These geometric changes in these potential deep water reservoir units have important implications for hydrocarbon exploration and development in deep-water slope settings. Turbidity currents appear to have flowed though such depressions without ponding.

The five styles of channel systems are based on observations of channel geometries in the study area (Fig. 7). Incised complexes (System 1) are common off Angola (Mayall & Stewart 2000; Kolla *et al.* 2001; Abreu *et al.* 2003). They can be several kilometres wide and 200–300 m deep. The fill of incised channel systems can consist of sand-rich, amalgamated channels and are therefore important deep water exploration targets (e.g. Abreu *et al.* 2003). Weakly confined complexes (system 2) are a little known, but important style of channel system. They form high-amplitude, broad depositional systems up to *c.* 5 km wide which can form stacked deposits within slope depressions and minibasins. Sinuous, leveed-channel complexes (System 3) are also recognized off Angola (Mayall & Stewart 2000). They can form deposits *c.* 5 km wide and

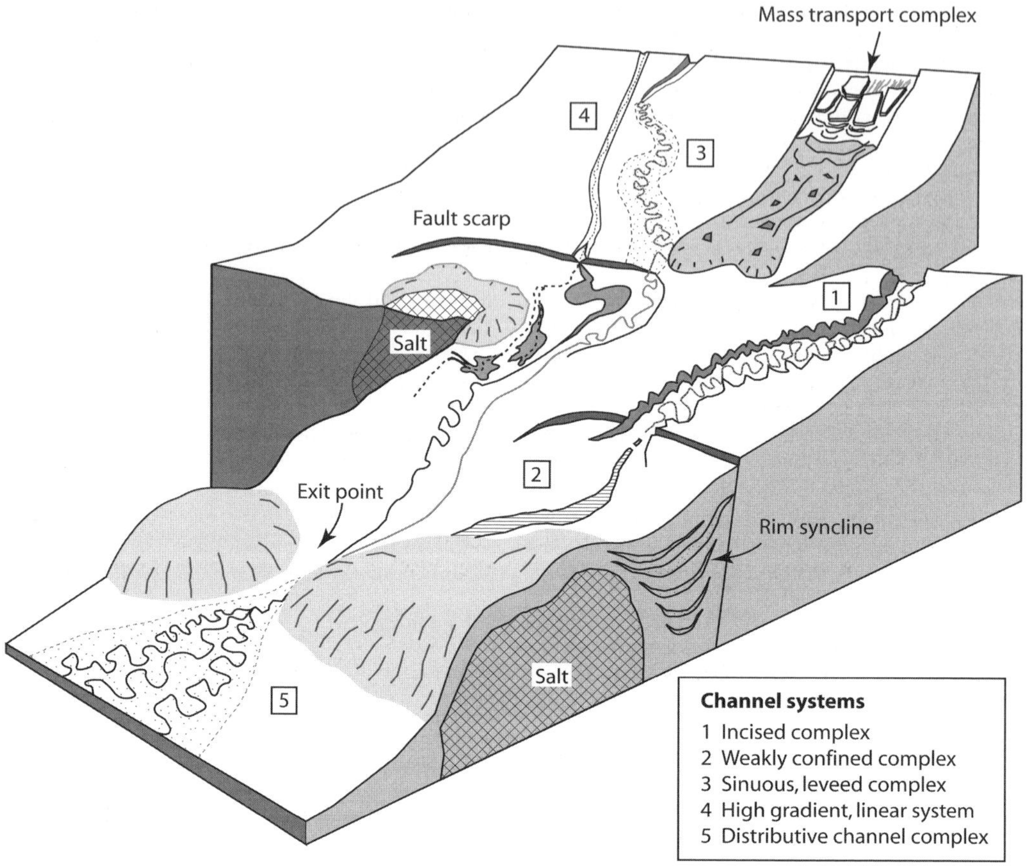

Fig. 7. Conceptual slope model illustrating range of channel types occurring within study area (after Gee & Gawthornpe 2006).

c. 200 m thick. High-gradient linear systems (System 4) have relatively small depositional volumes, but are numerous off Angola. They are typically a few tens of metres thick and can be associated with well-developed levees. These channels are often associated with abrupt transitions to larger depositional channels where they encounter slope depressions (Fig. 6). Distributive channel complexes (System 5) can occur where channels exit topographic constrictions. They may be several hundred metres thick and contain numerous sinuous channels which diverge and are associated with well-developed levees.

The flows that form the linear, high-gradient channels may be extremely sensitive to subtle changes in seafloor morphology compared with flows associated with sinuous and incised channels. Linear channels, which are typically less than 100 m wide, rapidly broaden and thicken as they approach the increased accommodation space and lower gradients provided by salt-withdrawal depressions. The resulting deposits can be up to several kilometres wide and tens of metres thick, having very unpredictable internal geometries. In planview, these deposits form distinctive S- or J-shaped morphologies. Several similar, but discrete, depositional bodies may form individually within slope depressions surrounding salt structures, recording lateral and radial shifts in early depositional facies. Such depositional systems are characterized by high amplitudes and form a recently and poorly defined slope reservoir class (often described as weakly confined channels) with relatively high net to gross sandstones (Mayall & Stewart 2000). The transitions between the two types of channel geometries described here (characterized by weaker seismic amplitudes) may result from erosion related to hydraulic jumps.

Evolution of straight and sinuous channels

Where channels have developed incisional valleys the planforms of channels, that appear responsible for the erosion, are often highly sinuous (e.g. Abreu *et al.* 2003, and this study). Even apparently linear channel systems are observed to have channel axes with low to moderate channel sinuosity. Straight channels observed in the study area appear to represent preserved immature channel systems characterized by narrow, thin channel sequences and higher gradients. A general model for the early evolution of submarine channels in a deep water setting is presented here, based on the range of linear, high gradient channels observed in the study area (Fig. 8; Gee *et al.* 2007). Stages 1 and 2 describe a simple channel 40 m wide and >15 m deep. These channels can form dense sub-parallel networks on the slope and in some cases appear to be associated with lobate depositional bodies which form on the slope. As channels become enlarged by flows the number of channels on the slope and their spacing decreases at a very early stage. Stages 3 and 4 typify the evolutionary stage where the channel broadens to around 60 m and a sinuous channel axis develops. Levees continue to grow laterally and vertically due to overbanking flows, often developing relatively continuous, relatively high seismic amplitude, regions along both channel flanks. Around this stage the valley containing the channel axis starts to develop a more complex geometry as the valley walls become unstable due to undercutting of a laterally and vertically eroding channel axis. Stages 5 and 6 describe a relatively mature channel system with a main channel *c.* 500 m wide and <100 m deep. Inner terraces may form due to intrinsic changes in flow depth and energy and considerable modification of the channel valley may result from the increasing confinement and focussed energy of turbidity currents. Channel sinuosity can reach quite high levels (2–3) and the out-of-channel flows erode lineations and grooves on levees and inter-channel regions.

Turbidity currents flowing in straight channels with higher gradients may be prone to surging (e.g. Friedmann & Beaubouef 2000), which means the rate of entrainment at the top of the flow is likely to be higher, leading to rapid flow thickening and channel overbanking (Parker *et al.* 1986; Pirmez 1994; Pirmez & Imran 2003). Straight channels may be

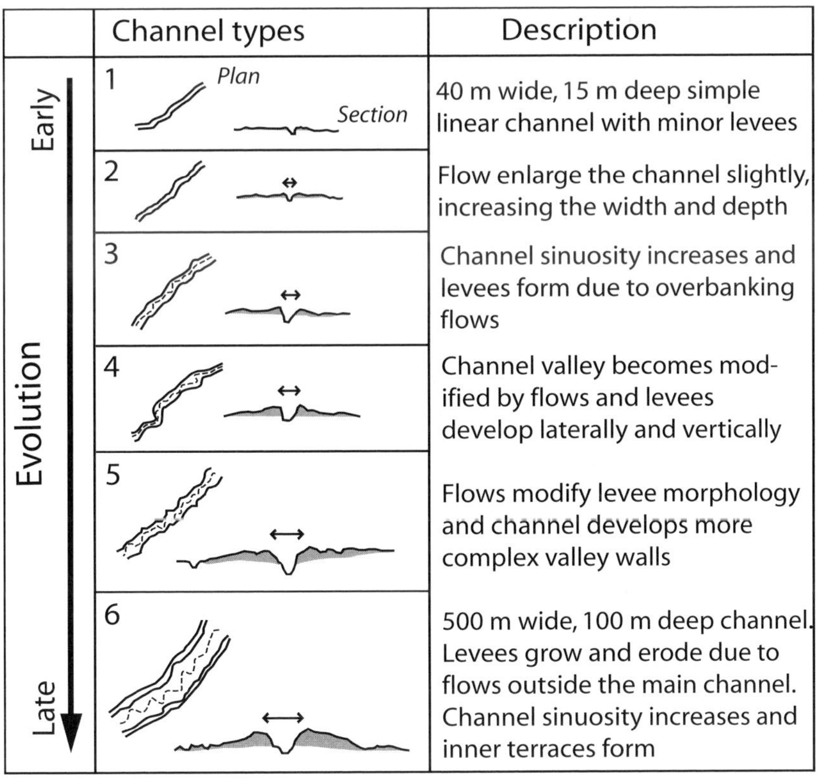

Fig. 8. Model illustrating the early evolution of a high gradient, low sinuosity immature system which describes the growth of a typical simple channel system (<40 m wide, <15 m deep) to a channel with developing sinuosity more complex geometry and an enlarged (500 m wide, 100 m deep), but still immature channel valley. Note the early onset of a sinuous channel axis and the growth and early erosion of channel levees.

preserved when they avulse. Without avulsion, sediment gravity flows would erode broader, deeper valleys in which a highly sinuous channel could develop.

On the Amazon Fan the base of channel sequences are associated with high amplitude, sheet like sands interpreted as base-of-levee depositional lobes (e.g. Flood & Damuth 1987). Young channel sequences reported here (Channel A, Fig. 6) have high amplitude, out-of-channel linear features that are comparable with the Amazon base-of-levee sands (see also Figs 2 and 4). These appear to result from turbidity currents which are not yet channel confined and are able to deposit frontal splays composed of sand- and silt-grade sediment. The linear, weakly incised channels associated with these systems are remarkable in that they appear truly straight over tens of kilometres. Potential analogues exist in the form of subaerial channels on uplifted and tilted slopes (Twidale 2004). These appear to develop though natural selection, with an initially high number of linear channels, of which only a few survive (Twidale 2004). A similar model for initiation and growth of submarine canyons has been suggested by Pratson & Coakley (1996). The observation here of a number of regularly spaced, out-of-channel lineations adjacent to a slightly larger linear channel (Figs 2 and 4) may form in a similar way to immature subaerial drainage patterns in unconsolidated subaerial steep substrates. Submarine lineations in close proximity to immature channels may indicate the passage of large unconfined turbidity currents capable of eroding the substrate. The exact mechanism by which a channel first establishes on the slope is not clear. Elliot (2000) described laterally extensive marine erosion surfaces associated with sandstone-filled channels from outcrop and speculated that channels were initiated by a single, high-magnitude, low-frequency turbidity current. Downslope oriented lineations (this study) have remarkably similar plan view geometries and locations to small established linear channels, indicating a common origin (Gee *et al.* 2007).

Studies of subaerial channels show that channels rarely maintain straight courses for more than ten or twenty times their channel width (Schumm & Khan 1972), unless influenced by regional structural control. Twidale (2004) reports that 'all (subaerial) channels are winding or sinuous' with even artificially straight channels such as canals tending to sinuosity. Submarine channels are observed to maintain straight courses several hundred times greater than channel width (this study), although the controls are not clear. This distance may be even greater since seismic resolution of the channel axis is limited in this study to *c.* 15 m vertically and tens of metres horizontally. Small offset faults which trend downslope are observed in the area, but they appear to have no relationship with linear channels.

Summary

The Angolan margin provides a unique opportunity to study the evolution of channel systems in areas of complex topography. High rates of deformation have resulted in immature channel systems being preserved in different stages of evolution. Abrupt increases in channel width, thickness and amplitude and planform geometry indicate that the flows that create high gradient, linear channel systems are very sensitive to seafloor topography compared to larger, more confined and aggradational channel systems. Large, low-frequency turbidity currents may be responsible for erosional lineations observed on levees and inter-channel regions. Some of these lineations appear to have developed into submarine channels whereas others clearly have not. Where channels encounter low gradients in salt-related slope depressions surrounding salt intrusions, abrupt transitions in channel geometry are observed. Channel structures in salt-related slope depressions (consisting of J- or Z-shaped morphologies in plan view) are deposited in discreet bodies and may be associated with avulsed feeder channels which record the lateral migration of sedimentary facies in relation to tectonic movements. A simple model of early submarine channel evolution which reconstructs temporal evolution using spatial observations can help to predict and understand complex and perhaps poorly resolved channel geometries from deep water reservoir sections.

We are grateful to Total Exploration and Production Angola, Norsk Hydro, Esso Exploration Angola (Block 17) Ltd, BP Exploration (Angola) Ltd, Statoil Angola Block 17 A.S. and Sonangol for permission to publish the 3D seismic data from offshore Angola. We particularly thank Norsk Hydro for supporting for this project, and the ongoing work at Petroleum Geoscience Centre, Manchester. Software used in this study was donated by Schlumberger and Paradigm Geophysical. We acknowledge G. Badalini at Manchester and C. Pirmez useful discussions about submarine channels and slope processes.

References

ABREU, V., SULLIVAN, M., PIRMEZ, C. & MOHRIG, D. 2003. Lateral accretion packages (LAPS): an important reservoir element in deep water sinuous channels. *Marine and Petroleum Geology*, **20**, 631–648.

AMARAL, J., BITEAU, J. J. & ZAROSLINSKA, P. 1998. Angola – the lower Congo basin Tertiary petroleum systems hydrocarbon distribution in relation with structural and sedimentary evolution (abstract). *AAPG International Conference and Exhibition, Extended Abstracts*, 924–925.

BADALINI, G. B. KNELLER, B. & WINKER, C. D. 2000. Architecture and processes in the late Pleistocene Brazos-Trinity turbidite system, Gulf of Mexico continental slope. *In*: WEIMER, P., SLATT, R. M. ET AL. (eds) *Deep-water Reservoirs of the World: Gulf Coast Section Society of Economic Palaeontologists and Mineralogists Foundation 20th Annual Research Conference*, Houston, TX, 16–34.

BEAUBOUEF, R. T. & FRIEDMANN, S. J. 2000. High resolution seismic/sequence stratigraphic framework for the evolution of Pleistocene intra slope basins, western Gulf of Mexico. *In*: WEIMER, P., SLATT, R. M. ET AL. (eds) *Deep-water Reservoirs of the World: Gulf Coast Section Society of Economic Paleontologists and Mineralogists Foundation 20th Annual Research Conference*, Houston, TX, 40–60.

DAMUTH, J. E., FLOOD, R. D., KOWSMANN, R. O., BELDERSON, R. H. & GORINI, M. A. 1988. Anatomy and growth pattern of Amazon Deep-Sea Fan as revealed by long-range side-scan sonar (GLORIA) and high-resolution seismic studies. *American Association of Petroleum Geology Bulletin*, **72**, 885–911.

DEPTUCK, M. E., STEFFENS, G. S., BARTON, M. D. & PIRMEZ, C. 2003. Architecture and evolution of upper fan channel belts on the Niger Delta slope and in the Arabian Sea. *Marine and Petroleum Geology*, **20**, 649–676.

ELLIOT, T. 2000. Megaflute erosion surfaces and the initiation of turbidite channels. *Geology*, 119–122.

FLOOD, R. D. & DAMUTH, J. M. 1987. Quantitative characteristics of sinuous distributary channels on the Amazon Deep-Sea Fan. *Geological Society of America Bulletin*, **98**, 728–738.

FONNESU, F. 2003. 3D seismic images of a low-sinuosity slope channel and related depositional lobe (West Africa deep-offshore). *Marine and Petroleum Geology*, **20**, 615–629.

FRIEDMANN, S. J. & BEAUBOUEF, R. T. 1999. Relationships between depositional processes, stratigraphy and salt tectonics in a closed intra-slope basin: E. Breaks area, Gulf of Mexico. *Extended Abstracts Volume, American Association of Petroleum Geologists Annual Conference*, A43.

GEE, M. R. J. & GAWTHROPE, R. L. 2006. Submarine channels controlled by salt tectonics: Examples from 3D seismic data offshore Angola. *Marine and Petroleum Geology*, **23,** 443–458.

GEE, M. J., GAWTHORPE, R. L., BAKKE, K. & FRIEDMANN, S. J. 2007. Seismic geomorphology and evolution of submarine channels from the Angolan continental margin. *Journal of Sedimentary Research*, in press.

KOLLA, V., BOURGES, P., URRUTY, J. M., CLAUDE, D., MORICE, M., DURAND, J. & KENYON, N. H. 1998. Reservoir architecture in recent and subsurface deep-water meandering channel and related depositional forms (abstract). *European Association of Geoscientists and Engineers/AAPG Third Research Symposium Extended Abstracts*, unpaginated.

KOLLA, V., BOURGES, P., URRUTY, J. M. & SAFA, P. 2001. Evolution of deep-water Tertiary sinuous channels offshore Angola (West Africa) and implications for reservoir architecture. *American Association of Petroleum Geologists Bulletin*, **85**, 1373–1405.

LAVIER, LUC. L., STECKLER, M. S. & BRIGAUD, F. 2001. Climatic and tectonic control on the Cenozoic evolution of the West African margin. *Marine Geology*, **178**, 63–80.

MAYALL, M. & STEWART, I. 2000. The architecture of turbidite slope channels. *In*: WEIMER, P., SLATT, R. M. ET AL. (eds) *Deep-water Reservoirs of the World: Gulf Coast Section Society of Economic Paleontologists and Mineralogists Foundation 20th Annual Research Conference*, Houston, TX, 578–586.

PARKER, G., FUKUSHIMA, Y. & PANTIN, H. M. 1986. Self-accelerating turbidity currents. *Journal of Fluid Mechanics*, **171**, 145–181.

PIRMEZ, C. 1994. *Growth of a submarine meandering channel levee system on the Amazon Fan*. Ph.D. thesis, Columbia University.

PIRMEZ, C. & IMRAN, J. 2003. Reconstruction of turbidity currents in a meandering submarine channel. *Marine and Petroleum Geology*, **20**, 823–849.

POSAMENTIER, H. W. 2003. Depositional elements associated with a basin floor channel–levee complex: case study from the Gulf of Mexico. *Marine and Petroleum Geology*, **20**, 677–690.

PRATSON, L. F. & COAKLEY, B. J. 1996. A model for the headward erosion of submarine canyons induced by downslope-eroding sediment flows. *Geological Society of America Bulletin*, **108**, 225–234.

ROBERTS, M. T. & COMPANI, B. 1996. Miocene example of a meandering submarine channel-levee system from 3D seismic reflection data, Gulf of Mexico Basin. *Gulf Coast Section Society of Economic Palaeontologists and Mineralogists Foundation Seventeenth Annual Research Conference*, 241–254.

SAMUEL, A., KNELLER, B., RASLAN, S., SHARP, A. & PARSONS, C. 2003. Prolific deep-marine slope channels of the Nile Delta, Egypt. *American Association of Petroleum Geologists Bulletin*, **87**, 541–560.

SATTERFIELD, W. M. & BEHRENS, W. E. 1990. A late Quaternary canyon/channel system, northwest Gulf of Mexico continental slope. *Marine Geology*, **29**, 51–67.

SCHUMM, S. A. & KHAN, H. R. 1972. Experimental study of channel patterns. *Geological Society of America Bulletin*, **83**, 1755–1770.

SIKKEMA, W. & WOJCIK, K. 2000. 3D visualization of turbidite systems, lower Congo Basin, Offshore Angola. *In*: WEIMER, P., SLATT, R. M. ET AL. (eds) *Deep-water Reservoirs of the World. Gulf Coast Section Society of Economic Paleontologists and Mineralogists Foundation 20th Annual Research Conference*, Houston, TX, 928–939.

TWIDALE, C. R. 2004. River patterns and their meanings. *Earth Science Reviews*, **67**, 159–218.

UCHUPI, E. 1992. Angola Basin: geohistory and construction of the continental rise. *In*: POAG, C. P. & DE GRACIANSKY, P. C. (eds) *Geologic Evolution of Atlantic Continental Rises*. Van Nostrand Reinhold, New York.

WESSEL, P. & SMITH, W. H. F. 1991. Free software helps map and display data. *EOS Transactions, AGU*, **72**, 441.

Seismic expression of turbidity-current and bottom-current processes on the Northern Mauritanian continental slope

A. M. SCHWAB, S. TREMBLAY & A. HURST

Department of Geology and Petroleum Geology, King's College, University of Aberdeen, Aberdeen AB24 3UE, UK (e-mail: a.schwab@abdn.ac.uk)

Abstract: The architectural framework of the Mauritanian continental slope is characterized by a complex mixture of gravity and bottom current deposits that modify the pelagic background sedimentation. Since the Neogene, the Mauritanian passive margin bottom-currents have been the main control on the construction and topography of the slope. On the lower slope, numerous sediment waves and contourites occur. Turbidity-current channel incisions and slope failures are responsible for the destructive remodelling of the slope. The upper-slope incisions have a dendritic pattern, and converge into highly sinuous main channels in the lower slope, where the channels become constructional with levees and lateral accretion packages. Subsurface data exhibit older slope failures, which guide the location of later turbidity-current channel pathways. The seabed and shallow seismic expression of deep-water slope deposits generated by variable gravity-transport and bottom-current processes in offshore Northern Mauritania can be compared with analogous settings in lower resolution, deeper subsurface seismic data to gain a better understanding of the slope environment and deposits.

The Late Miocene–Present Mauritanian deep-water muddy slope is not affected by tectonics, but is a depositionally complex margin (Tremblay 2005). This paper will document the seismic expression of the architectural elements that characterize the Mauritanian continental slope, i.e. the slope features deposited by bottom-currents and gravity-transport processes.

Bottom-current deposits mainly result from along- and upslope-flowing processes (i.e. contour-bottom currents; e.g. Stow *et al.* 2002), while gravity deposits result from all downslope processes (i.e. downslope turbidity-currents, mass-wasting and settling suspension; e.g. Stoker *et al.* 1998). Settling suspension leads to the deposition of a muddy/silty carpet of hemipelagic-pelagic sediment on the entire slope (e.g. Stow & Piper 1984). On seismic, these deposits are characterized by an homogeneous seismic facies with a concordant reflection configuration that drapes or onlaps slope features. Settling suspension deposits will not be addressed in this paper.

The study area is located offshore along the Northern Mauritanian passive margin, comprising a continuous transition from shelf (<200 m water depth) to upper slope (200–1300 m water depth) to middle slope (1300–1800 m water depth) to lower slope (+1800 m water depth) environments (Fig. 1). The Northern Mauritanian continental shelf extends over 100–120 km from the coast and is relatively flat. The upper slope is steepest (3–4°), grading into an average of 2.6° in the middle slope and 1–2° in the lower part of the middle slope.

Several incisions cut the slope seabed in the study area and exhibit dendritic patterns. Thirty-nine major tributary incisions occur along the shelf-break and upper slope, and converge into 17 tributary incisions on the middle-lower slope (Fig. 2). The Arguin tributary systems are located in the northern part of the study area and drain into the Cape Verde Abyssal Plain (Fig. 1b; Wynn *et al.* 2000). The Timiris tributary systems are located in the southern part of the study area and drain into the Gambia Abyssal Plain (Fig. 1b; Wynn *et al.* 2000). The Senegal River is the only currently active fluvial system in this region and is located over 300 km south of the study area (Fig. 1b). All of the currently active wadis on the continent adjacent to the study area end in sabkhas (Fig. 1b; Elouard 1975) and could not have connected to shelf and slope seabed channels since the Tchadian wet period (*c.* 10 ka; Elouard 1975).

The two- (2D) and three-dimensional (3D) reflection seismic examples shown in this paper are from the southern part of the study area (tan boxes in Fig. 2), and occur in water depths of 300–2000 m. These Late Miocene–Present slope features occur on the seabed and in the shallow subsurface (<1 s, TWT), and thus provide very good quality seismic examples (*c.* 60 Hz). The general slope architectural elements illustrated in this paper are: a tributary channel system, channel-levees, slope failures, sediment waves and contourites. Seismic examples

From: DAVIES, R. J., POSAMENTIER, H. W., WOOD, L. J. & CARTWRIGHT, J. A. (eds) *Seismic Geomorphology: Applications to Hydrocarbon Exploration and Production*. Geological Society, London, Special Publications, **277**, 237–252.
0305-8719/07/$15.00

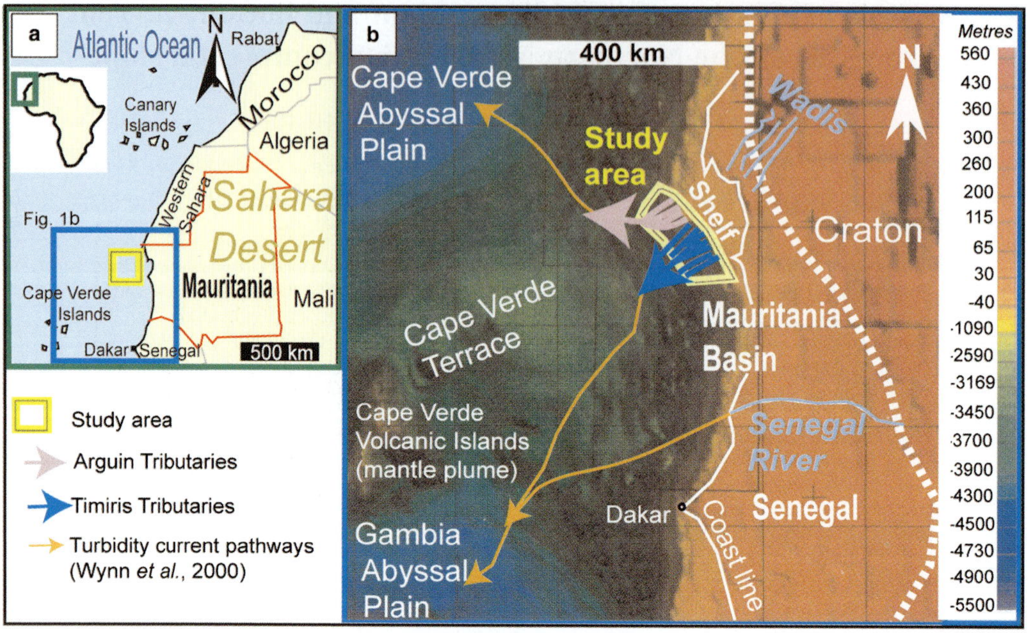

Fig. 1. Location map and regional bathymetric map of offshore Mauritania: (**a**) the yellow block shows the location of the study area. (**b**) The yellow outline shows the study area and the Arguin (violet) and Timiris (dark blue) tributary slope systems. Bathymetric image is from Jewell (2000).

of these slope features are described in the following sections and their depositional interpretations are presented in the discussion.

Morphology of a tributary slope channel system

In seismic profile, a channel is a negative relief feature that truncates reflections (incision). If two or more channels merge on the slope, then each of the channels can be called a tributary channel, with these channels merging downslope to form a tributary slope channel system. In the study area, the tributary incisions of the upper slope act as sediment-bypass conduits, whereas the middle to lower slope tributary incisions tend to be more depositional.

The seabed seismic examples in Figures 3 and 4 illustrate the downslope morphologic evolution of the southernmost Timiris tributary channel system (water depths of 750–2000 m; area I on Fig. 2). From the shelf-break to 750 m water depth, the tributary channel system comprises 'u' shaped channels that act as sediment-bypass conduits. The channels are 1–2 km wide, 150–400 m deep, and have steep channel walls (up to 34°) that are not terraced. The channels reach their maximum size on the upper slope. A strike seismic profile from the upper slope (750–1000 m water depth) shows three typical 'v'-shaped tributary incisions (A, B, C channels in Fig. 3a) that probably act as bypass/erosion conduits for sediment. The tributaries average 2.5–4 km wide, 600 m deep and have steep channel walls (up to 20°). The incisions clearly truncate underlying reflections and locally have terraced flanks, which were probably sculpted by the flow of turbidity currents during the early stage of channel incision. Fifteen kilometres downslope (water depths of 1000–1500 m), tributaries B and C merge and the tributary incisions have become more 'u'-shaped (Fig. 3b). The incisions continue to act as bypass conduits for sediment on the upper-middle slope. The tributaries are 2.5 km wide and 200–300 m deep. In this zone the tributaries have terraced flanks (resulting from rotation of slide blocks), and slope failure features on the tributary interfluve (Fig. 3d).

Thirty-five kilometres farther downslope, in the lower-middle slope (water depths 1500–1700 m), the tributary incisions merge into one depositional channel–levee system (A–B–C channel in Fig. 3c) with low sinuosity. Symmetrical levees are characterized by stacked 'wing-shaped' reflections (Fig. 3c) that wedge and thin by downlap away from the channel, or onlap onto the slope of the previous terrace. The levees are 1.5–4 km

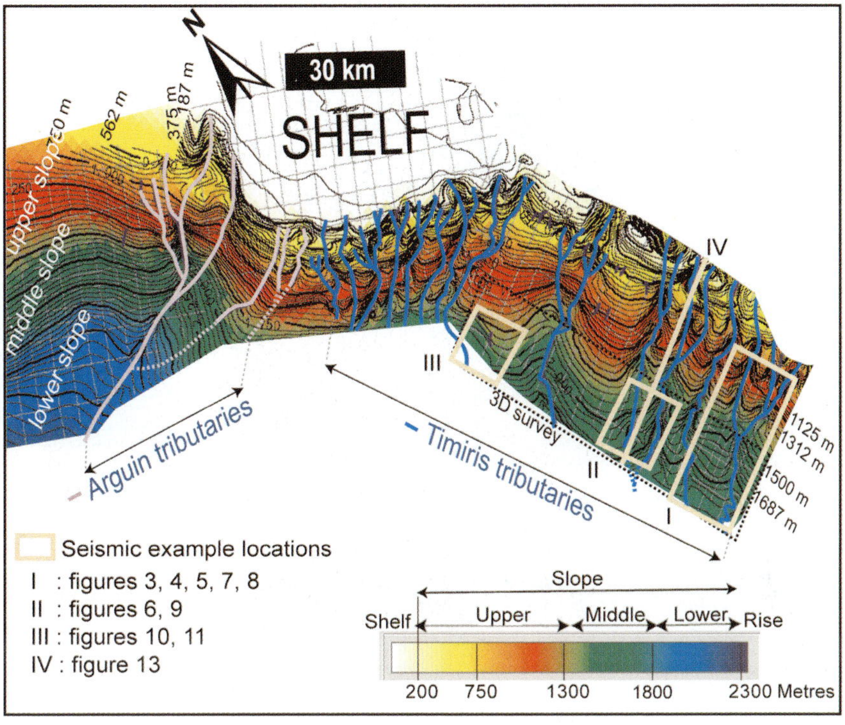

Fig. 2. A seabed bathymetric map of the study area showing the topography of the slope (produced from seismic data). The locations of the areas that contain the seismic examples described in the paper are shown. Area I contains the morphology of a tributary slope channel system (Figs 3–5), example II (Fig. 7), and example III (Fig. 8). Area II contains example I (Fig. 6) and example IV (Fig. 9). Area III contains example V (Figs 10 and 11). Area IV shows the location of the summary slope depositional seismic profile (Fig. 13).

wide, no more than 75 m thick, and confined to flat-floored terraces that result from previous erosion in the channel fairway. There is a decrease in average depth of the incisions downslope, from 630 m (upper slope; Fig. 3a) to 150 m (upper-middle slope; Fig. 3b) to 75 m (lower-middle slope; Fig. 3c).

In the lower slope environment, the channel A–B–C has become highly sinuous (below 1800 m water depth; Fig. 4a). The time slice in Figure 4(a) shows this sinuous channel with lateral accretion packages (LAPS of Abreu *et al.* 2003) on the insides of the bends. The lateral accretion packages can be seen both on seismic profiles as low-amplitude inclined reflections (Fig. 4b, c) and in plan view (Fig. 4a). Low-relief levee deposits (35 m) may occur on a terrace outside of the channel fairway (Fig. 4c).

Large slope failures, as much as 7 km long, occur on tributary–channel interfluves. An example of a slope failure in the upper-middle slope environment is shown in Figure 3(d), illustrating the listric detachment fault plane at the head of the failure, the extensional zone with associated slip (normal faults), the zone of accumulation and the compressional toe of the slope failure (with associated thrust faults). Ponded post-slope failure sediments have accumulated on the irregular topographic surface of the failure deposits; note the overlying high-amplitude reflections at the end of the extensional zone (Fig. 3d). A perpendicular seismic profile (Fig. 3b), confirms that this slope failure is located on a tributary–incision interfluve, covering an area of 25 km^2.

Control of a tributary slope channel system

An example of an older slope failure controlling a younger channel fairway location is shown in Figure 5. This example occurs in the middle slope environment of the upper Miocene. The travel time map of the Upper Miocene reflection (base yellow on Fig. 5a) shows the sharp change in palaeotopography (scarp) due to an underlying slope failure (detachment fault marked on the travel time map,

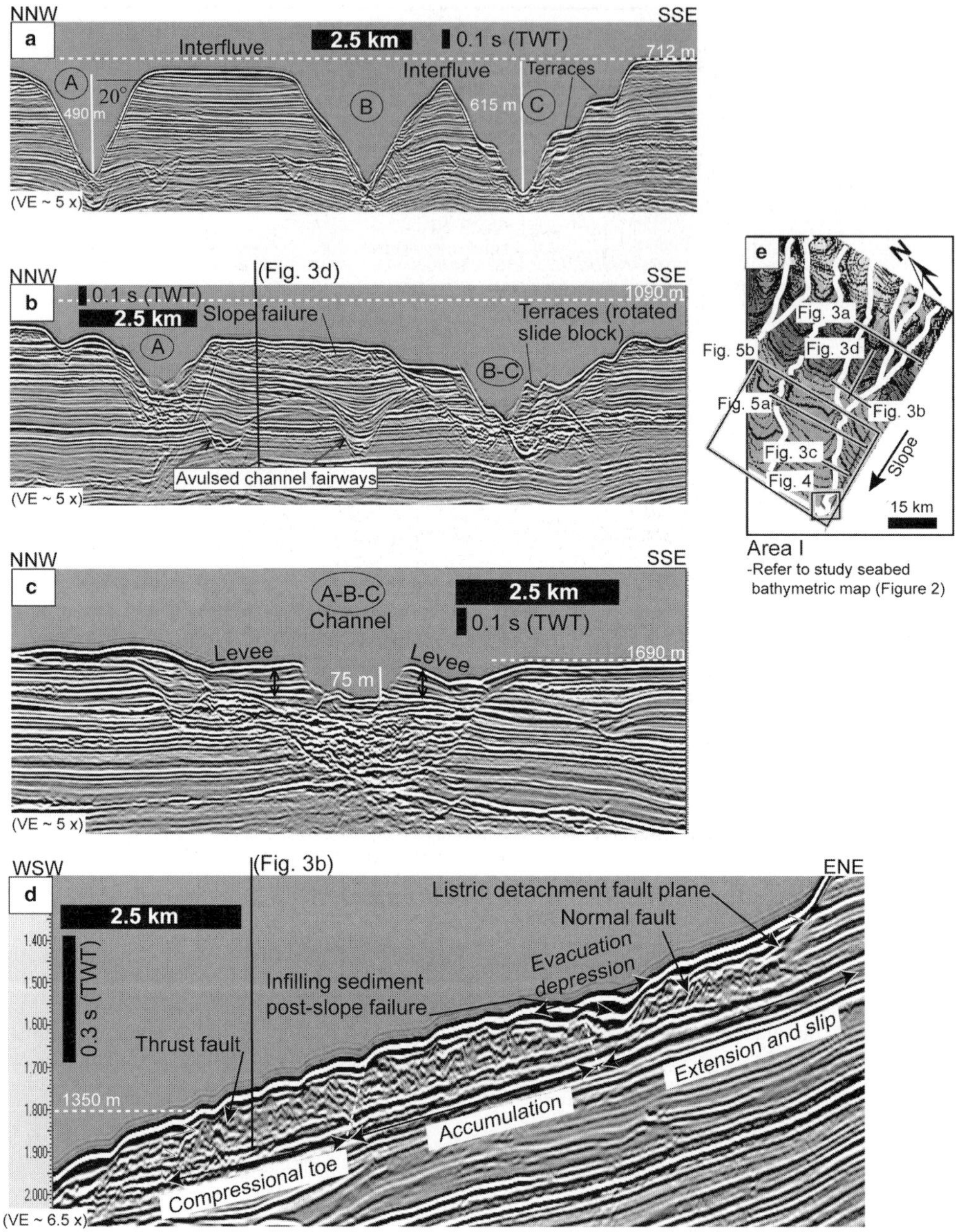

Fig. 3. Seismic profiles showing downslope variations in the morphology of the southernmost tributary slope channel system. (**a**) Strike seismic profile in the upper slope. (**b**) Strike seismic profile in the middle slope. (**c**) Strike seismic profile in the lower-middle slope. (**d**) Seismic profile in the upper slope showing a slope failure on a tributary interfluve. Note this is a dip profile, perpendicular to Figure 3(b). (**e**) Seabed bathymetric map of area I showing the seismic profile locations of Figure 3(a–d). Refer to Figure 2 for location of area I within the study.

Fig. 5b is the location of the head of the older green and violet slope failure deposits in Fig. 5a). This scarp modified the location of the subsequent channel fairway (yellow Channel BLMC E on Fig. 5a), confirming that older slope failures can guide later turbidity-current channel pathways.

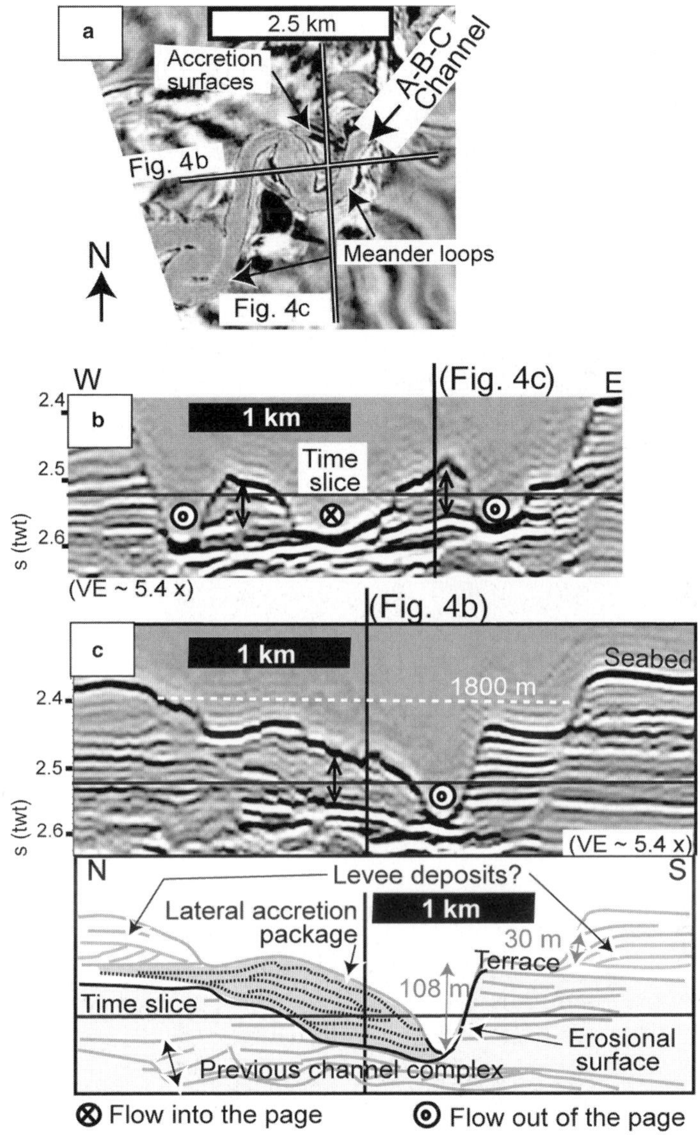

Fig. 4. Seismic expression of a channel meander loop in the lower slope. Refer to Figure 3(e) for location. (**a**) Time slice just below 2.5 s twt, i.e. horizontal line on the seismic profiles. Location of seismic profiles (b) and (c) are shown. (**b**) Dip seismic profile. (**c**) Strike seismic profile with interpretation showing the lateral accretion package and possible levee deposits.

Sediment waves

Sediment waves are identified as any seismic package that contains at least two wave-shaped units (horizontal and vertical) characterized by several undulating reflections. Sediment waves are depositional features generated from currents flowing across the seabed (Wynn & Stow 2002). They can be triggered either by downslope flowing turbidity currents (Kubo & Nakajima 2002; Lee *et al.* 2002; Normark *et al.* 2002), or by along-slope flowing bottom currents (Wynn & Stow 2002; Habgood *et al.* 2003). Usually sediment waves develop crests normal to the current flow, and their migration trends indicate the direction of current flow (e.g. Behrens 1994; Lee *et al.* 2002; Normark *et al.* 2002).

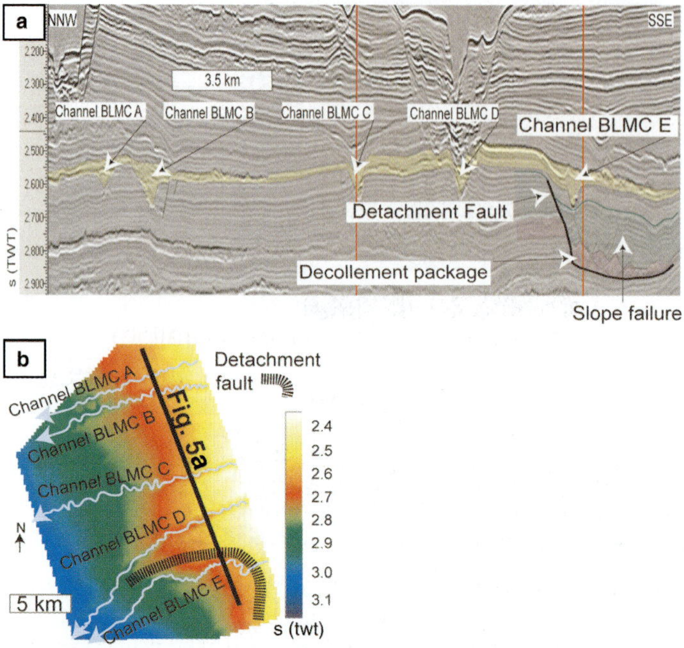

Fig. 5. Seismic example of an older slope failure controlling a younger channel fairway location in the middle slope environment of the Late Miocene. Refer to Figure 3(e) for location. (**a**) Strike seismic profile; note the yellow coloured Late Miocene deposits, in particular the channel BLMC E on the right, and the older green and violet-coloured slope failure deposits with the detachment fault directly underlying the edge of the BLMC E channel. (**b**) Travel time map of the base yellow seismic package. Note that the location of the underlying detachment fault is marked on the map by a hachured line and this seems to control the location of the meander of the younger channel (BLMC E).

Two examples of seismic sediment waves from different areas of the slope environment are described below. Each example has a different spatial extent (size) and is generated by a different process.

Example I

Example I illustrates a sediment wave field, located in the middle slope environment, covering an area of 25 km^2, in water depths of 1500–1700 m (area II on Fig. 2). These sediment waves are restricted to an area bordering the flank of a buried channel fairway. In a dip seismic profile (Fig. 6a) the sediment wave package is characterized by undulating reflection units (0.2 s twt thick) with a flat base and an undulating top. The seismic package contains several smaller, stacked seismic units (0.05 s twt thick) separated by undulating reflections (the interval between the two lines on Fig. 6a). Within these undulating units, moderate amplitude, angular reflections are inclined to the southwest. A time slice through the sediment wave package (Fig. 6b) shows that the linear sediment wave crests are perpendicular to the channel fairway, with a spacing of 250 m between crests. The waves show clear evidence of migration toward the southwest, basinward (Fig. 6a).

Example II

The second example of buried sediment waves comes from the middle slope environment, covering an area of 10.5 km^2, in water depths of 1500–1700 m (area I on Fig. 2). The wave field is elongate in the slope direction and is 7 km long by 1.5 km wide. In a dip seismic profile (Fig. 7a) the sediment wave package is characterized by low-amplitude undulating reflections (0.12 s twt thick). The sediment waves are located above a narrow, elongate pre-existing gully that incises a palaeoslope (seen only in strike profiles; Fig. 7b and 7c). The seismic package contains four waves that decrease in size downslope (purple package of Fig. 7a). The biggest wave is slightly asymmetrical, whereas the others have a symmetrical shape. The travel time map of a horizon within the seismic

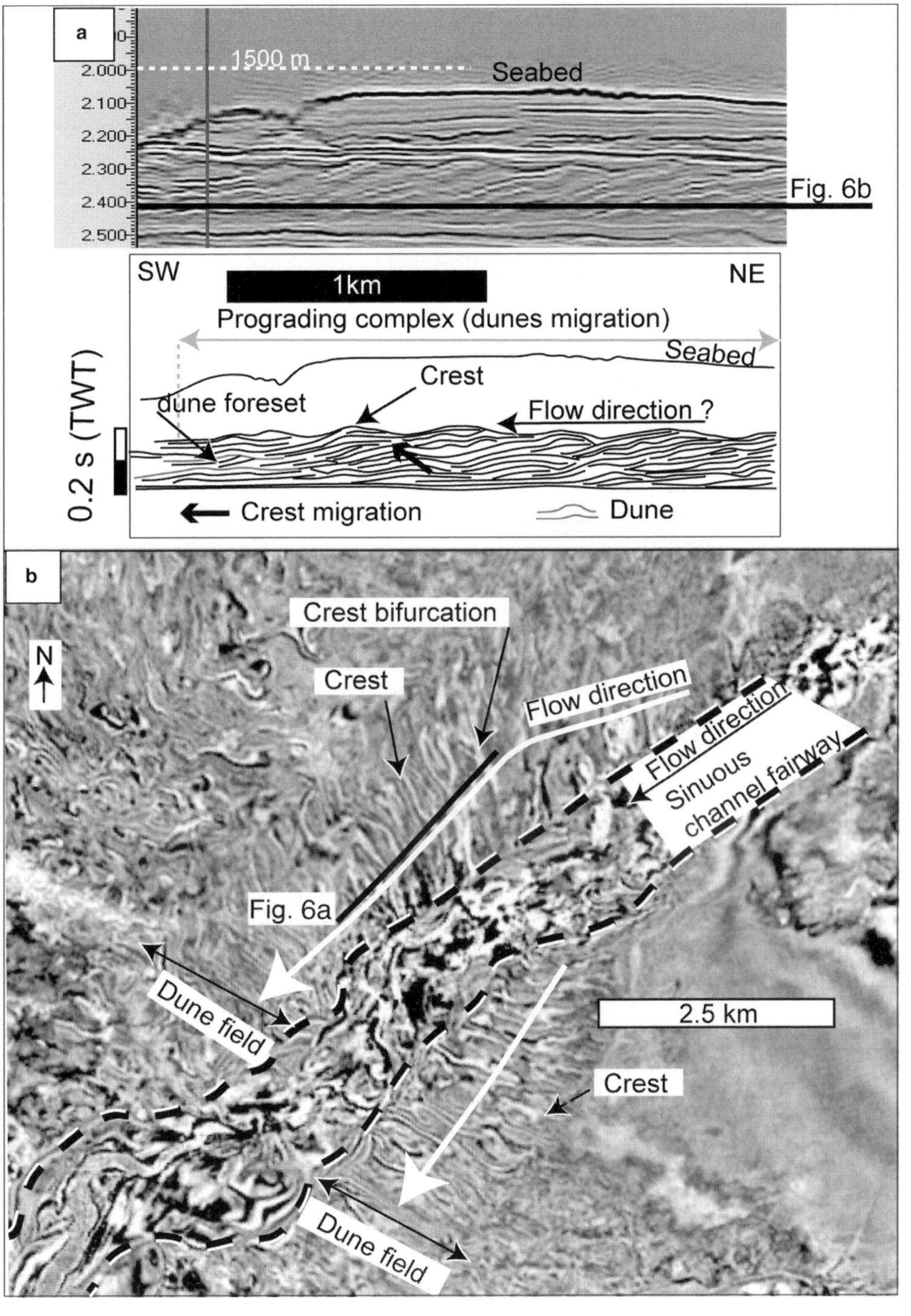

Fig. 6. Seismic example I – sediment wave field in the middle slope environment. Refer to area II of Figure 2 for the location. (**a**) Dip seismic profile through a sediment wave field interpreted as migrating dunes. (**b**) Time slice showing the sinuous channel fairway and the sediment wave fields on either side of the channel fairway. Refer to the line on seismic profile in (a) for the location of the time slice.

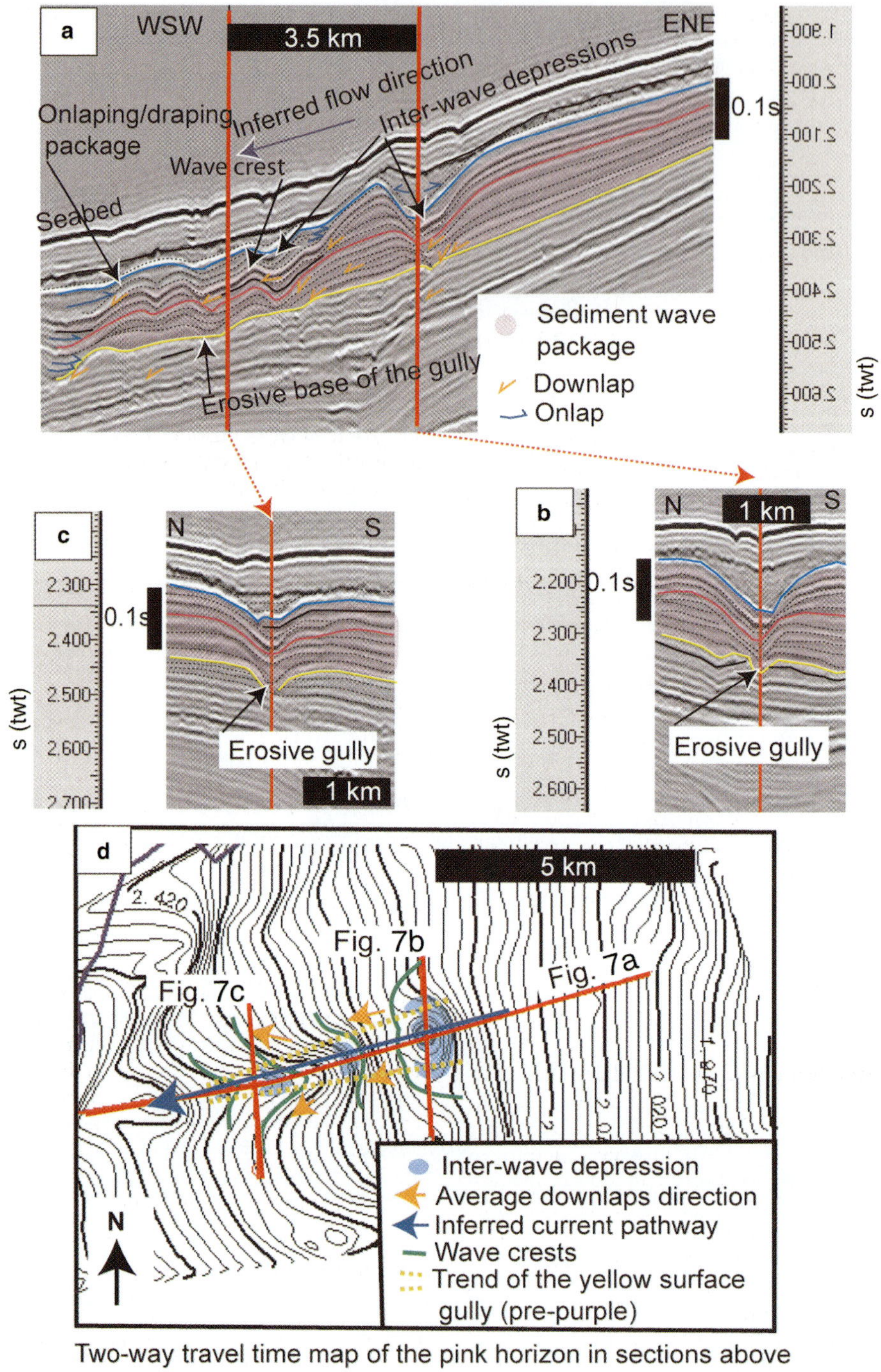

Fig. 7. Seismic example II – buried sediment waves in the middle slope environment above a gully on the palaeoslope. Refer to area I of Figure 2 for location. (**a**) Dip seismic profile through a sediment wave field showing the migrating waves, purple seismic package. (**b and c**) Strike seismic profiles showing the erosive gully located below the seismic sediment waves. (**d**) Travel time map of the pink reflection within the seismic sediment wave package, showing the wave crests (green lines), inter-wave depressions (blue), gully location (yellow dashed lines) and locations of the seismic profiles shown in (a–c).

waves (Fig. 7d, pink horizon on seismic profiles Fig. 7a–c) shows that the wave crests are 1–1.5 km apart, and are parallel to the strike of the slope (green lines on Fig. 7d). The waves are aggradational because their crests have migrated almost vertically (Fig. 7a).

Contourites

Contourites are sediments that have been deposited primarily by bottom-contour currents, although locally they can have downlsope, upslope or oblique-to-slope flow (Stow *et al.* 2002). Elongate-contourite drifts are common on the seabed of the middle to lower slope of the study area. For contourites, the term moat refers to the area where the bottom-current is channelized, and the moat pathway corresponds to the bottom-current pathway. Elongate-contourite drifts are mounded and elongate along the margin, in plan view (McCave & Tucholke 1986; Faugeres *et al.* 1999). They are divided into separated and detached drifts according to the nomenclature of McCave & Tucholke (1986). Separated drifts are separated from the margin by a moat pathway and contain along-slope and upslope sediment migration in an area downslope of the current flow, whereas detached drifts are not separated from the margin by a moat and contain downslope sediment migration in an area upslope of the current flow (McCave & Tucholke 1986; Faugeres *et al.* 1999). Drifts are the seismic packages corresponding to the sediment accumulation principally deposited by bottom-currents. Three examples of seabed-elongate-contourite drifts from the middle to lower slope environment will be shown. Each has a different spatial extent.

Example III

Example III illustrates separated and detached drifts deposited along a contour-current pathway on the seabed in the lower slope environment. The drifts cover an area of 30 km^2, in water depths of 1700–1800 m (area I on Fig. 2). In dip seismic profiles (Fig. 8a, b), the elongate drifts are separated by a narrow, sinuous moat, which is 40 m deep, 400 m wide, 10 km long and follows the 1760 m isobath (Fig. 8c). The moat truncates reflections on the upslope side (Fig. 8a, b). The depositional seismic package on the downslope side of the moat is termed a separated drift (Fig. 8a, c). It is characterized by convex-up low-amplitude, parallel reflections (0.1 s twt thick) that spill out of the moat in a downslope direction, but the separated drift migrates upslope (Fig. 8a).

Detached drifts occur on the upslope side of the moat, where three distinct protuberances (1 km^2 in area) occur along the smooth slope gradient. The protuberances are slightly elongate subparallel to the moat (Fig. 8c). In a dip seismic profile through these protuberances, the detached drift package is characterized by convex-up reflections with bi-directional downlap (Fig. 8b).

Example IV

Example IV illustrates a detached mounded drift deposited on the terrace of a turbidity-current channel on the seabed in the lower slope environment. On the map in Figure 9(b), the dark areas show the turbidity-current channels and the grey line shows the inferred contour-current pathway. The drift covers an area of 6 km^2 (area II on Fig. 2). In a dip seismic profile (Fig. 9a) the detached mounded drift is characterized by stacked convex-up reflections with bi-directional downlap (0.07 s twt thick, Fig. 9). The downslope migration of the detached drift, toward the channel fairway, is shown in Figure 9(a).

Example V

Example V illustrates a separated drift located on the seabed in the middle slope environment. The drift covers an area of 80 km^2, in water depths of 1250–1500 m (area III on Fig. 2). In dip seismic profiles (Fig. 10a, d), the drift is separated from the margin by a narrow, bent moat, 30–75 m deep, 800 m wide and 15 km long, which follows the 1400–1500 m isobath interval and bends toward the slope (Fig. 10e). This moat is bordered by truncated reflections on the upslope flank and a depositional seismic package on the downslope side (Fig. 10a, b, d). Truncation within the moat sculpts a dramatic submarine cliff (150 m) on the upslope flank (Figs 10 and 11). The depositional seismic package is characterized by convex-up low-amplitude, parallel reflections (0.07 s twt thick) that spill out of the moat in a direction perpendicular to the moat pathway (Fig. 10e). The separated drift migrates upslope (Fig. 10d), and sediment thickness within the drift decreases downslope. A 3D volume image (Fig. 11) shows possible small wave crests that occur on top of the separated drift.

Discussion and conclusions

The previous examples show the seismic expressions of deep-water slope architectural elements generated by variable gravity-transport and bottom-current processes in the slope environment of the offshore Northern Mauritanian passive margin. These processes interacted and modified, resulting in the deposition of a variety of degradational and constructional features. The schematic block diagram of Figure 12 shows the general processes, the

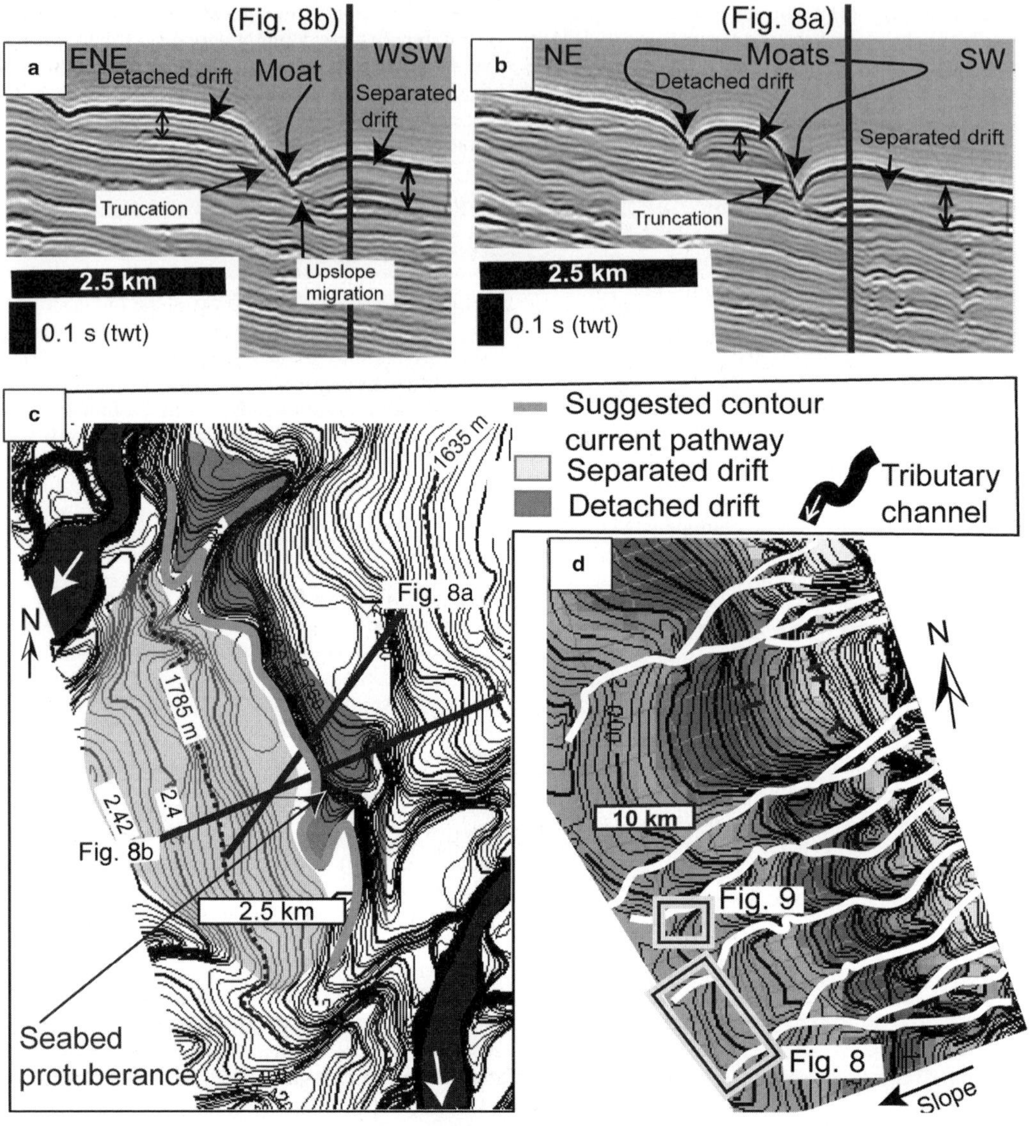

Fig. 8. Seismic example III – contourite drifts deposited along a contour-current pathway on the seabed between turbidity-current channels in the lower slope environment. Refer to area I of Figure 2 for location. (**a** and **b**) Dip seismic profiles showing the moat, detached drifts and separated drift. (**c**) Travel time map of the seabed showing the drifts, contour-current pathway, and seismic profile locations. (**d**) Seabed bathymetric map of area I showing the location of example III (Fig. 8), between two turbidity-current channels and the location of example IV (Fig. 9) alongside a turbidity-current channel.

resulting depositional features and their locations along this passive margin.

Turbidity currents create degradational dendritic tributary-incisions in the shelf and upper slope environments (Fig. 12a), which acted as sediment-bypass conduits (Fig. 3a, b). The terraced walls of the tributaries suggest several stages of incision. Slope failures occur on tributary interfluves (Figs 3d and 12b) and are probably due to slope instability generated by channel migration. Older destructional features, such as slope failures, can guide the location of later turbidity-current channel pathways (Figs 5, 12a). In the lower-middle slope environment the tributary-incisions merge into main channels (low sinuosity, almost straight) that become depositional with channel-levee deposits (Figs 3c

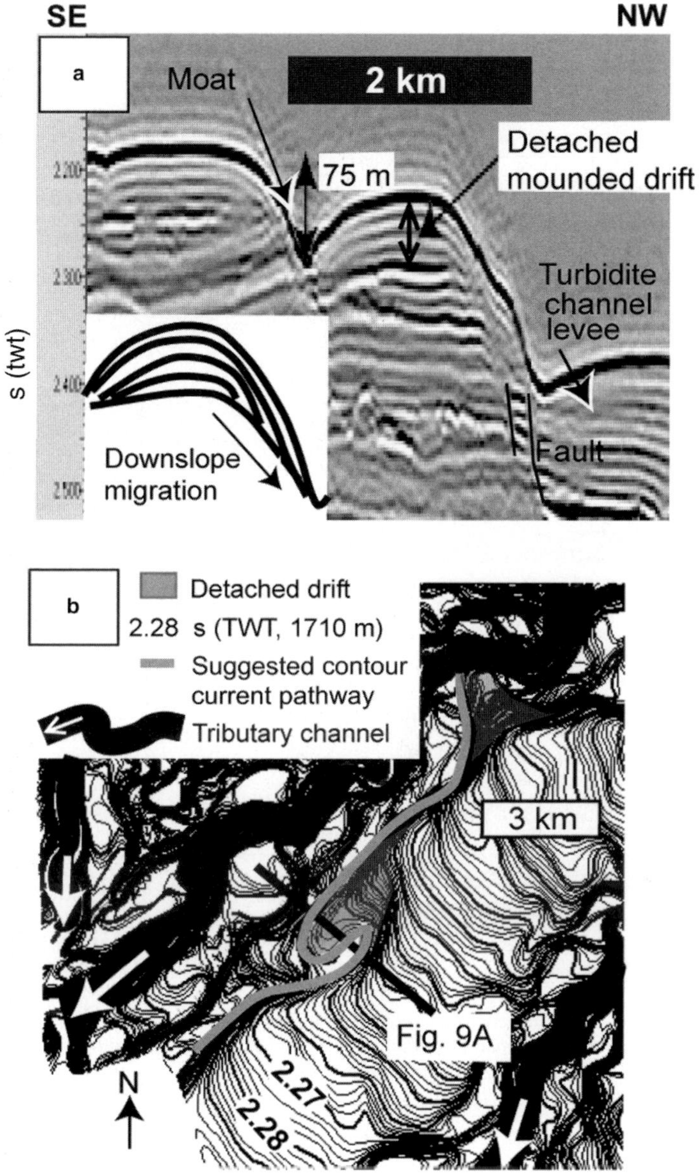

Fig. 9. Seismic example IV – detached mounded contourite drift deposited on the terrace of a turbidity-current channel on the seabed in the lower slope environment. Refer to area II of Figure 2 for location. (**a**) Dip seismic profile through the drift, moat and turbidity-current channel-levee deposits. (**b**) Travel time map of the seabed showing the mounded drift, contour-current pathway (grey line), turbidity-current channel (dark line) and seismic profile location. Refer to Figure 8(d) for actual location of example IV (Fig. 9).

and 12c), and then become sinuous with both erosion and redeposition within lateral accretion packages from channel migration (Figs 4 and 12d). The high sinuosity of the channel in the lower slope is probably due to a low gradient.

Downslope-turbidity-currents can also generate sediment waves in the middle slope environment (Fig. 12e, f). A large sediment wave field was deposited by a turbidity-current flowing downslope along the flank of a buried channel fairway, in example I (Figs 6 and 12e). The individual dunes and dune foresets imaged on seismic (Fig. 6a, b) indicate dune field migration to the southwest, basinward. Owing to the complex cut-and-fill seabed

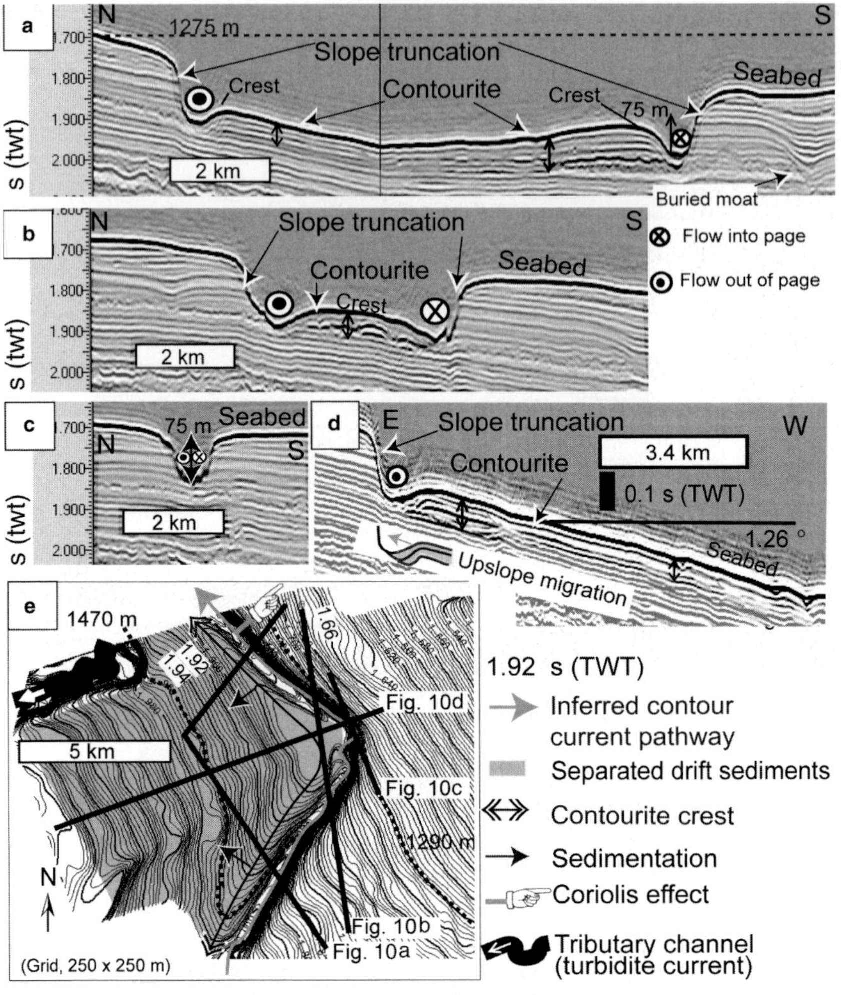

Fig. 10. Seismic example V – separated contourite drift deposited by a bottom current, influenced by Coriolis forces, in the middle slope environment. Refer to area III of Figure 2 for location. (**a–d**) Seismic profiles showing the moat and drift. (**e**) Travel time map of the seabed showing the extent of the drift, contour current pathway and seismic profile locations.

architecture of example I, it is difficult to determine whether the sediment wave deposits formed earlier or contemporaneously with the channel deposits. Example II shows localized sediment waves developed above a narrow erosive gully (Figs 7 and 12f). The depositional current flowed along an upslope/downslope-axis (blue arrow on Fig. 7d) because the crests are parallel to the strike of the slope. The wave size diminishes and some internal wave reflections downlap downslope, suggesting downslope turbidity-currents deposited this feature. The wave field developed above a narrow, straight erosive gully, probably controlling the growth of the waves by channelizing the downslope-flowing turbidity currents (Fig. 7a).

Contour bottom-currents create elongate-contourite drifts, which can be subdivided into separated and detached drifts. Moats are the contour bottom-current pathways that are associated with the drift deposits. Detached drifts occur upslope of the moat, and migrate in a downslope direction (Fig. 12g). In the study area, it is suggested that detached drifts are deposited by southerly flowing currents, which migrate basinward, away from the slope, due to the affect of the Coriolis Deviation preventing erosion of the margin (blue on Fig. 12).

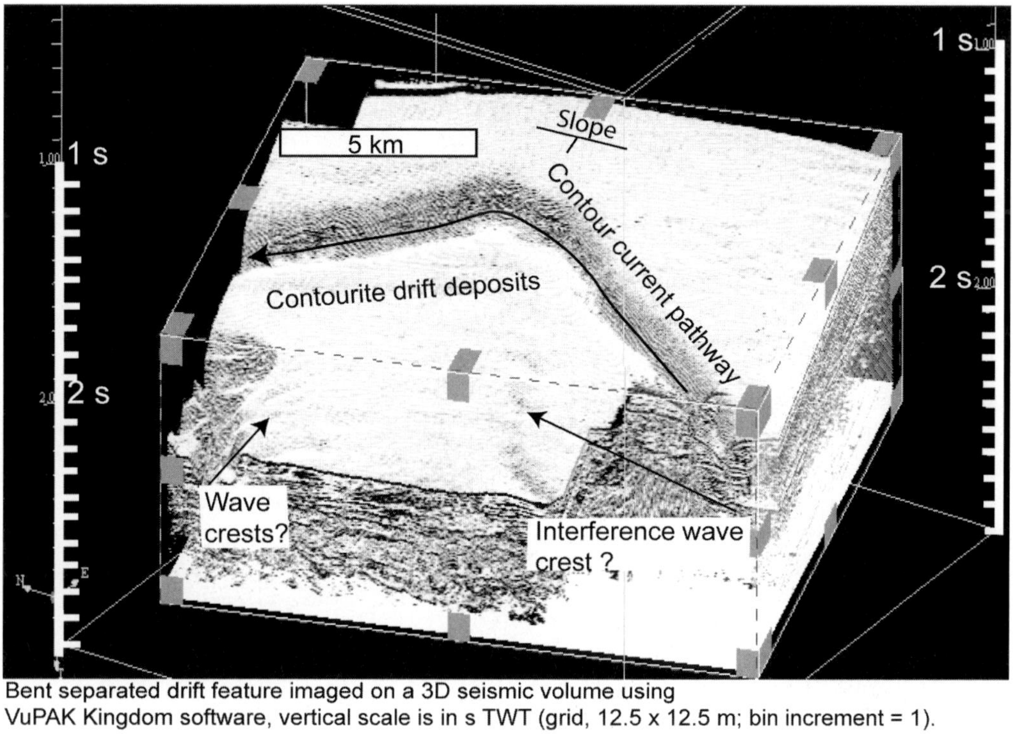

Fig. 11. Three-dimensional seismic volume of the separated contourite drift of example V (Fig. 10). Note the contour current pathway, moat (black line), the steep scarp on the upslope flank of the moat (150 m relief), and sediment waves on top of the drift.

Separated drifts occur downslope of the moat, and migrate in an upslope direction (Fig. 12h, i). In the study area, it is suggested that separated drifts are deposited by northerly flowing currents, which migrate landward, into the slope, due to the affect of the Coriolis Deviation, causing erosion of the margin and redeposition of the sediment in levee-like deposits on the downslope side of the moat (violet on Fig. 12).

Separated and detached drifts deposited along a contour-current pathway on the seabed are shown in example III (Fig. 8), where the larger separated drift is deposited on the downslope side and the detached drift is deposited on the upslope side of the pathway. The timing of deposition of the detached and separated drifts (Fig. 8b) is unclear in example III. Contour bottom-currents could have deposited the two drift types contemporaneously or sequentially. If the features were deposited contemporaneously, then example III represents terraced contourite drifts (Fig. 12g). Otherwise, an initial southerly flow could have first deposited the detached drift along the slope, while a later northerly flow could have eroded the slope, on its right (landward) side, and deposited the separated drift, on its basinward side (Fig. 12h).

In example IV, a localized detached mounded drift was deposited on top of a terrace on the flank of an earlier turbidity-current channel (Figs 9 and 12g). The correlation of numerous protruding detached drifts in the channel fairway Fig. 9b) and the migration towards the channel fairway (Fig. 9a) suggest a bottom current flowing toward the southwest in example IV. This example suggests that contour currents can follow the sculpted topography of a previous turbidite system (Fig. 12g).

The separated drift in example V is interpreted to be the result of a bottom-current that gently ascended from the 1500 m to the 1400 m isobath (Fig. 10e). The current is deviated to the right, triggering a dramatic erosion of the margin (Fig. 10e) and causing the drift to migrate upslope (Fig. 10d, e). The bent shape of the moat is probably due to an interaction between the slope-morphology (i.e. slope gradient and regularity), the current system and the intensity of the current. The

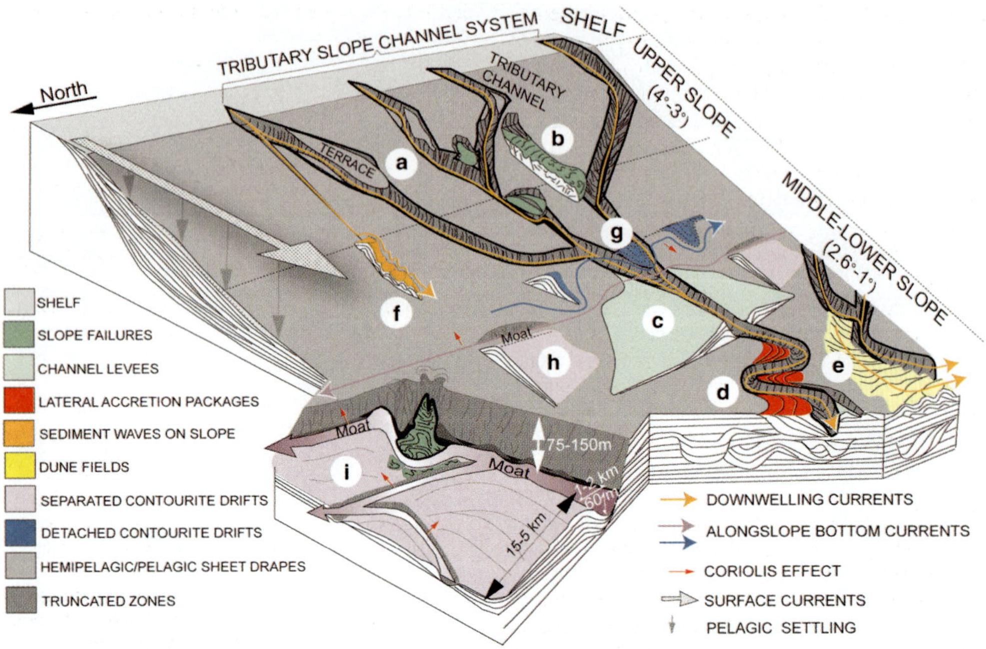

Fig. 12. Schematic block diagram of processes and resulting depositional features on the Northern Mauritanian passive margin, not to scale. (**a**) Turbidity currents create degradational tributary-incisions on the shelf edge and upper slope. Similar to Figure 3(a, b) (**b**) Slope failures occur on tributary interfluves (green). Similar to Figure 3(d). (**c**) Levee deposits occur on the middle and lower slope adjacent to straight tributary-incisions (light green). Similar to Figure 3(c). (**d**) Lateral accretion packages occur on the lower slope adjacent to sinuous tributary-incisions (red). Similar to Figure 4. (**e** and **f**) Turbidity currents associated with tributary channel systems generate sediment waves on the lower slope (yellow), or on the middle slope (orange). (**g**) Contour currents create detached drifts (blue). (**h** and **i**) Contour currents create separated drifts (purple).

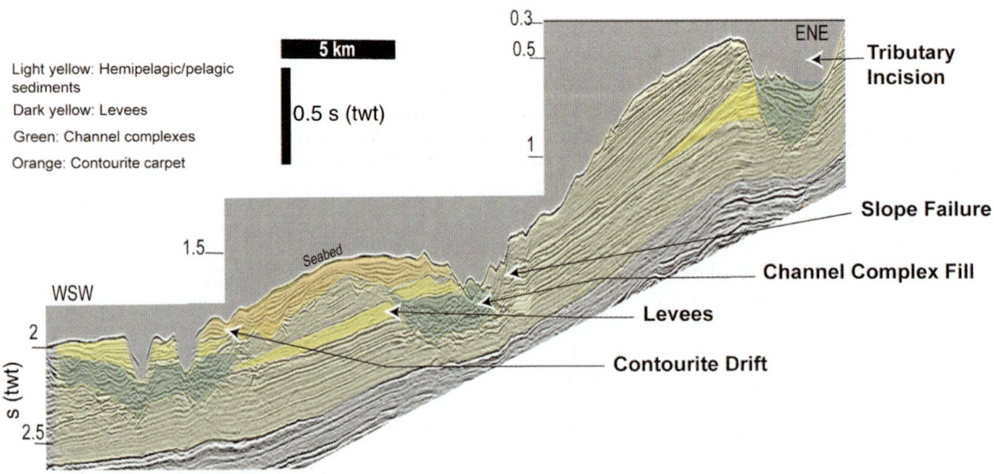

Fig. 13. Dip seismic profile covering the upper to middle slope environment in the southern part of the study area. Note the complexity caused by the stacking of the various deposits, interaction of multiple processes, complex cut-and-fill seabed architecture, and reworking of deposits by secondary processes. Refer to line IV on Figure 2 for location.

sediment eroded from the cliff is redeposited in the inner zone of the bent bottom-current pathway (violet on Fig. 10b). This type of bent separated drift is not elongated, but bent due to the bent shape of the current pathway. Example V shows that separated drifts are not necessarily elongate in shape as originally defined by McCave & Tucholke (1986). This example also shows how later secondary contour currents and gravity flows can rework drift deposits, i.e. the small sediment waves that occur on the surface of the drift (Fig. 11).

Constant, complex dynamic sedimentary processes have occurred and interacted with each other on the Mauritanian slope: downslope-flowing turbidity currents have produced tributary channel systems, channel levees, sinuous channels and sediment waves; mass-wasting has produced slope failures; and along-slope contour-bottom currents have produced contourites. Because of these multiple processes, the complex cut-and-fill seabed architecture, the stacking of depositional features, and the reworking of deposits by secondary processes it is difficult to determine the relative timing of depositional features and sometimes exactly which processes initially formed them. This complexity is illustrated in Figure 13, which is a typical dip seismic profile covering the upper to lower slope environment. Many of these depositional features have similar seismic geometries, but are created by different processes.

In particular, moats (i.e. bottom-current pathways; Figs 8–11) can be confused with erosive turbidity-current channels (Figs 3 and 9b). Contourite drifts deposited in channel pathways can be confused with turbidity-current channel-levees, which support Faugeres *et al.* (1999) conclusion that distinguishing between turbidity-current and bottom-current depositional features based mainly on 2D seismic characteristics is ambiguous (Figs 9b and 12g). When the deposits are buried, they become even less distinguishable due to the loss of seismic resolution with depth. The seismic examples of slope architectural elements and associated depositional processes from the muddy slope on the Northern Mauritanian offshore passive margin documented in this study can be applied to analogous settings in lower resolution subsurface seismic data to gain a better understanding of the slope environment and deposits. This complexity makes a seismic stratigraphic interpretation difficult and requires a good knowledge of the sedimentological features that make up the slope geomorphology in the area.

The authors wish to thank DANA Petroleum plc for providing data, support for the project and giving permission to publish, and to SMT for the Kingdom Suite University Licence Grant used in this study. C. Fulthorpe and T. F. Hentz are thanked for their constructive reviews which have improved the manuscript.

References

ABREU, V., SULLIVAN, M. ET AL. 2003. Lateral accretion packages (LAPs): an important reservoir element in deep water sinuous channels. *Marine and Petroleum Geology*, **20**, 631–648.

BEHRENS, E. W. 1994. Abyssal sediment waves in the Gulf of Mexico: an enigma. *Paleoceanography*, **9**, 1087–1094.

ELOUARD, P. 1975. Formations sedimentaires de Mauritanie atlantique. *In*: *Notice explicative de la carte geologique a 1/1 000 000 de la Mauritanie*. BRGM, Paris, 173–233.

FAUGERES, J.-C., STOW, D. A. V. ET AL. 1999. Seismic features diagnostic of contourite drifts. *Marine Geology*, **162**, 1–38.

HABGOOD, E. L., KENYON, N. H. ET AL. 2003. Deep-water sediment wave fields, bottom current sand channels and gravity flow channel-lobe systems: Gulf of Cadiz, NE Atlantic. *Sedimentology*, **50**, 483–510.

JEWELL, C. J. 2000. *Report on the processing of satellite gravity and topographic/bathymetric data of Mauritania*. Unpublished Dana plc report.

KUBO, Y. S. & NAKAJIMA, T. 2002. Laboratory experiments and numerical simulation of sediment-wave formation by turbidity currents. *Marine Geology*, **192**, 105–121.

LEE, H. J., SYVITSKI, J. P. M. ET AL. 2002. Distinguishing sediment waves from slope failure deposits: field examples, including the 'Humboldt slide', and modelling results. *Marine Geology*, **192**, 79–104.

MCCAVE, I. N. & TUCHOLKE, B. E. 1986. Deep current-controlled sedimentation in the western North Atlantic margin. *In*: VOGT, P. R. & TUCHOLKE, B. E. (eds) *The Geology of North American, Vol. M, The Western North Atlantic Region, Decade of North America Geology*. Geological Society of America, Boulder, CO, 451–468.

NORMARK, W. R., PIPER, D. J. W. ET AL. 2002. Variability in form and growth of sediment waves on turbidity-current channel levees. *Marine Geology*, **192**, 23–58.

STOKER, M. S., EVANS, D. & CRAMP, A. (eds). 1998. *Geological Processes on Continental Margins: Sedimentation, Mass-wasting and Stability*. Geological Society of London, Special Publications, **129**.

STOW, D. A. V. & PIPER, D. J. W. (eds). 1984. *Fine-grained Sediments: Deep-water Processes and Facies* [*sic*]. Geological Society of London, Special Publications, **15**.

STOW, D. A. V., FAUGERES, J.-C. ET AL. 2002. Bottom currents, contourites and deep-sea sediment drifts; current state-of-the-art. *In*: STOW, D. A. V., PUDSEY, V. C. J., HOWE, J. A., FAUGERES, J.-C. & VIANA, A. R. (eds) *Deep-water Contourite Systems: Modern Drifts and Ancient Series, Seismic and Sedimentary Characteristics*. Geological Society of London, Memoirs, **22**, 7–20.

TREMBLAY, S. 2005. *Geological evolution and depositional architecture of the Northern Mauritanian passive margin*. Ph.D. Thesis. University of Aberdeen.

WYNN, R. B., MASSON, D. G. *ET AL.* 2000. The Northwest African slope apron: a modern analogue for deep-water systems with complex seafloor topography. *Marine and Petroleum Geology*, **17**, 253–265.

WYNN, R. B. & STOW, D. A. V. 2002. Classification and characterisation of deep-water sediment waves. *Marine Geology*, **192**, 7–22.

Application of three-dimensional seismic data to documenting the scale, geometry and distribution of soft-sediment features in sedimentary basins: an example from the Lomre Terrace, offshore Norway

C. A.-L. JACKSON

Department of Earth Science and Engineering, Imperial College, Prince Consort Road, London SW7 2BP, UK (e-mail: c.jackson@imperial.ac.uk)

Abstract: Three-dimensional seismic data are used to document the geometry, scale and distribution of soft-sediment deformation features in the post-rift succession of the Lomre Terrace, offshore Norway. In the Cretaceous to Upper Oligocene succession a polygonal fault network, developed in the in response to compaction and dewatering of the interval, was mapped using dip and azimuth grid-based attributes. In the same stratigraphic interval a series of chaotic seismic reflection packages are developed which are visualized using a volume-based seismic coherency attribute and interpreted as the seismic expression of mobilized mud masses. Immediately overlying the mobilized mud masses are a series of fault-bounded depressions that are interpreted to have formed in response to deflation of the mobilized mud masses caused by loading of the overlying succession. A series of shallow, curvilinear erosion surfaces are present on the seismic horizon bounding the top of the Pliocene succession and represent iceberg-keel plough marks. This study demonstrates that interpretation and visualization of three-dimensional seismic data coupled with attribute analysis provide valuable insights into soft-sediment deformation features in sedimentary basins, in particular the scale, geometry and distribution of such features and their temporal and spatial inter-relationships.

Three-dimensional (3D) seismic data are increasingly used in basin analysis as they provide seismic interpreters with the ability to map structural and stratigraphic features only a few tens of metres in areal extent over thousands of square kilometres (see recent reviews by Davies *et al.* 2004; Cartwright & Huuse 2005). This has resulted in the identification and enhanced understanding of previously unrecognized structural and stratigraphic features (e.g. Cartwright 1996; Davies *et al.* 1999, 2002; Davies 2003; Trude 2004), and has helped refine the interpretation of complex structural settings (e.g. Nestvold 1992; Demyttanaere *et al.* 1993; Hesthammer 1998). In addition to providing very high spatial resolution of extremely large parts of sedimentary basins, an additional key benefit of 3D seismic data is that numerous spatially continuous attributes can be extracted or derived directly from the 3D seismic reflectivity cube (the data cube typically available to the seismic interpreter). For example, a secondary attribute termed variance or coherence (e.g. Marfurt *et al.* 1998; Brown 2003), which documents the spatial variations in seismic reflection continuity and amplitude in the primary data volume, can be used to map seismic facies or discontinuities in three dimensions. In addition, specific parts of the amplitude range within the seismic reflectivity cube can be rendered opaque such that high- (or low)-amplitude features of interest can be isolated and analysed in three dimensions (Dorn *et al.* 1995; Lees 1999; Bond 2001; James *et al.* 2004). Three-dimensional seismic data also allow spatially continuous seismic horizons to be rapidly mapped (Herron 2000). These can then be analysed using various grid-based attributes (e.g. dip, azimuth, etc.), which can enhance the 3D geometry of structural and stratigraphic features along the mapped horizon (Fig. 1; e.g. Dalley *et al.* 1989; Bahorich & Farmer 1995; Hesthammer 1998).

Based on the benefits listed above, 3D seismic data are particularly useful for studying soft-sediment deformation features in sedimentary basin. Such features are typically smaller than the usual grid spacing (i.e. >1 km) of 2D seismic data and thus such features may not be imaged at all (Cartwright & Huuse 2005; their fig. 1). In addition, soft-sediment deformation features display complex geometries in 3D, an important aspect which reveals much about the genesis of these features that would not be captured by 2D seismic data (Huuse *et al.* 2004; Jackson 2006).

Furthermore, the ability to co-visualize seismic data and mapped seismic horizons allows the spatial inter-relationship between various soft-sediment deformation-related features to be examined. When

From: DAVIES, R. J., POSAMENTIER, H. W., WOOD, L. J. & CARTWRIGHT, J. A. (eds) *Seismic Geomorphology: Applications to Hydrocarbon Exploration and Production*. Geological Society, London, Special Publications, **277**, 253–267.
0305-8719/07/$15.00

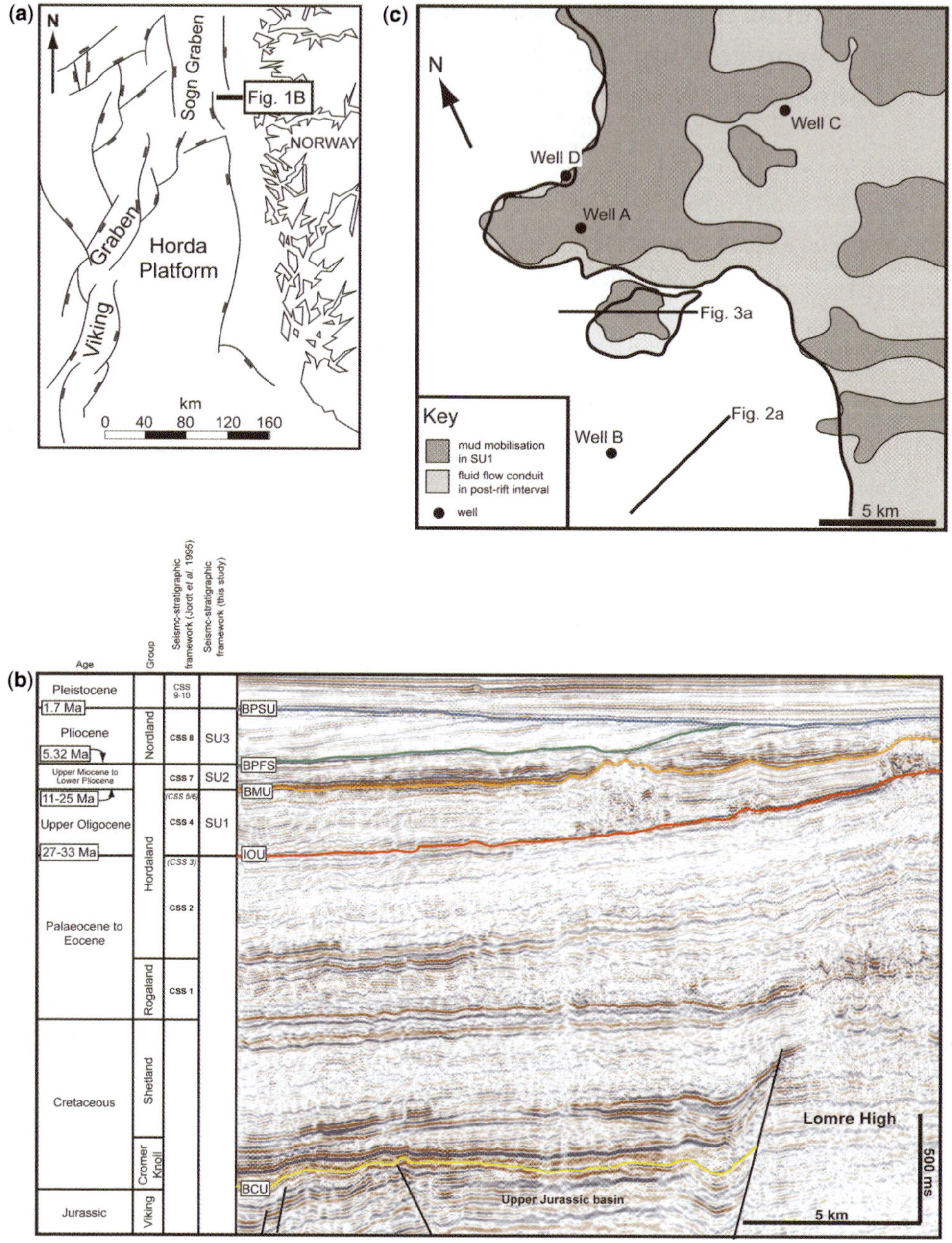

Fig. 1. (**a**) Simplified map showing the general structural setting of the Lomre Terrace study area along the eastern margin of the Sogn Graben. The locations of Figures 1(b), 2(a) and 3(a) are shown. (**b**) Seismic section across the study area illustrating the stratigraphic setting and seismic expression of the studied succession. A simplified geological timescale is shown as is the seismic-stratigraphic framework used in this study and its relationship to that of Jordt *et al.* (1995). Note that seismic packages of Jordt *et al.* (1995) not developed in the present study area are bracketed. The ages of key stratal surfaces and seismic horizons mapped in the study are given. The location of the seismic section is shown in Figure 1(a). (**c**) Use-Location map of the study area. Wells used in the study are marked. Locations of seismic sections shown in Figs. 2a and 3a are marked. Areas with thin bounding lines and dark grey shading represent areas of middle to Upper Oligocene strata affected by large-scale mud mobilisation. Areas with thick bounding lines and light grey shading represent areas where fluid flow conduits can be found.

combined with standard seismic–stratigraphic techniques, important insights into the temporal relationships between soft-sediment deformation-related features can also be gained from 3D seismic data. In this paper the scale, geometry and distribution of soft-sediment deformation features in the post-rift succession of the Lomre Terrace, offshore Norway, is documented through the interpretation and visualization of 3D seismic data. Key seismic-stratigraphic features and the relationships between them were analysed using various industry-standard volume and grid-based seismic attributes. It is demonstrated that the post-rift succession has undergone numerous phases of soft-sediment deformation related to a series of linked fluid flow, dewatering, compaction, and loading processes.

Dataset

The seismic dataset for this study consists of one 3D seismic reflection survey that partially covers North Sea exploration blocks 35/8 and 35/11 (Fig. 1a). The survey has a total areal extent of 782 km^2 and an inline and crossline spacing of 12.5 m. Although the survey is not depth-converted, by using velocity information from wells it is possible to convert all depth and thickness measurements from two-way travel time to metres. Data quality in the survey at the studied stratigraphic interval is excellent. The survey is zero-phase processed with normal polarity, thus a negative trough event (blue or black seismic reflection on seismic sections) represents an increase in acoustic impedance with depth and a positive peak event (red or yellow seismic reflection on seismic sections) represents a decrease in acoustic impedance with depth. The studied stratigraphic interval within the seismic dataset used is penetrated by numerous wells, four of which are used in this study to establish the age of the seismic packages and calibrate them to lithology (Fig. 1c). All of the wells contain a full suite of wireline data, biostratigraphic data and lithological cuttings for the stratigraphic intervals of interest.

Geological setting

The study area is located on the Lomre Terrace along the eastern flank of the Sogn Graben in the northern part of the North Sea Basin (Fig. 1a). The Viking Graben formed in response to crustal extension which commenced in the late Middle Jurassic and continued in some places until the latest Jurassic–earliest Cretaceous. During the main Late Jurassic rift phase subsidence was relatively rapid and the Viking Graben and the proximal flanks were characterized structurally by large-displacement normal faults which controlled a series of graben and half-graben basins (Fig. 1b; Badley *et al.* 1988; Yielding 1990; Færseth *et al.* 1997). The post-rift phase commenced in the Early to Middle Cretaceous and was characterized by a marked decrease in subsidence rate associated with a switch from fault to thermally controlled subsidence and a transition to a more ramp-type basin geometry (e.g. Joy 1993; Jordt *et al.* 2000; Gabrielsen *et al.* 2001). Although the post-rift period was generally a period of relatively tectonic quiescence, major uplift of the eastern and western margins of the basin occurred sporadically during the Tertiary although the cause of this uplift is still debated (see review by Faleide *et al.* 2002).

During the post-rift period a smectite-rich (see Aplin *et al.* 1995) mudstone-dominated succession up to 5 km thick was deposited which can be divided into four lithostratigraphic groups; the Shetland, Rogaland, Hordaland and Nordland Groups (Fig. 1b; Isaksen & Tonstad 1989). Although mudstone-dominated, two regionally developed sandstone-bearing units are present in the post-rift interval: the Upper Oligocene Skade Fm and the Upper Miocene to possibly Lower Pliocene (see Piasecki *et al.* 2002) Utsira Fm, which are developed in the upper part of the Hordaland Group and lower part of the Nordland Group, respectively. The Skade Fm is interpreted to have been deposited in a series of slope and basinfloor channels and fans, whereas the Utsira Fm has been interpreted to have been deposited either in a large shallow marine shelfal shoal (Galloway 2002) or a series of submarine fans (Gregersen *et al.* 1997). Sand-rich glaciogenic units are also developed in the uppermost part of the Nordland Group within the Pleistocene succession (e.g. Sejrup *et al.* 1991, 1996). Well data indicate that a number of regionally significant key stratal surfaces (i.e. flooding surfaces, sequence boundaries) are developed within the post-rift succession of the study area (Martinsen *et al.* 1999), which manifest on seismic data as regionally-mappable seismic reflections that are associated with truncation of seismic reflections below and/or onlap of seismic reflections above (Fig. 1b).

Seismic-stratigraphic framework and geomorphology

This paper adopts the seismic-stratigraphic framework established by Jordt *et al.* (1995) for the post-Cretaceous succession of the central and northern North Sea Basin (Fig. 1b). This scheme indicates that 10 seismic-stratigraphic units can be defined based on the identification of seismic reflection terminations (e.g. truncation, onlap, downlap) between the sequences and internal seismic characteristics of individual sequences (e.g. seismic facies). This seismic-stratigraphic framework is calibrated to biostratigraphic studies of Norwegian, Danish and British wells, is placed within the absolute geological timescale of Harland *et al.* (1990) and Gradstein & Ogg (1996) and uses the formal lithostratigraphy for the central northern North Sea established by Isaksen & Tonstad (1993). This combined seismic-stratigraphic and chronostratigraphic framework has been previously applied to the present study area by Clausen *et al.* (1999) during their study of intraformational faults in the Upper Oligocene succession (see description of Seismic Unit 3 below).

In the following sections detailed descriptions and interpretations of four seismic horizons and four seismic packages within the Cretaceous to present succession are provided in stratigraphic order (Fig. 1b). The relationship of each seismic package defined here with the existing seismic-stratigraphic framework of Jordt *et al.* (1995) is also highlighted so that the units can be placed in the existing regional stratigraphic framework. Note that for each seismic horizon and seismic package particular attention is paid to the seismic attribute used to investigate the geometry, scale and distribution of seismic-stratigraphic features of interest.

Intra-Oligocene unconformity

The intra-Oligocene unconformity (IOU) bounds the base of SU1 and is defined on seismic data by subtle truncation of seismic reflections below and marked onlap of seismic reflections above (Fig. 1b). Biostratigraphic data indicates a 6 Ma erosional hiatus associated with the IOU which spans 27–33 Ma (i.e. late Early Oligocene to early Late Oligocene; Fig. 1b). Based on these observations the IOU is interpreted as a sequence boundary formed by a fall in relative sea-level during the late Early Oligocene related to uplift of the eastern margin of the basin in combination with a major eustatic sea-level fall (Martinsen *et al.* 1999). Within the study area the relative sea-level fall associated with the formation of the IOU resulted in a minor influx of sandstones and siltstones into the basin which are time correlative to sandstones of the Skade Fm identified further westwards in the Gullfaks area (e.g. Martinsen *et al.* 1999; Løseth *et al.* 2003). Biostratigraphic data of Martinsen *et al.* (1999) indicate that the unconformity merges basinwards to the west with a slightly older Early Oligocene age unconformity which is associated with the basinward pinchout of an Early Oligocene age seismic-stratigraphic unit (CSS-3 of Jordt *et al.* 1995); this unit is below seismic resolution in the study area and is therefore not mapped in this study.

Seismic sections across the study area indicate that the Upper Cretaceous to Upper Oligocene succession is deformed by numerous faults that are planar to slightly listric in cross-section, have up to 30 m displacement and typically have their upper tips truncated by the overlying base-Miocene unconformity (BMU; Fig. 2a). Subtle folds are developed adjacent to many of the faults, with footwall and hanging wall bedding displaying slight upward and downward flexure respectively (Fig. 2a). Owing to the low-displacement of the faults, time-structure maps of the IOU do not clearly display the geometry or distribution of the faults in map-view (Fig. 2b). As the faults have well-defined, closely spaced hanging wall and footwall cut-offs the seismic reflection autotracking software picks the fault plane itself, thus a time-dip map of the IOU provides a much clearer indication of the geometry and distribution of the faults. This map indicates that the faults have a distinct polygonal distribution, and increase in density westwards (Fig. 2c). Individual fault segments are up to 2.5 km in length as mapped at the IOU, closely spaced (i.e. 150–900 m), are either straight to strongly arcuate, and display dominantly orthogonal intersections, although oblique (i.e. 40°) intersections are locally identified (Fig. 2c, d). The time-dip map also indicates that the axes of folds in the immediate footwall and hanging wall of the faults trend parallel to the faults and merge to define dish-shaped, fault-bounded polygons (Fig. 2d). A combined dip-azimuth map of the IOU provides spatial and directional information regarding the faults and the structural polygons that they bound and indicates that on a local scale the faults have no preferential orientation or dip direction (Fig. 2e).

Based on their cross-sectional and map-view seismic expression, the fault system developed within Paleocene to Upper Oligocene succession is interpreted as a polygonal fault network that formed in response to 3D volumetric contraction of the mudstone-dominated succession during burial and compaction (e.g. Cartwright & Dewhurst 1998). The subtle fault-parallel folding adjacent to the faults (Fig. 3d, e) is interpreted as reverse-drag folding that reflects the decrease in displacement of stratigraphic units with increasing distance from the fault zone (cf. Schlische 1995). The mixture of straight and curved, well-connected map-view fault traces and dominance of orthogonal fault intersections allows the fault system to be further classified as an irregular polygonal fault system based on the terminology and criteria of Lonergan *et al.* (1998). Although within the area of study the polygonal faults have no preferred orientation, a study by Clausen *et al.* (1999) on a more regional scale indicated that the faults have a dominant NW–SE trend which is related to the formation of the faults under a regional stress field.

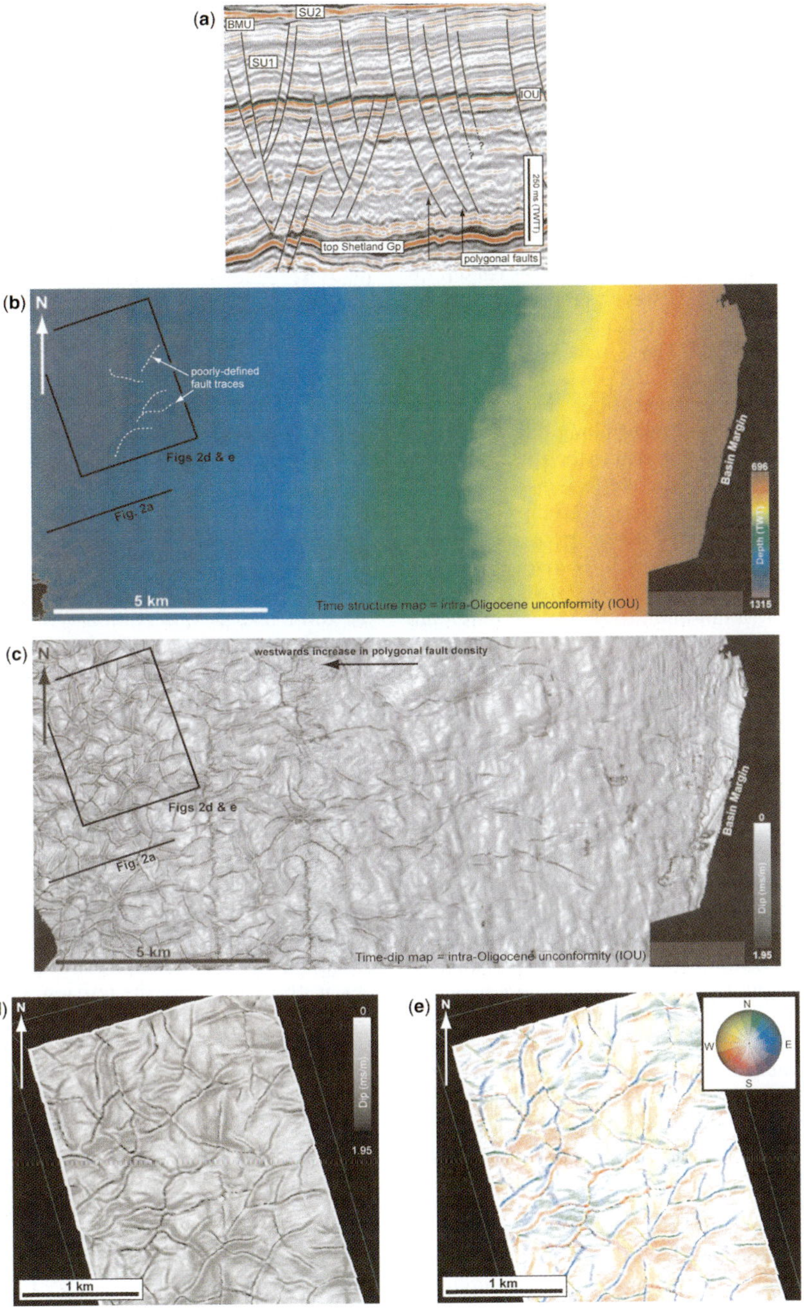

Fig. 2. (**a**) Seismic section through the Paleocene to Upper Oligocene of the study area illustrating the cross-sectional geometry of polygonal faults developed within the interval. IOU, intra-Oligocene unconformity; BMU, base-Miocene unconformity. (**b**) Time–structure map of the IOU. A westwards dip towards the basin centre is observed in addition to a series of poorly-defined fault traces in the west of the area. The locations of (a, d and e) are shown. (**c**) Time-dip map of the IOU. A polygonal fault network is developed in the western parts of the basin. (**d**) Close-up of the time-dip map shown in (c) illustrating the detailed geometry of the polygonal fault network developed along the IOU. (**e**) Combined dip-azimuth map of the IOU illustrating the geometry and variable dip directions of faults within the polygonal fault network. The map also highlights the development of flexures defined by pastel shades in the immediate hanging walls and footwalls of individual fault segments. Location of map shown in Figure 2(c).

Seismic Unit 1 – middle to Upper Oligocene

Seismic Unit 1 (SU1) is bound below by the IOU and above by the BMU, is up to 495 m thick and is dated as Upper Oligocene in age (Fig. 1b; Martinsen *et al.* 1999). SU1 corresponds to seismic-stratigraphic unit CSS-4 of Jordt *et al.* (1995) (see also Clausen *et al.* 1999). Well data indicate that SU1 is dominated by mudstone with thin beds of siltstone and sandstone. Internally, SU1 is dominated by low to moderate amplitude, parallel reflections (marked PR in Fig. 3a). Locally however, zones of variable amplitude, highly chaotic reflections (marked CR in Fig. 3a) are developed. These units have sharp, subvertical margins against the surrounding parallel reflections, and have mounded tops that result in the development of marked relief along the BMU (Fig. 3a–c, e). Immediately adjacent to the zones of chaotic reflections, the parallel reflections are locally highly deformed exhibiting steep-dips and rare reverse faults that verge away from the chaotic reflection packages (Fig. 3e). Towards the bases of chaotic reflection packages, steeply-dipping, discontinuous high-amplitude reflections are observed which have a distinct V or W geometry in cross-section (Fig. 3d, e).

Owing to their distinctly chaotic seismic expression set against a background of dominantly parallel reflections, the overall spatial distribution of chaotic reflection packages can be mapped by using a variance cube, with the chaotic reflection packages emerging as areas of high variance and parallel-stratified reflections as areas of low variance. By making areas of low variance transparent, the distribution of the chaotic reflection packages can be observed, revealing they form a series of circular to ovate bodies, 4.2×10^9–7.2×10^{10} m^3 in size (orange areas in Fig. 3b). The distribution of the high amplitude features developed within the chaotic reflection packages can be illustrated by 'sculpting' the seismic volume using the IOU and BMU seismic horizons which underlies and overlie respectively the features of interest, and rendering all low amplitudes within the resultant cube transparent. The results of this for one of the chaotic reflection zones is shown in Figure 3(d) and indicates that the anomalies are best developed at the base of the chaotic reflection packages, with voxel picking and surface mapping of these amplitude anomalies, indicating they have a clear conical geometry in 3D (Fig. 3e).

The chaotic reflection packages are located immediately above features interpreted as fluid flow conduits or gas chimneys emanating from the underlying Upper Jurassic basins (Fig. 1c; see full discussion by Jackson & Stoddart 2005). Based on their chaotic seismic expression, mudstone-dominated lithology and location above interpreted fluid flow conduits, the chaotic reflection packages are interpreted as the seismic expression of zones of mud which became mobilized and highly deformed in response to gas and water migrating through the poorly consolidated, mudstone-dominated succession (Løseth *et al.* 2003; Jackson & Stoddart 2005). Amplitude anomalies within the mobilised zones may represent the seismic expression of thin gas-charged siltstones and sandstones that were injected upwards from the Skade Fm immediately overlying the IOU.

Base-Miocene unconformity

The base-Miocene unconformity (BMU) bounds the base of SU2 and is defined on seismic data by truncation of seismic reflections within the underlying SU2 and onlap and/or downlap of reflections in the overlying SU3 (Fig. 1b). Biostratigraphic data from within the study area indicates a 14 Ma erosional hiatus associated with the BMU which spans 11–25 Ma (i.e. Late Oligocene to Late Miocene). Based on the seismically defined truncation below the BMU and the observation of an erosional hiatus in nearby wells, the BMU is interpreted to represent a sequence boundary formed during a fall in relative sea-level. The origin of this sea-level fall is not well constrained but may be related to major uplift of the eastern basin margin and associated shallowing of the North Sea Basin (Clausen *et al.* 1999; Martinsen *et al.* 1999; Jordt *et al.* 2000). Irrespective of its origin, the BMU is associated with a major influx of sands which are lithostratigraphically ascribed to the Utsira Fm (see description of SU3 below; Gregersen *et al.* 1997; Galloway 2002).

Three-dimensional visualization of the BMU time-structure horizon indicates that major topography is observed along the horizon which is spatially related to zones of large-scale mud-mobilization developed in the underlying Upper Oligocene units of SU2 (Fig. 3c). This topography is characterized by discrete, isolated sub-circular mounds and more amorphous mounds which protrude up to 257 ms above the essentially flat surrounding surface (Fig. 3c) and have maximum flank dips of 8°. The origin of the mounded topography developed along the BMU can only be interpreted after the seismic-stratigraphic patterns within the overlying units are considered.

Seismic Unit 2 – Upper Miocene to Lower Pliocene

Seismic Unit 2 (SU2) is bound below and above by the BMU and BPFS, respectively, and is dated as Late Miocene to Early Pliocene, broadly corresponding to seismic-stratigraphic unit CSS-7 of Jordt *et al.* (1995) (Fig. 1b). Lithostratigraphically,

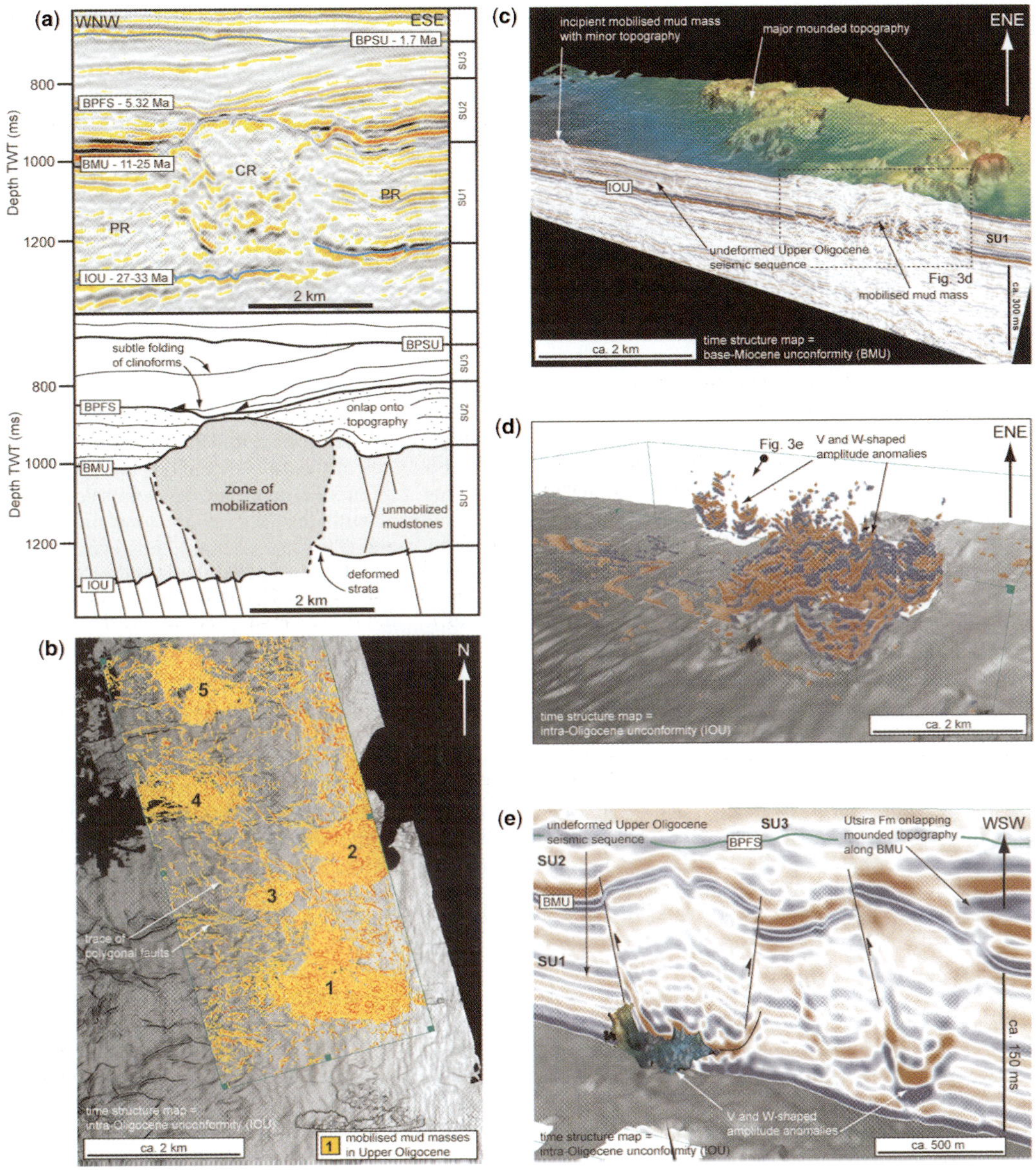

Fig. 3. (**a**) Seismic section through and geoseismic interpretation of a chaotic reflection package (CR) developed in SU1. Note the polygonal faults in parallel seismic reflections (PR) lateral to the chaotic reflection packages, the onlap of SU2 onto the top of the package and folding of clinoform units SU3 above the package. (**b**) Variance cube illustrating the geometry and distribution of chaotic reflection packages (numbered 1–5) in SU1. The volume has been rendered such that only areas of high variance are shown for clarity. (**c**) 3D visualization illustrating the development of marked topography along the BMU related to chaotic reflection packages in SU1. (**d**) 3D visualization of V- and W-shaped amplitude anomalies developed at the base of the chaotic reflection packages in SU1. The volume has been sculpted on the IOU and BMU and rendered such that only high amplitudes are visible. (**e**) 3D visualization showing detail of V- and W-shaped amplitude anomalies at the base of the chaotic reflections packages in SU1. Note the relationship between the steep-dipping limbs of the anomalies and reverse faults at the margins of the chaotic reflection packages.

SU2 corresponds to the Utsira Fm (Gregersen *et al.* 1997; Galloway 2002). SU2 is 0–248 m thick, with marked local variations in thickness observed within the central and eastern parts of the study area. All the studied wells indicate that SU2 is extremely sand-rich. SU2 is composed of moderate to high amplitude, parallel to mounded reflections, which onlap the BMU on the steep flanks of the mobilized mud masses developed in the Upper Oligocene age SU1 (Fig. 3a, e). Locally, steep-sided, intra-formational erosional furrows are observed within SU2.

Although the internal seismic character of SU2 has not been studied in detail, it may be speculated that the locally mounded reflections and erosional furrows are indicative of the depositional environment represented by the unit. If a basinfloor fan depositional model is preferred for the Utsira Fm as advocated by Gregersen *et al.* (1997), the mounded geometries and furrows could represent submarine fan lobes and channels respectively. Alternatively if a shelfal sand model as suggested by Galloway (2002) is preferred, the mounded units and erosional features may represent convex-up shelfal sand bars and tidal rip-current channels respectively. Regardless of the depositional setting for the Utsira Fm, thinning and onlap of the unit onto the marked topography developed above the mobilized mud masses in SU2 suggest that: (1) mobilization occurred prior to deposition of the Utsira Fm; (2) the mobilization processes created significant topography at the contemporaneous seafloor (i.e. mounded features developed along the BMU); and (3) this topography controlled the spatial distribution and thickness of the Utsira Fm.

Base-Pliocene flooding surface

The base-Pliocene flooding surface (BPFS) bounds the top of SU2 and is defined on seismic data as a major regional downlap surface (Fig. 1b). Well data indicates that the BPFS represents a condensed section which is dated as Early Pliocene (5.32 Ma) in age (Martinsen *et al.* 1999). Based on its seismic expression as a regional downlap surface and the correlation of this surface with a period of sedimentary condensation, the BPFS is interpreted as a maximum flooding surface formed in response to a major relative sea-level rise (Martinsen *et al.* 1999).

Mapping of the BPFS bounding the top of the SU2 indicates the development of a series of sub-circular depressions, 3–6 km in diameter and 23–58 m deep (Fig. 4a–c). The distribution and geometry of these features can be observed on a time structure map which has a compressed colour-scale to accentuate the features, and three-dimensional visualization of the BPFS indicates that the depressions occur above the crests of the interpreted mobilized mud features identified in the underlying SU3 (Fig. 4a). As the depressions result in a pronounced change in both dip and azimuth of the BPFS, details of deformation associated with the craters can be illustrated with time-dip (Fig. 4b) and dip-azimuth maps (Fig. 4c). The maps clearly indicate that the margins of the depressions are deformed by a series of curvilinear, convex-into-the-footwall, low-displacement (up to 40 m) normal faults that dip in towards the centre of the depressions.

The spatial relationship between the depressions and the underlying interpreted mobilized mud masses in SU2 (Fig. 4a) suggests a causative relationship between the two features, and two potential mechanisms are proposed for their linked formation. Firstly, the depressions could have formed as erosional features at the palaeo-seafloor due to bottom currents. The circular nature of the depressions (Fig. 4a), coupled with the lack of erosional truncation beneath them (Figs 3a and 4a), however, argues against an erosional origin for the features. An alternative interpretation, based on the inference that mobilization of the Upper Oligocene succession occurred due fluid flow through the interval, it is possible that the depressions formed in response to dissipation of fluid from and deflation of the underling mobilized mud masses, possibly augmented by loading from the overlying Pliocene units. Deflation of the gas-charged mobilized mud masses could have been enhanced by differential compaction between the mobilized mud masses and the surrounding, sand-rich Utsira Fm.

Seismic Unit 3 – Pliocene

Seismic Unit 3 (SU3) is bound above and below by the BPFS and BPSU respectively and corresponds to the Pliocene age seismic package CSS-7 of Jordt *et al.* (1995). SU3 is characterized by low to moderate amplitude, westward-prograding clinoformal reflections that downlap the BPFS but *do not* onlap topography associated with the depression features developed along this horizon (Figs 1b and 3a). The clinoforms dip 1–1.5° (not corrected for compaction), and are subtly folded above the depression features developed along the BPFS with the fold amplitude decreasing upwards towards the BPSU (Fig. 3a). A seismic timeslice that intersects SU3 at 800 ms indicates subtle lateral variations in the width (and accordingly thickness) of unconformity-bounded seismic packages (units 1–5 in Fig. 5a).

Based on their scale and westward progradation away from the Norwegian margin, clinoforms within SU3 are interpreted as continental margin, slope accretion clinoforms. Subtle variations in the width of seismic packages within this unit revealed by seismic timeslices (e.g. Fig. 5a) are inferred to

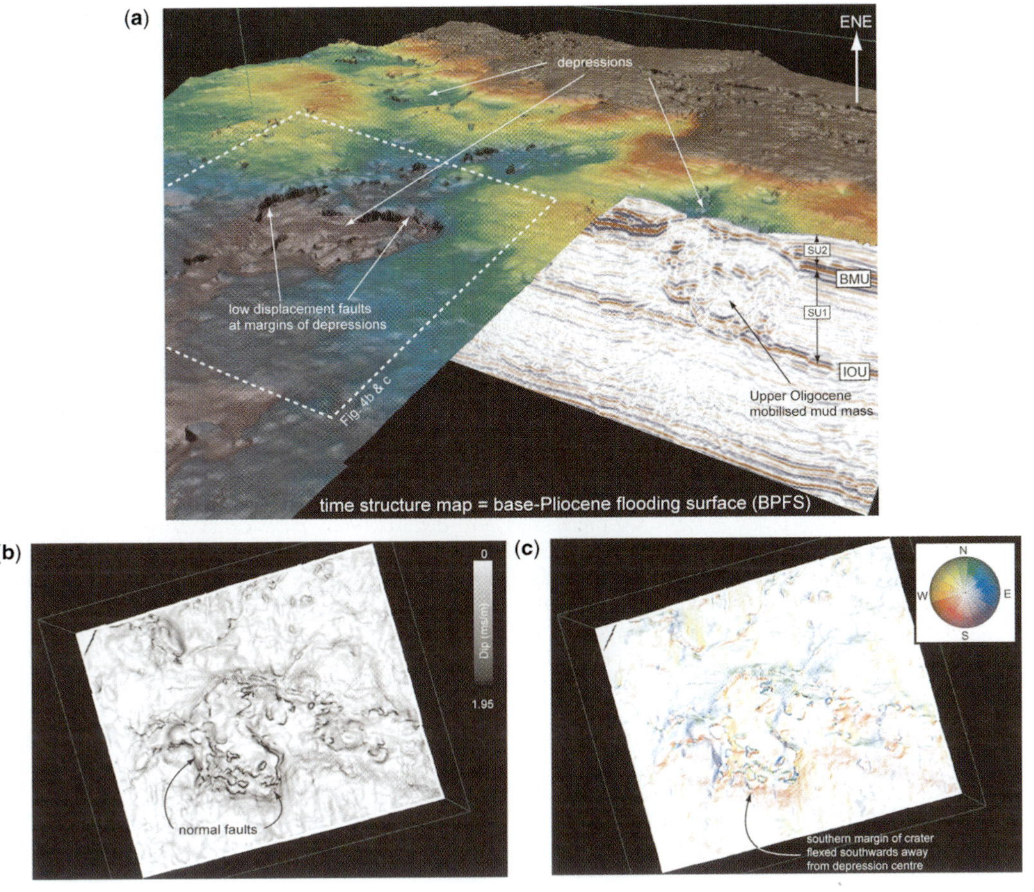

Fig. 4. (**a**) 3D visualization of the BPFS and intersecting seismic section showing the geometry of depressions developed along this surface and their spatial relationship to the interpreted mobilized mud masses developed in SU1. The locations of (**b** and **c**) are shown. (**b**) Time-dip map and (**c**) dip-azimuth map over one of the largest depression developed along the BPFS illustrating the deformation associated with the depressions. Note the development of low-displacement curvilinear faults at the margins of the depressions. See (**a**) for the location of maps.

reflect temporal and spatial, lateral variations in sediment supply and accretion to the shelf margin clinoform front.

Base-Pleistocene unconformity

The Base-Pleistocene unconformity (BPSU) is defined on seismic data as a major angular unconformity with marked truncation of underlying seismic reflections below the horizon. Seismic reflections within the overlying seismic package (not discussed here but correlative to CSS-9 and 10 of Jordt *et al.* 1995) are moderate to high amplitude, parallel and conformable to the BPSU. Biostratigraphic data from wells within the study area indicates that the BPSU is associated with an erosional hiatus which is dated at ca. 1.7 Ma and bounds the base of the Pleistocene succession (Fig. 1b). In the Troll area of the Norwegian margin, a shallow borehole indicates that this surface correlates with a major lithological change from open marine mudstones and sandstones to glaciomarine and fully glacial till deposits (see Sejrup *et al.* 1991, 1996 for full details). A boundary of a similar age with an angular relationship to underlying seismic packages is observed 300 km further north in the Haltenbanken area (Rokoengen *et al.* 1995; their figs 6 and 7). Based on the switch to glacial sediments above the surface and the regionally developed nature of the unconformity, this surface is interpreted to have formed in response to encroachment of Fennoscandian ice-sheets onto the Norwegian margin continental shelf (Sejrup *et al.* 1991, 1996).

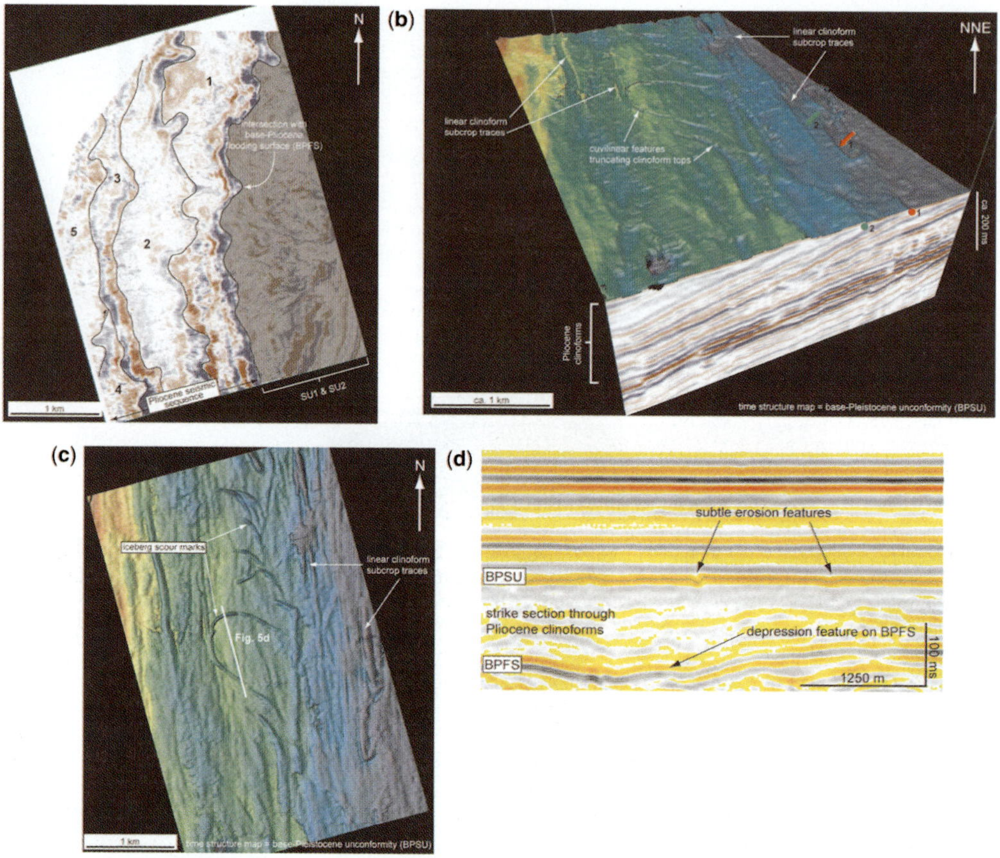

Fig. 5. (**a**) Timeslice at 800 ms through the three-dimensional seismic volume showing the planform seismic expression of clinoform units developed in SU3. Units numbered 1–5 refer to unconformity-bounded seismic packages within SU3. See text for full discussion. (**b**) 3D visualization of clinoforms in SU3 and the BPSU bounding the top of the unit. 1 and 2 refer to clinoform subcrop points beneath the BPSU that can also be traced in planview on the BPSU (arrowed). Clinoform subcrops traces are erosively cross-cut by semi-discontinuous, curvilinear features. (**c**) Map view image of the BPSU showing details of the erosive, curvilinear features identified on this surface. These features are interpreted as basal iceberg scours. The location of (d) is shown. (**d**) Seismic section through SU3 and the overlying Pleistocene seismic package showing the cross-sectional expression of the basal iceberg scour features developed on the BPSU. The location of the seismic section is shown in (c).

Figure 5(b, c) illustrates time structure maps of the BPSU that erosively truncate the upper parts of the Pliocene clinoforms. Where the clinoformal reflections subcrop the BPSU they affect the autotracker software, causing the seismic pick to 'skip' down to the lower clinoform reflection. As a result of this, these maps clearly illustrate the dominant NNW–SSE-trending, linear orientation of the underlying clinoform units (Fig. 5b). A series of semi-discontinuous, curvilinear to arcuate features are also identified on the BPSU (Fig. 5b, c). These features display no preferential orientation, are up to 75 m wide with individual segments up to 3 km long. Seismic sections across these features indicate they form extremely subtle, slightly erosive discontinuities that are 10–15 m deep and erode into the tops of the underlying Pliocene clinoforms (Fig. 5d). Features with a similar seismic expression to these semi-discontinuous, curvilinear erosive features are identified on three-dimensional seismic data from the Pleistocene successions of the Barents Sea, Northern Norway (e.g. Solheim *et al.* 1988; Rafaelsen *et al.* 2002) and the Måløy Plateau, Mid-Norway (e.g. Nygård *et al.* 2004). In these localities these features are interpreted as iceberg plough marks formed by the basal keels of drifting icebergs. Based on their development at the base of the Pleistocene section in the present study area and similarities in geometry, a similar interpretation is favoured for the erosive, curvilinear features identified here.

Summary

Interpretation, visualization and attribute analysis of three-dimensional seismic data reveal that the Paleocene to Present succession of the Lomre Terrace has been subject to several phases of deformation caused by a combination of fluid flow, dewatering, differential compaction and loading, superimposed on regional-scale crustal tectonics. The development of these deformation features and the temporal and spatial inter-relationships between them as discussed above is summarised below and shown schematically in a summary tectonostratigraphic model (Fig. 6a–f). Although this model does not consider in detail basin-scale tectonics, in particular the exact location and controls on uplift of the eastern basin margin, the uplift events which were important for key stratal surface development and the delivery of sands to the basin that subsequently became remobilized during burial are discussed.

During the Cretaceous to Late Oligocene, a smectite-rich mudstone-dominated succession was deposited in a period characterized by relative tectonic quiescence. Mudstone deposition was interrupted during the Late Oligocene by deposition of a silty sandstone unit (the Skade Fm), which is interpreted to have been due to a relative sea-level fall and sequence boundary formation (i.e. the IOU) linked to a period of uplift of the eastern margin and/or a eustatic sea-level fall (Fig. 6a). Compaction and dewatering of the Cretaceous to Upper Oligocene succession occurred sometime between the Late Oligocene and Late Miocene, resulting in the formation of a polygonal fault network (Fig. 6b). It is not possible to further constrain the timing of polygonal faulting as any growth sequences deposited at the upper tips of the polygonal faults were subsequently eroded during formation of the BMU (Fig. 6c).

A second major relative sea-level fall, possibly related to a period of uplift of the eastern basin margin, led to the development of another major sequence boundary (i.e. the BMU), marked shallowing of the basin and a major influx of sands of the Utsira Fm into the basin (Fig. 6c, d). Immediately prior to or synchronous with deposition of the lowermost parts of the Utsira Fm, the upwards migration of light hydrocarbons and associated formation water from the underlying Upper Jurassic basins triggered large-scale mobilization of poorly consolidated mudstones in the Upper Oligocene succession (Fig. 6d). This led to the formation of significant (>100 m) seafloor topography as indicated by folding of the BMU, and onlap of the Utsira Fm onto this horizon. Owing to the lack of well data it is not possible to investigate whether this pre-existing topography influenced facies variability within the Utsira Fm, although it may be speculated that wave and/or tidal processes in the basin may have been influenced by the prominent seafloor topography.

Deposition of the Utsira Fm ceased during the Early Pliocene due to a relative sea-level rise which resulted in the formation of the BPFS. Subsequent to the Early Pliocene flooding event, major basinward progradation of the Norwegian shelf-margin occurred (Fig. 6e). Regional data indicate that shelf-margin progradation was broadly synchronously along the length of the southern and mid-Norwegian margins, thereby suggesting that sediment supply to the progradational margin must have been equally distributed along the length of the margin. This observation has led several authors to suggest that widespread glacial erosion of the Norwegian margin was the key control on sediment supply to the Norwegian continental margin (Rokoengen *et al.* 1995; Martinsen *et al.* 1999). Local lateral variations in shelf-margin sediment supply are indicated, however, by the occurrence of subtle unconformity-bounded seismic packages in the Pliocene clinoform units. Progradation of the Pliocene shelf-margin over, and consequent deflation of gas-filled mobilized mud masses in the Upper Oligocene succession, caused the formation of a series of depressions along the BPFS (Fig. 6e). This deflation also caused low-relief folding of the Pliocene clinoforms (Fig. 6e, f). It is speculated that the formation of the depressions and folding of the Pliocene succession may have been augmented by differential compaction between the gas-filled mobilized mud masses and the adjacent sandstone-dominated Utsira Fm.

Encroachment of the Fennoscandian ice-sheet onto the shelf during the Late Pliocene to Early Pleistocene led to truncation of the underlying stratigraphic units, and a switch from fully marine to glaciogenic sedimentation (Fig. 6f; Sejrup *et al.* 1991, 1996). Eventual retreat of the main ice-sheet from the shelf during the Pleistocene and partial break-up of the frontal part of the ice-sheet resulted in the development of isolated icebergs. These drifted across the shelf and scoured the upper parts of the underlying Pliocene clinoforms resulting in the formation of the curvilinear iceberg-keel scour marks (Fig. 6f).

Conclusion

This study demonstrates that the interpretation and visualization of three-dimensional seismic data combined with volume and grid-based attribute

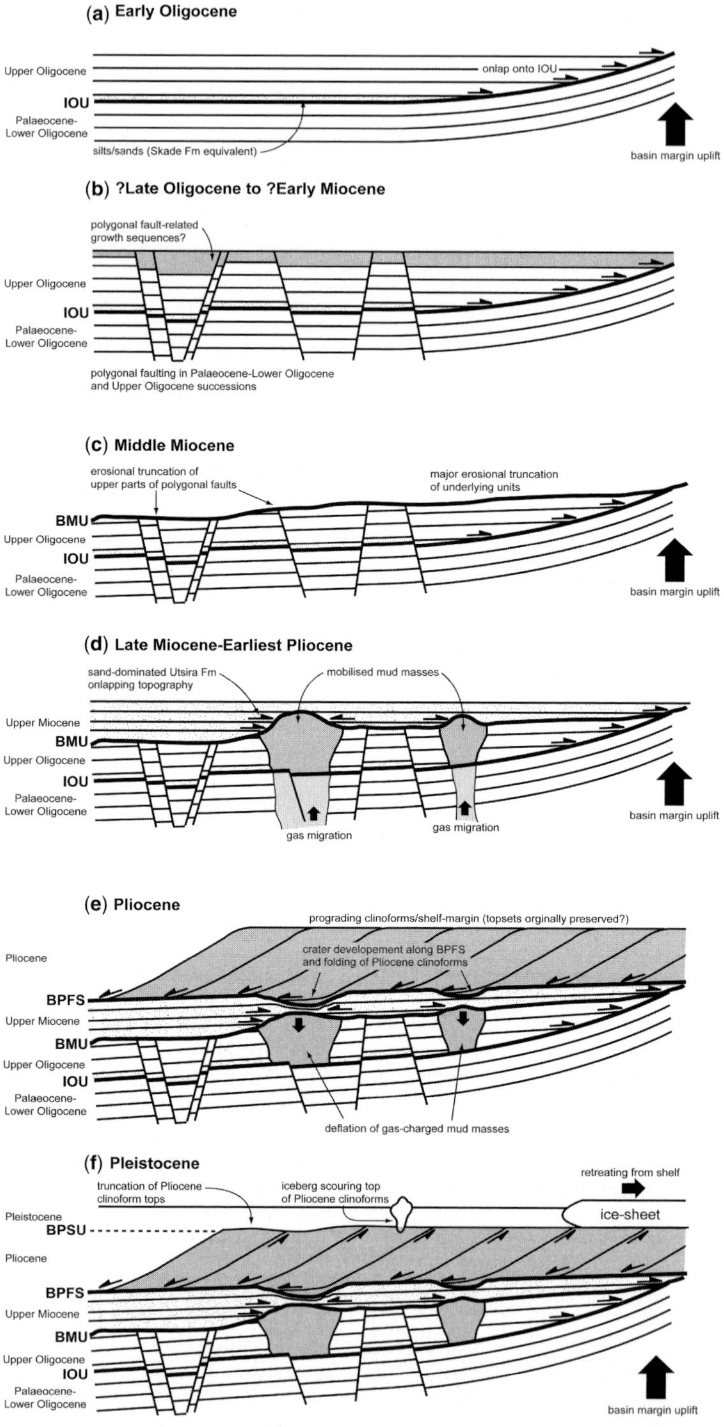

Fig. 6. (**a–f**) A series of summary diagrams illustrating the development of soft-sediment deformation features in the Lomre Terrace area, offshore Norway during the Early Oligocene to Present. The diagrams are not to scale and the vertical exaggeration has been applied to clearly illustrate some of the structural and stratigraphic features and their inter-relationships. See text for full discussion.

analysis can provide important insights into the scale, geometry, distribution and inter-relationships between of soft-sediment deformation features in sedimentary basins. Furthermore, through the application of standard seismic-stratigraphic techniques it is possible to elucidate the timing of deformation. Many of the features described here would not either identified or confidently interpreted using widely-spaced 2D seismic data due to their scale and highly variable distribution. With respect to the regional geological setting of the eastern North Sea Basin, this study has indicated that the Oligocene to Present succession of the Lomre Terrace has been subject to several phases of soft-sediment deformation caused by a combination of basinal fluid flow, dewatering, compaction, and loading processes.

Thanks to Norsk Hydro ASA and partners for permission to publish the results of this study, and to critical reviews by H. Løseth and B. Clarke which helped to improve the clarity of the manuscript. Thanks also to S. Corfield, T. Dreyer, D. Stoddart, M. Charnock, P. Spencer and A. Graham for discussions which helped clarify many of the ideas presented here. Liz Chellingsworth of Foster Findley Associates (FFA) is thanked for assistance with Sea3D™ software, which was used to generate some of the seismic volumes used in this study.

References

APLIN, A. C., YUNLAI, Y. & HANSEN, S. 1995. Assessments of the ß, the compression coefficient of mudstones and its relationship with detailed lithology. *Marine and Petroleum Geology*, **12**, 955–963.

BADLEY, M. E., PRICE, J. D., RAMBECH DAHL, C. & AGDESTEIN, T. 1988. The structural evolution of the northern Viking Graben and its bearing upon extensional modes of basin formation. *Journal of the Geological Society*, **145**, 455–472.

BAHORICH, M. & FARMER, S. 1995. 3-D seismic discontinuity for faults and stratigraphic features: the coherence cube. *The Leading Edge*, **14**, 1053–1058.

BOND, R. 2001. Volume-based visualisation and interpretation of three-dimensional seismic data. *First Break*, **19**, 472–478.

BROWN, A. R. 2003. *Interpretation of three-dimensional seismic data*, 6th ed. American Association of Petroleum Geologists, Memoirs, **42**.

CARTWRIGHT, J. A. 1996. Polygonal fault systems: a new type of fault structure revealed by 3-D seismic data from the North Sea Basin. *In*: WEIMER, P. & DAVIES, T. L. (eds) *Applications of 3-D seismic data to exploration and production*. AAPG Studies in Geology, **42**, and SEG Geophysical Developments Series, **5**, 225–230.

CARTWRIGHT, J. A. & DEWHURST, D. N. 1998. Layer-bound compaction faults in fine-grained sediments. *Geological Society of America Bulletin*, **110**, 1242–1257.

CARTWRIGHT, J. A. & HUUSE, M. 2005. 3D seismic technology: the geological Hubble. *Basin Research*, **17**, 1–20.

CLAUSEN, J. A., GABRIELSEN, R. H., REKNES, P. A. & NYSÆTHER, E. 1999. Development of intraformational (Oligocene-Miocene) faults in the northern North Sea: influence of remote stresses and doming of Fennoscandia. *Journal of Structural Geology*, **21**, 1457–1475.

DALLEY, R. M., GEVERS, E. C. A., STAMPFLI, G. M., DAVIES, D. J., GASTADI, C. N., RUIJTENBERG, P. A. & VERMEER, G. J. O. 1989. Dip and azimuth displays for 3-D seismic interpretation. *First Break*, **7**, 86–95.

DAVIES, R. J. 2003. Kilometer-scale fluidization structures formed during early burial of a deepwater slope channel on the Niger Delta. *Geology*, **31**, 949–952.

DAVIES, R. J., CARTWRIGHT, J. A. & RANA, J. 1999. Giant hummocks in deep-water marine sediments: evidence for large-scale differential compaction and density inversion during early burial. *Geology*, **27**, 907–910.

DAVIES, R. J., BELL, B., CARTWRIGHT, J. A. & SHOULDERS, S. 2002. 3D seismic imaging of dike-fed submarine volcanoes. *Geology*, **30**, 223–226.

DAVIES, R. J., STEWART, S. A., CARTWRIGHT, J. A., LAPPIN, M., JOHNSTON, R., FRASER, S. I. & BROWN, A. R. 2004. 3D seismic technology: are we realising our full potential? *In*: DAVIES, R. J., CARTWRIGHT, J. A., STEWART, S. A., LAPPIN, M. & UNDERHILL, J. R. (eds) *3D Seismic Technology: Application to the Exploration of Sedimentary Basins*. Geological Society, London, Memoirs, **29**, 1–9.

DEMYTTANAERE, R. R. A., SLUIJK, A. H. & BENTLEY, M. R. 1993. A fundamental reappraisal of the structure of the Cormorant Field and its impact on field development strategy. *In*: PARKER, J. R. (ed.) *Petroleum Geology of Northwest Europe: Proceedings from the 4th Conference*. Geological Society, London, 1151–1157.

DORN, G. A., COLE, M. J. & TUBMAN, K. 1995. Visualization in 3-D seismic interpretation. *The Leading Edge*, **14**, 1045–1050.

FÆRSETH, R. B., KNUDSEN, B.-E., LILJEDAHL, T., MIDBOE, P. S. & SODERSTROM, B. 1997. Oblique rifting and sequential faulting in the Jurassic development of the northern North Sea. *Journal of Structural Geology*, **19**, 1285–1302.

FALEIDE, J. I., KYRKJEBO, R., KJENNERUD, T., GABRIELSEN, R. H., JORDT, H., FANAVOLL, S. & BJERKE, M. D. 2002. Tectonic impact on sedimentary processes during Cenozoic evolution of the northern North Sea and surrounding areas. *In*: DORE, A. G., CARTWRIGHT, J. A., STOKER, M. S., TURNER, J. P. & WHITE, N. (eds) *Exhumation of the North Atlantic Margin: Timing, Mechanisms and Implication for Petroleum Exploration*. Geological Society, London, Special Publications, **196**, 235–269.

GABRIELSEN, R. H., KYRKJEBO, R., FALEIDE, J. I., FJELDSKAAR, W. & KJENNERUD, T. 2001. The Cretaceous post-rich development in the northern North Sea. *Petroleum Geoscience*, **7**, 137–154.

GALLOWAY, W. E. 2002. Paleogeographic setting and depositional architecture of a sand-dominated shelf depositional system, Miocene Utsira Fm, North Sea Basin. *Journal of Sedimentary Research*, **72**, 476–490.

GRADSTEIN, F. M. & OGG, J. 1996. A Phanerozoic time scale. *Epsisodes*, **19**, 3–6.

GREGERSEN, U., MICHELSEN, O. & SØRENSEN, J. C. 1997. Stratigraphy and facies distribution of the Utsira Fm and the Pliocene sequences in the northern North Sea. *Marine and Petroleum Geology*, **14**, 893–914.

HARLAND, W. B., ARMSTRONG, R. L., COX, A. V., CRAIG, L. E., SMITH, A. G. & SMITH, D. G. 1990. *A Geological Time Scale 1989*. Cambridge University Press, Cambridge.

HERRON, D. A. 2000. Horizon autopicking. *The Leading Edge*, **19**, 491–492.

HESTHAMMER, J. 1998. Evaluation of timedip, correlation and coherence maps for structural interpretation of seismic data. *First Break*, **16**, 151–167.

HUUSE, M., DURANTI, D. *ET AL.* 2004. Seismic characteristics of large-scale sandstone intrusions in the Palaeogene of the South Viking GRABEN, UK and Norwegian North Sea. *In*: DAVIES, R. J., CARTWRIGHT, J. A., STEWART, S. A., LAPPIN, M. & UNDERHILL, J. R. (eds) *3D Seismic Technology: Application to the Exploration of Sedimentary Basins*. Geological Society, London, Memoirs, **29**, 263–278.

ISAKSEN, D. & TONSTAD, K. 1986. *A revised Cretaceous and Tertiary lithostratigraphic nomenclature for the Norwegian North Sea*. Bulletin **5**, Norwegian Petroleum Directorate.

JACKSON, C. A. L. in press. The geometry, distribution and development of clastic injections in slope systems: seismic examples from the Upper Cretaceous Kyrre Formation, Måløy Slope, Norwegian Margin. *AAPG Memoir on Clastic Injections* (in press).

JACKSON, C. A. L. & STODDART, D. P. 2005. Temporal constraints on the growth and decay of large-scale mobilized mud masses and implications for fluid flow mapping in sedimentary basins. *Terra Nova*, **17**, 580–585.

JAMES, H., BOND, R. & EASTWOOD, L. 2004. Direct visualisation and extraction of stratigraphic targets in complex structural settings. *In*: DAVIES, R. J., CARTWRIGHT, J. A., STEWART, S. A., LAPPIN, M. & UNDERHILL, J. R. (eds) *3D Seismic Technology: Application to the Exploration of Sedimentary Basins*. Geological Society, London, Memoirs, **29**, 227–234.

JORDT, H., FALEIDE, J. I., BJØRLYKKE, K. & IBRAHIM, M. T. 1995. Cenozoic sequence stratigraphy of the Central and Northern North Sea Basin: tectonic development, sediment distribution and provenance areas. *Marine and Petroleum Geology*, **12**, 845–879.

JORDT, H., THYBERG, B. I. & NOTTVEDT, A. 2000. Cenozoic evolution of the central and northern North Sea with focus on differential vertical movements of the basin floor and surrounding clastic source areas. *In*: NOTTVEDT, A. (ed.) *Dynamics of the Norwegian Margin*. Geological Society, London, Special Publications, **167**, 219–243.

JOY, A. M. 1993. Comments on the pattern of post-rift subsidence in the central and northern North Sea Basin. *In:* WILLIAN, G. D. & DOBB, A. (eds) *Tectonics and Seismic Sequence Stratigraphy*. Geological Society, London, Special Publications, **71**, 123–140.

LEES, J. A. 1999. Constructing faults from seed picks by voxel tracking. *The Leading Edge*, **18**, 338–341.

LONERGAN, L., CARTWRIGHT, J. A. & JOLLY, R. 1998. The geometry of polygonal fault systems in Tertiary mudrocks of the North Sea. *Journal of Structural Geology*, **20**, 529–548.

LØSETH, H., WENSAAS, L., ARNTSEN, B. & HOVLAND, M. 2003. Gas and fluid injection triggering shallow mud mobilization in the Hordaland Group, North Sea. *In:* VAN RENSBERGEN, P., HILLIS, R. R., MALTMAN, A. J. & MORLEY, C. K. (eds) *Subsurface Sediment Mobilization*. Geological Society, London, Special Publications, **216**, 139–157.

MARFURT, K. J., KIRLIN, R. L., FARMER, S. L. & BAHORICH, M. S. 1998. 3-D seismic attributes using a semblance-based coherency algorithm. *Geophysics*, **63**, 1150–1165.

MARTINSEN, O. J., BØEN, F., CHARNOCK, M. A., MANGURUD, G. & NØTTVEDT, A. 1999. Cenozoic development of the Norwegian margin 60–64°N: sequences and sedimentary response to variable basin physiography and tectonic setting. *In*: FLEET, A. J. & BOLDY, S. A. R. (eds) *Petroleum Geology of Northeast Europe: Proceedings of the 5th Conference*, 293–304.

NESTVOLD, E. O. 1992. 3-D seismic: is the promise fulfilled? *The Leading Edge*, **11**, 12–19.

NYGÅRD, A., SEJRUP, H. P., HAFLIDASON, H., CECCHI, M. & OTTESEN, D. 2004. Deglaciation history of the southwestern Fennoscadian Ice Sheet between 15 and 13 ^{14}C ka BP. *Boreas*, **33**, 1–17.

PIASECKI, S., GREGERSEN, U. & JOHANNESSEN, P. N. 2002. Lower Pliocene dinoflagellate cysts from cored Utsira Formation in the Viking Graben, northern North Sea. *Marine and Petroleum Geology*, **19**, 55–67.

RAFAELSEN, B., ANDREASSEN, K., KUILMAN, L. W., LEBESBYE, E., HOGSTAD, K. & MIDTBØ, M. 2002. Geomorphology of buried glacigenic horizons in the Barents Sea from three-dimensional seismic data. *In*: DOWDESWELL, J. A. & COFAIGH, O. (eds) *Glacier-influenced Sedimentation on High-Latitude Continental Margins*. Geological Society, London, Special Publications, **203**, 259–276.

ROKOENGEN, K., RISE, L., BYRN, P., FRENGSTAD, B., GUSTAVSEN, B., NYGAARD, E. & SÆTTEM, J. R. 1995. Upper Cenozoic stratigraphy on the Mid-Norwegian continental shelf. *Norsk Geologisk Tidsskrift*, **75**, 88–104.

SCHLISCHE, R. W. 1995. Geometry and origin of fault-related folds in extensional settings. *AAPG Bulletin*, **79**, 1661–1678.

SEJRUP, H. P., AARSETH, I. & HAFLIDASON, H. 1991. The Quaternary succession in the northern North Sea. *Marine Geology*, **101**, 103–111.

SEJRUP, H. P., KING, E. L., AARSETH, I., HAFLIDASON, H. & ELVERHØI, A. 1996. Quaternary erosion and depositional processes: western Norwegian fjords, Norwegian Channel and North Sea Fan. *In*: DE BATIST, M. & JACOBS, P. (eds) *Geology of Siliciclastic Shelf Seas*. Geological Society, London, Special Publications, **117**, 187–202.

SOLHEIM, A., MILLIMAN, J. D. & ELVERHØI, A. 1988. Sediment distribution and sea-floor morphology of Storbanken: implications for the glacial history of the northern Barents Sea. *Canadian Journal of Earth Sciences*, **25**, 547–556.

TRUDE, K. J. 2004. Kinematic indicators for shallow level igneous intrusions from three-dimensional seismic data: evidence of flow direction and feeder location. *In*: DAVIES, R. J., CARTWRIGHT, J. A., STEWART, S. A., LAPPIN, M. & UNDERHILL, J. R. (eds) *3D Seismic Technology: Application to the Exploration of Sedimentary Basins*. Geological Society, London, Memoirs, **29**, 209–218.

YIELDING, G. 1990. Footwall uplift associated with Late Jurassic normal faulting in the northern North Sea. *Journal of Structural Geology*, **147**, 219–222.

Index

Note: page numbers in *italics* refer to Figures, while those in **bold** denote Tables.